D0119179

ENCYCLOPEDIA OF THE
BIOSPHERE

Humans in the World's Ecosystems

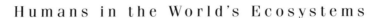

ENCYCLOPEDIA OF THE
BIOSPHERE

VOLUME 5: Mediterranean Woodlands

Project Director
Ramon Folch

Assistant Project Director
Josep M. Camarasa

GALE GROUP

Detroit
San Francisco
London
Boston
Woodbridge, CT

ENCYCLOPEDIA OF THE
BIOSPHERE

Encyclopedia of the Biosphere is an 11 volume work that treats the bioclimatic zones of the planet Earth and their corresponding biomes, and covers the settlement and use of these areas and systems by humans, as well as the problems that this has led to. This work has been planned in accordance with the principles of UNESCO's MAB (Man and Biosphere) Programme, under whose patronage it has been prepared.

ENCYCLOPEDIA OF THE
BIOSPHERE

Project Director

Ramon Folch
UNESCO/FLACAM Professor of Sustainable Development
Secretary-General of the Spanish Committee of the UNESCO/MAB

Assistant Project Director

Josep M. Camarasa
Member of the Spanish Committee of the UNESCO/MAB Programme

Editorial Advisory Committee

Francesco di Castri
Head of Research of the CNRS [Montpellier]
Former Assistant General Director of UNESCO's Environmental Coordination Programmes [Paris]

Mark Collins
Director at the World Conservation Monitoring Centre [Cambridge]

Ramon Margalef
Professor emeritus of Ecology of the University of Barcelona

Gonzalo Halffter
Director of the Institute of Ecology [Xalapa, Veracruz]

Pere Duran Farell
Founder member of the Club of Rome
President of the Spanish Chapter of the Club of Rome [Barcelona]

Alpha Oumar Konaré
Former President of the International Council of Museums [Bamako]

The original Catalan edition of this work was accomplished (1993-98) with the conceptual assistance and logistics
of the United Nations Educational, Scientific, and Cultural Organization (UNESCO).
The positions held by the Authors, the Project Director, the Assistant Project Director and the members of the Editorial Advisory Committee refer to the period
when the series was first prepared.

Catalan-language edition (volume 5): 1993
Biosfera. Els humans en els àmbits ecològics del món
Enciclopèdia Catalana

English-language edition (volume 5): **2000**

Editor: **ERF - Gestió i Comunicació Ambiental, SL** (Barcelona)
Director: **Ramon Folch**
Chief Editor: **Caterina López**
Editorial Team: **Josep M. Palau, Marina Molins**
Updating: **Josep M. Camarasa**

Publisher: **The Gale Group** (Farmington Hills, MI)
Art Directors: **Cynthia Baldwin, Martha Schiebold**
Editorial Coordinators: **Christine Jeryan, Pamela Proffitt**

Translation
Trevor Foskett
Graduate in Biology

Revision
Vernon Heywood
Ph. D., D. Sc., Professor emeritus, The University of Reading

English-language edition distributed to all markets worldwide by The Gale Group
27500 Drake Rd.
Farmington Hills, MI 48331-3535
U.S.A.

Printed by PRINTER, Indústries Gràfiques, S.A. (Barcelona)

ISBN 0-7876-4506-0 (complete set)
ISBN 0-7876-4511-7 (Volume 5)

5

Mediterranean Woodlands

Josep M. Camarasa
Marcos del Castillo
Montserrat Comelles
Graham Drucker
Lluís Ferrés
Ramon Folch
Teresa Franquesa
Cristina Junyent
Juan Pablo Martínez-Rica
Àngels Puig
Jordi Ruiz
Adolf de Sostoa
Ramon Vallejo
Marta Vigo

and

Margarita Arianoutsou
John S. Beard
Josep Canadell
Santiago Lavín
Esteve Masagué
Louis Trabaud

The authors and collaborators - volume 5

Margarita Arianoutsou
Professor at Ecology Dept,
University of Athens

John S. Beard
Researcher at "Vegetation Survey
of the Western Australia" project

Josep M. Camarasa
Member of the Spanish Committee
of the UNESCO/MAB Programme

Josep Canadell
Professor of Ecology
at the Universitat Autònoma de Barcelona

Marcos del Castillo
Professor of Zoology
at the University of Barcelona

Montserrat Comelles
Graduate in Biology

Graham Drucker
Researcher at the World Conservation Monitoring Centre
[Cambridge, United Kingdom]

Lluís Ferrés
Doctor of Biology

Ramon Folch
Secretary-General of the Spanish Committee
of the UNESCO/MAB Programme

Teresa Franquesa
Doctor of Biology

Cristina Junyent
Graduate in Biology

Santiago Lavín
Professor at the Universitat Autònoma de Barcelona

Juan Pablo Martínez-Rica
Researcher at the Instituto Pirenaico de Ecología,
CSIC [Jaca, Spain]

Àngels Puig
Doctor of Biology

Jordi Ruiz
General Direction of the Natural Environment
of the Generalitat de Catalunya

Adolf de Sostoa
Professor of Zoology at the University of Barcelona

Louis Trabaud
Research Director
at the Centre d'Ecologie Fonctionelle et Evolutive,
CNRS [Montpellier, France]

Ramon Vallejo
Professor of Plant Biology at the Universitat de Barcelona

Marta Vigo
Graduate in Biology

EDITORIAL TEAM

DIRECTOR: **Ramon Folch**, Doctor of Biology
ASSISTANT DIRECTOR: **Josep M. Camarasa**, Doctor of Biology
CHIEF EDITOR: **Montserrat Comelles**, Graduate in Biology
EDITOR: **Cristina Junyent**, Graduate in Biology
ART DIRECTION: **Rosa Carvajal,** Graduate in Geography, **Mikael Frölund**
SCIENTIFIC ASSESSMENT: **Jaume Bertranpetit, Francesco di Castri**
DESIGN AND PAGE-MAKING: **Toni Miserachs**
ADMINISTRATIVE SECRETARIES: **Maria Miró, Mònica Díaz**

EDITORIAL DIRECTOR: **Jesús Giralt**
PUBLICATION MANAGER FOR MAJOR PROJECTS: **Josep M. Ferrer**
HEAD OF PRODUCTION: **Francesc Villaubí**

Presentation

The Mediterranean world exists. It might seem banal to point this out, but it is not. The Mediterranean world exists for everyone in a geographical sense, but the idea is more open to debate if the cultural dimensions are considered. It is by no means clear that there is a single Mediterranean culture, which would have to be European and African, Jewish, Christian and Muslim, Roman and Greek, as well as Carthaginian, Byzantine, and Iberian. And it is even less clear that there is a biogeographical Mediterranean world, not only around the Mediterranean Sea, but also overseas, in the four corners of the planet. This ecological Mediterranean world exists, although the fact is little known, and thus it is not a commonplace to state it.

This presentation is not intended to be a summary of the text it introduces, so it would be inappropriate to explain the characteristics and the geographical distribution of the Mediterranean areas. The aim of this introduction is to point out the relative novelty of the concept of ecological Mediterranean-ness. It is not a new idea to experts in biogeography, but it is new to the general public, even those who are considered well-educated. Concepts such as the rainforest or hot desert are generally accepted internationally, to a far greater extent than that of the Mediterranean. Whether it is in the Amazon, in Zaire, or in Indonesia, the tropical rainforest is universally considered to be a specific environment, and the fact that it is distributed over three continents does not make this idea any less valid. In the same way, the desert is the desert, whether it is in the Sahara, in Australia or in Arizona. How then should we approach the Mediterranean-ness of Chile or California?

The absence of a general habit of considering ecological Mediterranean-ness as a global phenomenon is clearly shown by the fact there is no word for it. It is necessary to use cumbersome phrases like the "Mediterranean world" or the "Mediterranean areas," which are provisional, inconvenient and not very practical solutions. This problem has been around for a long time because the "Mediterranean lands" take their name from a sea, and this sea takes its name from those lands; the Mediterranean sea is the sea that lies in the midst of the Mediterranean countries. In the face of all this confusion, the best solution is not to hide behind circumlocutions and so this volume talks unashamedly about the mediterraneans, creating a new noun by the simple method of generalizing what until now has been a particular, special world. Mediterranean, like the jungle or the desert, is the term used for one of the Earth's bioclimatic regions, or biomes. There is a Mediterranean in California, one in central Chile, one at the southern tip of the African continent, one in southern Australia, and around the Mediterranean itself. The solution we have adopted is a little startling, and both the new concept and the new term require a small, but worthwhile, effort to get used to. This volume will help.

It will help because it deals with the different Mediterraneans around the world, i.e. the Mediterranean biome. This volume is mostly written by Catalans for obvious reasons: *Encyclopedia of the Biosphere* is prepared and coordinated in Barcelona, and Barcelona is part of the Mediterranean world, in strictly geographical terms, as well as culturally, biogeographically, and ecologically. In the same way as the entire work avoids anthropocentric (and misanthropic) attitudes, this volume aims to consider the Mediterraneans as a whole, without giving undue attention to the area where it was written. In spite of this, the Mediterranean Basin enjoys a special treatment in this volume, because there is more information available, its area is objectively much larger than the other Mediterraneans and it has a longer historical background. We have tried not to exacerbate this bias with a subjective treatment.

In any case, it was in this basin that the biogeographical unity, and the specific climate and ecology of the Mediterraneans was first recognized. Since long ago, and especially since the Renaissance, it has been obvious to European botanists that there are conspicuous differences between the flora of the Mediterranean Basin (the flora the classical Greek and Roman authors had described) and that of central Europe. A.P. de

Candolle, born in Geneva and thus central European, expressed this clearly when he dedicating the third edition of the *Flore Française* (1805) to Lamarck, "..ces pays fertiles sont placés sous un ciel different du nôtre à bien des égards" (these fertile countries lie, from more than one point of view, under a different sky from ours). Augustin Pyrame de Candolle was the first to map a part, the French part, of "the space occupied by the type of plants that I would call *Mediterranean*, because they are found in almost all the countries around the Mediterranean." This map appeared in the second volume of the *Flore Française* which he published in 1805 with Lamarck. Later exploration of Algeria by the *Comission Scientifique de l'Algérie* (1840-1844) confirmed his intuition about the phytogeographic unity of the basin's southern and northern shores. The similarities in climate and vegetation between eastern Algeria and southern and insular Italy, between the region surrounding Algiers and Provence, between the area around Oran and both Murcia and Andalusia led to the scientific recognition that the Mediterranean Basin was a biogeographical unit. In the middle of the 19th century books on geography, botany, natural history, and even guide books, talked about the Mediterranean region, and while M. Wilkomm was exploring its western edges in the Iberian Peninsula, E. Boissier dealt with its eastern limits in Anatolia and Syria in his *Flora Orientalis* (1867).

The idea of the bioclimatic unity of the Mediterranean Basin with the Mediterraneans overseas is much more recent. Despite early contributions by A. Grisebach, V. Köppen and E. De Martonne (1927), the concept was not generally accepted until the 1950s when F. Bagnouls and H. Gaussen on the one hand, and L. Emberger on the other, provided a simple way of defining it. Using tables and indexes summarising many factors, they defined the Mediterranean climate and showed it was clearly associated with a specific type of vegetation. In the 1960s comparisons between the different areas with Mediterranean climates became common. In 1973 the compilation *Mediterranean Type Ecosystems*, coordinated by F. di Castri and H.A. Mooney was published, and in 1981 *Mediterranean-Type Shrublands*, compiled by F. di Castri, D. W. Goodall, and R. L. Specht was published.

The structure of this volume is similar to the other biomes treated in Volumes 2 to 9 of *Encyclopedia of the Biosphere*. Unlike the other volumes, however, much of the text has been written by one person, Lluís Ferrés, Doctor of Biology, former professor at the Universitat Autònoma in Barcelona, who now devotes most of his time to environmental education in the media. Lluís Ferrés was management assistant to one of us (R. Folch) in the production of "Mediterrània," coproduced by Televisió de Catalunya S.A. and Caixa de Barcelona under the patronage of UNESCO/MAB Programme (1988) and broadcast by this and other television stations. The editorial team has played a greater role in this than in other volumes, due to their personal knowledge of the subject.

The first section, by Lluís Ferrés and Ramon Vallejo, professor at the Biology Faculty of the Universitat de Barcelona, describes the main features that characterize the Mediterraneans in terms of climate, soils and geographical distribution. The second section consists of chapters describing the ecological functionality of the different Mediterraneans, their flora, their fauna, life in their fresh waters, their variability in time and their variability in space. The first, second, fifth and sixth of these chapters were written by Lluís Ferrés: the second includes contributions from Margarita Arianoutsou, professor at the Ecology Department of the University of Athens, John S. Beard, researcher with the "Vegetation Survey of Western Australia" project, Louis Trabaud, of the Centre d'Ecologie Fonctionelle et Evolutive in Montpellier, and Ramon Folch. The third chapter, dealing with the fauna and animal populations, was written by Juan Pablo Martínez-Rica of the Instituto Pirenaico de Ecología Experimental, in Zaragoza and Jaca, with contributions from Marcos del Castillo, professor at the Universitat de Barcelona. The fourth chapter, dealing with life in rivers and lakes, was written by Adolf de Sostoa, professor of Zoology at the Universitat de Barcelona, Montserrat Comelles, Lluís Ferrés and Àngels Puig.

The third section, dealing with human beings in the Mediterraneans, consists of four chapters. The first deals with human populations, and was written by Josep M. Camarasa. The second deals with the use of plant resources, and was written by Lluís Ferrés, with contributions from Josep M. Camarasa and Cristina Junyent. The third chapter deals with the use of animal resources, and was prepared by Jordi Ruiz, of the Direcció General del Medi Natural of the Generalitat of Catalonia, Esteve Masagué, of the Barcelona Beekeepers Association, Marta Vigo, a biology graduate from the Universitat de Barcelona and Santiago Lavín, of the General Veterinary Pathology Department of the Universitat Autònoma de Barcelona. The fourth chapter on environmental problems and their management, was mainly written by Josep M. Camarasa with contributions from Lluís Ferrés, John S. Beard, Josep Canadell of the Ecology Unit of the Universitat Autònoma de Barcelona, and Jordi Ruiz.

The fourth section, dealing with the Biosphere Reserves exemplifying the biome, is the work of Graham Drucker of the World Conservation Monitoring Centre, Cambridge (U.K.), and Teresa Franquesa. The volume contains over twenty inserts written by Josep M. Camarasa, Lluís Ferrés, Ramon Folch, Cristina Junyent, and Marta Vigo, with graphics by Toni Miserachs.

This volume would not have been possible without the invaluable advice of Francesco di Castri, one of the leading experts on the Mediterraneans of the world, and who has done much to further their understanding. We must also acknowledge the crucial guidance received from professor Jaume Terradas, professor of ecology at the Universitat Autònoma de Barcelona, and director of the Centre de Recerca Ecològica i Aplicacions Forestals (CREAF), and his collaborators.

We would also like to thank the more specific contributions from the persons and institutions that have provided documentation, or who have commented on sections of the volume. These were Gaspar Jaén (Elx Council), Oficina Tècnica d'Imatge (Barcelona Council), Archivio di Stato di Foggia, Archivo General de Indias, Archivo General de Simancas, Biblioteca de Catalunya, Arlene Fanarof (Library of South Africa), Centre de Documentació i Animació de la Cultura Catalana (Perpignan), Corporación Nacional Forestal (CONAF, Chile), Hans van Baren (International Soil Reference and Information Centre, Wageningen), Servei de Parcs Naturals (Diputació de Barcelona), Nina Cummings (Field Museum, Chicago), the Huntington Library, the Institut Botànic de Barcelona, Francesc Vives (Institut Tirant lo Blanc, Elx), Agencia de Medio Ambiente (Junta de Andalucía), Mateo Martinic (Instituto de la Patagonia, Chile), Laboratori de Referència de Catalunya, Phoebe Apperson Hearst Museum of Anthropology (Berkeley), Departament d'Educació (Museu d'Art Modern de Barcelona), Bernat Martí (Museu Arqueològic de València), M. Dolors Llopart (Museu d'Arts Indústries i Tradicions Populars de Barcelona), British Museum (London), Museu Marítim (Diputació de Barcelona), Albert Masó, Biblioteca General d'Història de l'Art (Museu Nacional d'Art de Catalunya), Ramon M. Planas (Museu del Perfum, Barcelona), the National Museum of South Africa, Centre de Recerca Ecològica i Aplicacions Forestals (CREAF, Universitat Autònoma de Barcelona), Francesc Calafell (Departament d'Antropologia, Facultat de Biologia, Universitat de Barcelona), Jacint Nadal (Departament de Zoologia, Universitat de Barcelona), Laboratori del Suro (Universitat de Girona), H. J. Deacon (Stellenbosch University), Martí Domínguez (Departament de Biologia Animal-Entomologia, Universitat de Valencia) and the Wildlife Society of South Africa.

Ramon Folch
Josep M. Camarasa
1993

Mediterranean-type sclerophyllous forests and shrub communities

Know you the land where the lemon-trees bloom? In the dark foliage the gold oranges glow. A soft wind hovers from the sky, the myrtle is still, and the laurel stands tall. Do you know it well? There, there, I would go, O my beloved, with thee!

Johann Wolfgang von Goethe
Wilhelm Meisters Lehrjahre (1795–1796)

1
Classical, friendly, and fragile

1. A biome and climate on a human scale

1.1 Dry summers, mild winters

The Mediterranean biome, which is present on all five continents, takes it name from the Mediterranean Basin, the largest of the areas forming this biome. It includes parts of the three continents encircling the Mediterranean, the "sea between the lands." This sea tempers the climate and has witnessed the appearance, rise and fall of many cultures and empires over the centuries. The Mediterranean Basin was, in fact, the birthplace of western civilization: the mild climate has formed landscapes where human life, without being extremely easy, is nonetheless possible.

Rather than being a frontier or division between the peoples and cultures living on its shores, the Mediterranean Sea has served to bring them together, so that people, goods and information have been crossing it for more than three thousand years. This is why most of the Mediterranean biome can be called classical. Together the landscape and mild climate, in which neither drought nor cold is serious enough to present major problems to agriculture and cattle raising, have allowed the development of different cultures that have something in common: an open-air lifestyle in streets and public spaces where goods and ideas can be exchanged. But the Mediterranean biome is also fragile, with a combination of poor soils and irregular, torrential rains, meaning that any damage to the landscape may become permanent, and that the transformation and exploitation of land may cause its progressive deterioration. To sum up, it is a classical, friendly, and fragile biome.

The Mediterranean climate can be defined as a transition zone between temperate and dry tropical climates. It is characterized by having fairly long summer dry periods, great variations in yearly rainfall, hot summers, and cool or moderately cold winters. It is a very varied climate, with annual rainfall varying from 4-98 in (100-2,500 mm), and average temperatures ranging from 41-64°F (5-18°C). The most important characteristic of the Mediterranean climate is that the hot period and the dry period coincide. This combination is unusual, because if we ignore very continental climates and very dry ones, in most climates the hot period is the wet season, due to the greater evaporation.

In summer, areas with Mediterranean climates show hot dry conditions like those that prevail in the neighboring subtropical deserts. During the winter, on the other hand, the strong subtropical anticyclones retreat to lower latitudes, exposing the areas to the influence of the cyclonic rains that normally dominate temperate areas in higher latitudes throughout the year. These conditions are related to the presence of cold oceanic currents that give rise to high humidity near the coast.

The Mediterranean climate is relatively recent, having appeared for the first time in the Pleistocene. Throughout its existence, it has been subject to major climatic fluctuations that have had an effect on the area dominated by this climate. This character of transition between two climates, together with other factors like its reduced area, the abrupt relief in many areas under its influence and the variety of geological substrates and soils present, have given rise to a great variety of environmental conditions within this biome.

1.2 Climatic and meteorological parameters

The duration of the hot dry period is of great importance in determining the nature of the landscape within the Mediterranean biome, as it is the most important environmental factor. This parameter allows us to clearly define the areas of transition towards drier or wetter climates, and serves as the basis of the classification of the different subclimates within the biome. Rainfall in the areas with a Mediterranean climate is character-

1 The blue of the sea, whitewash, and the green of the vine: these colors symbolize the humble but hospitable Mediterranean. They are the colors of a mainly coastal landscape, created by human toil, and they are present in towns and villages, like this back street in Cadaqués on the Costa Brava in Catalonia.
[Photo: Joan Biosca]

2 Ombrothermic (temperature and rainfall) diagrams of various locations with a Mediterranean climate representing the different areas of the biome. The dry summer period (from June to September in the northern hemisphere and from November to March in the southern hemisphere) is clear in all of them. Average annual temperatures range between 59-68°F (15-20°C), while total annual precipitation is around 500-700 l/m².
[Cartography: Editrònica]

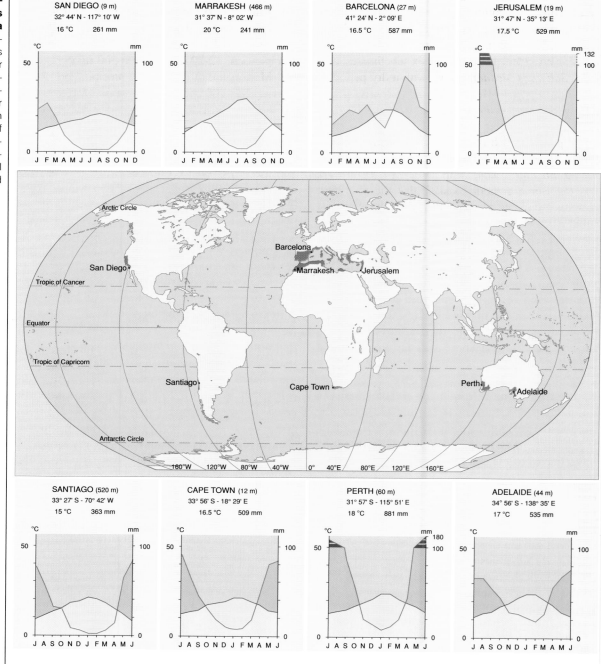

SAN DIEGO (9 m)
32° 44' N - 117° 10' W
16 °C   261 mm

MARRAKESH (466 m)
31° 37' N - 8° 02' W
20 °C   241 mm

BARCELONA (27 m)
41° 24' N - 2° 09' E
16.5 °C   587 mm

JERUSALEM (19 m)
31° 47' N - 35° 13' E
17.5 °C   529 mm

SANTIAGO (520 m)
33° 27' S - 70° 42' W
15 °C   363 mm

CAPE TOWN (12 m)
33° 56' S - 18° 29' E
16.5 °C   509 mm

PERTH (60 m)
31° 57' S - 115° 51' E
18 °C   881 mm

ADELAIDE (44 m)
34° 56' S - 138° 35' E
17 °C   535 mm

ized by its concentration in the colder periods and, above all, by its irregularity. Rains tend to be short and heavy, reducing the availability of water to vegetation. Variations from year to year also accentuate this climate's harshness, because when the rains become scarce, they also become more irregular. The total rainfall in areas with a Mediterranean climate is no lower than in many wet temperate areas, but its irregularity makes it clearly different. In Paris and Marseille, for example, total annual rainfall is similar (around 24 in/year or 600 mm/year), but in Paris the total annual rainfall is very close to this average in 30% of the years, while in Marseille this only occurs in 14% of the years, clearly showing this irregularity.

From semi-arid to subhumid

The Mediterranean climate includes a range of transitions, from those of wet climates to those found in the boundaries with desert climates: thus it extends from subhumid to semi-arid.

Many classifications have been proposed for Mediterranean subclimates, based on the use of different criteria for combining rainfall and temperature data to determine the dry period. Between four and six categories are normally accepted. The widest division considers the following six subclimates: *very arid (Saharan) Mediterranean*, with a dry period of 11-12 months; *arid Mediterranean*, with a dry period of 9-10 months; *semi-arid Mediterranean*, with a dry period of 7-8 months; *subhumid Mediterranean*, with a dry period of 5-6 months; *humid Mediterranean*, with a dry period of 3-4 months; and lastly the *super-humid Mediterranean*, with a dry period of 1-2 months. Some authors consider that the two most extreme examples correspond to different climates, subdesert and temperate, respectively. In this case, they propose a classification with four subclimates called *xeromediterranean*, *thermomediterranean*, *mesomediterranean* and *submediterranean*, which correspond approximately to the four intermediate classifications above.

Different criteria can be used to determine the duration of the dry period. Thus if we use diagrams combining rainfall and temperature (ombrothermic graphs), the biologically dry periods are those in which the temperature curve is higher than the rainfall one.

There are also more complex numerical criteria that combine rainfall and temperature to define coefficients whose values can be used to distinguish different subclimates. In each case an attempt is made to define the coefficient so that it is possible to calculate when rainfall is lower than potential evapotranspiration, and there is, thus, a lack of water. The simplest coefficient would be one directly relating precipitation and evaporation (*P/E*), but this is not feasible due to the great methodological difficulty of measuring potential evaporation or evapotranspiration. Therefore this parameter is usually substituted by temperature, which is much easier to measure and directly related to evapotranspiration. Emberger's coefficient, one of the most widely used, is based on a combination of the average of the mean maximum temperatures of the hottest and mean minimum of the coldest months, thus seeking to establish more significant values than the average annual temperature: $Q=P/[(M-m)(M+m)/2]$, which can be simplified as $Q=2P/[(M-m)(M+m)]$, where P is total annual rainfall, M is the mean maximum temperature of the hottest month and m is the mean minimum temperature of the coldest month. The term $(M+m)/2$ is an improvement on the mean annual temperature, while the term $(M-m)$ is a measure of annual temperature range closely connected to evaporation. To avoid negative values, all temperatures are expressed on an absolute scale in degrees Kelvin (°K). The drier the climate, the lower the value of this coefficient is, and we can thus clearly separate different subclimates. Classifications based on these values suggest the following subclimates: arid Mediterranean ($20<Q\leq30$, precipitation 12-20 in/yr or 300-500 mm/yr), semi-arid Mediterranean ($30<Q\leq50$, precipitation 20-27 in/yr or 500-700 mm/yr), subhumid Mediterranean ($50<Q\leq90$, precipitation 27-39 in/yr or 700-1,000 mm/yr), and humid Mediterranean ($Q>90$, precipitation >39 in/yr or 1000 mm/yr).

The usefulness of a coefficient like this is obvious when we combine it with temperature values, especially with the mean value for the coolest month. By plotting any pair of values (Q and m) on a system of axes, we obtain a separation of the different subclimates, ranging from subarid to subhumid, and from cold to hot for each of the categories of aridity.

Relatively hot, tolerably cold

In the same way that we have defined subclimates by the length of the dry period, we can also define them by temperature. If we use the mean temperature of the coldest month, we can make the following division: a hot Mediterranean climate ($m\leq45°F$ or 7°C), where there are no frosts; temperate Mediterranean (37°F or $3°C<m\leq45°F$ or 7°C), where there are occasional frosts; cool Mediterranean (32°F or $0°C\leq m<37°F$ or 3°C), where there are frequent frosts; and lastly cold Mediterranean ($m<32°F$ or 0°C), with long periods of freezing weather.

Minimum values are used to define subclimates because of the great importance of cold as a factor in the Mediterranean climate. In fact, mild winters are often a defining characteristic of this climate, but because the coldest season is also the wettest season and, thus, the most favourable for vegetation, restrictions imposed by low temperatures are very important. Temperatures around 32°F (0°C) or slightly below do not cause serious problems for the functioning of the vegetation, which is adapted to support them without being damaged, but they do impose serious restrictions on its activity.

It is also very important to bear in mind that minimum temperatures can occasionally reach very low values in certain conditions. Irruptions of cold, dry air from the polar regions may cause temperatures to fall as low as -4°F (-20°C) near the coast. These temperatures are very infrequent, separated by intervals of several decades, but they are of great importance in determining the distributions of many long-lived plant species, as they may cause serious dam-

3 The mild Mediterranean climate allows the population to enjoy many open air activities, even in the fairly mild winters, such as these fishmongers who have set up market on this fishing wharf on the Bosphorus in Istanbul, Turkey. Throughout the Mediterranean Basin commercial activities and public meetings have developed in the open air, with practically no need for shelter. The wharves of ports and squares in towns and villages in the Mediterranean are famous as the settings for diverse activities, including commercial exchanges of merchandise, revolutionary movements, and schools of philosophy, like those of classical Athens.
[Photo: Joan Biosca]

age to populations with little resistance to cold. They are also of great economic importance, because of their disastrous effects on both native and introduced tree crops. The terrible frosts in the Mediterranean Basin at the beginning of 1956 killed thousands of olive trees, some of them centuries old, thus showing how rare such severe frosts are. The exceptionally cold period in the Mediterranean Basin in the 1984-1985 winter, with minimum temperatures reaching -4°F to -22°F (-20°C to -30°C), greatly reduced the ranges of some species introduced from other areas of the biome, such as the Australian acacias, although the most affected were succulent plants, such as the Hottentot fig (*Mesembryanthemum* spp.), which died out over much of the Mediterranean Basin.

Maximum temperatures vary greatly, the highest values normally occurring in the most arid areas, bordering on subdesert areas. Where mean temperatures for the hottest months vary between 72°F and 109°F (22°C and 43°C). The difference between the mean temperature of the hottest month and that of the coldest month, the temperature range, is between 23 and 81 Farenheit degrees (13 and 45 Centigrade degrees), depending on how continental the area is. In fact, a temperature range of between 36 and 54 Farenheit degrees (20 and 30 Centigrade degrees) is usual. The highest values are found in the most continental areas, that is, in zones that are transitional to arid climates.

Overall, however, temperatures in Mediterranean climates oscillate between limits that do not impede biological activity. The combination of high temperatures and the lack of water is the most important limiting factor. Low temperatures only act as a limiting factor in transitional areas, although we should not forget the limitations imposed by the occasional very cold spells caused by the irruption of polar air masses in winter, when vegetation is still active and thus very sensitive to low temperatures.

4 Mist and fog frequently appear inland in the Mediterranean. This is due to the condensation of atmospheric water vapor by saturation and direct cooling, such as in this early morning picture of the Plana de Vic in Catalonia. This phenomenon is most common in winter, and provides a supplement the rainfall, allowing the growth of vegetation that is surprisingly hygrophilous for this area. [Photo: Tavisa]

Equinoctial rains

Although in the Mediterranean biome rainfall is normally concentrated in the cooler seasons, it may also vary in its annual distribution. It may be concentrated in winter, that is to say around the solstice, or it may show two annual peaks associated with the spring and autumn equinoxes.

In Mediterranean areas, rain is normally associated with high latitude storms moving towards the Equator as strong sub-tropical anticyclones withdraw, which normally occurs between the autumn and spring equinoxes. The cold currents affecting these coasts create conditions of high humidity that favor winter rains.

Maximum precipitation may occur in winter, spring, or autumn depending on the characteristics of each area. The more oceanic areas, that is the coastal areas influenced by cold currents,

show maximum rainfall in midwinter when the depressions are most active and reach lower latitudes. On the other hand, in those areas furthest from the oceanic coasts, such as the western part of the Mediterranean Basin, the storms that arrive have already shed most of their rain, meaning that winter rains are scarce. In these conditions the dominant rains are equinoctial and linked to irruptions of rising cold air, and are therefore derived from unstable air masses and are not caused by the fronts associated with depressions.

Frontal rains tend not to be very intense, but rather continuous and gentle, whereas equinoctial rains are usually violent. The sudden destabilization of air masses due to irruptions of rising cool air or irruptions caused by the interaction of cold air with warm sea water at the beginning of autumn, gives rise to torrential storms in which a year's average rainfall may fall in a single day.

Dry riverbeds and autumn floods

Mediterranean rivers have a regime that is highly influenced by the irregular distribution of rainfall over the year and their torrential nature. Only rivers with a large drainage area and headwaters high up in the mountains have a more or less constant flow. The high water retention capacity of a very large catchment area, together with the higher rainfall and the possible local accumulation of snow in the headwaters are necessary to ensure a constant, if slight, flow throughout the dry summer period. Even so, the reduction of water level in the summer is considerable, and the river flow in the summer is considerably less than in the rest of the year.

In most Mediterranean rivers no water flows for much of the year, resulting in their being little more than dry stony beds. These river beds can easily be crossed on foot, and they are covered with typical terrestrial vegetation. This absence of water leaves the bridges crossing them looking for the time rather absurd and pointless. Yet any Mediterranean village or city close to a river has a collective memory of the "year of the flood," often reflected in some kind of stain or mark on the wall of a building, showing the highest level the water reached. To sum up, these rivers either lack water or suffer from an excess.

La Rambla, Barcelona [Lluís Ferrés]

The situation is different in rivers with smaller catchments that occur entirely within areas with a Mediterranean climate, that is, without the headwaters reaching high up in the mountains. In this case, the fluctuations in the flow are even greater, and the river may run dry in the summer. The little remaining water accumulates in pools, which get smaller and smaller as the summer progresses. This has forced the species living in them to develop adaptations that allow them to survive these conditions. In some Mediterranean rivers, when the water is at its summer low, the fish are concentrated in the few remaining pools, and may reach densities of up to 30 to 40 fish per square meter (1 square meter=10.8 square feet).

In the most extreme cases, these rivers only flow during the short period that the torrential rains last, and they are not even known as rivers. The name "rambla" is given to riverbeds that are dry for most of the year. When the river does have water, its flow may be surprisingly high indeed, as hundreds of liters of rain may fall on each square meter in a torrential storm in just a few hours. If rainfall is heavy and rapid, the change from a dry riverbed to a river may take place abruptly, with the water forming a mass that advances like a wall; this is totally unlike a normal river's increase in flow, which is gradual and depends on the intensity of the rain and the steepness of the relief. This is why ramblas that are almost always dry are often quite wide and they may change substantially after very intense spates. In fact, in Mediterranean areas we can find all the possible gradations between large permanent rivers and immense desert "wadis" in which water flows only at intervals of many years.

Sign indicating danger of flooding, NE of the Iberian Peninsula [Ernest Costa]

Their highly irregular nature means that people often forget that part of the land they live on belongs to these fickle rivers. The frequent location of towns and cities in valley bottoms and near to this type of watercourse in Mediterranean coastal areas, may give rise to somewhat picturesque scenes that may easily turn into tragedies. In cities, it is not uncommon to see bridges over dried-out riverbeds that are asphalted and developed in various ways, or even tree-lined promenades that have become the local main traffic route. When torrential rains occur, damage to property is inevitable and accidents to people are not uncommon.

Even larger and more devastating than these surges of floodwater (freshets) are the large floods that affect considerable areas. Local freshets are usually the result of violent local storms. When intense rainfall is widespread, associated with periods of general instability, even the most constant and regular rivers may show large freshets. When this occurs, the rivers once again come to occupy their extensive flood basins. These basins often house human settlements, whose inhabitants suffer the effects of the freshet; they may even aggravate them by impeding the flow of the water, as some urban features, such as roads or bridges, may act as effective dams and cause the water level to rise.

Sign indicating danger of flooding, Arenys de Mar, Catalonia, NE of the Iberian Peninsula [Lourdes Sogas]

Flood in Australia [Matt Jones / Auscape International]

The larger the flood, the less frequent it is, and this low frequency makes them even more devastating. People seem to find it easy to forget that water flows down some ramblas once or twice a year, and even easier to forget that certain areas occasionally act as flood basins at intervals of several decades. Settlements on these flood basins, on lands which to some extent form part of the river, helps to turn this type of flood into a catastrophe. It is only necessary to think of the 1,000 people who died in the Vallès district of Catalonia (Spain) in the floods in September 1962, or the damages caused in Italy by the Arno and the Po, both permanent rivers with flow regimes highly influenced by the irregular nature of the Mediterranean climate. They both broke their banks after torrential rains in November 1966, flooding 800 municipalities and destroying 12,000 buildings. This flooding caused the death of 120 people, the evacuation of an other 45,000, and the loss of 165,000 tons (150,000 metric tons) of crops.

5 Summer hail on the streets of Barcelona. In periods of meteorological instability, large vertical cloud formations develop (cumulonimbus). The top of the cloud is usually freezing, and layers of ice particles form, which grow as they fall through the storm cloud. If these grains of ice are larger than 2-5 mm, they are called hailstones, otherwise they are called hail. Both can seriously damage crops.
[Photo: Ernest Costa]

The violent nature of this rainfall makes it highly erosive in effect, and its concentration in a short period of time means that the water is of little use to the vegetation.

These rains become even more torrential when the climate is dry, and may lead to greater aridity even though this is not shown by the figures for total annual rainfall. This figure may easily reach 23-27 in/yr (600-700 mm/yr) but does not represent a continuous supply of water even in the wet season.

Water, snow, and hail

Most precipitation in Mediterranean regions is in the form of water, as corresponds to a climate characterized by mild winters. Solid precipitation is relatively rare, but it is important for its effects. There are two types of solid precipitation: snow and hail.

Snow consists of small ice crystals formed when the temperature of the clouds drops below 32°F (0°C), and is typical of winter. It is uncommon in Mediterranean areas and is associated with irruptions of cold humid air. It may have considerable negative effects on the plant cover, not so much due to the low temperatures but because of the mechanical action of snow settling on trees and shrubs that are in leaf. The weight of the snow that accumulates on the branches of evergreens

may cause the breakage of branches and other wounds that jeopardise further development. The heavy snowfalls in the west of the Mediterranean Basin during the winter of 1986-1987 caused serious damage to woods of holm oak and cork oak, especially to older stands consisting of trees with large crowns, and isolated specimens.

Hail is also a form of solid water but has different characteristics and causes. Water falls in the form of transparent or translucent grains of ice, from less than an inch to 2 in (2-5 mm) in diameter, although it may form much larger grains, called *hailstones*. Hail is typical of unstable situations that cause the formation of large storm clouds that develop vertically (cumulonimbus). At the top of these clouds, about 5 mi (7-8 km) above the Earth's surface, temperatures are below 32°F (0°C) and ice particles are formed. As these fall through the cloud, they cause the supercooled water freezes, forming hailstones. Depending on their size and weight, they can cause serious damage to vegetation.

Hailstorms are typical of summer and early autumn, when contact between cold air from higher latitudes and the air heated by contact with the sea or the land gives rise to highly unstable conditions, with associated phenomena of rising masses of hot air and the formation of vertically developing clouds. They are very rare during the winter as there are no masses of hot air to cause these violent instabilities.

2. Reddish and brown poor soils

2.1 The formation of Mediterranean soils

The Mediterranean conditions affecting soil formation are its low levels of dissolved nutrients and leaching (little excess water percolates through and escapes the soil and the plant), its strong contrasts in humidity, and its high erosion potential. One highly characteristic factor in the Mediterranean Basin is the fact that humans have manipulated the soil and vegetation intensively and extensively for millennia, often with major effects. A typical example of this influence is soil salinization and destruction caused by crop irrigation in ancient Mesopotamia. This long and intense human influence is probably one of the factors distinguishing soils in the Mediterranean Basin from the world's other areas with a Mediterranean climate.

Soil evolution

The end result of the operation of these soil formation factors in the Mediterranean area is that most soils are little developed, i.e. they are not very different from the rocky substrate; and they are shallow in high areas that do not receive sediments. The most soluble salts, such as chlorides and sulfates, have generally been washed out of the soil profile, while carbonates have only been partially mobilized. Thus soils which have developed over

6 **The Mediterranean soils and landscapes show strong anthropic influence**, as they have been modified by human activity for many centuries, as can be seen in these terraced fields on these hills in the Maestrat area, near Morella, at the southern end of the Iberian Range, which marks the southwest of the Ebro Depression, in the northeast of the Iberian Peninsula.
[Photo: Ramon Vallejo]

7 **Red Mediterranean soils are often palaeosoils formed under conditions more arid than the present**, and are often exposed by the strong rains, because of the poor plant cover. This is clearly shown by these slopes suffering gulley erosion in la Noguera, in the southern foothills of the Pyrenees, in the northeast of the Iberian Peninsula.
[Photo: Josep Maria Barres]

a calcareous substrate usually contain carbonates. Soils developed on non-calcareous parent materials show moderate alteration of the primary minerals (silicates), weak washing and acidification. This may sometimes lead to precipitation of calcium carbonate within the soil, derived from calcium released by silicates. These relatively immature soils often retain some characteristics of the unmodified rock, such as its color or particle size. Iron, the element that almost always colors the soil, is released in moderate quantities in these conditions and remains in an amorphous or crypto-crystalline state that is normally associated with organic material and with clays. This produces the darkish coloration characteristic of what were known as the *brown earths* in earlier soil classifications. To sum up, in keeping with the current soil formation conditions, the dominant soils in the Mediterranean area are not very differentiated due to the limited climatic weathering, and they have often only developed for relatively short periods due to the intensity of erosion processes.

A Mediterranean soil has no absolute age limit that can be fixed in the recent Quaternary, unlike the soil in areas at high latitudes or high altitudes. In these areas, the preexisting soils were swept away by glaciers only recently in geological terms (some 10,000 years ago). The Mediterranean countries thus contain a mosaic of old, even relict soils beside recently-formed soils with well-defined characteristics. Old soils persist on stable surface formations, such as extensive karstic limestone platforms or river terraces.

The case of red soils

The most typical mature soils of Mediterranean areas used to be called *Mediterranean red soils*, with a reddish brown color produced by the presence of hematite-type iron oxides, normally associated with the surface of clays. These soils are not exclusive to Mediterranean regions, as they are also found in tropical areas with a dry season, since the predominance of hematite (Fe_2O_3) as the mineral form of iron only occurs when there is an annual dry period. The formation of these soils from, for example, calcareous rock involves the following phases. First, carbonates are dissolved by carbonic acid in rain water, enhanced in the soil by the action of living organisms, and by the action of organic acids released by roots. Secondly, there is a period when carbonates are washed out of the upper horizons, or out of the profile, in the form of bicarbonates, generally calcium bicarbonate. Thirdly, there is a period of rubifaction (reddening), i.e. soils turn red due to the formation of hematite from the free

iron already present in the sedimentary rock, or from iron produced by the alteration of primary minerals (biotite, for example). Fourthly, there may be migration of clays with their associated iron oxides from the surface horizons, and their accumulation in subsurface *argillic* horizons. This difference in texture between a surface horizon that has lost its clay, and is therefore more sandy, and a lower clay horizon has led to these soils being known in Australia as *duplex soils*. This completes the description of the maturation process of red soils. However, aging processes may also occur. This would constitute a fifth phase, with carbonate enrichment of the soil profile due to later percolation of water saturated with calcium bicarbonate and the appearance of various types of calcium carbonate precipitates. As a result the soil would partially lose the initial bright red color. Also frequently seen in the Mediterranean Basin are sediments from the breakdown of nearby red soils that retain some of their original properties.

If red soil develops directly from hard limestone, the soil is usually shallow and patchy, even if it has been developing for a long time. This soil was typically called *terra rossa* (red earth), and is widely distributed throughout the Mediterranean Basin. The solid matrix of these soils is derived from the silicates contained in the limestone and any possible input of wind-borne particles. As these limestone rocks usually only contain small quantities of silicate residue, as little as 5% or less, enormous volumes of rock must be dissolved to produce small quantities of soil. It is thought that these soils have required very long periods to develop, and probably did so in more aggressive climates than those of the contemporary mediterraneans.

In the current landscape, red soils are found on stable surfaces that may have lost some or most of the top layers, or be in the form of paleosols, often fossilized under more recent sediments. In river terrace systems, the higher, older terraces have red soils, sometimes showing migration of clays and reprecipitation of carbonates, while in more recent terraces these characteristic are less clear. Time is therefore an important factor in this process. The different degrees of carbonate loss, acidification, and rubifaction of these soils clearly vary with respect to the current rainfall regime: soils in subhumid areas are redder, lack carbonates in the fine soil and may be slightly acid, while in semiarid conditions carbonate loss is incomplete, the color is poorly developed, and the pH is around 8. Everything seems to indicate that there has been a marked gradient in the evolution of soil formation in the Mediterranean area that has persisted for a long time.

The deepest and most mature red soils are found overlying loess materials and in high terraces. On consolidated substrates, they are found on rocks that allow easy moisture movement, that is with good drainage, such as hard limestones, sandstones, conglomerates, granites, and schists.

Red Mediterranean soils fit within the general scheme of soil formation in hot countries, and vary according to the level of hydrolysis of the primary minerals. Where gently changing, subtropical or Mediterranean conditions occur, the primary minerals have altered little, there was limited leaching of silica and bases, and iron oxides and various types of clays have accumulated, which, in their most developed state, migrate to a subsurface (argillic) horizon and give rise to Mediterranean red soils, called *Luvisols* in the FAO-UNESCO classification. Under tropical conditions, where the alteration of primary materials and the washing of silica and bases is greater than for *Luvisols*, simple clay minerals (kaolinites) are formed that coexist with sesquioxides of iron and a residue of relatively unchanging primary minerals, and also with clays mobilized, resulting in the formation of *Acrisols* and *Lixisols*. Finally, under equatorial conditions of high rainfall and temperature, profound alteration of primary minerals and the leaching of bases and silica occur, and the residue is enriched by more insoluble secondary compounds, sesquioxides of iron and aluminium. These conditions give rise to what are typically called *lateritic soils* (called *Ferralsol* soils in the FAO-UNESCO classification), and are also called bauxites when they are found as paleosols/rocks. These soils represent the most altered and transformed state of the rocks and are not very fertile. This series can also be considered as different phases of the same single pathway of change, which can occur over time in equatorial countries.

2.2 Mediterranean soil types

An area's prevailing climate broadly determines soil development, but differences in rock substrate, topography, activity of living organisms (including humans) and how long the soil has developed all combine to create a wide variety of soil types. For example, 70% of the 28 principal groups of soils defined by FAO-UNESCO in their world map of soils are present, to some extent, in the areas with Mediterranean climates though they occupy only a small part of the total land surface.

The different regions with a Mediterranean climate show many similarities in soil types. The differences

8 **Red Mediterranean soil on the Catalan coast (NE of the Iberian Peninsula)**, enriched in carbonates at depth and formed on an alluvial plain. This is an example of an old soil which has undergone various developmental cycles: carbonate loss, rubifaction, movement of clays and subsequent carbonate enrichment.
[Photo: Jordi Vidal]

9 Red soils (such as in this photo, on the island of Minorca) can also be formed on karstic limestone platforms. The discontinuous and fissured red soil is often colonized by communities of deeply-rooted shrubs, such as the mastic tree (*Pistacia lentiscus*) in this photograph. [Photo: Ramon Vallejo]

are basically due to the different distribution of the rock substrates and their particular geomorphological history in the Quaternary, which greatly influenced the present day mosaic of soils. The most notable of these differences is the predominance of calcareous rocks in the countries around the Mediterranean Basin compared to the other continents. Also the existence of large areas covered with residual materials derived from ancient, profound changes that have taken place in Australia.

Poorly developed soils

As already mentioned, Mediterranean soils have been conditioned by slow soil formation, potentially high erosion, greater evaporation than percolation, and their uninterrupted formation in the Quaternary. All this means that the majority of soils are poorly developed.

In mountain areas, plateaus, and on steep slopes that are often the result of degradation by human action, *Leptosols* are found. These soils are shallow and rest on a hard substratum. They show minimal development and are generally very stony. These soils predominate, for example, on karstic limestone formations: poorly developed or eroded red soils, where deep-rooted plant communities grow well. *Regosols*, which are more restricted in occurrence, show deeper profiles on unconsolidated parental material, but there is little soil differentiation. They develop on very recent sediments, such as lava and other pyroclastic materials from active volcanoes (Italy), or very stony and/or sandy sediments, for example dune materials.

Also included within this general group of not very differentiated soils are the *Fluvisols*. They are formed from recent alluvial material, as in fluvial terraces, deltas, or alluvial plains. They are deep soils of sedimentary origin with a profile characterized by alternating levels of different particle size, and showing little soil differentiation. Although these are immature soils, paradoxically they are generally very fertile: the most extreme examples are almost hydroponic media. The depth accessible to the roots is large, particle size is often well balanced, nutrient content is high, and the water regime is very favourable. In fact, the most productive agricultural areas in the Mediterranean are on these soils.

Well-developed soils

Moving on to soils with more developed profiles, *Cambisols* show a horizon that is clearly differentiated from the rocky substrate, altered and with a clear soil structure. They have a minimum depth of more than 12 in (30 cm). These are the most common soils in Mediterranean areas. Their properties are usually highly influenced by the nature of the parent material. On siliceous rocks, they may include brown-colored soils with moderate acidity (*dystric* type) or insignificant acidity (*eutric* type). On calcareous rocks they include the decarbonated red soils showing little migration of clays in the profile (*chromic*), and the soils that retain significant quantities of carbonates, normally on argillic materials (*calcaric*). This group includes soils with distinct soil formation processes that occupy a large part of the surface of areas with mesic conditions in the Mediterranean areas: they are the main soil resource for unirrigated farming.

Calcisols have very similar characteristics to those described above, but are distinguished by their strong secondary deposition of carbonates. They are thus impregnated with calcium carbonate, incorporated in solution; in extreme cases, reprecipitation of carbonates may form a hard crust that restricts the soil depth available for root growth. This is a clear example of soil senescence, with loss of productive potential.

Luvisols represent the greatest level of maturity found in soils in Mediterranean climates. They are characterised by the formation of a horizon of accumulated clays (argillic *B* horizon) and reddish colors (red-brown soils). They have a pH between neutral and moderately acidic, usually occupy very stable, discontinuous surfaces, and are generally considered relict or ancient soils. In many cases they have lost their surface horizon, after losing the clays that accumulated in the *B* horizon. They do not occupy very large areas in the Mediterranean Basin except in Greece. In the arid limits of the Mediterranean areas in Australia, red soils show transitions to alkaline soils (*Solonetz*). This is because of the abundance of sodium ions, and that causes highly alkaline soils (pH over 8.5) and the loss of soil structure.

Marginal soils

Other types of soils present in the Mediterranean areas occupy marginal locations and small areas. The most important, due to their influence on plant growth, are saline soils (*Solonchaks*). They are characterized by high levels of salts more soluble than gypsum, such as sodium chloride or sulphate. The high salinity produces osmotic problems for the roots of plants not adapted to it, and makes water uptake difficult even in soils saturated in water. Saline soils appear on the arid border of Mediterranean regions, in low-lying areas that water flows into but cannot flow out of (*endorheic*), and in coastal marshlands in any type of climate. Saline soils have been spreading in areas with a water deficit, as a result of introducing irrigation in soils with poor drainage, with salts at depth, or with saline irrigation water. Soil salinization is now the main overall cause of declining soil productivity.

Gleysols are associated with anaerobic conditions caused by soil waterlogging over long periods. As well as the soil's mottled, rusty and blue-grey colors, the reducing conditions change the soil's chemical activity, with the loss of nitrogen by denitrification, the presence of sulphides, and showing the mobility of ferrous (Fe^{2+}) salts. Many plants adapted to these soils pump oxygen to their roots and produce localized oxidation of the surrounding soil.

Finally, *Vertisols* show the relationship between Mediterranean areas and tropical areas with a dry season. They form in depressions over materials with a high clay content. Their characteristics include saturation in basic cations (positively charged ions), and particularly the abundance of swelling clays, that change their volume enormously according to their level of hydration, leaving large cracks in the dry season. This soil mobility creates problems for the construction of buildings and roads. These soils are typical of the Sahel.

2.3 Poor soils or impoverished soils?

Many Mediterranean plants show adaptations to drought that also allow them to conserve nutrients and use them more efficiently. These adaptations have been related to the poverty of Mediterranean soils, although this interpretation would have to be refined before extending it to the entire Mediterranean biome.

10 On the border of a **Mediterranean climate with a semiarid climate**, such as in Australia's Flinders Ranges, the accumulation of sodium in red soils leads to a loss of soil structure and consequently of the soil's physical fertility.
[Photo: Fritz Prenzel / Bruce Coleman Limited]

1. CLASSICAL, FRIENDLY, AND FRAGILE

Nutrient availability

Of all the essential mineral nutrients, *nitrogen* is consumed in the largest quantities by plants. The nitrogen that plants use comes almost exclusively from organic sources: in principle it enters the plant-soil system through the symbiotic fixation of atmospheric nitrogen. Its subsequent availability is regulated by the breakdown of organic material in the soil. *Nitrogen* deficiency is associated with very young or highly degraded soils, as well as with the slow breakdown of organic material. Mediterranean soils have a low organic material content compared to soils in temperate environments, despite the fact that leaf litter input into the soil is much the same in both types of ecosystem. The low levels of organic material are due to rapid breakdown, typical of hot environments. It is thus to be expected that the nitrogen cycle should be relatively efficient. Net losses of nitrogen are further limited by the small amount of water lost from the soil profile by drainage.

Potassium is another element that plants absorb in large quantities. Potassium levels may be low in soils developed from, for example, calcareous rock. However, its rapid recycling in the plant and between the plant and soil, and the soil's effectiveness at retaining potassium in an assimilable form, make it unlikely that plant communities will suffer from deficiencies.

Phosphorous is a more complex nutrient. It can be very scarce in highly leached soils and in sediments derived from ancient weathering (as in Australia), or in soils formed from limestone. In addition to these potentially low levels of phosphorous, its absorption

11 **Levels of phosphorous and nitrogen** (total, in percentages) in soils in the five regions with Mediterranean climates.
[Diagram: Editrònica]

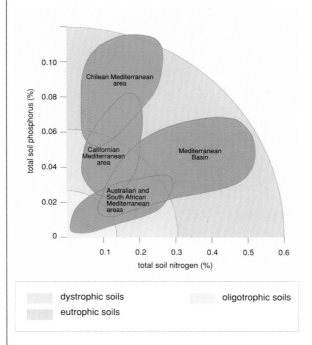

total soil phosphorus (%)

total soil nitrogen (%)

dystrophic soils
oligotrophic soils
eutrophic soils

is highly conditioned by the low solubility of its most common salts. Its availability in more altered soils is low because it is sequestered in the form of sesquioxides of iron and aluminium. In carbonate-rich soils availability is limited by the low solubility of calcium phosphate. This low availability explains the variety of adaptations plants have evolved to increase their phosphorous uptake. Some Mediterranean areas are thus naturally phosphorous-deficient.

Water, pH and depth

Soil nutrient availability and water availability are not independent, as nutrients are absorbed in solution. This means that it is sometimes difficult to dis-

tinguish the relative importance of the two limitations. Furthermore, the quantity of water required for the transpiration involved in carbon dioxide uptake, especially in *C3* plants, is much greater than is necessary for nutrient absorption. This suggests that in Mediterranean climates, in general, water shortage is more important than nutrient deficiency.

The chemical environment of Mediterranean soils is usually favorable to plant growth with a more or less neutral pH, except in extreme situations (alkaline soils). High levels of calcium in many areas leads to specific characteristics of nutrient availability and soil structure. Apart from calcium's effect on phosphorous availability (mentioned above), it also greatly reduces the solubility of some elements, especially iron. Calcicolous plants have evolved specific mechanisms to overcome these low levels of solubility. Many cultivated non-calcicoles, however, suffer from extreme iron deficiency (chlorosis) and require iron to be added in the form of an organic complex.

The depth of useful soil, that is to say, the soil that can be colonized by roots, is one of the most important factors in determining fertility. In many cases in dry farming (without irrigation), it has been possible to correlate the depth of the soil with the harvest produced. The properties of the soil surface are also very important, as they determine water uptake, and if that is insufficient, runoff and potential erosion. The physical properties of the soil are often its most fragile feature, and the factor with the most serious consequences in ecosystems affected by water shortage. The restoration of degraded soils usually presents more serious problems of physical fertility than of nutritional fertility.

Soil impoverishment

Soil impoverishment has been occurring in Mediterranean areas as a whole, and particularly in the Mediterranean Basin, for centuries. It can be attributed to the effects of human activities and to the fragility of the Mediterranean ecosystems, which can border on desert conditions. Fires almost always represent a net loss of nutrients, especially of nitrogen. Farming in mountain areas runs a well-known risk of eroding the surface horizons. Excessive grazing in recent historical times may also have impoverished the plant cover and, indirectly, the soil. Slightly or strongly degraded soils are common in the current mosaic of soils; although the reasons for the degradation are not always clear. In any case, soil degradation by human activity is a relevant factor when interpreting soil fertility in Mediterranean areas.

12 **The degradation following the abandonment of fields of marginal cultivation**, such as in this one in a dried-up stream in Granyena, in Catalonia (NE of the Iberian Peninsula), has been very important in the Mediterranean for many years. Other factors, such as fire or grazing, have also contributed to the increase in erosion.
[Photo: Ramon Vallejo]

3. The Mediterranean regions of the world

3.1 A fragmented and scattered biome

The climatic conditions that define the Mediterranean biome occur in only a small part of the biosphere, occupying only 1.2% of the Planet's land surface. This limited space is scattered over all the continents, and consists of five separate areas between 30° and 40° latitude, in both the northern and southern hemispheres. The boundaries of each of the biome's areas depend upon the criteria used to define the Mediterranean climate.

The Mediterranean Basin is the largest of these five areas. It occupies the coastal areas around the Mediterranean Sea, including parts of the European, Asian, and African continents. The biome's other four areas are more uniformly situated: they are on the western coasts of the continents at about 35° latitude. They include small areas of California, central Chile, the extreme southwest tip of South Africa, and the south and southwest of Australia. The differences among them in size and location are caused by the climatic differences between the two hemispheres, and by the geographical features of each continental mass.

3.2 The limits of the Mediterranean biome

The Mediterranean Basin

The Mediterranean Basin, the land around the Mediterranean Sea, is the area from which the Mediterranean biome takes its name. It comprises a strip of land, of varying width, surround the Mediterranean Sea in the latitudes between 30° and 45°N, and is approximately 1,056 mi (1,700 km) from north to south and almost 3,105 mi (5,000 km) from east to west.

13 Sea, mountains, sparse vegetation and buildings are the elements common to all areas in the Mediterranean biome, as illustrated by this view of Amalfi on the Salerno Gulf in Italy. The Mediterranean Basin began to adopt its current form 60 million years ago when, among other tectonic movements, the plate bearing Africa moved north. At the same time, due to the great pressure involved, the crust surrounding the new sea started to fold, leading to the formation of the mountain ranges surrounding the coasts, such as the Sierra Nevada, the Pyrenees, the Alps, the Balkans and also the many islands scattered throughout the *Mare Nostrum*. The abrupt and uneven terrain, with deep valleys and high plains, has led to the creation of many different microclimates and to a very diverse vegetation that is the result of adaptation to the general climatic conditions. *[Photo: J. Enric Molina]*

14 The distribution of the Mediterranean biome around the world. Apart from the Mediterranean Basin itself, which stretches about 2,485 mi (4,000 km) from Portugal to Iraq, other areas of the planet have similar climates: the rainfall coincides with the cold season, while the hot season is dry. There is a region with these climatic characteristics in the Cape area at the southwestern tip of Africa. Other such regions occur on the Californian coast of North America and the Pacific coastline of Chile, South America. There are two regions with Mediterranean climates in the south and southwest of Australia. There are significant variations within each of these five regions due to their local relief.

[Cartography: Editrònica]

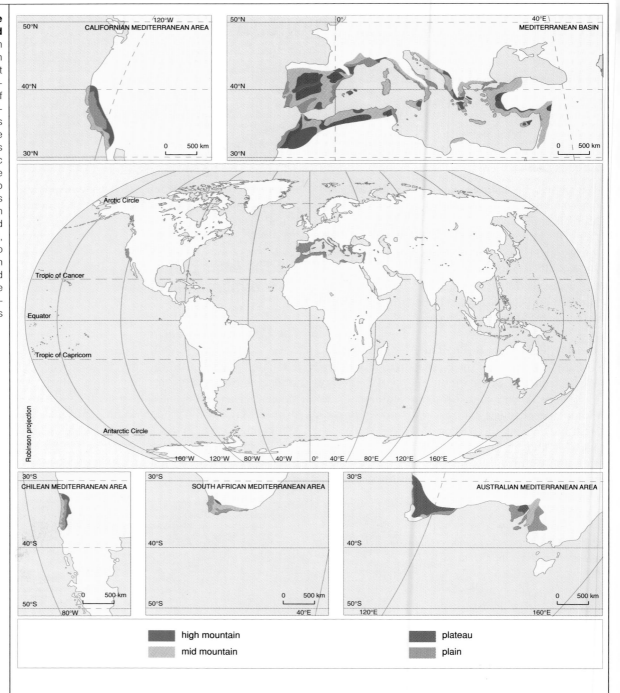

The highest latitude that this biome reaches is in the French Riviera at around 43°N. One must bear in mind that this apparently vast area contains a sea, the Mediterranean, and the areas that form this biome consist only of a narrow strip around the sea. The largest area is in the Iberian Peninsula, occupying almost all of Spain and Portugal, except the north which has a temperate climate, although much of the Atlantic coast is clearly Mediterranean in climate. This coastal strip continues through the south of France, Italy, the former Yugoslavia, Albania, Greece, Bulgaria, Turkey, Syria, Lebanon, Israel, Jordan, Tunisia, Algeria, and Morocco. The biome also includes all the islands, but excludes the coastal regions of Egypt and much of Libya which have distinctly desert climates. Overall, this area represents 60% of the biome's total area, the largest areas being in the Iberian Peninsula, Turkey, Morocco, and Italy. The presence of a small closed sea between large continental masses favors the entry of the Mediterranean biomes deep into the interior of the continents, unlike the situation in the other areas.

The climatic conditions conform to the general Mediterranean climate scheme mentioned before, with a dry summer caused by the powerful Azores

15 The topography and climate in California have allowed a rich and varied flora to develop. Mediterranean sclerophyllous trees (such as the oaks, strawberry trees and *Ceanothus*, in the photograph) mix with conifers and New World succulents (cacti and agaves) in an extremely broad botanical miscellany. The concentration of rainfall in winter and the extremely dry summers make California the area with the world's most typically Mediterranean climate.
[Photo: Larry Minden / Minden Pictures]

anticyclone, and a wet winter caused by the passage of cyclones or depressions that are not blocked by this anti-cyclone after its retreat south. In most of the basin the summer is relatively dry, with an almost total lack of rain, at least during high summer. Rainfall peaks may coincide with the winter period or with the equinoxes, depending on where the area is in the basin. The Mediterranean Sea's presence has significant effects, as its water is warm after the summer and when it comes into contact with the cold air masses descending from higher latitudes violent equinoctial storms occur. In certain places, rains caused by easterly winds, saturated with humidity after crossing large areas of open sea, are much more important than rains caused by cyclones, associated with the general wind circulation from the west.

The Mediterranean Basin is climatically less homogeneous than the other areas of the biome, due to the large area it covers and the effect of the sea it encloses. The 25,000 mi (40,000 km) of coastline and hundreds of islands contrast greatly with the markedly continental areas, such as the Iberian Peninsula or the interior of Turkey. Within this area, we can find all the thermal subclimates and all the variations in rainfall distribution mentioned above, with Morocco's climate being most typically Mediterranean. The western area of the basin normally receives some rainfall during the summer months, while the eastern area receives almost none, as in California. Overall, the Mediterranean Basin is the biome's most continental area, which is expressed in occasional summer rains, wide temperature variations and winters cold enough to be a limiting factor.

The Californian mediterranean

The Californian mediterranean occupies a narrow coastal strip between 42°30'N and 30°N, from Cape Blanco in the United States to Punta Baja in Mexico, with its approximate center in San Francisco. To the east, the strip only reaches 62-124 mi (100-200 km) inland, as it is limited by the Cascade and Sierra Nevada Ranges. In Baja California, the Mediterranean area is limited to the peninsular territory. Although it is a small area (approximately 10% of the biome's world total), this is the zone with the largest continuous areas

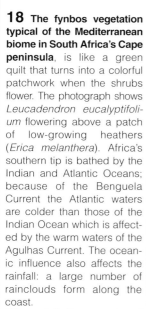

16 Evergreen broadleaf matorral landscape on the coastal range in the Chilean mediterranean, showing the dominant species *Peumus boldus*. The small area with a Mediterranean climate shows a series of climatic transitions as it is greatly affected by the Andes. The orientation and layout of mountain chains and plains in the Chilean mediterranean is similar to that of the Californian mediterranean, which means they coincide in many features.
[Photo: Lluís Ferrés]

17 The limited height of Australia's mountains, as shown in this photo of the Flinders Ranges in South Australia near Adelaide, and the influence of the surrounding oceans, mean that seasonal changes are not very drastic in Australia. Two areas on this continent can be considered to have a relatively mild Mediterranean climate, with rainfall mainly in the winter and with summers that are dry, and the characteristic vegetation is of sclerophyllous forests.
[Photo: Oriol Alamany]

18 The fynbos vegetation typical of the Mediterranean biome in South Africa's Cape peninsula, is like a green quilt that turns into a colorful patchwork when the shrubs flower. The photograph shows *Leucadendron eucalyptifolium* flowering above a patch of low-growing heathers (*Erica melanthera*). Africa's southern tip is bathed by the Indian and Atlantic Oceans; because of the Benguela Current the Atlantic waters are colder than those of the Indian Ocean which is affected by the warm waters of the Agulhas Current. The oceanic influence also affects the rainfall: a large number of rainclouds form along the coast.
[Photo: Colin Paterson-Jones]

of unaltered Mediterranean landscape, due to its relatively recent settlement by humans.

The area's climate is typically Mediterranean with a dry summer, due to high temperatures and an almost total lack of rainfall, and wet winters. Rainfall is mainly in the winter period, accounting for 85% of the yearly total. There are none of the equinoctial rains typical of some parts of the Mediterranean Basin, although in the coastal areas environmental humidity remains high throughout the summer period due to the influence of ocean air cooled by sea currents coming from the north. This why the Californian area is considered as the classic example of the Mediterranean climate, unaffected by the continental influences present in some parts of the Mediterranean Basin, such as the oceanic influences found in Chile, and the tropical influences in South Africa and Australia.

The Chilean mediterranean

The Chilean mediterranean consists of a narrow coastal strip between latitudes 30°S and 37°S, approximately coinciding with the location of the coastal cities of La Serena and Concepción, that is to say, central Chile. The inland city of Santiago de Chile is in the heart of the area. The strip is only about 62 mi (100 km) wide, bordered by the Andes, which in some places reach an altitude of 22,960 ft (7,000 m) within only about a 62 mi (100 km) of the coast. It accounts for less than 5% of the biome's total area.

The region's climate is greatly influenced by the cold Humboldt Current which flows along the Chilean coast and helps to mitigate the Mediterranean area's summer drought, and its daily and seasonal variations in temperature. Even so, there is a short period without any rainfall, but this drought is attenuated by the relatively low temperatures (5.4 Farenheit degrees or 3 Centigrade degrees lower than in the equivalent latitudes in the northern hemisphere). The rainfall peak is in the winter period, and is relatively high. Altogether, the Chilean mediterranean has a relatively cool climate, due to the pronounced oceanic influence.

The Australian mediterraneans

The Australian mediterranean is the second largest of the biome's five areas, accounting for just over 20% of its total area. It consists of two disjunct areas situated in the southwest (the region of Perth) and south (the region of Adelaide) of the continent. They lie between 37°30'S (southern region) and 27°S (southwestern region), the latter limit being the closest latitude to the Equator in the entire Mediterranean biome. In the southwestern region, the southern limit is further to the north than in the other region, around 33°30'S, as it borders an area with high rainfall at a latitude which, in other areas, is at the heart of the biome. The width of the area from east to west is considerable, about 1,550 mi (2,500 km), but there is an intermediate arid area separating the two regions. These areas' overall relief is gentle in comparison with the others, giving rise to a much more uniform landscape, due to lack of variation in height and orientation. The smooth relief has important effects on the river systems, which have slightly different regimes to those in the other mediterraneans.

The climate is typically Mediterranean, but with tropical influences, as in the South African region. The abundant rain mainly falls in the winter months, peaking at the winter solstice rather than at the equinoxes. There is a dry summer season, but there is no period without any rainfall. The edaphic factor, related to poor soils is also of great importance.

The South Africa mediterranean

The South African mediterranean, at the southwestern tip of the African continent, is the smallest of all the mediterraneans: it covers a small area, accounting for only about 3% of the total biome. It lies between latitudes 30° and 35°S (the latter representing the southernmost extreme of the African continent), and occupies a coastal strip that stretches along the western and southern edges of the continent. Its approximate center is Cape Town and it continues for another 186 mi (300 km) to the east, reaching the arid zones of the interior. This area, unlike the others, does not make contact with temperate climates at higher latitudes, as it lies at the southernmost tip of the African continent.

In general terms, its climate conforms to the scheme outlined above. Rainfall is mainly in the cold season, with a clear maximum in winter, as in Chile, but with lower total precipitation. Rain is uncommon and irregular during the dry summer period, but there is no period in which it is totally absent. This is due to the influence of tropical climates with summer rains. Edaphic factors are very important, as more than half the area consists of very nutrient-poor soils.

The Cape kingdom: geraniums and gazanias

The gardener sent by the Royal Botanic Gardens, Kew, arrived in Cape Town in 1772, on board the ship *Resolution* under the command of the famous Captain Cook. His mission was to collect South African plants for use in gardening. He returned to Europe with many different plants, including fifteen species of *Pelargonium*: the era of the geranium, perhaps the most internationally widespread ornamental plant, had begun.

Francis Masson (1741-1805) [Library of South Africa (Cape Town)]

Like geraniums, many of the most commonly used garden plants are from South Africa. Masson was the first person who systematically exploited the decorative possibilities of the plants found in the Mediterranean region at the southernmost tip of Africa. In recognition of his work, botanists named a species of Cape heather after him, *Erica massoni*. In fact, there were plenty to choose from: compared to the dozen or so species of heathers that are native to Europe, the South African flora lists over 500. On the Cape of Good Hope itself (or in short, the Cape peninsula), in an area covering a few dozen square kilometers, there are over a hundred species of heather. Not only are there many species of plant on the Cape, but they are so distinct from those found in the rest of the world that they are generally accepted as an independent floristic region, called the Cape kingdom.

Spring flowers in Namaqualand, South Africa [Colla Swart / ABPL]

B. J. Burchell (1781-1815) [Library of South Africa (Cape Town)]

Geranium (*Pelargonium cuculata*) [*Flowers of South Africa*, Auriol Batten, Southern Book Publishers, Johannesburg (1988)]

William J. Burchell explored the area in depth, travelling in a well-equipped and picturesque wagon, between 1810 and 1815. Shortly after beginning his research, he wrote: "To give some idea of the botanical wealth of the region, I will say that in the course of a single mile I have managed to collect 105 distinct species of plants, even at this unfavourable season." Later, as irrefutable proof, he published a floristic catalogue listing 8,700 species. Even so he did not find them all: we know today that the Cape flora (Cape Floristic Region) contains different 8,600 species of which 7,396 are found only on the Cape Peninsula. There is probably no other area in the world with such a high degree of botanical diversity, not even the tropical rain forests.

Appropriately enough for plants from a Mediterranean region, Cape species do not require a great deal of water and tolerate poor soil conditions. This fact, together with their beauty, explains why they have been so successful as garden and pot plants. We have already mentioned geraniums and heathers (*Pelargonium*, *Erica*), but there are also gazanias and agapanthus (*Gazania*, *Agapanthus*), commonly planted in masses in flower beds, and gladioli and calla lilies (*Gladiolus*, *Zantedeschia aethiopica*), two of the most important cut flowers, although the horticultural varieties are quite distinct from their wild Cape ancestors. Some succulent plants from the Cape have also prospered, such as the remarkable little plants of the genus *Lithops*, which look just like pebbles (why they went unnoticed for such a long time, until Burchell discovered the first *Lithops* species, *L. turbiniformis*). Another that has spread widely is the Hottentot fig (*Mesembryanthemum* [=*Carpobrotus*] *edule*), so-called because of its edible fruits which were used by the Hottentots as food and which resemble somewhat those of the prickly pear (*Opuntia*). It is an extremely efficient colonizer of arid slopes in temperate climates, and has even become quite invasive. And of course, there are also the famous Proteaceae, (*Protea*, *Leucospermum*, *Leucadendron*, etc.), which feature in many arrangements of dried flowers, or the delicate Cape shamrocks (*Oxalis*), small plants but impossible to eradicate from fields and roadsides. With all these and many other species to display, it is easy to understand why the botanical gardens in the heart of the Cape kingdom, especially the renowned National Botanic Gardens at Kirstenbosch, present an unrivalled explosion of colour and beauty.

The reasons for this surprising floral diversity are still far from clear. The Cape flora obviously has certain similarities with the neighbouring tropical flora, and also with that of Mediterranean Australia, but it is nonetheless clearly separate from both of them. It would seem that it is a relict floristic assemblage, the remains of the flora that occupied a much greater area when the land surfaces we know as Africa, southern America, and Australia formed the Gondwana continent and the climate was moister. The poverty of the soils has also contributed to the appearance and consolidation of this diversity, as always happens in adverse environments. Whatever the case, it is quite remarkable to find so many thousands of different and singular species in an area only 621 miles (1,000 km) long and barely 62 miles (100 km) wide.

The German botanist Rudolf Marloth, at the beginning of the twentieth century, was the first to talk of the Cape kingdom. This stretches from Port Elizabeth on the shores of the Indian Ocean to Port Nolloth, on the Atlantic coast in Namaqualand, and has perhaps the largest and most beautiful display of spring blooms in the world. The fascinating Cape kingdom, with its "karoo" and "fynbos" formations and dominated by Table Mountain (Taffelsberg in Afrikaans), is the pearl of the unrivalled Cape Peninsula, the queen of the kingdom.

2
Life in sclerophyllous formations

1. Mediterranean sclerophyllous formations

1.1 The trick of sclerophylly

Each combination of environmental factors may be tackled with different strategies, but one of these may be optimal and end up being dominant. In the case of plants, climatic factors determine the structure and duration of the leaves, two basic responses to the distribution of temperatures and rainfall over the year. In environments where temperature is never a limiting factor and where water is always abundant, plants are evergreen and have relatively large leaves with little thickening, and are termed malacophyllous. These types dominate in wet temperate and tropical biomes, where, for example, tropical rainforest and laurisilva (temperate rainforest) develop. When cold or dry conditions are limiting in one period of the year, the malacophyllous vegetation is deciduous, which means that it sheds its leaves in the unfavorable period. This is what happens to the summer deciduous plants in sub-desert climates, and winter deciduous plants in cold temperate climates. When the unfavorable period is very long, this strategy does not work, because there is not enough time to make new leaves and pay back the energy invested during the short period of vegetative growth, and it is necessary to adopt a strategy based on resistance. In these cases, the vegetation is dominated by plants with acicular, or needle-shaped, ever-

green leaves, such as conifers, whose leaves can resist the long periods of excessive cold in an inactive state, as happens in the taiga. In most of the Mediterranean biome a set of factors have led to the appearance of a different strategy, that of the evergreen sclerophyllous plants. This strategy is a response to an environment that is never clearly limiting, but is often close to the limit, as well as being highly irregular and unpredictable.

Functional features

The limiting environmental factors in the Mediterranean biome are the dry summer period and the cold winter. This means that apart from the warm, wet periods in the spring and autumn, Mediterranean ecosystems have to deal with restrictive environments that slow down their activity, although they are rarely harsh enough to cause biological activity to stop completely. The evergreen nature of typical Mediterranean vegetation is a response to these factors, as it allows activity to begin as soon as environmental conditions cease to be limiting, which may happen in any period of the year due to the biome's irregular climate. In laboratory experiments, it has been shown that sclerophyllous plants can respond rapidly; some can begin to photosynthesise within 10 minutes of being watered

19 **The possession of small, hard, shiny leaves**, such as those of this splendid *Protea cynaroides* from the Cape scrub in South Africa, are characteristic of Mediterranean vegetation. This is called sclerophylly, and is a conspicuous characteristic of the Mediterranean biome.
[Photo: Nigel Dennis]

20 **The leaf structure of the holm oak** (*Quercus ilex*), a typical sclerophyllous plant. Its leaves are dark green and leathery, and have a thickened cuticle and parenchyma to resist dehydration, although they are more tender when they are young. They usually have a high specific weight, low nitrogen content, and relatively low photosynthetic rates.
[Photo: Ernest Costa]

21 Daily changes in plant activity in a sclerophyllous plant on a summer's day. As the dry season progresses, the assimilation phase shortens because the water shortage becomes limiting more and more quickly. In the most adverse conditions, total respiration may be greater than assimilation, which may even fall to zero.
[Drawing: Editrònica]

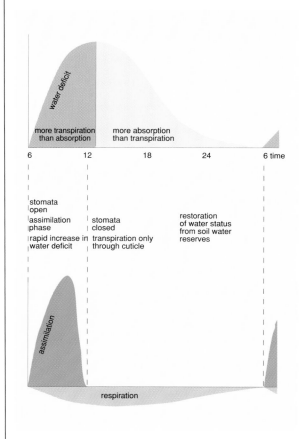

following a really severe dry period in which they were totally inactive. This statement is generally valid, but must be interpreted with caution due to the biome's different subclimates, as both the length of the dry period and the intensity of the cold period show wide variations and only in certain highly continental regions, such as central Syria, are both drought and cold conditions harsh.

In these conditions there is no option but to maintain the leaves throughout the year in order to take advantage of the short favorable periods that may occur at any moment, and this explains the evergreen nature of most of the biome's vegetation. On the other hand, the leaves must support dry and cold periods, and this forces them to adopt an especially resistant structural design. Sclerophyllous leaves typically have thick cuticles, a highly developed palisade parenchyma, a high content of strong structural materials, such as cellulose and lignin, and a low number of sunken stomata that are partially covered by hairs. These leaves loose little water through evaporation from the cuticle, and control transpiration effectively by opening and closing their stomata. They also show good mechanical resistance to the loss of water, and can withstand temperatures below 32°F (0°C) without suffering damage. They are more like an all-terrain vehicle, a

jeep, than a family car or a limousine, with all the advantages and disadvantages this implies.

Transpiration and catching sunlight

Two factors contribute to control water loss, as mentioned before. The thickened cuticle acts as insulation and is combined with a low density of stomata. In the leaf of a malacophyllous plant the area of the stomata may represent 0.5-1.5% of the total leaf area, while in a sclerophyllous leaf they may only represent 0.2-0.5% of the area. This rigid structure allows significant water loss without withering that would lead to irreversible mechanical damage: while in malacophyllous plants the loss of 30% of leaf water content begins to produce damage, many sclerophyllous plants can withstand loss of up to 70%. This structure also allows them to withstand temperatures close to 14°F (-10°C). However, such a resistant structure cannot function very quickly, which results in a relatively low rate of photosynthesis. The barriers that prevent water loss also mean that it is difficult for carbon dioxide to reach the chloroplasts, and the high content of structural materials means that active proteins, such as photosynthetic enzymes, are only present in low proportions. The leaf of a malacophyllous plant may photosynthesise between 15 and 40 mg of carbon dioxide per square decimeter per hour and its photosynthetic system is not saturated at high light intensities, while a sclerophyllous leaf has to make do with values of between 4.5 and 16 mg of carbon dioxide per square decimeter per hour, and is saturated at light levels below the environmental maximum.

In this biome, however, total annual fixation of carbon dioxide by summer-deciduous plants and sclerophyllous plants is similar, since although the former show clearly higher photosynthetic rates, the latter compensate for this by prolonging their activity. Overall, the problem is the same as that of all Mediterranean vegetation: too much light and not enough water.

To understand the advantages of sclerophyllous morphology, we can compare the behaviour of two typically Mediterranean trees, the holm oak (*Quercus ilex*), an evergreen with sclerophyllous leaves, and *Quercus humilis*, a winter-deciduous tree. *Q. humilis* begins to close its stomata when leaf water content is below 90% of saturation, and totally closes them when this reaches 70%, but it continues to loose water by transpiration through the cuticle at a rate equivalent to 50% of maximum transpiration. The holm oak, however, starts to close its stomata when water content reaches 85% of saturation, and closes them totally at 70%, in much the same way as *Q. humilis*, but with the difference that transpiration from the cuticle only represents 3% of the total. This, together with its ability to support greater water loss without suffering damage, means that the holm oak can resist dry periods up to 15 times longer than *Q. humilis*.

Structural cost

The robust nature of sclerophyllous leaves results in a high specific weight, i.e. they contain a high amount of organic material per unit area. While many malacophyllous plants have specific weight of less that 10 mg per square centimeter, sclerophyllous leaves may reach 20-40 mg per square centimeter. A resistant structure may last longer, but it must last longer because it costs more to produce and is harder for it to show a net gain, especially bearing in mind that this structural resistance imposes restrictions on the maximum photosynthetic rate. To make 0.035 oz (1 g) of sclerophyllous leaf represents an energetic cost equivalent to 0.059 oz (1.7 g) of glucose, much more than is necessary to construct 0.035 oz of malacophyllous leaf, and also more needed to make needle-shaped leaves in the taiga, which can withstand extremely low winter temperatures but which only require a cost equivalent to 0.045 oz (1.3 g) of glucose.

When something is expensive and not very productive it must last a long time, and this is why sclerophyllous plants are evergreen, that is to say they keep their leaves in winter and the leaves last more than one year, the minimum necessary to qualify as evergreen. The index of sclerophylly, the ratio of lignin and cellulose weight to that of proteins, is between 230-450 for sclerophyllous plants, in comparison with values of 80-180 for Mediterranean winter-deciduous plants.

Sclerophyllous leaves are thus relatively poor in nutrients because of their low protein content, and so their construction is not expensive in nutritional terms. Yet they are also resistant to decomposition, so the few nutrients they contain are released only with difficulty. This characteristic may be connected to their evolutionary origin, which many authors locate in former oligotrophic environments that existed before the appearance of the Mediterranean climate. Whatever their origin, this set of advantages and disadvantages has become an excellent solution to the environmental problems raised by the Mediterranean climate, making sclerophylly one of the most notable characteristic of the biome's landscapes.

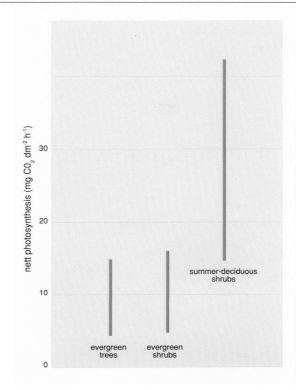

22 **Photosynthetic capacity of plants from Mediterranean climates**, measured mainly on live plants. *[Drawing: Editrònica]*

2. LIFE IN SCLEROPHYLLOUS FORMATIONS

1.2 Never at a standstill, always ticking over

The ecosystems of the Mediterranean biome are as a whole not very lush or productive in comparison with warmer biomes, such as the tropical ones, or colder, such as the temperate ones, because they suffer from continuous environmental restrictions, whether due to water shortage or low temperatures, that forces them to function at a slow rate.

Biomass

The quantity of organic material accumulated in an ecosystem, termed *biomass*, is a parameter indicating its level of complexity. Environmental factors, basically climatic factors such as temperature and water availability, and soil factors, such as nutrient availability, determine the maximum development of terrestrial ecosystems. In the case of the Mediterranean biome, with the environmental limitations mentioned above, this development is less than that of temperate forest systems and greater than that of deserts, which is in accordance with its situation of transition between the two biomes.

In general, when rainfall exceeds 16-20 in (400-500 mm) a year, the ecosystems that develop are dominated by trees, or if not then the vegetation is dominated by shrubs. In this case, the accumulated biomass

23 Holm oak woodlands are typical of the Mediterranean biome, and are lush communities that are rich in species with persistent, coriaceous, dark-green leaves, as can be seen in this photograph of the Collserola hills, near Barcelona. The leaf biomass of holm oak woods is usually greater than that of deciduous forests and smaller than that of coniferous woodlands. The biomass of a holm oak wood usually contains a large amount of nutrients, such as calcium, followed by nitrogen and potassium, and in lesser quantities, magnesium and phosphorus. [Photo: Ernest Costa]

is relatively modest, with values ranging from less than 1 t/ha (metric tons per hectare) in shrub formations in transition zones to deserts, to 50-60 t/ha in the most developed ecosystems, in the transition to forest formations. It is very difficult to assign values to each type of shrub ecosystem in the biome, because although soil and climatic factors play a very important role, other factors such as recurrent fires or its management history may play an even more determining role. In shrub ecosystems that have not been affected by fire or human management, biomass never reaches much more than 100 t/ha.

In Mediterranean forests these values may be greater, although if they have been exploited or are regenerating after a fire, the values may be less than the maxima cited for shrubs. In the Mediterranean biome it is very rare to find woods that have not been subjected to continuous exploitation that has maintained them in a state far from their maximum possible development, and this strengthens the idea that the biome is sparse in woods.

In the few remaining well-preserved woods, it has been shown that biomass may reach 300 t/ha, similar to the values for many forest ecosystems in other biomes. This is true for some of the magnificent holm oak woods in sparsely populated areas of the Mediterranean Basin, such as the mountains in Sardinia, or the remains of the thick Australian jarrah woods, whose trees can reach 130 ft (40 m) and maintain their activity in summer thanks to their deep roots. Environmental restrictions thus lead to slow renewal of biomass rather than low absolute values. To put it another way, Mediterranean woods may reach high biomass values, but their production is low, meaning that rates of renewal and accumulation are relatively low, clearly lower than those found in a temperate woodland or in a rainforest.

All the biomass values cited so far are in metric tons of dry material accumulated in the aerial parts of the plant, and ignore water content. Data for animal biomass is scarce, but never represents a quantitatively important proportion, although its role may be of great importance. Data for underground root biomass are also scarce because of the methodological difficulty of studying it, although quantitatively it may represent a large proportion of the total biomass. In a biome where the summer dry period is a determining environmental factor,

24 **In the Western Australian mediterranean**, near Perth, there are dense woodlands of jarrah, dominated by the species (*Eucalyptus marginata*). The undergrowth is rich in vegetation, especially in the spring. The photographs shows an *Acacia*, and *Kennedia coccinea* in flower. This eucalyptus can live for 200 years and reach a height of 131 ft (40 m) and forms woods so dense that they do not allow sufficient light to pass through for the leaves of the side branches to photosynthesize. As a result these leaves are shed and the plant is left with a naked trunk topped by a tuft of leaves. [Photo: Jean-Paul Ferrero / Auscape International]

2. LIFE IN SCLEROPHYLLOUS FORMATIONS

one may expect to find well-developed underground parts to ensure an adequate and constant water supply. Subterranean biomass may be greater than aerial biomass in the case of shrub communities of the driest areas of the biome. In most forest formations, the ratio of subterranean biomass/aerial biomass is less than one. Underground biomass cannot be ignored, since in a well-developed woodland like a holm oak forest, that has an aerial biomass of 170 t/ha, the roots may account for 70 t/ha, which represents 40% of the aerial biomass and 30% of the ecosystem's total biomass.

Water economy

Low winter temperatures impose restrictions on the functioning of Mediterranean ecosystems, but the most important factor and the one that has caused the most conspicuous adaptations, is the combination of dryness and high temperatures in the summer period. Saving water and making the best use of it, is a problem that species in this biome have had to resolve. Animals, in general, obtain their water from their food or ingest it directly, and so do not face serious supply problems, as they are mobile enough to find it for themselves. In fact, most animals are more independent of climatic conditions than plants, due to their structure, their highly regulated vital functions and their mobility. With the exception of those animals that are less adapted to terrestrial life, such as some crustaceans, molluscs, and amphibians, the lack of water in the Mediterranean biome does not reach the extremes found in deserts and which have favoured the appearance of physiological, ethological, and morphological adaptations in most species that live there. The soil fauna is without doubt the fauna that shows the most marked response to seasonal water shortage, and its members show three main types of strategy to deal with this problem. The first strategy is to remain deep in the soil all the time at levels that never dry out and where plant roots, abundant in the Mediterranean biome, keep the soil spongy and ensure a supply of organic material. The second strategy is practised by animals that always live close to the soil surface, and respond to the dry period in two different ways: by lethargy or by producing reproductive resistance structures to avoid water shortage. On the other hand, the animals in this group that remain active have developed protective structures, such as thick cuticles complemented by physiological mechanisms to save water and to consume the materials forming the hard sclerophyllous leaves. The third strategy

25 The preferred habitat of the five species of tortoise found around the Mediterranean Basin is the coast, mainly low areas with gentle relief (the picture shows *Testudo graeca* next to a clump of mastic tree [*Pistacia lentiscus*] in Mallorca, the largest of the Balearic Isles). Mediterranean tortoises like all reptiles, are cold-blooded, and so their activity changes greatly over the course of the year. In the winter period they reduce their activity, resting underground. When the spring arrives they bask in the sun, and are very active, moving around, feeding and seeking a sexual partner. In the summer, when they have to tolerate the dryness of the climate, more and more individuals rest in the shade, and eating and movement are concentrated in the first hours of daylight.
[Photo: Javier Andrada]

consists of migration within the soil, either vertically to lower soil levels or obliquely towards depressions with deeper, moister soils.

Plants are very sensitive to water shortage and, as they dominate terrestrial ecosystems, this factor is essential to understanding their modes of functioning. In the Mediterranean biome, the basic plant types are winter-deciduous malacophyllous plants, evergreen sclerophyllous plants and summer-deciduous malacophyllous plants. The first group are restricted to the wetter, cooler parts of the biome and their strategy is aimed more towards adapting to the cold winter, rather than to the summer water shortage, which is limited and not very determining in these environments. The other two strategies are responses to the water shortage, and only to a lesser extent to the temperatures. In conditions of summer water shortage and moderately cold winters, which are average or typical for the biome, the dominant plants are evergreen sclerophylls, and where the dry period is harsher, and the winter cold is less, as in the warmer areas of the biome, the dominant vegetation is summer-deciduous plants. In the first case, the dry period can be dealt with by a series of morphological and physiological adaptations, while in the second case the rigours of the dry period combined with the mild winter temperatures favors the concentration of activity in the winter, making the summer period a rest period.

In ecosystems dominated by evergreen sclerophyllous plants, saving water in the dry summer period is achieved by having an effective stomatal closure mechanism that regulates water loss. Thus, activity is limited to the coolest periods of the day, and ceases during the hotter periods in order to avoid water loss that cannot be compensated from the soil reserve. The result of this is that transpiration over the summer period is up to 20% less than in the winter period. This is done without reducing leaf-area, but by using physiological mechanisms to control transpiration.

In the case of summer-deciduous plants, transpiration is reduced to almost nothing in high summer, by the drastic or total reduction of leaf surface; their activity is concentrated into the rest of the year. There are intermediate strategies that consist of either a partial reduction of leaf area, by losing some of the leaves, or a reduction by replacing the winter leaves with more xeromorphic (adapted to an arid environment) ones. This seasonal dimorphism combines two forms of regulation: transpiration is reduced by diminishing leaf area and by increased physiological control of transpiration. In such cases, the reduction of the transpiring mass in the summer dry period may be so great as to leave only 15% of the winter transpiring mass.

Mediterranean ecosystems thus have two mechanisms allowing them to use the water available thoroughly: it is essential to optimize use of this limiting factor. A holm oak woodland, with an annual rainfall of about 23 in (580 mm), shows evapotranspiration of 80% of total rainfall, that is to say about 18 in (464 mm), while in conditions where rainfall is 34 in (870 mm), evapotranspiration is 54%, which corresponds to 18.5 in (469 mm), which shows the minimum requirement of water needed to function. When rainfall is less than 16 in (400 mm), or the dry period is very long, the landscape is dominated by shrub formations with lower water requirements. In Californian chaparral, water consumption by evapotranspiration in a single shrub community may vary from 10-14 in (250-350 mm) a year, values which correspond respectively to rainfall of 12-51 in (300-1,300 mm) a year, showing both the Mediterranean climate's irregularity and the importance of low and average rainfall values in determining the characteristics of the landscape. An excess of light and a shortage of water means that Mediterranean forests and matorral leave little water for the biome's rivers.

Production

The limiting nature of water supply in Mediterranean ecosystems has the effect of overall productivity being low, and that rates of biomass renewal and increase are also low. Mediterranean woods and shrublands function slowly, like low-yield, low-consumption engines.

Net aerial primary production in a tropical rainforest is around 20 t/ha per year dry weight, and that of a deciduous or coniferous temperate woodland is around 12 t/ha per year. In Mediterranean ecosystems, average production values are around 5 t/ha per year in woods and around 3 t/ha per year in shrub formations, although in especially favourable conditions this figure may reach 10 t/ha per year, and in extremely dry conditions it may not even 1 t/ha per year. It should be borne in mind that many Mediterranean woodlands show a production that is lower than that of the taiga, which produces around 8 t/ha per year in especially harsh conditions caused by low temperatures, and this shows the determining role of water availability in terrestrial ecosystems.

Decomposition

Decomposition of dead organic material plays a very important role in the flow of energy and the cycling of nutrients in ecosystems, as it determines the return of the elements to their inorganic forms, i.e. the form in which they are available again to the vegetation. Decomposition is a very complex process in which fungi and soil bacteria play a very important role, as do all the soil fauna that feed on these remains. In Mediterranean ecosystems, decomposition is relatively slow as it faces two major problems: the shortage of water, which limits decomposer activity, and the structural resistance of sclerophyllous litter.

The summer dry period causes intense desiccation of the upper soil layers and this greatly limits the decomposing activities of fungi, bacteria and the many members of the soil fauna that also play a very important role in decomposition, as they

26 One strategy to reduce excessive water loss in summer is to shed the leaves in the driest period, which is what summer-deciduous plants do. The plant in the photograph (*Aesculus californica*) is following this strategy by shedding its very tender broad leaves in summer. This is due to physiological mechanisms that have evolved to avoid water loss in the period of minimum environmental humidity and the highest temperatures, when evaporation is greatest.
[Photo: Ramon Folch]

27 Leaf litter, consisting of the leaves that have fallen from trees, and other animal and plant materials, such as the rhytidome (outer bark) of eucalyptus, is reused by organisms through the action of decomposers that close the cycle of materials. Decomposers are very important in the material cycles in ecosystems. Their activity, which is mainly oxidative, is also influenced by climatic conditions. In the mediterraneans, the decomposition cycle reflects the summer dry period, when lower water availability reduces the process. In the winter, the cold also reduces this return of materials. At the equinoxes—in spring and autumn—decomposition starts again. The photo shows *Eucalyptus*, and a clump of mastic tree (*Pistacia lentiscus*) on the left. [Photo: Josep Pedrol]

fragment the leaf litter making it easier for bacteria and fungi to attack. In the most arid areas of the Mediterranean biome, the soil fauna shows a winter peak in density and a summer minimum, due to the reduction of soil fauna populations and their migration to deeper soil layers maintaining some humidity.

In the cooler, more humid areas of the biome there are two yearly maxima, in the spring and in the autumn, as the winter cold also acts as a limiting factor. The bacteria show a succession, or progressive replacement of some species by others, as soil desiccation increases. Microbial activity is greatly reduced when midsummer arrives, and this stops decomposition almost completely. Bacterial populations may be activated by water, giving rise to peak density and activity within a month of the first rains at the end of the dry period; or they may also be activated by temperature in the spring coinciding with the increasing temperatures. Thus, decomposition is restricted to the wetter periods of the year, which are also the coolest, giving rise to a discontinuous decomposition process with two annual peaks separated by an excessively dry period and a relatively cold one. The simultaneous occurrence of heat and humidity that is so favourable to decomposition, since most fungi and bacteria show maximum activity between 77°F and 95°F (25°C and 35°C), does not happen in Mediterranean climates.

The sclerophyllous nature of many Mediterranean plants is a second problem that slows down the process of decomposition. The lignin and nitrogen content of plant remains strongly influences their rate of decomposition. Sclerophylly is accompanied by another special feature of the leaves an especially high proportion of structural materials, such as lignin, together with a low proportion of essential nutrients like nitrogen, meaning that these leaves are more similar to those of conifers than to those of broad-leaved trees. The leaves of a typical Mediterranean sclerophyllous plant like the holm oak (*Quercus ilex*) contain little more than 1.1% nitrogen, while the leaves of a broadleaf from a wet temperate climate, like the beech (*Fagus sylvatica*) are close to 2%. This means that in the same period (18 months) that it takes the leaf litter of a temperate climate oak forest to lose 40% of its weight, the leaves of a community dominated by sclerophylls, such as chaparral, only lose 20% of their weight. Since leaves, which are renewed more quickly than stems, represent more than 70% of the litter produced annually by the vegetation, its characteristics determine the rate of decomposition of the leaf litter as a whole. The sclerophyllous nature of the vegetation thus causes a slowdown in the return of nutrients to the inorganic fraction, a delay that may lead to relative impoverishment of the soil, but this is compensated by the frequent fires that mineralize the accumulated organic material.

The circulation of nutrients

The overall circulation of nutrients in an ecosystem occurs partly within the ecosystem and partly externally. The internal nutrient cycle consists basically of absorption of nutrients by plants, their translocation and re-mobilization from old organs to new parts, retention in the newly created structures and then their return to the leaf litter and subsequent decomposition. The external nutrient cycle includes the inputs to and outputs from the system, most of which are related to rainfall and drainage of water, although the inputs derived from dry deposition and the direct absorption of nitrogen gas may also be important. The rate of nutrient circulation is closely linked to that of organic material, because high productivity or rapid decomposition lead to intense mineralization of nutrients and speed up their cycles.

We have already mentioned the low productivity of Mediterranean systems, caused by the low availability of water, and this causes the cycles to be, in general, slow. The limited speed of decomposition reinforces this slowness, meaning that the general rate of nutrient turnover in these ecosystems is slow.

In some conditions, the availability of nutrients, or a particular essential element, such as phosphorus or nitrogen, may be a limiting factor. This happens in all biomes with poor soils, but it is relatively common in the mediterraneans, and especially in Australia and in the Cape region. In these conditions, ecosystems bring into play mechanisms that ensure nutrient supply and control by strengthening some parts of the cycle, such as retranslocation. Thus, by withdrawing essential nutrients from the structures that are going to be shed, such as old leaves, nutrient accumulation in the leaf litter is avoided, as is loss by leaching.

Mediterranean plants adapted to these conditions, normally small-leaved sclerophyllous plants such as members of the Ericaceae or Epacridaceae, have leaves with a low level of nutrients, and translocation allows them to recycle 70% of the essential nutrients during leaf senescence, reducing even further the rate of nutrient circulation. Some plants, such as *Banksia ornata*, which grow on poor soils, may withdraw as much as 90% of the phosphorus and 30% of the nitrogen during leaf senescence. Long-lived tissues and structures, such as stems and roots, can store water and nutrients, increasing plant control over nutrient cycles. Large rootstocks consisting of a specialized tissue called a lignotuber, have been interpreted as an adaptation to fire, although they also

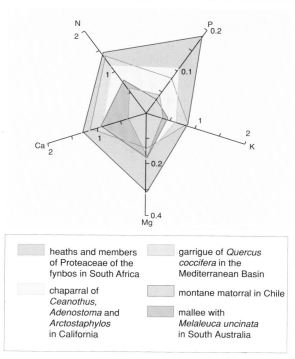

28 Nutrient content of leaves (*N* nitrogen, *P* phosphorus, *K* potassium, *Mg* magnesium, *Ca* calcium) in the four most representative types of shrublands in the Mediterranean biome, expressed as a percentage of total leaf dry weight. The acidic Cape fynbos is clearly poorer in nutrients than the calcicolous Chilean matorral. [Drawing: Editrònica, based on Kruger, 1983]

heaths and members of Proteaceae of the fynbos in South Africa

chaparral of *Ceanothus*, *Adenostoma* and *Arctostaphylos* in California

garrigue of *Quercus coccifera* in the Mediterranean Basin

montane matorral in Chile

mallee with *Melaleuca uncinata* in South Australia

play an important role as a nutrient and water store; they are highly developed in species living on oligotrophic soils, such as heathers and the Australian jarrah, which can form highly developed forests on what are considered to be the poorest forest soils in the world. In these cases of marked oligotrophy and slow decomposition, fire mobilizes nutrients by mineralising the layer of accumulated leaf litter, and occasional fires appear to be essential in speeding up the functioning of these ecosystems.

Oligotrophic soil conditions are not general in the Mediterranean biome, but it must not be forgotten that most of the biome's flora comes from species adapted to nutrient-poor environments, and thus may show adaptations related to nutrient economy that may seem excessive under current conditions. On the other hand, dry conditions cause relative poverty in nutrients, as they cannot be absorbed from the soil because there is not enough water to dissolve them. It is thus not surprising that most Mediterranean plants show characteristics that are related to saving nutrients, such as low leaf nutrient levels, and related to controlling nutrient cycles, such as their high levels of retranslocation.

Nor is it surprising that they shed the old leaves when they grow the new ones, and store high levels of nutrients in structures like bark or wood, from which they can be mobilized to cover the needs of the moment.

1.3 The biogeography of sclerophylly

Sclerophyllous formations in the Mediterranean Basin

In the Mediterranean Basin there are several types of sclerophyllous formation, dominated by shrubs or trees depending on the soil and climatic conditions of the particular site. When rainfall exceeds 16 in (400 mm) a year, the landscape is usually dominated by woodlands and where it is lower shrub communities dominate. Soil conditions also affect this broad scheme, as on a calcareous substrate or on a shallow soil with a steep slope, even rainfall exceeding 16 in (400 mm) a year may not be sufficient for the growth of a tree community, and a shrub community would then form. The division between tree and shrub formations is not very clear, as intermediate states may be found, such as poorly developed forests and shrub communities with a sparse tree layer.

Evergreen oak and pine forests

The most typical and best developed sclerophyllous woodlands are formed by different evergreen species of the genus *Quercus*. The wetter areas of the Mediterranean are dominated by thick, exuberant forests with a tree layer typically dominated by the holm oak (*Quercus ilex*). These woodlands can reach a considerable height, with trees 82 feet (25 m) tall. The densest and richest holm oak woods are found in coastal areas, since those that develop

in mountain areas are floristically poorer, as they lack thermophilous shrub species.

In areas with more continental climates, such as the interior of the Iberian Peninsula and in areas of transition towards more arid subclimates, woods are dominated by *Quercus ballota* (= *Q. rotundifolia*) —*carrasca* in Spanish—a species closely related to the holm oak. These woodlands must withstand harsher environmental conditions, and are less dense than the holm oak woodlands in coastal areas, as the tree layer is not highly developed and the shrub and herbaceous levels are rather thin.

In the western Mediterranean Basin there is a third type of oak woodland, dominated by the cork oak (*Quercus suber*). Cork oak woodlands occupy humid siliceous areas, and do not attain such high altitudes as the montane holm oak forests. The dominant tree species is different, but the accompanying shrub and herbaceous plants are no different from those found in coastal holm oak woods.

In the eastern Mediterranean Basin there are tree formations that are dominated by *Q. calliprinos*, similar to the kermes oak (*Q. coccifera*) but larger, which is often accompanied by other evergreen species, such as the mastic tree (*Pistacia palaestina*) and the Phoenician juniper, (*Juniperus phoenicea*). These often very open tree formations are common in Syria, Lebanon, Israel, and Jordan where they occupy an ecological niche similar to the holm oak in the west.

29 At medium altitudes and in shady places, communities of holm oak (*Quercus ilex*) mingle with deciduous oaks (*Quercus cerrioides*) forming a humid variant of the coastal holm oak woodland, as can be seen in this photograph of the Natural Park of Montseny, Catalonia (NE of the Iberic Peninsula). These mixed woodlands have lost much of their shrub layer, and show clear differences in color between the two species throughout the year.
[Photo: Oriol Alamany]

In areas with sub-Mediterranean climates, that is to say areas with a less pronounced summer dry period and with slightly colder winters, holm oaks and *Q. rotundifolia* are replaced by deciduous or semi-deciduous woods dominated by other species of the genus *Quercus*. In general, they are trees with marcescent leaves, that is to say the withered leaves stay on the branches for much of the winter, a trait that is halfway between the broadleaves and the evergreens. The genus *Quercus* has several species that occupy different ranges; *Q. humilis*, the small-leaved gall-oak (*Q. faginea*), *Q. cerrioides*, the Pyrenean oak (*Q. pyrenaica*) and the Canary oak (*Q. canariensis*) are all found in the western Mediterranean Basin, while *Q. macrolepis, Q. ithaburensis, Q. infectoria, Q. boissieri, Q. frainetto, Q. trojana, Q. aegilops* and *Q. humilis* are all found in the eastern sector. In many places they may form mixed stands of evergreen and marcescent species.

They are also tree masses dominated by conifers, for example the Aleppo pine (*Pinus halepensis*) and the Calabrian pine (*P. brutia*), which is restricted to the eastern sector. The role of these species may vary considerably as they may form pure stands, or mixed ones with different species of *Quercus*, or they may form an open tree layer in shrub communities. Their role is often considered to be secondary, in that they are forest masses that replace disturbed evergreen or semi-deciduous forest. There are other species of conifers that play a lesser role in the landscape, such as the stone pine (*P. pinea*), the pinaster, or cluster pine, (*P. pinaster*), the Italian cypress (*Cupressus sempervirens*), the arär (*Tetraclinis articulata*), and the Phoenician juniper (*Juniperus phoenicea*), which form relatively limited stands.

In mountain environments under clearly Mediterranean climatic conditions, and sharing an area with deciduous or evergreen broadleaves, there are forests consisting of different subspecies of a typically Mediterranean pine, the black pine (*P. nigra*), which in some areas gives way to a clearly extra-Mediterranean species, the Scots pine (*P. sylvestris*). The many subspecies of black pine are the result of the isolation of their populations in the mountainous habitats of the Mediterranean, a phenomenon which in other cases this has given rise to clearly different species. This is the situation in the cedars (*C. atlantica, C. libani* and *C. brevifolia*) and the firs (*Abies maroccana, A. pinsapo, A. nebrodensis, A. cephalonica, A. borisii regii, A. equi-trojani, A. bormulleriana* and *A. cilicica*), whose distributions are all very restricted, but of great interest as they are endemic species with relict distributions. Some of these species are in danger of extinction because of their dwindling populations. The most dramatic example is the Nebrodean fir (*A. nebrodensis*), with only ten to twenty specimens still surviving in the wild.

Maquis, matorral and thyme scrub

Shrub formations may show very different physiognomies, depending on the soil and climate. The most developed ones are called maquis, and are known as *macchia* in Italian, *maquis* in French, *xerovuni* in Greek, and *choresh* in Hebrew. They are very dense, almost impenetrable, formations that may reach 6.5 ft (2 m) in height, and are dominated by shrub species that can reach the size of a small tree, such as the wild olive (*Olea europea* var. *sylvestris*), the carob tree (*Ceratonia siliqua*), the mastics (*Pistacia lentiscus, P. palaestina*), the myrtle (*Myrtus communis*) and the strawberry trees (*Arbutus unedo, A. andrachne*), among others. Lower shrub formations (about one metre in height), dominated by the kermes oak (*Quercus coccifera*, called *garric* in Catalan), are known as garrigues. In many cases these shrub formations may show an open conifer layer, mainly of Aleppo pine (*P. halepensis*) and the Calabrian pine (*P. brutia*), giving them a forest-like appearance. They are of great importance in the landscape, as they occupy many degraded forest areas.

There are also more simply structured shrub communities, not exceeding two meters in height and much sparser, allowing sunlight to reach the ground. These are called matorral and are dominated by different species of cistus (*Cistus*), heathers (*Erica*), brooms (*Genista*), and labiates, such as rosemary (*Rosmarinus officinalis*). The dominant species are often not strictly sclerophyllous, since they have some malacophyllous features, have very small leaves or lack them entirely.

Low shrub formations, typical of degraded or very dry places, are dominated by thyme (known as *tomillares* in the Iberian Peninsula, *phrygana* in Greece, and *batha* in Palestine), which are rich in labiates, such as thyme, lavender and Jerusalem sage (*Thymus, Lavandula, Phlomis*), and members of the Cistaceae, or rockrose family, such as *Helianthemum*. Degradation of these communities leads to the implantation of dry meadows that are the least developed form of Mediterranean vegetation.

Finally, Mediterranean high mountain areas, with a clearly colder climate but fully exposed to the dry summers, show very characteristic oromediterranean shrub formations, dominated by spiny cushion-like plants such as *Erinacea anthyllis* and

30 Schematic diagram of the structure of a typical holm oak woodland showing the modest size and simplicity of the tree layer consisting of holm oaks (*Quercus ilex*) and some Aleppo pines (*Pinus halepensis*), the virtual absence of a herbaceous layer, and the rich shrub and liana layer. In the foreground are some of the most typical climbing or trailing plants, such as honeysuckle (*Lonicera implexa*), madder (*Rubia peregrina*), smilax (*Smilax aspera*) and ivy (*Hedera helix*) as well as butcher's broom (*Ruscus aculeatus*). [Drawing: Anna Maria Ferrer]

Lonicera implexa

Smilax aspera

Rubia peregrina

Ruscus aculeatus

Hedera helix

31 On coastal plains and small raised areas near the sea, like much of the Greek coastline, the climate is very dry and hot, and the predominant vegetation is maquis. In the Mediterranean Basin woody vegetation dominates over herbaceous vegetation right up to the shore, since it requires more water.
[Photo: Jaume Altadill]

Vella spinosa). They are similar in morphology to plants growing on wind-beaten coastal areas, and are adapted to withstand the intense sunshine, and the cold, dry conditions, which are accentuated by the desiccating effect of the wind.

Californian sclerophyllous formations

In the Californian mediterranean there is a similar distribution of formations with sclerophyllous woodlands, dominated by trees of the genus *Quercus* in wetter areas, and shrubland in drier areas.

Savannah oak and other woodlands

These woodlands contain a wide range of species, as do the shrub formations, because there were few extinctions, if any, due to the glaciations. There are deciduous species, such as the valley white oak (*Quercus lobata*), blue oak (*Q. douglasii*), Engelmann oak (*Q. engelmannii*), black oak (*Q. kelloggii*) and the Oregon white oak (*Q. garrayana*), alternating in some places with pines, such as *Pinus sabiniana* and *P. coulteri*. In drier areas, the evergreen tree species of *Quercus* are dominant, such as the canyon live oak (*Q. chrysolepis*), the coast live oak (*Q. agrifolia*), the scrub interior live oak (*Q. wislizenii*) and *Q. tomentella*. These tree formations vary greatly in physiognomy, as the different species associate to

form mixed stands. In many cases, the tree masses are not dense but open, and give rise to a pasture landscape with trees, known as savannah oak woodland, that is clearly anthropogenic and caused by grazing and recurrent low-intensity fires that do not affect the larger trees. Formations dominated by evergreen species may show a shrub structure in some limiting environmental conditions, resulting in an imprecise boundary between woodlands and scrub.

The woods of the Californian mediterranean include a remarkable tree, the coast redwood (*Sequoia sempervirens*) which forms forests on some parts of the coast, in the heart of the Mediterranean climatic area. These forests consist of trees as tall as 164 ft (50 m), and are an exception in a biome where forests are usually much more modest in appearance. They are similar to the forests of giant redwood (*Sequoiadendron giganteum*) in the mountainous north of the Californian mediterranean, but the development here is the result of clearly different climatic conditions. The determining factor in the case of the coast redwoods is the abundant fog formed by onshore ocean winds, which reduces water loss and provides extra water through condensation on the vegetation.

Overall, conifers play as variable a role as they do in the Mediterranean Basin, because they may form replacement forests or a tree layer over shrub vegetation. Characteristic species include *Pinus attenuata*, *P. muricata*, the Monterey pine (*P. radiata*), and several species of the genus *Cupressus*, such as the Monterey cypress (*C. macrocarpa*). Many of these species are clearly pyrophytic, which is to their advantage in environments where fires are common.

Californian chaparral

The most famous of the Mediterranean tree formations of California are dominated by shrubs and are generically known as chaparral. Chaparral thrives in areas with a rainfall of less than 16-20 in (400-500 mm) a year, but is remarkably rich in species. It is dominated by species of the following genera: *Adenostema*, such as chamise (*A. fasciculatum*); *Arctostaphylos*, such as manzanita (*A. glauca*); *Ceanothus*, such as the Californian lilac (*C. greggii*); *Heteromeles*; and *Rhus* (*R. diversiloba*), together with some species of *Quercus*, such as the California scrub oak (*Q. dumosa*).

Several different types of chaparral have been classified, depending on their environmental conditions and dominant species. These include low woodland formations, or woodland chaparral dominated by the California scrub oak (*Q. dumosa*), dense shrub communities dominated by *Ceanothus*, and therefore known as *Ceanothus* chaparral, as well as manzanita chaparral at the highest altitudes, dominated by species of *Arctostaphylos*. Drier areas with poor soils dominated by (*Adenostema*) form what is called chamise chaparral.

The driest and stoniest areas, exposed to desert or marine influences, have low shrub communities dominated by deciduous species of the genera *Artemisia* and *Salvia*, known as coastal sage scrub, very similar in appearance to the matorrals and Spanish thyme scrub of the Mediterranean Basin.

Finally, dry meadows are often degradation stages of tree or shrub communities, mainly due to fire.

32 Woodlands with trees of the genus *Quercus* only appear in the northern sector of the Californian mediterranean since the distribution of oaks is determined by the north-south rainfall gradient. They are also mixed with other broadleaf species. The current floristic diversity of the American west is remarkable; these communities are very species-rich because the flora is still very similar that of the Pliocene, as it did not suffer appreciable impoverishment during the Pleistocene. [Photo: Lluís Ferrés]

33 California's forests of the coast redwood are a relict of 65 million years ago, when dinosaurs still roamed the earth. The coast redwood, (*Sequoia sempervirens*), such as those in the Muir Woods, grows on humid soils and may live 2,000 years (although their closest relatives, the giant redwoods [*Sequoiadendron giganteum*], may live twice as long). The tallest known tree in the world is a sequoia 367 ft (112 m) tall, although, given time, it will probably grow even taller. [Photo: Teresa Franquesa]

34 Chilean coastal matorral, Punta Molles, in the Los Vilos area. The photo shows the formation has an unusual appearance for sclerophyllous vegetation, as it includes succulents and cacti. The limited natural area covered by these formations has been even further reduced by human action, and is now very small today.
[Photo: Xavier Ferrer & Adolf de Sostoa]

Chilean sclerophyllous formations

The Chilean mediterranean also shows a wide range of sclerophyllous landscapes. There are different tree and shrub formations, including encinal, coastal chaparral, and espinal.

Encinal

The woody formations, which may reach 33-49 ft (10-15 m) in height, and occupy the wettest areas, such as shaded valley bottoms, are dominated by sclerophyllous species such as "litre" (*Lithrea caustica*), soap bark (*Quillaja saponaria*), "boldo" (*Peumus boldus*), and "peumo" (*Cryptocarya alba*). The endemic Chilean wine palm (*Jubaea chilensis*) grows in a very small area to the northeast of Valparaiso. In fact, encinales occupy only a small part of the modern landscape.

Matorral and espinales

In drier areas, or due to the degradation of these tree formations, a taller shrub formation appears, called matorral. Often the species composition of matorral is very similar to the encinales, and the clearest difference is in their stage of development. In addition to the species already mentioned, we must add *Kageneckia oblonga* and *Colliguaja odorifera*, both sclerophyllous shrubs. There are also some malacophyllous and summer-deciduous plants, such as *Baccharis linearis*, *B. rosmarinifolius*, *Collettia spinosissima*, *Proustia pungens* and *Trevoa trinervis*. In drier areas, there are also columnar cacti, such as

Trichocereus chiloensis and bromeliads, such as *Puya berteroniana* and *P. coerulea*. In coastal areas, sclerophyllous species are less important and succulent and malacophyllous species dominate, the latter behaving as summer-deciduous plants. Examples include *Fuchsia lycioides*, *Proustia pungens*, and *Adesmia arborea*, in addition to those mentioned above, and they give rise to formations known as coastal matorral.

A characteristic savannah-like community grows in the central valleys, dominated by "espino" (*Acacia caven*), and called espinal. These small trees may reach 10-13 ft (3-4 m) in height and develop a canopy that is 16-20 ft (5-6 m) in diameter. They show great developmental adaptability and may be winter-deciduous, evergreen or summer-deciduous, depending on climatic conditions. Shrub cover may vary between 5 and 15%.

In the transition between espinal and matorral, the shrub layer becomes denser and is enriched in typical matorral species. The herbaceous stratum of these communities is dominated by introduced species of Eurasiatic origin that withstand grazing much better and have displaced the native herbaceous species. These savannah-like formations are considered to be degraded matorral, although in places that are too dry to support a shrub formation they may represent the climax vegetation.

On the slopes of the Andes, above 5,900 ft (1,800 m), the typical matorral is substituted by a

35 The South-African fynbos includes ericaceous and restionaceous plants, but is dominated by shrubs of the protea family; the photograph shows scrub with *Leucospermum cordifolium* in flower. [Photo: Nigel Dennis / ABPL]

scrub formation that is also xerophytic and dominated by species such as *Colliguaja integerrima*, *Kageneckia angustifolia* and *Guindilia trinervis*. At greater heights, spiny scrub formations appear, with abundant cushion-form scrub species. Conifers are not very important in the Chilean mediterranean but there are some forests of the alerce or Chilean cedar (*Libocedrus* [=*Austrocedrus*] *chilensis*).

Cape sclerophyllous formations

In the South-African mediterranean tree formations are absent, as the few tree species, such as the silvertree (*Leucadendron argenteum*) have very restricted distributions. The landscape is dominated by shrub formations. There are only a few remains of the former wet temperate forests, which have diminished greatly since European colonization.

There are considered to be four types of shrub vegetation, called "stranveld," "fynbos," "renosterveld," and "karoo." This classification is based on criteria derived from the agricultural and forestry uses each type can be put to, which is related to soil and climate conditions, and it clearly reflects the characteristics of the natural vegetation.

Broadly speaking, the stranveld formations occur in arid coastal areas; coastal fynbos and mountain fynbos occur in intermediately arid areas with relatively poor soils; coastal renosterveld and mountain renosterveld are found on very poor soils and in transition towards more arid areas; finally, karoo formations are found in the most arid areas, in transition towards deserts.

Fynbos and stranveld

Fynbos, which may reach 6-10 ft (2-3 m) in height, is dominated by members of the heather and restio families (Ericaceae and Restionaceae), accompanied by sclerophyllous shrubs (of the genera *Protea*, *Leucadendron*, *Olea*, *Euclea*, *Rhus*), summer-deciduous malacophyllous plants (*Zygophyllum*) and succulents (*Euphorbia*). The relative abundance of aphyllous species (restionaceous type), small-leaved sclerophylls (ericaceous type), and broad-leaved sclerophylls (proteaceous type), depends on soil fertility, and this gives rise to a vegetation mosaic showing a spectrum ranging from scrub or heath type formations to formations that are more similar to maquis.

The stranveld mainly occurs in coastal dunes with more fertile granitic soils and the dominant species are sclerophyllous, giving it an appearance that is very similar to the maquis of the Mediterranean Basin; it includes the same accompanying shrub species as the fynbos.

Renosterveld and karoo

Renosterveld grows in degraded coastal areas and in the inland areas of transition towards the Karoo

36 **The Australian jarrah woodlands** are dominated by a eucalyptus species typical of the Western Australian mediterranean, *Eucalyptus marginata*, which forms the tree layer, while the shrub layer is formed by *Acacia lateriticola*. Both species are characteristic of climates with winter rainfall and summer drought, such as that of Western Australia.
[Photo: Jean-Paul Ferrero / Auscape International]

semi-desert, and is a loose scrub formation that looks very similar to the coastal sage in California, and which is dominated by the composite "renoster-bos" or rhinoceros bush (*Elytropappus rhinocerotis*), frequently accompanied by summer-deciduous shrubs, although in coastal areas there may be some sclerophylls.

The karoo formations that occur in drier areas are dominated by succulents, giving this area of transition towards inland desert a very unusual appearance.

Australian sclerophyllous formations

The Australian mediterraneans show very similar characteristics to those of the Cape region, as the mediterranean climatic conditions are combined with markedly nutrient-poor and siliceous soils.

The vegetation is dominated by species of *Eucalyptus*, whose common names have been given to the formations they dominate, although there are also many species of the Proteaceae in the shrub layer and also members of the Epacridaceae that play the same role as heathers in Europe, that is to say they grow on the most nutrient-poor soils. There are some floristic peculiarities, such as the arborescent monocotyledons of the genera *Kingia* and *Xanthorrhoea*, and the cycad *Macrozamia*.

Jarrah woods, wandoo and the western mallee
In the rainiest areas of the southwestern Australian mediterranean there are forests of jarrah (*Eucalyptus marginata*), which like the other tree species in its genus is not strictly sclerophyllous, as it has coriaceous leaves. These evergreen woodlands may easily reach 66-82 ft (20-25 m) in height, and occur in areas where annual rainfall is over 23 in (600 mm).

Where rainfall exceeds 47 in (1,200 mm) a year, as in the southern tip of western Australia, the Australian jarrah forests are replaced by woodlands of karri (*E. diversicolor*) with less coriaceous leaves.

In the transition between jarrah and karri there are forest formations dominated by marri (*E. calophylla*), another eucalyptus that encroaches into jarrah areas giving rise to mixed woodlands.

In slightly drier areas, with rainfall between 20 and 23 in (500 and 600 mm) per year, the dominant plant is the wandoo (*E. redunca* var. *wandoo*) which forms a more open woodland. Areas that only receive 10-20 in (300-500 mm) of rainfall a year, as happens in most of the southern Australian mediterranean, open shrub formations called mallee appear. This name is derived from an indigenous language and designates a community dominated by more than a hundred different species of *Eucalyptus*. Both wandoo and mallee have been greatly modified and converted into pastures and cereal fields.

In even drier places (less than 16 in [400 mm] annual rainfall and a seven month dry period) and places with extremely poor sandy soils, there are shrub formations dominated by members of the protea family, with some species of *Eucalyptus* and bull oak (*Casuarina*). Although they are not very productive communities, their floristic richness is remarkable.

The southeastern mallee

The above scheme is valid in the Western Australian region, but not in the other Australian mediterranean, located in South Australia. Only 1.1% of this region has an annual rainfall greater than 23 inches (600 mm), and it therefore lacks forests and is dominated by shrubs or open woodlands.

These open tree formations contain some of the 18 species of tree-type eucalyptus found in the region, such as *Eucalyptus obliqua*, the sugar gum (*E. cladocalyx*), or the blue gum (*E. viminalis*). There are also areas with stands of casuarina (*Casuarina luehmannii*), or the cypress pine (*Callitris columellaris*).

Here, the mallee shrub formations are dominated by a large number of gums (*Eucalyptus*) and acacias (*Acacia*), accompanied by species of the genera (*Callitris*) and (*Casuarina*).

Seven main types of mallee are recognized, divided into two subcategories and dominated by different species, depending on their climatic and soil conditions. The first group consists of three types with small-leaved shrubby plants in the undergrowth, the first dominated by the white coastal mallee (*E. diversifolia*), the second by *E. incrassata* and broombush (*Melaleuca uncinata*), while the third is dominated by *E. incrassata* and *Triodia irritans*. The second group embraces the areas dominated by herbaceous plants or chenopods: one type by white coastal mallee (*E. diversifolia*), the second by white mallee (*E. behriana*), the third by red mallee (*E. socialis* and *E. dumosa*), and the last type is dominated by *E. oleosa* and *Triodia irritans*. The first group is typical of poorer soils and shows greater ecological similarities with shrub systems of oligotrophic soils in other biomes than with Mediterranean ones. In fact, on the Australian con-

37 *Eucalyptus redunca* **dominates the wandoo community**. Australia's plants reflect its evolutionary history. The differentiation of the genus *Eucalyptus* into 600 species is due to many factors including soil diversity, and the barriers represented by the central sea and by the aridity of the center of the continent. Because of these factors, the animals and plants of the eastern and western coasts have evolved separately for a long time. [Photo: Eric Crichton / Bruce Coleman Limited]

tinent there are mallee-type communities on poor soils in areas with temperate and tropical climates.

In the most arid areas or areas with poorer soils, low-growing (less than 6.5 ft or 2 m) shrub formations occur that include several members of the protea family (of the genera *Banksia*, *Hakea*, *Melaleuca*, *Leptospermum*) accompanied by some species of the genera *Eucalyptus*, *Casuarina* and *Xanthorrhoea* among others. There are also areas dominated by herbaceous vegetation with grass species, such as cutting grass (*Gahnia trifida*), and thatching grass (*G. filum*), or members of the lily family (Liliaceae) such as iron grasses (*Lomandra dura* and *L. effusa*).

2. Flora and vegetation

2.1 The origins of Mediterranean vegetation

The Mediterranean bioclimate, as has already been pointed out, has only come into existence relatively recently. Although there is evidence of the existence of climates with characteristics similar to the mediterraneans towards the end of the Miocene (seven million years ago) and in the middle of the Pliocene (four million years ago), the truly Mediterranean climate, with its characteristic period of summer drought, appeared in the Pleistocene after the first great glaciation (two million years ago). By then the five areas making up today's Mediterranean biome were already clearly separated from each other, and thus their floras evolved independently.

The flora of the mediterraneans of the northern hemisphere (California and the Mediterranean Basin) evolved from Tertiary elements belonging to hot moist tropical or to temperate environments. In the Chilean mediterranean something similar happened, as the pre-existing Tertiary elements belonged to hot moist tropical environments or to moist temperate Antarctic elements. In the South African and Australian mediterraneans, the flora evolved exclusively from the tropical elements that had surely been common in both regions during the Mesozoic. The presence in these areas of plants that grow during the summer, a typically tropical habit, shows the vegetation's origin, although tropical climatic influences which cause some rainfall during the dry season have contributed to maintaining its characteristics.

These two Mediterranean regions in the southern hemisphere share other features in common, since the both the African and Australian continents do not extend far into high latitudes, and have given rise to Mediterranean areas that are more isolated than the other areas of the biome, making them more vulnerable to invasions by non-native plants. The Monterey pine (*Pinus radiata*) from the Californian mediterranean has been planted widely for timber in all areas of the biome, as it grows very well in all of them, but it is only invasive in the Australian regions and in the Cape. In the Cape region, even some Australian species, such as the shrub *Hakea sericea* behave invasively.

Isolation between different areas, and between different parts of the same area due to complexities of the relief, together with the repeated climatic changes associated with the glaciations (which caused the Mediterranean areas to expand and contract) have given rise to a complex flora that is the result of intensive speciation and the survival of relict species in favourable environments.

The flora's high level of endemism is the result of speciation processes operating in a territory that is divided into valleys and mountains, and thus functions like an archipelago in terms of biological isolation.

In the South African mediterranean region this process of floristic differentiation has gone much further than in the biome's other areas, and has given rise to a flora in which 68% of all species and 19.5% of the genera are endemic. Furthermore, the South African mediterranean flora is so diverse that this small area's species density is greater than any other tropical or temperate areas of comparable size—this small area boasts 8,600 species of vascular plant. The South African mediterranean region's flora is so special that it forms a floristic kingdom in its own right, the Cape Floristic Kingdom, whose area is tiny in comparison with those of earth's other five accepted floristic kingdoms. The Holarctic Kingdom corresponds to Eurasia and North America: the Neotropical occupies all Central America and almost all South America; the Paleotropical covers almost all the African continent, except the Cape region; the Antarctic Kingdom occupies the southern tip of South America and the sub-Antarctic islands; and the Australian Kingdom corresponds to the continent of the same name.

Thus, the current Mediterranean flora evolved from the set of tropical and temperate species that were able to adapt to the new climatic conditions that prevailed. It should be borne in mind that

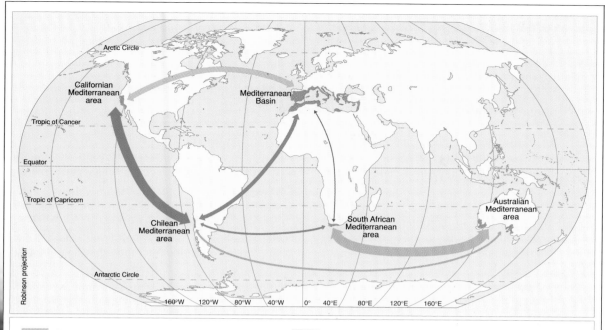

38 There are differences and similarities between the planet's disjunct areas of Mediterranean-type sclerophyllous vegetation. These are basically due to topographic and geological features, patterns of convergent evolution, phylogenetic relationships, and the parallel effects of human action. Phylogenetic relationships account for the similarities between the two regions in the western hemisphere (California and Chile) and between the regions that used to form part of Gondwana (South Africa and Australia).
[Drawing: Editrònica]

Californian Mediterranean area - Mediterranean Basin

- Geological connections
- Pleistocene affinities
- Some phylogenetic lines in common
- Shared continental Mediterranean climates
- High fire risk
- Shared land use models
- Invasions of species, especially herbaceous ones, from the Mediterranean Basin

Californian Mediterranean area - Chilean Mediterranean area

- Pleistocene affinities
- Similar tectonic structure
- Discontinuous geological connections through the Isthmus of Panama and the Andes
- Shared oceanic Mediterranean climates
- Anthropic biogeographic connections (Spanish colonization, the Californian gold rush)

Chilean Mediterranean area - Mediterranean Basin

- Pleistocene affinities, but without geological connections
- Spanish colonization
- Shared models of land use, but with less risk of fire in Chile
- Major impact of grazing and deforestation
- Invasion of plants, mainly herbaceous, from the Mediterranean Basin to Chile

Chilean Mediterranean area - South African Mediterranean area

- Paleoantarctic connections and phylogenetic lineages
- Temperate climate with tropical influences in South Africa
- Abrupt relief

Chilean Mediterranean area - Australian Mediterranean area

- Paleoantarctic connections
- Some ancient phylogenetic lineages in common
- Insular evolutionary conditions, especially in Australia
- Temperate climate with tropical influences in Australia

Mediterranean Basin - South African Mediterranean area

- Some biogeographical connections, mainly along the mountain ranges of southern Africa
- Invasions of species, especially conifers from the Mediterranean Basin (and from California)

South African Mediterranean area - Australian Mediterranean area

- Gondwanan affinities
- Geologically ancient leached substrate
- Large areas of oligotrophic soils
- Mediterranean climate with tropical influence
- High species diversity
- Very high risk of fire
- North European colonization
- Mutual invasions of species

---

sclerophyllous species were already present in these environments as an evolutionary response to water shortage and nutrient-poor soils. This water deficit might have been due to topographic factors creating rain shadows, while floods during the hot or rainy season may have forced the plants to grow during the warm but dry winter. In fact, the Mediterranean flora's ancestors were already adapted to conditions of water and nutrient scarcity and occupied a marginal space in their respective environments, whether these were tropical or temperate. The existence before the Pleistocene of periods of climatic conditions similar to the mediterraneans, with a mild cool period and a hot dry period (although they were less markedly arid), also favoured the appearance of plants especially adapted to water scarcity.

2.2 Sparse, slender, yellowish grasses

The Mediterranean biome is dominated by communities of woody plants, while herbaceous formations are relatively scarce. There are many species of herbaceous plants but they play a secondary role in the landscape. The long summer drought makes life difficult for herbaceous plants, whose short roots cannot reach the deeper levels of the soil, the only levels where any moisture is left in the hot summer period. Thus, herbaceous plants are obliged to use the water in the surface layers, and these completely dry out in summer. This is why Mediterranean meadows show their greatest development in the winter and spring, while in summer they enter their period of rest or marcescence. These dry meadows are dominated by tough-leaved grasses, which are often rolled up to avoid excessive water loss; they are almost always yellowish, and quite different from the lush, green appearance usually associated with the term "meadow."

Mediterranean herbaceous vegetation

When conditions are extremely dry, it is very difficult for perennial herbaceous plants to survive and meadows are absent from the landscape. In these conditions, all that can grow are meadows of small annual plants; these tiny herbaceous plants are called therophytes and complete their life-cycle in the few weeks of the wet season, and then pass the dry season as seeds. These small patches also occupy the spaces between herbaceous perennials in dried out meadows and the gaps left between the small shrubs that can resist the extremely dry conditions. This is why, in the Mediterranean biome, the destruction of the shrub communities in the driest areas does not lead to the

39 Dry Mediterranean meadow on the coast of the Iberian Peninsula showing the thin, dry covering of grass (*Brachypodium retusum*) and the presence of bulbous plants (geophytes), whose aerial parts grow in the favourable season, in particular the dwarf iris (*Iris chamaeiris*) and the asphodel (*Asphodelus cerasifer*). [Photo: Ernest Costa]

40 The most characteristic feature of an orchid is its flowers, which reveal the plant's close relationship with its pollinating insect. The meadow and forest species found in Europe produce small flowers. This photo, taken in Vacarisses in Catalonia, shows the yellow bee orchid (*Ophrys lutea*). The labellum (the flower's lowest tepal) mimics the form of the female of the wasp that pollinates the flower. This yellow orchid flowers in March and April in the warmer sites near to the Mediterranean coastline on shallow soils that are very sunny, as it does not tolerate shade.
[Photo: Oriol Alamany]

appearance of meadows, but to almost empty sites occupied only by patches of ephemeral plants. This is unlike the situation in wetter climates, where the disappearance of woody communities almost always leads to the growth of relatively lush replacement meadows.

Mediterranean herbaceous plants therofore play a marginal role in the landscape, occupying relatively dry sites and areas at levels above the tree line. In certain situations, fire also favors the installation of herbaceous vegetation, especially in the first phases of succession after a fire, before the shrubs that can sprout after burning have reoccupied the space, and pioneer and ephemeral herbaceous species can appear and take advantage of the temporary reduction of competition. In sites where there are repeated fires there are many geophytes, herbaceous plants with underground resistance structures, mainly bulbs, that allow them to grow and flower rapidly after a fire, such as asphodel (*Asphodelus*), orchids (*Serapias*, *Ophrys*) and gladiolus (*Gladiolus*).

The scarcity of good pasture in the Mediterranean environment means that livestock is often reduced to grazing the lower branches of trees and shrubs. Naturally, this prevents the proper development of forest communities, especially if the lack of tender shoots makes shepherds set fire to the woody vegetation to induce the growth of more palatable shoots, and may even lead to the appearance of grass on burnt sites that have been enriched by the nutrients in the ashes. The herbaceous communities that fleeting-

ly appear in these cases are never more than poor, very dry meadows, except perhaps for their first few weeks of existence. In any case, actions of this type mean that herbaceous vegetation now occupies a greater area in the Mediterranean area than it would do spontaneously, in the Mediterranean Basin as well as in the Chilean, South African, and Australian mediterraneans. This has also caused the replacement of many native species by other sub-spontaneous or introduced plants, as the herbaceous plants from these three regions are not at all adapted to being grazed, and the shepherds in the overseas Mediterranean regions, almost always of European descent, have had no hesitation about importing species suitable for their livestock. The open shrub and tree formations of these regions thus have an autochthonous (native) woody flora and a herbaceous flora that is largely non-native, forming landscapes that are clearly human-influenced, such as the "savannah oak" in California, "espinal" in Chile, and "wandoo" formations in Australia. The Chilean mediterranean's flora, for example, consists of about 3,000 species of which about 500 are introductions; about 85% of the introduced plants are herbaceous species. This dominance by introduced species is also found in the other areas of the biome.

Herbaceous species in the Mediterranean regions

In the Mediterranean Basin, the most abundant herbaceous perennials are grasses, such as false bromes

(*Brachypodium phoenicoides, B. retusum*), esparto grass (*Stipa tenacissima*) and albardine (*Lygeum spartum*), as well as various species of the genera *Avena, Bromus* and *Festuca*; some elements of tropical origin are also abundant, such as *Hyparrhenia hirta*. In the Californian area, the native species of *Aristida, Poa,* and *Stipa* have been largely replaced by introduced species of the genera *Avena, Bromus,* and *Festuca*.

In the Chilean mediterranean, perennial herbaceous plants formerly dominated annuals, but introduced herbaceous species have now displaced the native ones, so that in some pasture communities 25% of the species are not native. Some non-native species may be very abundant, such as *Erodium cicutarium, E. moschatum, Capsella bursa-pastoris, Lolium multiflorum, Stellaria media,* and *Vulpia bromoides*. Native plants of note, because of their abundance, include geophytes belonging to the Liliaceae, Amaryllidaceae, and Iridaceae, as well as ferns of the genera *Adiantum, Blechnum, Notholaena,* and *Pellaea*.

In the Cape region, geophytes such as *Watsonia* and *Gladiolus* normally grow in sites that are repeatedly burnt, and are so abundant that there are 350 species of them; also frequent are some annual species (*Dimorphotheca pluvialis, Senecio elegans*), originally characteristic of degraded areas, which produce spectacular flowering in the spring. In this mediterranean area, many herbaceous perennials, such as *Themeda triandra*, have been displaced by introduced annual herbaceous plants, although some native plants such as the geophyte *Oxalis*, maintain an important presence. The problem of invasive species in this unusual, extremely fragile and sensitive floristic region is becoming critical: while in the biome's other areas, introduced species may be successful in disturbed conditions, in the Cape region introduced species can establish themselves in all conditions. Thus, plants like *Brachypodium distachyon, Briza maxima, Polygonum aviculare* and *Rumex angiocarpus* only invade disturbed sites, but other species like *Fumaria muralis, Lolium perenne, Spergula arvensis, Spergularia media, Silene gallica,* and *Salsola kali* occupy all types of biotope. The South African mediterranean's susceptibility to invasions by foreign plants has put the very survival of some native species in doubt, yet the Cape region has exported herbaceous plants to gardens throughout the world (especially to the other Mediterraneans), such as the many species of geranium (*Pelargonium),* arum lilies (*Zantedeschia)* and clivias (*Clivia*).

The Australian landscape is now dominated by herbaceous plants introduced from Europe, especially species of the genera *Lolium* (Poaceae) and *Trifolium* (Leguminosae). The original native species of herbaceous plants belong, amongst others, to the Poaceae, Cyperaceae, Asteraceae, Liliaceae, Orchidaceae, Restionaceae, and Xanthorrhoeaceae. Among the most abundant native species on oligotrophic soils are the hummock grasses of the genus *Triodia*.

Australia's grass trees, "black boys" and "black gins"

Special mention should be made of Australia's species herbaceous plants which have an arborescent habit. Although these herbaceous trees are typically Australian, their growth form is not really that strange, as it is similar to other woody arborescent monocotyledonous plants, such as the palms and the screw pines, *Pandanus*. The main difference is that Australia's herbaceous or grass trees have very narrow, shiny, fragile leaves that are unlike the large, broad and fleshy leaves of the palms and the pandans. This sclerophyllous characteristic is a typical adaptation found in Australia's Mediterranean conditions.

The grass trees called "black boys" (*Xanthorrhoea*) and "black gins" (*Kingia, Dasypogon*) in Western Australia, are a typical feature of the moister areas of the Mediterranean landscape of Australia, although they also grow occasionally in the desert. Where they grow they are so common that it is hard not to notice them, and they seemed so strange to the European colonists that they were accepted as yet another unique member of Australia's flora and fauna, like the many species of *Eucalyptus* and kangaroos. Their strange and primitive appearance led to the legend that they are a relic of some ancient race that has survived in isolation in Australia.

Anatomy and development
The slow growth of grass trees, their height and their generally primitive appearance led some people to believe that some specimens could be up to 6,000 years old. The latest scientific estimates for the ages of the largest specimens of *Xanthorrhoea preissii* are 350-400 years, and 600-800 years for the largest specimens of *Kingia australis*.

A typical grass tree has an erect trunk, 8-12 in (20-30 cm) in diameter, with a dense tuft of narrow leaves at the top. The old dead leaves remain

41 Kingia and Xanthorrhoea
(*K. australis* in the photo on the left, and *X. glauca* in the photo on the right) are the best known of Australia's grass trees. These arborescent herbaceous plants (grass trees) grow in the south of the continent. Their strange growth form has led to *Kingia* being called "black gins" and *Xanthorrhoea* being called "black boys," in Western Australia. These plants show extreme sclerophyllous adaptations, as they have narrower, shinier, and more fragile leaves than closely related species. They are so well adapted to the frequent fires affecting the areas they grow that they stimulate flower production.
[Photos: Reg Morrison / Auscape International and Jaime Plaza Van Roon / Auscape International]

Proteaceae: at the service of Hephaestus or of Poseidon?

Mosaic of Proteus Thessalonica, Greece (1st century A.D.) [The Ancient and Architecture Collection (California).]

According to Greek mythology, Proteus was a god in the service of Poseidon. Living on the island of Pharos near Egypt, he was the sea-god responsible for herding Poseidon's aquatic flocks. When confronted by an enemy, he had the remarkable gift of being able to adopt different shapes and appearances, similar to the way that proteins made of amino acids can have different forms, which is why proteins are named after him. The proteas, members of the Protea family (Proteaceae), are also remarkable for their variety of forms, and so they too were also named after this Greek god. The members of the Proteaceae, however, are not too keen on water and are often closely linked to fire, and seem to prefer Hephaestus, the Greek god of fire, to Poseidon, the Greek god of the sea (or Vulcan to Neptune for those who follow Roman mythology).

The members of the Proteaceae do not, of course, change shape. It is just that different species vary greatly in appearance. Certainly, some are clearly pyrophytic (fire-loving), fitting in well with this symbolic affinity towards the fire gods. Furthermore, many prefer dry places, or at least places which are not too wet. This is true of the members of the Proteaceae that grow in the Mediterranean biome, an environment where many of the family's most remarkable species are very much at home.

Leucadendron discolor [*Flowers of South Africa*, Auriol Batten, Southern Book Publishers, Johannesburg (1988)]

The Proteaceae are basically southern hemisphere plants. Their distribution area covers Central and South America, the southern half of Africa and Asia, as well as Australasia. With a distribution of this kind, it is understandable that there are nemoral (woodland) species associated with the rain forests, such as the Chilean firebush (*Embothrium coccineum*), that grow in the so-called Valdivian forests, and the well-known Asiatic and Australian genus *Grevillea*. The latter has about 200 species, the best-known of which is the silky oak (*G. robusta*), a large tree that has been widely planted in gardens in temperate climates. Nevertheless, most of them prefer dry conditions with low humidity, and this is true of members of the Proteaceae from the South African and Australian mediterraneans.

In the Australian mediterranean, there are many banksias (*Banksia*), a genus with about fifty species. The members of the Proteaceae normally produce tubular flowers with four stamens and a protruding style in an inflorescence comparable to the capitulum of composites, an arrangement that is particularly clear in the banksias. The most spectacular must be *Banksia coccinea* with its inflorescences containing up to a thousand perfectly aligned, fiery red,

Grevillea robusta [Erich Crichton / Bruce Coleman Limited]

tubular flowers. Its very reproductive structures loop back on themselves, making the whole thing look rather like a particularly ingenious piece of wickerwork. The equally conspicuous inflorescences of some *Banksia* species are, however, a vivid sulphur yellow. Even so, their hard, evergreen sclerophyllous leaves remind us that they are Mediterranean plants.

Not only their leaves but also their well-know pyrophytic nature remind us that *Banksia* are Mediterranean plants. Their fruiting bodies are like heavily reinforced pine cones that can withstand very high temperatures. In fact, high temperatures are

Banksia coccinea [Jaume Altadill]

required to "persuade" them to release their seeds. This is why *Banksia* seedlings, which grow into large shrubs, dominate so completely after a fire. Here, Hephaestus is clearly winning the battle against Poseidon.

Australia's mediterraneans have many more genera of the Proteaceae, such as *Telopea*, *Hakea* or the famous *Macadamia integrifolia*, which produces an fatty edible nut, the Macadamia or Queensland nut, that is highly appreciated—which is why it is grown on a large scale both in Australia and Hawaii. However,

Protea cynaroides [Jules Cowan / Bruce Coleman Limited]

the *Protea* family is most abundant in the Cape region. Indeed, the South African mediterranean has a hundred or more species, the most representative members of the family, especially of the genus *Protea*. Most are trees and shrubs that are tolerant of poor soil conditions and they are highly prized by gardeners.

There are about a hundred species of the genus *Protea*, all of which grow in sandy siliceous soils. They are characterized by a collar-like ring of large, sometimes brightly colored, bracts that enclose the inflorescence, making them look rather like an artichoke. *Protea*

Macadamia integrifolia [Reg Morrison / Auscape International]

cynaroides, the most characteristic species and national flower of South Africa, received its specific name because its flower is similar to an artichoke (*Cynara scolymus*). Incidentally, the very first Cape plant to appear in a botanical book was a *Protea*, *P. neriifolia*, in a Flora by C. Clusius published in 1605.

On the other hand, the species of the genus *Leucospermum* lack floral bracts, but their very prominent styles explain why they are common called pincushions; there are some fifty species, all of which have yellow or red flowers, such as *L. cordifolium*, *L. lineare*, and *L. cuneiforme*. *Leucadendron* is also an important genus with some eighty species that are dioecious (the individual plants bear either male inflorescences or

Leucospermum cordifolium [Carol Hughes / Bruce Coleman Limited]

female inflorescences). The female inflorescences are like a sort of pine cone, while the male ones are surrounded by eye-catching pink or golden yellow bracts. And there are more genera in the Proteaceae, such as *Serruria* with its laciniate (jagged) leaves and bluish flowers, *Mimetes* with a bract under each flower, and *Aulax*.

Many of these South African and Australian members of the Proteaceae are highly valued as ornamental dry flowers, because the floral bracts and the "cones" of some genera are brightly-colored, long-lasting structures, and living up to their family's reputation, they come in a very wide variety of shapes.

attached to the stem, forming a "skirt" below the crown until it is destroyed by the next fire. The relatively fire-resistant leaf-bases protect the growing points from the locally frequent fires and also form an external sheath that insulates the trunk. Old specimens can reach 26 ft (8 m) in height. Most species never produce branches, but the oldest specimens of some species may do so.

Like other woody monocotyledons, the stems of Australia's grass trees have a central cylinder with relatively soft, juicy vascular bundles, surrounded by a harder layer that consists mainly of densely packed clusters of xylem and phloem vessels that are crossed by the leaf-bases. In *Xanthorrhoea*, all this is surrounded by a cambium-like layer known as *desmium*, which regularly gives rise to internal secondary vascular groups and is protected externally by a bark-like layer of periderm. The rings of secondary vascular groups are grouped into light circular areas that alternate with dark circles, giving the appearance of annual growth rings. Of course this process does not lead to increase in the stem's thickness, but causes cracks to form in the outermost sheath of the dead leaf-bases surrounding the epidermis. *Kingia* and *Dasypogon* lack this desmium and also lack the related secondary growth, so the stem cannot increase in thickness, but simply becomes longer. One interesting detail of *Kingia* is that the stem continuously produces roots about an inch (a few centimeters) below the growing point. These grow downwards through the persistent leaf-bases that surround the trunk.

The ways in which the flowers grow and develop is interesting. In *Xanthorrhoea* many small white flowers are borne on a very long, densely packed terminal spike that emerges from the top of the plant and grows rapidly to 10 or more feet (3 or more meters) in length. (Daily growth of 3-4 in [7-10 cm] has been reported!) Fire, although it is not essential, does stimulate flowering; the appearance of a grass tree in this state, a black trunk with a tuft of leaves at the top and a conspicuous inflorescence, reminded the first European colonists of Western Australia, of the figure of an Aborigine hunting with a spear, and this is the reason for the name "black boy." In *Kingia* and *Dasypogon*, each plant bears between 4 and 100 much smaller inflorescences among the leaves at the top of the stem. The individual flowers are borne on globose flower heads at the end of each inflorescence. The resulting shape recalls an Aboriginal woman with her head covered with decorations, the origin of

the name "black gins." Some people still believe that individuals of *Kingia* are the females and those of *Xanthorrhoea* the males.

Functional adaptation and ecological behaviour

Both *Xanthorrhoea* and *Kingia* have tough sclerophyllous leaves that are quadrangular in cross-section, with relatively large amounts of sclerenchyma bundles. The leaves break easily when bent by hand, but resist the wind. The leaves of *Kingia* look silvery as they are covered with silky hairs. The leaves curve outwards from the growing point, and eventually hang down to form the typical skirt. On the other hand, in *Dasypogon hookeri* the not very sclerified leaves are broad—up to 6 in (14 cm) wide—with a concave channel on its upper side. Anatomical examination shows the absence of sclerenchyma bundles near the upper concave surface of the leaf: it should be noted that this species grows in the moist sclerophyllous forest in the furthest southwest tip of Western Australia, where the sclerophyllous character of the vegetation is generally less marked than in other places.

Xanthorrhoea and *Kingia* are markedly sclerophyllous and thus typical of much of Australia's mediterranean vegetation, more so than in comparable regions of other continents. Even the Australian deserts are characterized by sclerophyllous cushion plants of the genera *Triodia* and *Plectrachne*. Likewise, the Australian grass trees are a unique biological growth form. This phenomenon of sclerophylly is attributed to the nutrient deficiencies common in Australia's soils, especially phosphate deficiency. Fire-resistance is another adaptation that many Australian plants, and specifically the "black boys," have had to acquire.

Phyletic origin and systematics

The taxonomy of the Australian grass trees is unclear and controversial. There are three accepted genera (*Xanthorrhoea*, *Kingia*, and *Dasypogon*), but there is dissent over which family they belong to. They were first considered to be primitive members of the Liliaceae, but in 1829 they were placed in a new family the Xanthorrhoeaceae. Since then *Dasypogon* and *Kingia* have been placed in a new family, the Dasypogonaceae. Seven of the thirty species of Australia's endemic genus *Xanthorrhoea* are isolated and restricted to the southwestern tip of Australia, a further species (*X. thorntonii*) grows sporadically on the western desert, while all the rest grow in the continent's eastern coastline. The monotypic genus *Kingia* contains a single species, while the genus *Dasypogon* has three species, all found in southwest Australia.

Their habitat is quite variable: moist forests, dry forests, scrub, and marshes. Not all the species in these genera are arborescent; some form perennial tussocks with a strong underground base but no trunk. The arborescent species of *Xanthorrhoea* are *X. australis* (Queensland and Tasmania), *X. arborea* (New South Wales to south Queensland), *X. johnstonii* (Queensland and northern New South Wales), *X. quadrangulata* and *X. tateana* (south Australia) and *X. thorntonii*, *X. preisii* and *X. reflexa* in Western Australia. At least another nine species produce short stems with a maximum height of 3 ft (1 m), and another thirteen normally lack any aerial stem. *Kingia australis* is arborescent, but there is only one similar species of *Dasypogon* (*D. hookeri*), making a total of ten species of "grass trees" in Australia.

2.3 Plenty of shrubs and small trees

The Mediterranean biome, as has already been pointed out, is dominated by woody plants, that is to say, by trees and shrubs that are often very small but whose roots are deep enough to give them access during the dry period to the accumulated water deep in the soil. They are phanerophytes, the name for the growth form of plants whose perennating (they survive from year to year) buds are more than 8-16 in (20-40 cm) above the soil, or chamaephytes, plants forming small clumps with perennating buds between 8-16 in (20-40 cm) above the soil surface. Anyway, the distinction between trees and shrubs is often unclear in the Mediterranean regions, as many low-growing tree species may grow in the form of shrubs under limiting environmental conditions, both in terms of size and general morphology. Thus, it is frequent to find in tree and shrub communities with a similar same species composition, or at least some shared dominant species. In any case, the Mediterranean biome's large number of tree and shrub species is due to the characteristics of the speciation process mentioned before (see also chapter 2.1 of this volume).

The trees of the Mediterranean Basin and of California

The dominant tree species in the Mediterranean Basin and the Californian mediterranean to the beech family (Fagaceae), and the pine family (Pinaceae). Most of the Fagaceae are members of

42 The acorns of *Quercus coccifera* (top), *Q. ithaburensis* and *Q. calliprinos* (in the center) and *Q. engelmannii* (bottom). Sclerophyllous adaptations are present throughout the plants, not just in their leaves. Their acorns, which are very large nuts in a basal cupule, ripen in autumn, and appear to have developed defenses against herbivores as well as against the dry conditions. The kermes oak (*Quercus coccifera*) of the western Mediterranean Basin produces spiny leaves and also bears its acorns in spiny cupules consisting of curved scales. Other species of *Quercus* such as *Q. ithaburensis* and *Q. calliprinos* of Upper Galilee, and the Californian *Q. engelmannii* show small variations to deal with Mediterranean climatic conditions.
[Photos: Ernest Costa, Jordi Bartolomé and Francesc Muntada]

43 **Aleppo pine (*Pinus halepensis*) and stone pine (*P. pinea*)** at the base of the Montseny range (Catalonia). The Aleppo pine is probably the Mediterranean Basin's most widespread and characteristic species of pine. It forms large forests from sea-level up to an altitude of 3,300 ft (1,000 m), especially on the coastline and on dry calcareous soils, although it can also grow in siliceous soils. In the past the Aleppo pine was almost certainly less abundant, as it has benefitted from intensive forestry planting and the increasing degradation of other plant communities in the areas most affected by humans. The stone pine probably originated in the western isles of the Mediterranean and was widely spread by the Romans who valued it highly.
[Photo: Ernest Costa]

the genus *Quercus*, and may be deciduous, such as the English oak, or evergreen, such as the holm oaks and cork oaks. Yet the genus *Quercus* also includes shrubby species such as the kermes oak (*Q. coccifera*) in the Mediterranean Basin and several species in the Californian chaparral. The members of the pine family are mainly species of the genus *Pinus*. Even so, there is greater species diversity in these genera in the Californian mediterranean than in the Mediterranean Basin, as in the latter the east-west mountain ranges acted as barriers duration the glaciations, thus increasing the rates of extinction. In fact, the Californian mediterranean's large number of species of pines and both deciduous and evergreen oaks reflects the high

specific diversity of these groups throughout North America, an area that has 40 species of *Quercus* and more than 80 conifers, 34 of which are species of *Pinus*.

Pines are the most abundant conifers in both the Mediterranean Basin and the Californian mediterranean. In the Mediterranean Basin the most abundant species are the Aleppo pine (*P. halepensis*), the Calabrian pine (*P. brutia*), the maritime pine (*P. pinaster*) and the black pine (*P. nigra*), while in the Californian mediterranean the most abundant species are the Monterey pine (*P. radiata*), the digger pine (*P. sabiniana*), *P. attenuata*, and *P. coulteri*. There are conifers belonging to families other than the Pinaceae,

44 The tree Californians call the canyon live oak (*Quercus chrysolepis*) is one of the most typical evergreen plants of the Californian mediterranean region. In English, the word oak refers to both evergreen and deciduous species of *Quercus*, requiring the use of an adjective to indicate the evergreen nature of a species. This is because English arose in higher latitudes, where only deciduous species of *Quercus* grow, and thus does not have two words to distinguish between deciduous and evergreen species.
[Photo: James Simon / Bruce Coleman Ltd]

such as the cypress family (Cupressaceae) and the yew family (Taxaceae) such as: different species of cypresses (*Cupressus sempervirens* in the eastern Mediterranean Basin and *C. macrocarpa* on the Californian coast), and the junipers and savins (*Juniperus*). Other important Californian conifers include *Pseudotsuga macrocarpa* and the redwoods (*Sequoia sempervirens* at low altitudes, and *Sequoiadendron giganteum* in the mountains). The conifers in the higher areas of the Mediterranean Basin include yews (*Taxus baccata*), cedars (*Cedrus atlanticus*, *C. libani*, *C. brevifolia* in the Atlas Mountains, Lebanon and Cyprus respectively), and even firs (*Abies pinsapo*, *A. maroccana*, *A. nebrodensis*, *A. cephalonica*, *A. cilicia* in Andalusia, Morocco, Sicily and Calabria, Greece, and Turkey respectively), although they are not very important elements in the landscape. Most of the broadleaf trees and shrubs in California and the Mediterranean Basin belong to the genus *Quercus*. The absence of evergreen species of *Quercus* in the English-speaking areas of Europe means the English language lacks the terminological distinction found in Latin languages between deciduous species (*robles* in Spanish, *roures* in Catalan, and *roveri* in Italian) and evergreen species (*encinas* in Spanish, *alzines* in Catalan, and *lecci* in Italian). In English they are all oaks, and likewise in French the word *chêne* is used for both deciduous and evergreen species, because in the area where French evolved, north of the Occitan area, evergreen oaks are also absent. To resolve this problem, the evergreen species are called *chêne vert* (green oak), namely the holm oak (*Q. ilex*), the carrasca (*Q. ballota*), the cork oak (*Q. suber*), the

kermes oak and other similar species (*Q. coccifera*, *Q. calliprinos*), and the evergreen species from California (*Q. agrifolia*, *Q. chrysolepis*, *Q. engelmannii*, *Q. wislizenii* and even the clearly shrubby *Q. palmeri* and *Q. morpheus*). The deciduous species naturally present no problem, but they are very scarce in the area (*Q. faginea*, *Q. pyrenaica*, *Q. humilis* in the Mediterranean basin, and *Q. lobata*, *Q. douglassii* and *Q. kelloggii* in California). In general, deciduous oaks have leaves that are more-or-less deeply lobed (or even incised, like *Q. kelloggii*), while evergreen species have entire leaves, often with slightly spiny margins. The leaves of the tanbark oak (*Lithocarpus densiflora*) are also entire; this unusual Californian oak belongs to a genus whose distribution is basically Asiatic. Obviously, not all trees in California or the Mediterranean Basin landscape are conifers or members of the Fagaceae. It would be unforgivable to omit two domesticated trees that have become symbols of the Mediterranean Basin: the carob or locust tree (*Ceratonia siliqua*) and the olive (*Olea europaea*), cultivated forms of the wild olive (*Olea europaea* var. *sylvestris*). Nor should we forget the Mediterranean deciduous trees that, like the deciduous oaks, grow in the coolest spots in the area or in the mountains, the best examples of which were, for example, the maples (*Acer opalus*), the black maple (*A. monspessulanum*), the bigleaf, or Oregon, maple (*A. macrophyllum*) and the Californian horse-chestnut (*Aesculus californica*). Other deciduous trees, such as walnuts, willows, and poplars, grow on the banks of permanent watercourses in the Mediterranean and in California, and are discussed in the section on aquatic ecosystems.

45 **The Chilean wine palm** (*Jubaea chilensis*) grows in the Chilean mediterranean area, such as this example photographed in La Campana, making it the palm with the southernmost distribution. The spongy nature of its stem (or false trunk) and the texture of its protective covering make it highly fire-resistant. It does not flower until it is 60 years old and it can reach a height of 98 ft (30 m), while its "stem" can reach a diameter of 3 ft (1 m). The sap yields a sugary liquid called *miel de palma* (palm honey) and its leaves are used in basketwork.
[Photo: Ramon Folch]

Tree of the Chilean mediterranean

The Chilean region's tree species are not related to those of the Mediterranean Basin and California. Chile has a set of species that are members of different families, such as the Anacardiaceae (*Lithrea caustica*), Rosaceae (*Quillaja saponaria*), and Auraceae (*Cryptocarya alba*). Some of these species, such as the Chilean *Lithrea caustica*, which produces fever and irritation of the skin on contact, the soap bark tree (*Quillaja saponaria*), whose saponin-rich bark is used to make soap, and *Cryptocarya alba*, may dominate both tree and shrub communities, clearly showing the Chilean region's lower level of specialization, probably due to its greater biogeographical isolation. Like the trees of the Mediterranean Basin and the Californian region, all these species have the same form of tree with a dark green crown at the top of a not very tall twisted stem (that is more useful for firewood than for timber), as a result of their difficult growing conditions.

The Chilean mediterranean's trees include an endemic palm (*Jubaea chilensis*) that grows in a very small area and is the most southerly of all palms.

46 *Leucadendron argenteum,* **a member of the Proteaceae,** is the Cape region's only native tree species. The Proteaceae is one of the most ancient of the angiosperm families and was already present on the continent of Gondwana 300 million years ago. The species of the Proteaceae only grow naturally in the southern hemisphere, as tropical and equatorial regions have acted as a climatic boundary to their dispersal.
[Photo: Colin Paterson-Jones]

Broad-leaved trees are not abundant in the truly Mediterranean region and are more characteristic of the zones of transition towards mountainous or more southerly climates, and are represented by the southern hemisphere's most typical members of the Fagaceae, the genus *Nothofagus,* commonly known as in English as southern beeches. Like most of South America, conifers are virtually absent, but the slopes of the Andes are the home of the Chilean incense cedar (*Libocedrus* [=*Austrocedrus*] *chilensis*) as well as other species belonging to the genera *Araucaria, Podocarpus,* and *Fitzroya,* which are not truly Mediterranean as they grow in cooler and wetter climates. The Chilean mediterranean has 57 tree species, 35 of which are endemic, while the others are more widely distributed.

Trees of the South African mediterranean

The South African mediterranean region has only a single native species that forms a true tree (*Leucadendron argenteum*), a member of the Proteaceae. Its natural distribution is restricted to the moist slopes of Table Mountain, near Cape

47 **The leaves and fruits of a *Eucalyptus*,** the genus that dominates the tree and shrub flora of the Australian continent. These trees show an interesting leaf dimorphism as the adult leaves are sickle-shaped and the same on both sides, while the rounded juvenile leaves are opposite, and a different, more glaucous, green color. This curious difference is considered to be an adaptation, because the adult leaves avoid overheating by hanging vertically, while the juvenile leaves have adapted to the reduced sunshine they receive in the undergrowth.
[Photo: Joaquim Reberté & Montserrat Guillamon]

The trees of the Australian mediterraneans

The tree and shrub flora of the Australian mediterraneans is dominated by a large number of species of Leguminosae (*Acacia*) and Myrtaceae, in particular the genus *Eucalyptus*. Species of *Eucalyptus* are found throughout the Australian continent in every ecological niche, thus illustrating the special biogeographical characteristics of a continent that has been isolated for such a long time.

The most abundant trees include the jarrah (*Eucalyptus marginata*), found in the typical Australian mediterranean, where it plays the same role as deciduous oaks, evergreen oaks, *Lithrea caustica*, and the soap bark tree in the other mediterraneans. The wandoo (*E. wandoo*) grows in zones with slightly drier climates, in the transition to Mediterranean shrub communities. Southern Australia has 18 tree species of *Eucalyptus* which form pure or mixed forests. In the southwestern Australian mediterranean (Western Australia), but not in the southern one, zones with potentially very heavy winter rains are dominated by karri (*E. diversicolor*), a large tree that can reach 279 feet (85 m) in height and which forms highly developed, dense forest masses. It is sometimes accompanied by another large species, marri (*E. calophylla*). Together with the Californian redwoods, these three species of *Eucalyptus* are the giants in a biome with so many small trees that would only be considered to be a shrub in other biomes.

Although *Eucalyptus* species dominate this region of the biome, there are other trees, especially species of she oak (*Casuarina*). Conifers are uncommon and restricted to the genus *Callitris*, such as the cypress pine (*C. columellaris*). And it is worth mentioning again the unusual monocotyledonous grass trees belonging to the genera *Xanthorrhoea* and *Kingia*, known as "black boys" and "black gins" due to the form of their inflorescence which give them a masculine or feminine appearance. Also noteworthy are their blackened trunks which bear witness to the many fires to which they have been exposed and which they have survived, due to their adaptations. (See also page 68.)

The shrubs of the Mediterranean Basin and the Californian mediterranean

As regards shrubs, all the areas have many more species of shrubs than trees, but there are also considerable differences between the different areas. In the

Town, but it is now abundant as it has been widely planted in parks and gardens as an ornamental. Other species of the Proteaceae are on the borderline between trees and shrubs, although they are generally considered as large shrubs rather than as trees, because they never grow to a large size.

The South African mediterranean region's soil and climatic conditions allow the development of forest in many parts of the biome, but paradoxically the presence of native tree vegetation is very restricted and almost a matter of anecdote. Some non-native tree species, such as the Monterey pine (*Pinus radiata*), grow very well in dense plantations and even behave as invasive species and form forests, thus highlighting the surprising lack of native trees. There are several hypotheses to explain the absence of trees, but the most widely accepted one relates it to the origins of the Cape flora. The potential tree flora of the Cape region would have been derived from tropical African trees growing on mountains, and this group's lack of species preadapted to poor soils and repeated fires was the cause of this surprising absence. It is not at all clear why species of this type have not appeared, as the evolutionary process could have given rise to the development of physiological adaptations in the species already present, but the total lack of pre-adaptations and the isolation of the South African mediterranean might explain this phenomenon.

48 **Branch with flowers and fruit** of the cypress pine (*Callitris columellaris*), one of the few members of the cypress family in Australia's flora, specifically, Australia's mediterranean flora. *[Photo: Lluís Ferrés]*

2. LIFE IN SCLEROPHYLLOUS FORMATIONS

Mediterranean Basin there are more than a hundred species belonging to 76 genera and 29 different families. There is a general difference between east and west: the former is dominated by members of the Lamiaceae, Rhamnaceae and Hypericaeae, while the west is dominated by members of the Leguminosae, Cistaceae and Ericaceae. Some genera have particularly large numbers of shrubs: thus there are 22 shrubby species in the genus *Genista*, 20 in *Sideritis*, 12 in *Cistus*, *Cytisus*, *Salvia* and *Satureja*, 10 in *Thymus*, 7 in *Erica* and 4 in *Ulex*, to cite just a few examples.

These shrubs vary greatly in size, from small trees 10-13 ft (3-4 m) high to bushes less than 8 in (20 cm) high. Species such as the wild olive (*Olea europaea* var. *sylvestris*) and the carob, or locust, tree (*Ceratonia siliqua*) are very important in terms of their importance in the landscape. Although previously discussed as cultivated trees, they often grow no larger than a shrub and may be considered as such. Other very important species are the mastic tree (*Pistacia lentiscus*), the strawberry tree (*Arbutus unedo*), and the kermes oak (*Q. coccifera*) because they dominate the taller shrub communities; while heathers (*Erica*) and cistus (*Cistus*) play an important role in shrubby matorral of medium height; and species of thyme (*Thymus*) dominate the communities of smaller plants, known in Spanish as "tomillares" (from the Spanish *tomillo* or thyme). There is also a species of palm (*Chamaerops humilis*), that is the only widespread native European member of the palm family.

These shrubs show enormous variation in leaf morphology: some have typically sclerophyllous leaves, like the kermes oak (*Quercus coccifera*), or are malacophyllous like the cistus (*Cistus*), while others have very small leaves, such as the heathers (*Erica*). There even are leafless plants, which are often extremely spiny, such as the brooms and gorses (*Genista* and *Ulex*).

The Californian mediterranean has a richer flora, especially considering that it covers a much smaller area, and this is because of the factors mentioned when discussing the tree flora. Its floristic composition is very different from that described for the Mediterranean Basin, but most families, and some genera, are the same. Very many shrub species occur that belong to the Fagaceae (*Quercus*), Ericaceae (*Arbutus*, *Arctostaphylos*), Rosaceae (*Adenostoma*, *Heteromeles*), Anacardiaceae (*Rhus*), and Rhamnaceae (*Ceanothus*). The diversity of the flora is shown by the number of species in some genera, such as *Ceanothus* (25 species) and *Arctostaphylos* (19 species). The dominant species in the typically Mediterranean communities are chamise (*Adenostoma fasciculatum*) with ericoid leaves, the typically sclerophyllous scrub oak (*Quercus dumosa*), several species of *Ceanothus*, and the manzanitas (*Arctostaphylos*), all of which also have a typically sclerophyllous morphology. In the eastern parts of the Sierra Nevada, where the winters are colder and the desert influence is stronger, the shrub *Artemisia tridentata* is very abundant. In the areas of transition towards warmer and drier environments there are even abundant succulent plants and members of the cactus family (Cactaceae).

The shrubs of the Chilean mediterranean

In the Chilean region, as has been mentioned, some species may behave as trees or shrubs, for example *Lithrea* and soap bark tree. Other common shrubs include two typically sclerophyllous species, *Colliguaja odorifera* and *Kageneckia oblonga*, and summer deciduous malacophyllous plants, such as

49 The main trees and shrubs of the Chilean matorral and the Californian chaparral. Most of the flora of these two Mediterranean regions consists of sclerophyllous species. The use of different names to describe them already suggests there are differences (the Californian mediterranean is dominated by shrub species of the genus *Quercus*, while not a single species of this genus is found in the Chilean mediterranean), yet there are similarities between the two regions, as can be seen in the drawing. It is interesting to note that plants from very different families have adopted similar strategies. The Chilean species drawn are *Colliguaja odorifera*, a member of the Euphorbiaceae; *Cryptocarya alba*, a member of the Lauraceae; *Lithrea caustica*, a member of the Anacardiaceae; and soap bark tree (*Quillaja saponaria*). The Californian species are manzanita (*Arctostaphylos glauca*), a member of the Ericaceae; coastal live oak, (*Quercus agrifolia*), a member of the Fagaceae; chamise (*Adenostoma fasciculatum*), a member of the Rosaceae; and *Ceanothus leucodermis*, a member of the Rhamnaceae. [Drawing: Eugeni Sierra]

Colliguaya odorifera

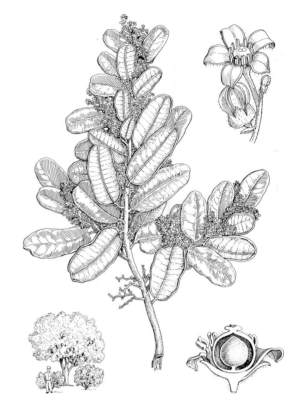

Lithraea caustica

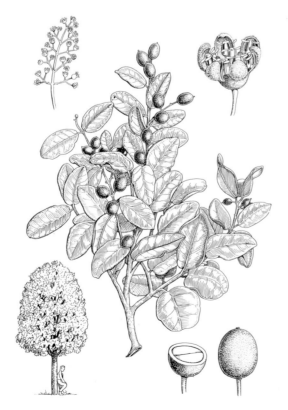

Cryptocarya alba

Quillaja saponaria

Arctostaphylos glauca

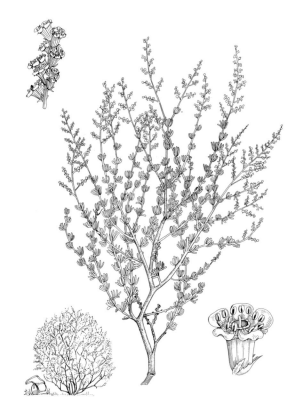

Adenostoma fasciculatum

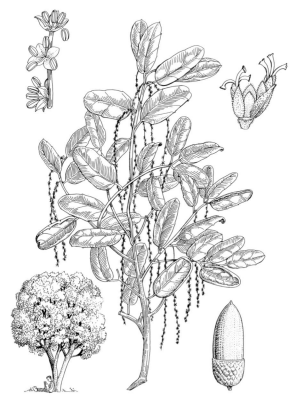

Quercus agrifolia

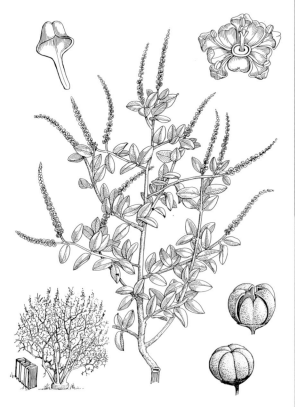

Ceanothus leucodermis

50 *Trichocereus chiloensis*, **near Valparaíso, Chile**. This columnar cactus, typical of arid areas, thrives in the Chilean mediterranean, reaching 10 ft (3 m) or even 13 ft (4 m) in height. It arrived from neighbouring semi-desert regions where it is abundant. The presence here and in the Californian mediterranean of these species so typical of arid areas gives the shrub vegetation of the American mediterraneans its distinctive appearance.
[Photo: Adolf de Sostoa & Xavier Ferrer]

Baccharis linearis, *B. rosmarinifolius*, *Trevoa trinervis*, *Satureja gilliesii*, *Proustia cuneifolia*, *P. cinerea*, and *Flourensia thurifera*. Mention should also be made, because of its abundance, of the "espino" (*Acacia caven*), a small tree whose distribution extends beyond the Chilean mediterranean.

Another distinctive feature of this mediterranean are the columnar cacti, such as "quisco" (*Trichocereus chiloensis*), some specimens of which can reach 13 ft (4 m) in height. As well as these cacti there are other succulent plants of the genus *Puya*, which help to give the Chilean mediterranean its distinctive appearance that is so unlike the other mediterraneans, except for the driest areas of California where similar plants are also abundant.

The shrubs of the Cape mediterranean

The southern tip of the African continent, the Cape region, has a surprisingly rich flora, as shown by the more than 600 species of the genus *Erica* and the more than 100 species of the genus *Protea*. The region's poor soils favor the abundant shrubby plants that are morphologically well adapted to these extreme conditions, with characteristics such as small leaves (*Erica*), or even no leaves as seen in members of the Restionaceae that have green photosynthetic stems instead. *Elytropappus rhinocerotis*, a member of the Asteraceae, is very abundant; it is a small evergreen cypress-like shrub whose growth is being adversely affected by excessive grazing and other environmental changes.

The drier sites are occupied by leafless succulents (such as *Euphorbia mauritanica*), summer-deciduous species (such as *Zygophyllum morgsana*) and plants with succulent leaves, such as *Aloe ferox*. Some sclerophyllous shrubs reach a considerable size, such as *Olea africana* and various species of the genera *Rhus*, *Euclea*, *Pterocastrus*, and *Sideroxylon*; but the most abundant species belong to various genera of the Proteaceae, such as *Protea*, *Leucadendron* and *Leucospermum*.

The shrubs of the Australian mediterranean

The Australian mediterranean's shrub flora is dominated, like its tree flora, by species of the genus *Eucalyptus*. The genus contains about twenty shrub species, all similar in morphology and known as mallee. They are all large shrubs or small trees with many stems growing from a single stock, and they lack the single main stem found in trees. They are accompanied by many species, including members of the Myrtaceae (*Melaleuca*, *Baeckea*, *Leptospermum*), Casuarinaceae (*Casuarina*), Cupressaceae (*Callitris*), Leguminosae (*Acacia*), Epacridaceae (*Leucopogon*, *Astroloma*), Myoporaceae (*Eremophila*), Rhamnaceae (*Cryptandra*, *Spyridium*), Chenopodiaceae (*Atriplex*, *Maireana*) and Proteaceae (*Banksia*, *Hakea*, *Grevillea*). They all contribute to forming a landscape of great floristic diversity.

2.4 Structure, competition, and adaptations

Mediterranean shrub communities are relatively simple in structure, consisting of a more-or-less dense shrub layer and a herbaceous layer. Even the most developed shrub formations, such as maquis, matorral, and chaparral, which are so dense they are almost impenetrable, are very similar in structure to what has been described.

Stratification of holm oak forests

Mediterranean forests are a special case. They are in the wettest areas of the biome and are only important features within the landscape in the Mediterranean Basin (forests of holm oak, kermes oak, cork oak and the deciduous oaks) in the southwest of Australia (jarrah) and in the Californian mediterranean (oak woodlands); they are very rare, or are lacking altogether in the other areas. The most complex structure is found in holm oak forests, the most developed of all Mediterranean forests. The tree layer may reach a height of 66-82 ft (20-25 m), but only rarely does it reach this height because of its intense exploitation and the fact that the dense crowns create a cool, moist environment within the forests. Underneath this protective canopy there is a dense and very rich shrub layer that can reach 10-16 ft (3-5 m) in height, with plants like the laurustinus (*Viburnum tinus*), *Rhamnus alaternus*, *Phillyrea latifolia*, mastic tree (*Pistacia lentiscus*) and the strawberry tree (*Arbutus unedo*). There is also a low shrub layer (1.6-3 ft or 0.5-1 m) with lower growing plants, such as butcher's broom (*Ruscus aculeatus*), asparagus (*Asparagus acutifolius*), and wild madder (*Rubia peregrina*). The herbaceous layer lacks light and is thus species poor, with only a few species like spleenwort (*Asplenium onopteris*), wall germander (*Teucrium chamaedrys*), or white violet (*Viola blanca*).

Yet the holm oak forest's most remarkable feature is the presence of a well-developed liana layer, especially in the coastal forests of typical holm oak (*Quercus ilex*), which are the most developed, but also in the dry continental forests of carrasca (*Q. ballota*). In fact, trees and shrubs are linked together by climbers such as like smilax (*Smilax aspera*), clematis (*Clematis flammula*), mediterranean honeysuckle (*Lonicera implexa*), rose (*R. sempervirens*), and even ivy (*Hedera helix*). They, and especially the foliage of the clematis hanging from the crowns of the trees, all give the forest as a whole a surprisingly tangled appearance. These different strata are full of lianas and give rise to very dense, almost impenetrable woods, with a forest-like appearance; woods that are the biome's most complex plant formations.

Adaptation to unfavorable circumstances

The plants of the biome face two basic problems: water shortage in the dry period and cold weather in the winter. And, although it is not a generalized problem, the nutrient poverty of certain soils, especially certain soils of the Australian and Cape regions, should be taken into account. The winter cold is an adverse environmental factor that plants must withstand and which can limit their growth, but water and nutrients are resources that have to be obtained from the environment, and this leads to competition between species or between members of the same species. Summer drought, cold weather, and oligotrophic soils are three problems with different impacts and occurrences in the differing areas, and they have given rise to differing strategies, given that each problem can be solved in more than one way. In addition, in a climate in which water is scarce, plant growth is relatively slow and the laboriously synthesised structures

51 Snow damage to Aleppo pines (*Pinus halepensis*) in Ordal, Catalonia. The Mediterranean occasionally experiences intense cold spells. When air from Siberia reaches the Mediterranean Basin, which happens sporadically, weather conditions change greatly and cause severe disturbances, since the Mediterranean is little accustomed to severe cold. These situations cause loss of crops and damage to the native vegetation, as well as disrupting normal life in the large cities. Mediterranean plants are not adapted to bearing the weight of snow on their branches, which may break.
[Photo: Josep Maria Barres]

are often protected from herbivores, for they can destroy in no time at all, what the plant has taken months to construct. Thus, in addition to the adaptations mentioned above, there are others intended to protect against grazing.

Winter cold and vegetative activity

The Mediterranean winters are only moderately cold. This is an affirmation that is constantly repeated when talking about the biome's climate, but the truth is that the cold does create problems for plants, as they are unable to control their temperature. The most thermophilic Mediterranean species, such as the carob, (*Ceratonia siliqua*) and the wild olive (*Olea europaea* var. *sylvestris*), have distributions restricted to warm areas, while other more cold-resistant species such as the holm oak (*Q. ilex*) can live in areas with lower temperatures. It is worth mentioning the close relationship between increase in temperature and reduction in rainfall, as this is the reason why many Mediterranean species that occupy colder sites cannot live in the warmer areas; it is because of problems of water economy not because of the high temperatures, which they can tolerate perfectly well.

Cold poses two different problems for plants: on the one hand, it limits their metabolism and on the other hand, it may cause physical damage if temperatures drop below a particular value. Many Mediterranean plants keep photosynthesising throughout much of the winter, but they cannot perform other activities, such as coming into growth, as the low temperature acts as a limiting factor. Most Mediterranean species cannot

grow when the temperature is below 50°F (10°C) and the most thermophilic species need temperatures of more than 62°F (17°C) to shoot. With regard to the possible damage caused by excessive cold, such damage is due to the formation of ice crystals within the tissues, especially serious for evergreen plants as they have to maintain their leaves throughout the winter, and these relatively unprotected structures with a high water content are vulnerable to the cold.

Many Mediterranean plants resist the cold very well; the holm oak, for example, can resist temperatures of 10°F (-12°C) without suffering damage to its leaves and as low as -4°F (-20°C) without damage to its trunk. Other species are less resistant. The leaves of the cork oak (*Q. suber*), kermes oak (*Q. coccifera*), and *Rhamnus alaternus* tolerate temperatures to only 12°F (-11°C), while those of myrtle (*Myrtus communis*) and oleander (*Nerium oleander*) are resistant to 18°F (-8°C). Carob leaves tolerate temperatures to 19°F (-7°C). The cold resistance of a single species or individual varies over the year, and increases in the winter due to the relative dehydration which concentrates the cell and vacuolar sap, thereby lowering their freezing point. The olive (*Olea europaea*) for example, can withstand temperatures below 14°F (-10°C) in mid-winter, but only 23°F (-5°C) in spring and autumn. For the holm oak, the figures are -13°F (-25°C) in winter and 23°F (-5°C) in summer for the new shoots, and 5°F (-15°C) in winter and 21°F (-6°C) in summer for the leaves. On the other hand, a winter-deciduous tree such as the oak (*Quercus humilis* [=*Q. pubescens*]) can easily withstand temperatures of -31°F (-35°C) when without leaves. As a result of these seasonal physiological adaptations, the first frost or late frosts, which are rare in, but not absent from, the Mediterranean, may cause serious damage to the vegetation and limit the distribution of certain long-lived species, even if they only occur with a very low frequency.

Water scarcity, nutrient poverty and the root system

Its irregularly distributed and moderate rainfall means the Mediterranean biome's vegetation combines species adapted to almost sub-desert conditions with other species that need more moisture; the biome includes plant types as diverse as succulents, summer-deciduous plants, evergreens, and winter-deciduous plants. Comparable variation is present in their root systems, which have been studied much less because of the methodological problems involved. Yet, it is logical to hope that some general tendencies can be detected in the development and morphology of root systems, as they are so closely related to water uptake.

The underground parts reproduce the morphological differences that are found in the aerial parts. In Chilean matorrals on soils more than 3 ft (1 m) deep, the sclerophyllous species like *Lithrea caustica* produce a large number of roots throughout the soil profile, while the summer-deciduous shrubs like *Satureja gilliesii* only root in the top 16 in (40 cm) of soil, and the succulent plants, such as "quisco" cactus (*Trichocereus chiloensis*) develop roots only in the top 6 in (15 cm). This shows the coexistence of three different strategies under analogous environmental condition: the succulent plants make use of surface water when it rains, and then store it in their tissues; summer-deciduous plants do not need to root at great depth, as they are active during the periods of abundant moisture, although they need to exploit a wider soil layer, since they are not able to accumulate water reserves; finally the evergreens, which are active throughout the year, root at greater depth in object to ensure their supply of water during the dry season.

These major interspecific differences correspond to very different strategies, but there are also considerable differences between species that are morphologically more similar. For example, one might suppose that evergreens are always deeply rooted, but there is in fact great variation between different situations. In the Chilean matorral, for example, the "colliguay" (*Colliguaja odorifera*) does not root below 3 ft (1 m), while the *Lithrea caustica* sends roots down to 16 ft (5 m) and the roots of the soap bark tree (*Quillaja saponaria*) can reach down to 26 ft (8 m). The advantages of being able to root deeply appear to be obvious enough. This habit is known, directly or indirectly, in many Mediterranean species such as *Quercus calliprinos*, *Q. ilex*, *Q. dumosa*, *Adenostoma fasciculatum*, and *Pistacia lentiscus*, among others, that can easily root down to 26 ft (8 m). The most deeply rooted plants are trees: the roots of the valley white oak (*Q. lobata*) are said to grow down to 65 ft (20 m), those of the scrub interior live oak (*Q. wislizenii*) and the blue oak (*Q. douglasii*) to 85 ft (26 m), and the roots of the jarrah (*Eucalyptus marginata*) to 164 ft (50 m).

The depth of the soil and the nature of the substrate are clearly very important factors, as they may ease or prevent the roots' penetration of the soil, and at the same time may make it necessary or useless. Shallow soils prevent deep root growth, but the roots of plants on a very fragmented calcareous substrate may be able to penetrate through the cracks to greater depths, which results in behaviorial and morphological differences between individuals of the same species, depending on the soil and substrate they are growing on. Three out of seven independent studies of the depth of the roots of the chamise (*Adenostoma fasciculatum*) describe the root systems as superficial and concentrated in the top 3 ft (1 m) of soil, while the other four emphasize the plant's capacity to root down to 13 ft (4 m). Root systems show great developmental adaptability because they can adjust to very different soil conditions, allowing the plants to cope with the arid conditions of some sites.

There are few data on the weights and surfaces of the root hairs, the ones that absorb water and nutrients. The relation between the functional root surface (absorbing surface) and the leaf surface (the transpiring surface) may be a very good guide to the strategies of different species, but not enough is known to allow us to make generalizations. It is known that only 10% of the weight of the roots corresponds to the root hairs, and only a small part of this 10% is fully functional. In general the area of soil occupied by the root system is greater than that of the tree's canopy, and varies depending on the depth of the soil. In the case of the chamise (*Adenostoma fasciculatum*), values for the ratio of the root surface area to the surface area of the canopy may be two or three in deep soils and may reach seven in shallow soils; this clearly shows the developmental plasticity mentioned earlier when discussing rooting depth. In any case, in spite of the lack of information about this, it is worth pointing out the relative abundance of adaptations in root systems that favor very efficient absorbtion of nutrients, such as the abundance of fine

52 The base of the stem of a holm oak (*Quercus ilex*), showing the thick moss-covered bark and the roots growing towards the soil, as well as ivy (*Hedera helix*), a hemi-parasitic creeper. The arrangement of the plant's roots seeks to make best use of water resources, just as the architecture of the aerial parts seeks to make best use of light. Mediterranean trees, usually have long roots so that they can reach down to the lower levels that contain enough usable water in the dry season. In fact, studies of the concentration of cell sap show the plant's root system has as many adaptations to drought as the aerial parts. In the dry season the water balance changes very little and the hydration of the cell protoplasm hardly alters.
[Photo: Josep Pedrol]

roots in the top layer of soil and in the leaf litter, the presence of mycorrhizae and symbiosis with nitrogen-fixing bacteria. These adaptations are present in all the biomes, but they are especially abundant in conditions of edaphic oligotrophy, and are a response to the Mediterranean biome's relative abundance of poor soils.

Another good indicator of a species' strategy is the ratio between the biomass of its roots and that of its aerial parts. Different Mediterranean plants show very different ratios, as they may be greatly influenced by the repeated destruction of the aerial parts by repeated fires or felling. Values for this ratio are generally between 0.2 and 0.9 for trees, and between 1 and 6 for shrubs. However, there are very many exceptions, and thus in holm oak forests the values may vary between 0.4 and 1.2, depending on whether the site is dominated by fruiting trees or by regenerating specimens. Much higher values have been found in some trees, for example 2.3 in the *Cryptocarya alba* and 5 in *Lithrea caustica*. It should be pointed out that these very high values are doubtless the result of repeated destruction of the aerial parts and they are not exclusive to the Mediterranean biome. For example, in the temperate American forests deciduous shrubs such as scrub oak (*Quercus ilicifolia*) may show values greater than 6 in periods of regeneration after fire. In any case, the combination of summer drought with occasional fires mean that it can be expected that the Mediterranean biomes will frequently, but not always, show high values for underground biomass. The coexistence of different rooting strategies allows more

thorough use of the soil's nutrient and water resources and also helps to reduce competition, as the different species exploit different layers of the soil system.

Many Mediterranean shrubs and trees with the ability to produce new shoots have semi-underground structures in the form of thick rootstocks formed by a tissue called lignotuber, that stores water and nutrients and can produce new shoots when the aerial parts have been destroyed by fire, grazing, or felling. Thus, these structures are related to fire as well as to the plants' water and nutrient economy, and they play a very important role in the ecology and functioning of Mediterranean systems.

The summer drought and the leaf system
Mediterranean plants have to withstand prolonged dry periods precisely in the hot season, giving rise to a series of adaptations collectively known as *xerophytism*. There are two basic ways to solve this problem, depending on the length of the dry period. The first strategy is to maintain the leaves and reduce water loss by controlling stomatal opening, while the second is to shed the leaves in the dry period and limit activity to the cool, moist period. The adoption of one or other strategy is a function of the economy of resources: thus, when the quantity of carbon necessary to maintain a leaf over the dry period is less than the cost of making a new leaf, the best strategy is to have evergreen leaves. Obviously, the costs of making and maintaining a leaf depend on its structure and are not fixed quantities. In general, however, when the dry period is

longer than 100 days, the summer deciduous strategy is adopted, and when it is less than 100 days, the evergreen foliage strategy is selected. In intermediate conditions, both strategies are found, but in extreme conditions one or the other predominates.

In addition to surviving drought, a leaf has to last long enough to recover its production cost, and this in turn depends on its structure and chemical composition. This means that values for the length of a leaf's life are very variable, as many summer deciduous plants recover the cost of leaf production in only 90 days, while many evergreens need one, two or even three years to recover this cost. As the rate of photosynthesis varies with the quantity of light received, the time necessary to recover production costs and the longevity of leaves differs within a single species and within a single plant, depending on the leaf's position within the tree's canopy or the plant's position in a sunny or shady site. Thus, for example, the leaves of the chamise (*Adenostoma fasciculatum*) last 2 to 3 years in shady sites but only 1 to 1.5 years in sunny ones.

Leaf persistence is also related to the availability of essential nutrients like nitrogen, as poor soils favor the long-lasting leaf strategy. In conditions of nutrient deficiency it is better to construct long-lasting leaves, that is to say, ones that are more resistant and less attractive to herbivores. It is worth pointing out that the response to summer drought is leaves that last for a little less than a year, while nutrient deficiency may lead to leaves lasting longer than one year. Furthermore, leaves produced in oligotrophic conditions contain a high percentage of structural materials, like those that have evolved in very dry conditions, since they have to avoid the mechanical deformations and damage that water loss may produce. Finally, to complicate the picture even further, the high *C/N ratio* of leaves rich in structural materials means they are especially resistant to decomposition, so that evergreen vegetation resulting from adaptation to summer drought conditions may give rise to relative nutrient poverty in the soil due to the slowness of leaf decomposition.

All this explains why part of the Mediterranean flora has been able to evolve from precursors adapted, not to summer drought, but to nutrient scarcity, a factor that may be present in any climate. This series of observations is also confirmed by the fact that most of the Chilean mediterranean plants growing on relatively rich soils show high rates of leaf renewal, that is to say, leaf longevity is relatively low and is determined by climate. The tendency in the Californian mediterranean is that the poorer the

53 Young shoots of the mastic tree (*Pistacia lentiscus*) near Banyoles, in the Pla d'Estany (Catalonia). The evergreen plants of the Mediterranean region sprout in spring. The new leaves in this case are paripinnate (pinnate and lacking a terminal leaflet); they can be easily distinguished as they are softer and paler green color than the leaves from the previous season. The mastic tree is usually a low-growing shrub, but it can grow into a tree. It grows in matorral, maquis, and open woodlands and is frequent throughout the western Mediterranean Basin.
[Photo: Ernest Costa]

54 Cross-sections of the leaves of the west Mediterranean heath (*Erica multiflora*) (top) and *Brachypodium retusum* (bottom). The leaves of Mediterranean evergreens show adaptations to dry conditions. The tendency to reduce water loss seems to follow a series of strategies, such as thickened cuticles, protected stomata, and, in some cases, the production of hairs to reduce evaporation. The leaves of heathers have a thick waterproof covering, and in summer they control the opening and closing of the stomata. The perennial leaves of *Brachypodium retusum* are an example of adaptation to dry conditions in the hot season, as they show several different strategies, such as hairiness and sunken stomata, to reduce water loss by transpiration.

[Photos: Unitat d'Ecologia, Universitat Autònoma de Barcelona]

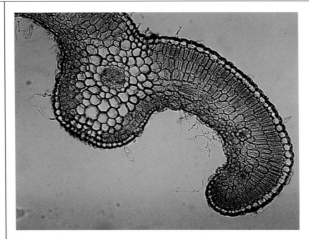

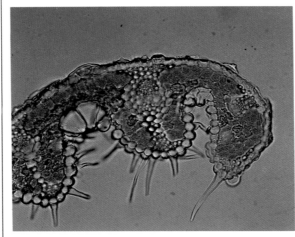

soils and the more extreme the drought conditions, the longer the life of the leaf. In South Africa and Australia, the extreme climatic effects are reinforced by the highly oligotrophic soils, giving rise to even greater values for leaf duration.

Leaf morphology is also very important as a factor of adaptation to the summer drought. The summer-deciduous plants are malacophyllous, that is to say, with relatively large leaves that are not very hard, as they avoid the problem of the summer drought by spending the summer in a state of repose or inactivity. Many malacophyllous plants have leaves covered in hairs to avoid excessive water loss: as they are not truly deciduous, they have to protect themselves from desiccation. In fact, some malacophyllous plants lose all their leaves, while in others leaf surface and activity are reduced in accordance with the severity of water shortage, showing relatively flexible response. This is what happens in the case of shrubs, like cistus (*Cistus*) and *Rhamnus lycioides*, that considerably reduce their leaf area and can even produce smaller, more leathery leaves in the summer, but never lose all their leaves.

Plants with evergreen leaves use a range of strategies to deal directly with the problem of the lack of water. Sclerophyllous plants have toughened leaves, rich in structural materials like cellulose and lignin, and are often covered by a hard cuticle. Their toughened leaves give structural resistance against desiccation, and their highly efficient control of their stomata means they can stop transpiring when water supply diminishes. Leaf water content is lower than in malacophyllous leaves, but is maintained at a more constant level due to their regulatory capacity. Sclerophyllous plants, which are very abundant in the Mediterranean biome, may have relatively large leaves; that is, they are broad-leaved, such as many species of *Quercus*, many members of the Proteaceae, and most of the Chilean mediterranean's trees and shrubs. Others have very small leaves that are covered with hairs, known as *ericoid* leaves, such as those of the members of the families Ericaceae and Epacridaceae. Also considered as sclerophylls are some perennial grasses, such as albardine (*Lygeum spartum*) and feather grasses (*Stipa*), which have long leaves that are rolled so tightly around the stem to avoid desiccation that they appear circular in cross-section.

Leafless xerophytes have lost their leaves and have photosynthetic green stems, allowing them to reduce water loss by having a very small transpiration surface. This is the case of the South African members of the Restionaceae and some shrubs of the Mediterranean Basin, such as Spanish broom (*Spartium junceum*). The most drastic strategy, which is uncommon in the Mediterranean biome, is the total loss of the leaves and the development of water-storage tissues to store the rain that falls in the wet period, or even the conversion of leaves into water-storage organs. Such plants, called *succulents*, are typical of drier and more irregular climates than the Mediterranean climate, such as the sub-desert environments. Yet some examples are present in the Mediterranean biome, ranging from herbaceous stonecrops (*Sedum*) to Chile's large columnar cacti (*Trichocereus chiloensis*), and include medium-sized shrubs, such as South Africa's euphorbias (*Euphorbia*) and the Californian Spanish bayonet (*Yucca whipplei*) whose succulent leaves have spines as sharp as the Spanish bayonets that gave them their name.

Defense against herbivory

A long-lasting leaf, produced at great expense in an environment where water is very scarce, is a precious commodity, something to be protected from herbivores capable of destroying the result of several month's activity in no time at all. The presence of spines is related to defence against herbivores. The accumulation of volatile essential oils in the stems and leaves makes them less palatable to herbivores, although some typically Mediterranean animals, such as the goat, are capable of eating almost anything, however bitter or aromatic it may be. Many Mediterranean species of Lamiaceae (*Thymus, Salvia, Rosmarinus, Lavandula*), Cistaceae (*Cistus, Helianthemum*), Pinaceae (*Pinus*), Cupressaceae (*Cupressus*), and Myrtaceae (*Myrtus, Eucalyptus*), to mention only the most aromatic ones, contain such a large quantity of essential oils that they give the Mediterranean landscape its typical intense smell, a smell that is much stronger in the hot dry months and comes not from the flowers but from the accumulated essential oils in the leaves and stems.

Essential oil accumulation may also be related to water economy, pyrophytism, and competition. When the essential oils evaporate, they increase the density of the air in the leaf cavities, raising resistance to the diffusion of water from the cell walls and thus reducing transpiration. The fact they increase the plant's inflammability can be interpreted as another pyrophytic adaptation.

Finally, in some cases it has been shown that the accumulated aromatic substances in plant organs enter the soil when the plants decompose and act to inhibit the germination of other species, thus avoiding competition for water and space. This phenomenon is typical of arid zones and has been demonstrated in eucalyptus and in some chaparral species, such as *Salvia leucophylla* and species of the genera *Arctostaphylos* and *Adenostoma*. Depending on the species producing them, these germination inhibitors last for a period ranging from less than one year to many years, ensuring the elimination of competition even if the plant's aerial parts are eliminated.

2.5 Fire and vegetation

Fire is a very important ecological factor in the world's biomes, not only in the mediterraneans. As an example, the northern conifer forests have for millennia experienced cyclic fires with a periodicity of around 200 years, as shown by the layers of ashes deposited in lake sediments. Prairies also suffer fires, although with a shorter periodicity, and this explains why herbaceous vegetation dominates in climatic conditions favourable to forest development. Yet fire is more important in the mediterraneans than in any other biome.

The causes of spontaneous fires

Fire has been a very important factor in the Mediterranean biome for a long time, as the hot season of the year coincides with the dry season and this favors fires. The very nature of sclerophyllous vegetation makes it especially inflammable, further encouraging the outbreak and spread of fires. Furthermore, fire rarely burns plant structures more than 0.8 in (2 cm) in diameter, leading to the accumulation of charred, dead biomass, further increasing inflammability until its disappearance by decomposition. The most important natural cause of fires is lightning, which is very frequent in the Mediterranean regions, especially during the short, violent summer storms caused by the local destabilization of air masses; these storms are accompanied by a great deal of lightning but here may be little or no rain, further encouraging the spread of fires. The Chilean mediterranean is an exception in that electric storms are very rare. In the Californian region, however, lightning is a major fire risk, and as many as 1,750 forest fires have been recorded in a single year (1972).

It is difficult to assess the frequency of fires in conditions of little or no human interference, but rough estimates speak of fires with a frequency of 15 to 100 years in the Australian mallee, of 6 to 40 years in the Cape fynbos, and 25 to 50 years in the Californian chaparral, although in the latter case the neighbouring drier formations of dry meadows may suffer regular annual fires. There is little information from the Chilean mediterranean about the frequency of fires before human settlement, but the rarity of electrical storms suggests this frequency was low. Some authors have estimated that in the Chilean mediterranean fires occurred at a frequency of 270 years, but, surprisingly, in this region there are many species that can produce new shoots after fire, and this is interpreted as an adaptation to fire, which makes one think that, in practice it is relatively important. In the Mediterranean Basin, the intense and ancient human settlement, combined with the landscape's structural complexity, makes it almost impossible to estimate the frequency of the natural fires, unless we resolve the problem by considering that since antiquity human beings have been the most important natural factor in this regard.

An environmental factor as regular and as drastic as fire has left its mark on the Mediterranean vegetation. One must not underestimate the effects of

55 Mediterranean pines burning in a forest fire. They are reduced to a skeleton consisting of the dead trunk and branches in next to no time. They contain a lot of inflammable substances and this means their foliage bursts into flames like a can of turpentine, often without burning the wood of the trunks.
[Photo: Ernest Costa]

fires in periods before human beings acquired the major role they have held for the last few millennia. Although fires must have been less frequent, the areas affected might have been much greater, since the fire acted on a landscape that was much more homogeneous than today's lanscape which is divided by barriers (crops, roads) that act as fire-breaks. Paradoxically, as will be discussed, in some Mediterranean areas there are actually many fires, but the frequency with which each area burns may be lower than before.

Adaptations to fire:
sprouting or germination en masse

Due to the frequent occurrence of fires, many of the Mediterranean biome's plants show adaptations allowing them to resist fire, although some have gone further and have developed life-cycles and reproductive mechanisms that give them clear advantages when a fire occurs. The first type of plants are called *passive pyrophytes*, while those in the second group are called *active pyrophytes*. The most extreme cases of pyrophytic adaptation involve a set of adaptations that make the plant highly inflammable, favouring the appearance and spread of fires.

Regeneration by sprouting: pyroresistance
One of the main mechanisms of resistance to the effects of fire (pyroresistance) is the development of thick, fire-resistant bark that insulates the plant from the heat, protecting the subcortical meristems that will sprout and produce new leaves. Some species of eucalyptus (*Eucalyptus*), such as the jarrah (*E. marginata*) and the marri (*E. calophylla*), and several species of *Quercus* show the same adaptation. The most notable example is that of the cork oak (*Q. suber*), a typical tree of the western Mediterranean Basin whose corky bark, a few centimeters thick, is highly fire-resistant. After a fire, branches 0.8-1.2 in (2-3 cm) or thicker can resprout and rapidly regenerate the crown. The temperature reached in the fire and the thickness of the bark determine the diameter of the branches whose sub-cortical tissues were sufficiently protected for them to resprout.

Bark is an effective insulator, and the thickness of bark is the most important factor in determining fire-resistance. In the Mediterranean Basin, the branches of the cork oak (*Quercus suber*) sprout abundantly when the fire has not been very intense (unless its bark has been stripped for

56 Examples of new shoots produced by Mediterranean sclerophyllous plants after a fire. The marri (*Eucalyptus calophylla*) (top), and the strawberry tree (*Arbutus unedo*) (bottom). Fire does not destroy mediterranean plants as they have had to live with it for millennia and have adapted. Resistance to flames takes different forms in different species. *Eucalyptus* species produce an insulating bark to protect the meristematic tissues that are able to produce new shoots after a fire. Other plants, such as the heathers (*Erica* spp.) and the strawberry tree produce new shoots from a rootstock. These underground rootstocks produce new shoots immediately after a fire, since a thick water— and nutrient— storage organ is formed at the transition from root to stem (the lignotuber) which can resprout when the aerial parts are damaged by fire, grazing, or felling.
[Photos: Ernest Costa]

cork), and within 2 or 3 years the tree has recovered its leaves. Many species of *Eucalyptus* have fire-resistant bark that allow the foliage to recover rapidly after fires. These trees usually have a thick bark that protects the cambium and makes the trunk fire-resistant. The new shoots with leaves sprout directly from the resting buds below the bark of the trunk. Some pines, such as the Canary pine (*Pinus canariensis*), also show this survival trait.

Many Mediterranean shrubs have half-buried, swollen lower trunks, that can produce new shoots after a fire: 50% of the dominant species in the Californian chaparral, 65% of those in the Cape fynbos, 70% of those in the Australian jarrah forests, and most woody species from the Mediterranean

57 Leaf and flower production by geophytes after a fire in the Cape scrub. The photo shows the orange flowers of *Watsonia borbonica*, the pink flowers of *Pillansia templemanii*, and the white inflorescences of *Lanaria lanata*. The bulbs of geophytes benefit from the fact that the soil insulates them from the heat of the fire, and so they do not suffer the effects of the high temperatures and can reconstruct their aerial parts immediately after the fire, at a time when there is little competition from other species.
[Photo: Colin Paterson-Jones]

58 Eucalyptus bark burning on the surface only, in South Australia. One mechanism plants use to protect themselves from fire is to develop a bark that reduces heat transfer and may have other characteristics to make it less inflammable. Under such conditions, the meristems beneath the bark may survive to resprout and form new leaves, as happens in many species of *Eucalyptus*, the cork oak, and other Mediterranean plants.
[Photo: Lluís Ferrés]

Basin, except the conifers, are capable of resprouting after fire. In all of these species the aerial parts may be totally burnt, but the rootstock survives due to the insulating effect of the soil that partially covers it. In the case of a fire reaching a temperature of 1,472-2,012°F (800-1,100°C), only 2 in (5 cm) below the soil surface the temperature is no more than 104°F (40°C), and 4 in (10 cm) below the surface the temperature has hardly changed. These rootstocks, consisting of a special tissue, called *lignotubers*, also store water and nutrients and ensure rapid sprouting even in dry summer periods, without needing to wait for the autumn or winter rains. Many plants possessing this structure develop it in the first few years of their life, when they devote nearly all their resources to the development of the rootstock and the roots. Within five years of germination, the jarrah (*Eucalyptus marginata*) already possesses a well-formed lignotuber to ensure it will survive a fire.

Most of the trees of Europe's matorrals, garrigues, and maquis, and those of the Californian chaparral produce sprouts from their rootstocks. Each species has a range of ways of resisting fire, and this depends on the fire's intensity. Also the ways of resisting fire vary over the year depending on the plant's phenological status and/or state of development. Thus, the holm oak (*Q. ilex*), which sprouts readily from its rootstock, can bear new shoots from buds under the bark when the fire is not very intense and the trees's bark is thick enough. Except for about fifteen species, all *Eucalyptus* have lignotubers; these woody growths start as small growths at the level of the axils of the cotyledons and of the plant's first leaves. As the plant grows so does the lignotuber, and it buries itself deeper and deeper into the soil. In shrubby species of *Eucalyptus* the lignotuber may grow continuously. This organ is a source of live buds that can survive a fire because they are underground. Similarly, many species of heather (*Erica*) have a similar rootstock that allows them to produce many new shoots and rapidly reconstruct the aerial part destroyed by fire.

In certain cases, the rootstocks are not half-buried, but are underground stems that are completely protected by the soil. This adaptation is present in some shrubs, such as the kermes oak (*Quercus coccifera*) and also in bulbous herbaceous plants, such as *Gladiolus*, *Asphodelus*, and *Ophrys*. These bulbous plants are *geophytes*, and they are very abundant in communities that are repeatedly affected by fires;

some only flower after fires, and if there are no fires they restrict themselves to vegetative growth. In some herbaceous plants, such as the grasses, the very rigid leaf-bases protect the plant, allowing it to sprout and regenerate rapidly.

One remarkable case of adaptation to producing new growth after fire is that of the Australian "black boys" or "yacca" (*Xanthorrhoea*), herbaceous plants with an arborescent growth-form that are well adapted to growth after fires. The seedlings of these arborescent species have contractile roots that drag the delicate growing points downwards. During a fire, a young plant loses its leaves, but these are replaced from below. The young plant's stem remains underground and many years may pass before an aerial stem appears above ground level. The base of the stem must first of all reach the diameter of an adult stem before the stem starts to elongate. When this happens, the growing point is protected from the fire's heat by the leaf bases, which are tightly-packed making them less inflammable than the tuft of exposed leaves. The leaf bases are resinous and thus inherently inflammable, but as most scrub fires are brief, they normally resist the fire long enough.

The growth of the shoots in any species that sprout after a fire is always very rapid, as they have a well-developed root system to take up water and nutrients. They can reach 3 ft (1 m) in the time that a plant growing from seed would take to grow a few centimeters. This growth may be very active for the first four or five years, slowing down to nothing after about 25-30 years. The morphology of the sprouts is varied and affects the soil's cover, and thus its protection from erosion. Most plants that sprout only from rootstocks give rise to a group of vertical shoots that only cover a small area. Plants that sprout from both rootstocks and roots or from subterranean stolons give rise to dense ground cover, although the shoots are not as high as those produced from rootstocks. Plants sprouting from both rootstock and roots, such as the kermes oak (*Q. coccifera*), show very rapid regeneration and *auto-succession* ensures the survival of the community even despite repeated fires.

The ability to produce new shoots might be an ancient adaptation that only came into operation after the destruction of the foliage by an external agent. Living foliage inhibits buds from becoming active, but its disappearance removes this inhibition and the resting buds then sprout. The passing through of a fire may stimulate resistant plants to

59 Black gin (*Kingia australis*) flowering after a fire in Western Australia. Australia's grass trees produce new shoots after fires, giving them a very strange appearance, with their fire-blackened trunks and a tuft of leaves at the top. They respond vigorously after fires, growing a complete new set of leaves and producing their conspicuous inflorescences. The subsequent seed production ensures effective colonization of the space left free by the fire. Leaf regeneration is extremely rapid, since it is due to the growth of the leaf bases which are not damaged by the fire. Flowering has been experimentally shown to be induced by the combination of increased light intensity (due to the destruction of the tree canopy) and the ethylene contained in the smoke. The aerial roots at the leaf bases allow the uptake of the additional nutrients needed for this rapid regeneration. Yet between fires, the black gins lead a quiet life, as they do not flower, and they grow very slowly. As these grass trees do not produce annual growth rings they cannot be used to calculate a plant's age. The first studies on normal growth rates, i.e. between fires, estimated the age of individual specimens 6-10 ft (2-3 m) tall as thousands of years, and this height is quite common. The mystery of these absurd and impossible ages was resolved when it was realized that the growth of the new leaves occurs precisely when the old leaves have been destroyed by fire and continues very slowly in the intermediate period. The black gins only wake up after a fire.
[Photo: Jan Taylor / Bruce Coleman Limited]

60 **The seeds of *Cistus salviifolius* in the Montseny cordillera (Catalonia)**. One way of persisting after a fire is to produce fire-resistant seeds. In a fire the entire adult plant of the *Cistus* is destroyed but the seeds survive, germinating abundantly with the first rains after the fire. This is why they are abundant under pine forests that suffer frequent fires.
[Photo: Jordi Vidal]

produce inflorescences; this is exceptional in the plants of the Mediterranean Basin, although it has been observed in monocotyledons and has been studied in *Xanthorrhoea australis*.

Regeneration by germination: anthracophytism and pyrophytism

Some plants die if their aerial parts are burnt. This is true for some small shrubs, many herbaceous plants and some trees, such as pines. In this case, the dead individuals are replaced by new individuals that have germinated from seed, which find themselves in a competition-free space, fertilized by the ashes of the fire. This strategy favours species that produce many seeds with a high dispersal capacity. Likewise, sun-loving species are those that benefit most, as the intense sunshine that results from the destruction of the plant cover does not favor germination. The destruction by combustion of the germination inhibitors produced by some plant remains during decomposition may also favor the appearance of new individuals from seed.

Some authors recognize two strategies with respect to regeneration by seeds so as to distinguish clearly what pyrophytism is. *Anthracophytes* are simply opportunist species with a high dispersal ability whose seeds rapidly colonize burnt areas from areas unaffected by the fire, as happens in the case of *Galactites tomentosa*, which produces a large quantity of wind-dispersed fruits. On the other hand, plants that reproduce by seed and whose germination is actually helped by fire are called *pyrophytes*. Such pyrophytes typically show high production of seeds protected by hard coverings that germinate with difficulty unless exposed to a heat shock, and

the plants accumulate volatile substances that make them highly inflammable, although this can also be interpreted as an adaptation to water shortage.

A good example of this strategy is the genus *Cistus*. Plants of this genus are heliophilous shrubs that produce a large quantity of small seeds protected by a hard covering that germinate in large numbers after a fire. In a year without fire, germination of *Cistus* may be of the order of 10-20 individuals per square meter, but after a fire this value may reach 300-400 individuals per square meter. The adult plants die in the fire and so possible recolonization depends on the frequency of fires, which must be separated by intervals that are long enough to allow the new individuals to grow and flower.

Most pyrophytic shrubs, however, can produce seeds within a year of germinating. This is the explanation for their dominance in repeatedly burnt landscapes and their gradual disappearance if fires do not occur. For example, in the Californian mediterranean, *Lotus scoparius* behaves as a germination pyrophyte. It is very abundant until 5 years after the fire while after 20 years there are only a few scattered individuals, although the seeds that have accumulated in the soil ensure it will once more be dominant after the next fire. Germination pyrophytes are thus plants that favor the appearance and spread of fire, and plants that rapidly colonize the burnt areas by their abundant germination. Their only serious problem arises if fires occur too frequently, since, in that circumstance, the plants do not have time to go through their entire life cycle between fires and thus cannot produce new seeds. In the most extreme cases of pyrophytism, the seeds are protected by structures that insulate them from high temperatures, and they remain on the plant until released by the action of fire. This is the case with the pyrophilous plants we will be discussing later.

Germination experiments show that the response to heat depends on both the temperature and the length of exposure. Long periods of exposure to relatively low temperatures may have the same effect as short exposures to high temperatures, and this effect may be germination or seed death. The result, however, also depends on the species and, therefore, on the seed's physical and physiological characteristics. Thus, the seeds of *Cistus monspeliensis* show peak germination after exposure to temperatures between 194°F and 302°F (90°C and 150°C). Even so, fifteen minutes at 194°F (90°C) is needed to achieve this

rate of germination, while at 302°F (150°C) one minute is enough. Above 302°F (150°C), the temperatures are too high and kill the embryo. These effects have been observed in Australian sclerophyllous plants, in Californian chaparral species, and in plants in the Mediterranean Basin.

Most species of pines (*Pinus*) are unable to sprout after a fire, and can only recover by germinating from seeds. Their entire strategy is adapted to turning this fact to their advantage. The seeds of the Aleppo pine (*P. halepensis*), a tree that forms extensive pine forests in the western and central regions of the Mediterranean Basin, are mechanically dispersed when the tree's cones burst violently during the fire or afterwards. Thus, after a fire, pine forests are colonized by a large number of pine seedlings that recreate the forest. Given the ability of these seeds to germinate successfully in the mineral soil of areas totally exposed to light (where competition is lowest and where mineral nutrients are abundant), the pine forest may perpetuate itself indefinitely, as long as the fires are not so frequent that the trees do not have enough time to produce fertile cones.

In nature, the response of the soil seed bank depends not only on its own properties, but also on those of the fire, that is to say, on the fire's intensity, the heat penetrating the soil, the time exposed to fire, and on the depth at which the seeds are buried. After a fire, many annual or biennial herbaceous species appear, flower, and disappear again within two or three years. This behavior has been observed in many different regions: California, Australia, the Cape region, Israel, France, etc. This phenomenon has been thoroughly studied in Californian chaparral, and the conclusion reached is that the abundant germination and growth of herbaceous species is due to the suspension of the tree-induced chemical inhibitors acting on the seeds in the soil. What is preventing germination in this case is not limiting factors, such as light, water, or nutrients, but what is called *allelopathy*. Extracts derived from woody species inhibit the germination of seeds, while exposure to a temperature of 176°F (80°C) for an hour raises germination rates. Heat may not be necessary for germination, but does stimulate it; fire, as it destroys woody species, removes the source of the inhibitors. As soon as the woody species have reestablished, germination inhibition reappears, although the seeds remain viable until another fire occurs. In the Mediterranean Basin this phenomenon of allelopathy has not yet been demonstrated,

61 **Flower production by the South African *Haemanthus caniculatus* after a fire.** Fires shape the landscape in the Cape province, too. This member of the Amaryllidaceae produces its capsule-like flower buds after fire, and their colorful flowers add color to the ashes. Naturally, the life of the fauna is also affected by the repeated fires. Next to the plant there is a small frog completely camouflaged by the ashes covering the soil.
[Photo: Colin Paterson-Jones]

62 Cones of *Banksia serrata* burning (above), and the burnt cones of a species of the same genus (*B. grandis*), together with shoots produced from the trunk after a fire (below), in Australia. The shrubs of the genus *Banksia* are pyrophytic because they need fire to reproduce. The hard seeds are protected inside highly lignified fruits, and need fire to germinate. Germination immediately after a fire has the advantage of having large quantities of readily available ash (and trace elements), and little competition for light.
[Photos: Wayne Lawler / Auscape International and Lluís Ferrés]

but there is harsh competition for light and space from perennial plants, making it very difficult for annual or biennial plants to persist.

Extreme cases of pyrophily

Pyrophytism, that is to say, adaptations going beyond the mere ability to withstand fire can reach extremes, and in some pyrophytes it has even become necessary for their reproduction. These are called *pyrophilous* plants, and they include many Mediterranean pine species and some Australian banksias (*Banksia*).

Pines (*Pinus*) accumulate a large amount of resins and volatile essences, including turpentine, a product that until recently was only obtained from the resin of some species of pine. These resins and essences are present throughout the plant, making it highly inflammable, and so fires are more devastating in forests dominated by pines than in the Mediterranean's typical sclerophyllous woodlands. This inflammability, however, is not accompanied by the ability to resprout after a fire; the fire causes the death of the trees and regeneration is only by seed. The problem of the too high frequency of fires has led to relatively short maturation times in Mediterranean pine species, in which the period that elapses from germination to production of fertile seed is shorter than that of other species of the same genus.

Some species of pine have cones, called serotinous cones, that only open after a fire or under the action of fire. These cones have scales glued together by a heat-sensitive resin or wax that melts with the heat of the fire, thus allowing the cones to open and drop their seeds. This phenomenon has been thoroughly studied in the United States, but the Aleppo pine (*Pinus halepensis*) might also behave this way in some areas of the Mediterranean Basin. The pine cones are in fact acting to protect the seeds from the fire. After a fire the cones look charred, but the seeds they contain have not been affected by the high temperatures. Furthermore, the heat helps to open the cones, thus freeing a large number of seeds to colonize the bare soil with relative ease. This behavior, which can be observed in several species of the Mediterranean Basin, such as the Aleppo pine (*P. halepensis*), the Calabrian pine (*P. brutia*), the maritime pine (*P. pinaster*) and the stone pine (*P. pinea*), is accentuated in the Californian area, where there is a whole group of pines (*P. attenuata, P. muricata, P. remorata, P. torreyana, P. contorta*) that only open their cones after a fire. In this case, fire is necessary to release the reproductive potential stored within the pine cones accumulated year after year on the branches. The precious pinenuts will only fall on ground that is totally free of competitors.

Many Australian shrubby members of the Proteaceae behave similarly, especially the banksias (*Banksia*). Their seeds are enclosed within dense woody fruits vaguely reminiscent of pine cones and function in exactly the same way (the most spectacular case is that of the bull banksia, a species that not only sheds its seeds as described above, but is also capable of

63 Undergrowth burning in a fire in the Vallès region (Catalonia, in the Iberian Peninsula). The combustible material of the undergrowth helps the fire to spread from tree to tree. If the leaf litter consists of pine or eucalyptus leaves—and both genera are typical of Mediterranean climates—it is even more inflammable because of the resins and oils they contain. The production of these compounds is considered to favor the fires that are already common in their habitats. These fire-resistant species, have protective bark or cones that do not open unless they are burned. They need fire to germinate. [Photo: Oriol Alamany]

sprouting after fire). Fire is responsible for releasing the seeds and if fire does not occur, the fruits accumulate forming enormous reserves of inflammable materials that eventually favor the occurrence of fire. The cones do not open until they are exposed to the heat of a fire, and the seeds do not germinate until fire has partially burnt the surface protective layers or has caused internal physiological changes. Many of these shrub species produce enormous quantities of seed; average values of ten million seeds per hectare per year have been recorded, and in the moist sclerophyllous forests of Western Australia it is not unusual to find more than a thousand seeds per square meter in the top layer of soil. After the fire releases the seeds they can germinate in the bare, ash-enriched, earth, while the adult individuals rapidly regenerate their foliage by sprouting, not only from their rootstock, but also from their relatively thin branches which have remained protected by the bark.

The genus *Eucalyptus* shows increased inflammability due to the volatile oils in the leaves and the enormous quantity of dry leaf litter. In addition to the large quantities of leaf litter, many species of *Eucalyptus* also shed patches of dry bark, while others have strips of bark hanging from their branches that help fire to spread from the undergrowth to the canopy. Such marked pyrophily can be explained as a further stage of pyrophytic adaptation: fire-resistant vegetation evolves until it becomes dependent on fire for its maintenance and regeneration.

In Victoria and Tasmania, both outside the Mediterranean biome, where sclerophyllous and hydrophytic forests coexist side by side, the hydrophytic species continually invade the *Eucalyptus* forests unless fires occur, and would eventually displace them if the process lasted long enough (at least 200 years without fires would be necessary). If a fire occurs, the hydrophytic species are eliminated from the forest and succession will start all over again. In Mediterranean sclerophyllous forests, analogous phenomena would occur if there were never any outbreaks of fire. Thus, fire, an apparently and genuinely devastating environmental factor, has been so effectively integrated into the life-cycles of some Mediterranean plants that it is now necessary for their continued survival.

Adaptive strategies

Some interesting questions are raised by the different responses of plants to fire. Some plants regenerate vegetatively while others do so by means of obligate sexual reproduction. Plants that resprout strongly seem to have a considerable advantage. When a species of this type colonizes a space, it occupies it for a long time. However, sexual reproduction appears to be clearly less advantageous. A plant that reproduces sexually must have an unoccupied space to germinate successfully, and this is highly conditioned by the distance the seeds are dispersed and the rate of turnover of seedlings.

It is difficult to explain the coexistence of two such different strategies in plants of Mediterranean communities affected by fire, and it raises the question of what advantages each strategy offers the plant. In particular, it raises the question of the advantage for plants that obligatorily reproduce by seed, because a fire may destroy the entire population before the individuals have matured, flowered, and seeded again. The lapse between two fires has to be long in the case of plants that only reproduce sexually, and may be shorter for those that can reproduce vegetatively. Apparently, one type (vegetative reproduction) serves to recover space that was previously occupied, while the other (germination strategy) offers the possibility of invading unoccupied spaces. The coexistence of the two types must be due to the alternation of successive disturbances and rest phases, which have influenced selection pressure over the course of time.

The effects of fire on vegetation and the soil

The greater or lesser frequency of fires favors different strategies and it also influences the processes of selection and speciation. The coexistence of these processes leads to a greater abundance of "germinating" species than "resprouting" species, as the shorter life-cycle of the former means the process of speciation can occur more rapidly. Thus, in two genera typical of chaparral, *Arctostaphylos* and *Ceanothus*, there are (respectively) 16 and 12 species that resprout after fire, while there are 59 and 46 that germinate; and in two genera typical of fynbos *Leucadendron* and *Leucospermum*, there are (respectively) 7 and 4 species that resprout after a fire, while 73 and 36 germinate. This numerical frequency of species within the genus is quite distinct from abundance and dominance within communities, as in each case it is the frequency of fires that determines which type of species and strategy is favoured.

The overall effect of these adaptations is that Mediterranean vegetation has a high capacity to regenerate after fires: depending on the climate, Mediterranean vegetation after a fire shows production of between 1 and 4 t/ha per year, a quite respectable figure for this biome and higher than that of some Mediterranean formations unaffected by fires. Fire destroys the vegetation and starts a process called *auto-succession*, which is not the same as *secondary succession*. Secondary succession occurs when the plants completely disappear, including their underground parts and seeds, and is a long process from bare soil to the development of climax vegetation, by means of the colonization and successive replacement of species, from pioneering species to those corresponding to the final stages. Thus, if a site occupied by maquis is ploughed for cultivation and is then abandoned, a slow process begins that will lead to the regeneration of the maquis after passing through a series of different stages. Yet if the destructive agent is fire, the same type of vegetation regenerates rapidly from the sprouts and seeds of the species already present before the fire. The ability to sprout after fire means that even the same individuals continue to exist; in the low shrub formations such as chaparral, rootstocks estimated at more than 250 years old have been found, showing that a very old plant does not have to be a large tree. In these forests, regeneration by auto-succession is very rapid, since they comprise trees protected by thick barks that can resprout epicormically (i.e. from dormant buds) and from the rootstock and by small shrubs able to sprout from a rootstock, in a combination of strategies that corresponds to the intensity of the fire in each of the layers of the forest.

In some cases, fires appear to be necessary to keep the vegetation in a healthy state, as the absence of fire gives rise to a certain "decadence" with the accumulation of a great deal of dead biomass, which indirectly favours fire; this is what happens in the Californian landscapes dominated by chaparral and in the South African and Australian areas dominated by members of the Proteaceae. In Australia's jarrah forests fire appears to be necessary for the periodical mobilization of the nutrients immobilized in a leaf litter that decomposes very slowly, thus increasing the soil's nutrient poverty. Australia's prevailing climatic conditions mean this detritus decomposes very slowly and continues to accumulate for about 30-40 years until a balance is reached between build up and decomposition. This leaf litter accumulates in the soil at a rate of 1-3.5 t/ha per year in dry sclerophyllous forest and 2.5-11 t/ha per year in wet sclerophyllous forest.

Yet the Mediterranean sclerophyllous vegetation's high capacity for regeneration after fire does not mean the forest is totally immune from the effects of fire, as fire affects the fauna and, strangely enough, the soil. In fact, the most negative effect of fire is the erosion it causes, which is very intense when the torrential rains typical of the Mediterranean climate fall on soils unprotected by plant cover. It should be emphasized that during the dry summer period, regeneration after a fire can be very slow, and so the first, violent, autumn storms fall on the bare soil. In environmental conditions that give rise to little or no soil formation, erosion represents an almost irreversible loss, and puts the

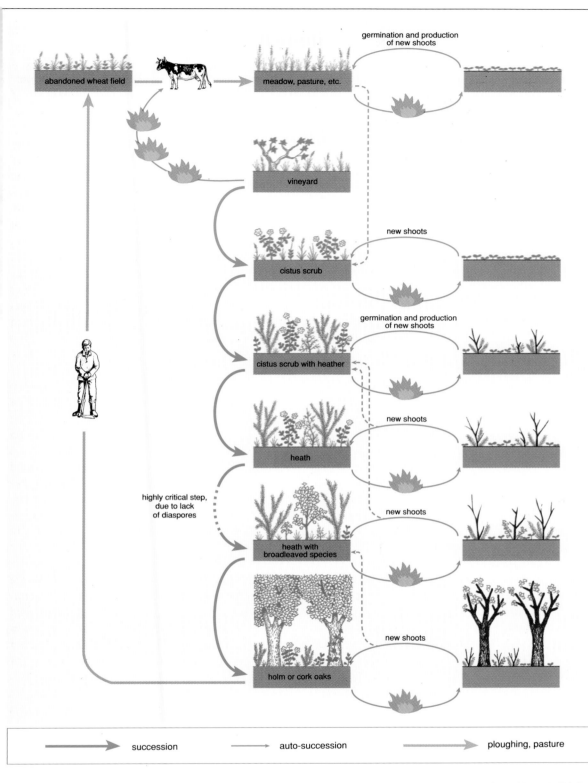

germination and production
of new shoots

abandoned wheat field

meadow, pasture, etc.

vineyard

new shoots

cistus scrub

germination and production
of new shoots

cistus scrub with heather

new shoots

heath

highly critical step,
due to lack
of diaspores

new shoots

heath with
broadleaved species

new shoots

holm or cork oaks

new shoots

→ succession   → auto-succession   → ploughing, pasture

64 Models of plant succession after fire, on peninsula of the Cape Creus (northeast tip of the Iberian Peninsula). If a vineyard is abandoned—for example, one in an area that had formerly been a forest—succession reestablishes a heath, and eventually a Mediterranean forest, as long as there are seed-producing trees nearby; the shortage or lack of these seeds may prevent the recovery of the native vegetation. On the other hand, after a fire the former community usually reestablishes itself, a phenomenon known as auto-succession.

[Drawing: Jordi Corbera]

future regeneration of the vegetation at risk. In some areas of chaparral it has been shown that erosion, which may cause the loss of up to 2.7 in (7 cm) of soil, increases for 8 years after a fire. This increase is due to the loss of plant cover and to the destruction of soil structure by the high temperatures, since in a fire that reaches a surface temperature of 2,012°F (1,100°C), the top centimeters of soil may reach temperatures of 482°F (250°C), which causes the destruction of all the organic material in the top layers, with the consequent loss of soil cohesion.

It is worth pointing out that the problem of soil destruction and nutrient loss caused by fire is especially serious when the burnt community was

65 Fire affects nutrient cycles, as it mobilizes the nutrients contained in living plants and deposits them in the form of ash in the same area. Yet some nutrients are lost in the smoke, and ash deposition on soil lacking plant cover leads to even greater nutrient loss due to the action of wind and rain after the fire. Nutrients are lost to other ecosystems, whether nearby or more distant.

[Drawing: Editrònica]

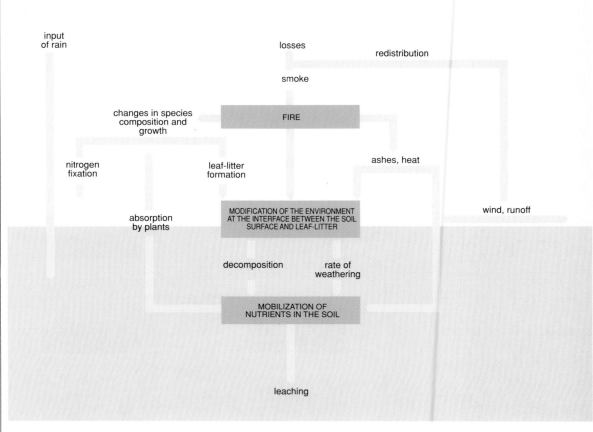

shrub, as often happens in the Mediterranean biome, because the surface accumulation of biomass means that higher temperature are reached in the upper layers of the soil. In a chaparral fire that reaches temperatures of 1,832°F (1,000°C) in the foliage, the soil surfaces reaches 1,292°F (700°C), while a fire in a conifer forest that reaches the same temperature in the branches only heats the soil to 572°F (300°C), and a grassland with less combustible material, reaches only 482°F (250°C). These high temperatures also cause the loss of essential nutrients, such as nitrogen, as volatile gases from both the top layers of soil and the biomass, leading to a progressive and very serious impoverishment, as these soils are frequently very poor to begin with.

Obviously, if fires produced by human activity are very frequent, this aggravates the problem of erosion and soil impoverishment and makes some of the adaptations discussed above less effective. Repeated fires also affect species able to resprout after fire, as a single rootstock can only sprout a limited number of times. Furthermore, on poor soils and in very dry conditions, a very high percentage (as high as 50%) of the individuals able to resprout may, in fact, die after the fire. Fire, an old acquaintance of the Mediterranean biome, has become in human hands an important factor in landscape degradation, as is discussed later.

Fire and the nutrient cycle

The ecosystems of Mediterranean regions are generally considered to be poor in nutrients, although between them the five areas with Mediterranean climates show differences in the quantity of soil nutrient reserves. During and after a fire, the soil humus subsystem is directly or indirectly affected by the action of the fire. As soil fertility is the critical factor that controls the structure and function of plant communities and ecosystems as a whole, it is important to understand the effects of fire on nutrient reserves.

Nutrients are stored in variable quantities in the plants, in the humus layer and in the soil. The vegetation's nutrient reserves basically depend on the age of the forest, the aerial biomass and its concentration of nutrients; nutrient reserves are also influenced by the relative amount of the biomass present in the form of foliage, because leaves contain more nutrients than woody materials. The older a forest, the greater the amount of woody material and the greater the risk of fire. The humus nutrient reserves are present in the layer in which dead, partially or

66 **A cork oak (*Quercus suber*) forest after a fire** on the peninsula of the Cape Creus (northeast tip of the Iberian Peninsula). The cork oak's thick, corky, bark means the trunks can resist the fire without their cambium (the delicate meristematic tissue) being damaged. The cork oak usually loses its leaves in a fire, but its branches and trunk are not killed, and therefore produce new shoots. Although it seems difficult to believe, these devastated trees will only take a few years to grow a new canopy. [Photo: Teresa Franquesa]

totally decomposed, organic material accumulates. The chemical composition of this material depends on the way the humus is deposited when it falls on the ground, on the rates of accumulation and of decomposition (strongly influenced by its chemical composition), on the volume of leaf material relative to that of wood, and on possible losses through erosion and leaching. The reserves of water-soluble nutrients (potassium, calcium, magnesium) are highly dependent on the water cycle. Finally, the soil's nutrient reserves depend on the local bedrock, the climate, the age of the soil and the vegetation, and of course, on the anthropogenic factors that affect the soil. In general, the soils of Mediterranean ecosystems are nutrient deficient, especially in nitrogen and phosphorus.

The mechanisms by which fire affects nutrient reserves and cycles are complicated. Fire mobilizes nutrients by incinerating the forest's biomass, humus and soil organic material and then depositing it in the form of ashes. Some nutrients are lost in the smoke, and these potential losses increase with rain and/or the action of the wind (leaching, runoff and wind action). Some nutrients transferred to the atmosphere are redeposited in the same ecosystem once the fire has gone out, while others may be deposited near to or far from the burnt area.

In addition to these abiotic responses there are the effects produced by biological processes caused or magnified by fire. In other words, the effects of fire on nutrient cycling and reserves depend to a large extent on the specific processes occurring in each ecosystem, which imply interactions between the prevailing climatic conditions, the type of soil, the type of vegetation, and human interference.

The burning of vegetation has many effects on nutrient reserves in the soil-plant system, and thus on nutrient cycles within the ecosystem. It is generally considered that fire is a rapid decomposer because it determines in most cases the amounts of nutrients available to the plants from the forest's biomass or humus. In any case, this ecological consequence acts in favor of ecosystems only under certain relatively clearly defined conditions of normal intervals between fires.

Frequent fires lead to nutrient depletion and the desertification of the burnt area, due to their cumulative effects on nutrient reserves. But if the natural interval between fires (which most scientists estimate is between 25 and 30 years) is maintained, this environmental risk can be considered as an agent that promotes the renewal and stabilization of ecosystems.

The effects of fire on nitrogen

Among the major nutrients, nitrogen deserves special attention because of its specific chemical properties and its scarcity in the soils of Mediterranean ecosystems. The direct effect of fire is the loss of much of the nitrogen as a gas due to the intense heat. It is estimated that only 8% of the energy released in the form of heat is absorbed by the soil and transferred to soil layers. Potential losses of nitrogen are closely related to the temperatures the fire reaches. At temperatures below 392°F (200°C) there are almost no detectable losses, while losses are greater at 572-752°F (300-400°C), 75-00%. At temperatures above 932°F (500°C) almost all the humus and plant nitrogen reserves may be lost. The soil's nitrogen losses depend to a large extent on its water content. Water has such a high specific heat capacity and latent heat of vaporization that the soil cannot reach a temperature of more than 100°C until the water has completely evaporated or moves to lower layers. The soil's texture also plays an important role in heat transfer during a fire.

Although most of the nitrogen is volatilized in a fire, the quantities of the available forms of this element are greater in burnt sites than in unburnt ones, due to the rapid breakdown and weathering of the humus and the consequent enrichment of the soil after a fire. In general, after a fire concentrations of ammonium-nitrogen (NH_4^+-N) are higher, while the rate of nitrification increases substantially. Even if the total quantity of nitrogen is reduced after a fire, the soil conditions after the fire favor and enormously increase the microbial activity of the soil. Fire promotes the fixation of nitrogen by free-living organisms, which is considered to be induced by the change in the soil pH towards neutrality.

In addition to the fixation of nitrogen by free-living organisms, there is also nitrogen fixation by symbiotic organisms, which is attributed to the spread of members of the Leguminosae during the stages immediately after the fire. The Leguminosae —one the world's most successful flowering plant families—has the highest number of genera in Mediterranean matorral (15 out of 76, i.e. 19.7%), and it accounts for approximately 8% of the flora of the countries of the Mediterranean Basin. This percentage may be an underestimate, because some of the countries involved have both Mediterranean and non-Mediterranean climates,

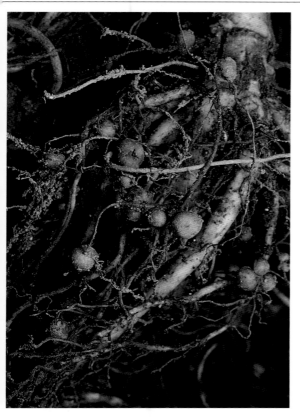

such as Algeria, Syria, Egypt, and the north of France. The proportion of Leguminosae in the total flora for other regions with Mediterranean climates varies between 3.8% (Chile) and 13.3% (southwest Australia). In Mediterranean ecosystems, legumines are considered an important component of the plant communities that flourish after a fire.

Available information clearly shows that after a fire there is a local enrichment of legumes in the flora. This enrichment may not be very grand, as is the case of the Californian chaparral, where the native members of the Leguminosae increase by only 7.3%, or as striking as in Greece where burned sites may show an increase in members of the Leguminosae of more than 50%, with many intermediate responses. The proliferations of members of the Leguminosae in burnt areas only occurs in the period immediately following the fire. Most are annual, or at most biennial, species belonging to the genera *Vicia*, *Lathyrus*, *Lotus*, *Medicago*, and *Trifolium*. Over time the number of species diminishes and their ground cover reduces until it disappears, leaving a seed bank ready for the next fire. Nitrogen fixation by these short-lived herbaceous legumines is

67 **After fires the plants whose roots bear nitrogen-fixing nodules (a symbiosis with a bacteria)** grow better than those lacking them. *Rhizobium*, the bacterium forming these root nodules on a runner bean (*Phaseolus multiflorus*) is the most common root symbiont among the members of the Leguminosae. *[Photo: Adrian P. Davis / Bruce Coleman Limited]*

68 **The effects of a fire on pine trees on old abandoned agricultural terraces** in Requena (Autonomous Community of Valencia) in the eastern Iberian Peninsula. Many species of Mediterranean trees and shrubs accumulate resins and volatile essential oils that are highly inflammable. This seems to be related to their being the first to occupy the space cleared by the fire, thanks to their protective bark or fire-resistant seeds, etc. After the fire, the vegetation rapidly produces new shoots, and plants such as cistus, rosemary, mastic tree, and gorse grow under the scorched plants, covering the affected soil and protecting it from being washed away by the runoff.
[Photo: Jordi Vidal]

still being investigated, but the quantity of nitrogen that returns to the soil subsystem in the form of humus ready for decomposition should be considered as an important additional source of this element.

The effects on phosphorus

Phosphorus does not appear to be much affected by fire; all authors are in agreement on this. Yet available forms of phosphorus increase after a fire, a fact of great importance if we bear in mind that the life of phosphorus is very short in the soils of Mediterranean ecosystems.

Effects on cations

Although some losses of cations have been detected immediately after a fire, the availability of most cations increases. The ashes deposited on the soil surface after the fire contain relatively high quantities of potassium, calcium, and magnesium. Yet the increase in the availability of some of these cations, such as potassium, which are relatively more mobile, is reduced in the stage before the fire in unburnt areas, possibly due to leaching. A change in soil pH is often observed as a result of the cations released by the ashes of the burnt plants and organic material in the soil. The size of this change and the time it lasts depend on several factors, mainly the original pH, the quantity of ash produced, its chemical composition, and the rainfall regime.

Effects on soil microbial activity

The change in soil pH after a fire affects microbial activity directly or indirectly. Highly activated bacterial populations show an immediate response. Increases in the rate of ammonification and nitrification after a fire have been reported from many places in the Mediterranean. Fungi seem to be affected negatively to begin with, because in general they do not thrive in acid conditions. Nitrogen fixation by symbiotic and free-living organisms has already been described. Very little is known about the effect of fire on mycorrhizal propagules in the soil conditions following the fire. It is supposed that, as in other groups of fungi, those that form mycorrhizal associations are negatively affected by the fire, especially by the change in pH to more acid levels, but also by the increase in soil temperature although this rarely reaches lethal values. The means used, the speed and the degree of cover of the mycorrhizal associations of the roots of most Mediterranean plants are very important, because they play an essential role in the establishment and survival of plants in adverse soil conditions, as they act as specialized roots that branch their way through large volumes of soil, translocating nutrients to the plants much faster than roots lacking mycorrhiza.

2.6 The annual cycle

Plant activity is conditioned by seasonal fluctuations in the resources they need, that is to say, light, water, and nutrients. These seasonal variations give rise to adaptive changes in the activities of the organisms, which may use the level of a specific resource (such as soil moisture) as a direct indicator or may make use of an indirect indicator (for example, variation in daylength as an indicator of the arrival of the cold or warm period). In any case, organisms begin these activities as soon as the resources become available. In the case of the Mediterranean climate, there is a clear seasonal fluctuation in temperature, moisture, and nutrient availability, which is directly dependent on water availability.

Temperature shows very regular and predictable fluctuations, and during a part of the cold period acts as a factor limiting growth. On the other hand, rainfall is irregular and unpredictable, although it tends to fall outside the hot season. The activity of Mediterranean vegetation is limited by the relatively low temperatures in winter, and by the lack of water and nutrients in the dry period, and growth is concentrated in the autumn and spring; the factors responsible for this are the water availability and the temperature. This scheme is only a broad generalization, as different plants have evolved different mechanisms to deal with the same problems, so that some species may be active when others are inactive.

It should be noted that there are clear differences between the general behavior of the vegetation in, on the one hand, the Mediterranean Basin, California and Chile, and on the other hand, Australia and South Africa. Those in the first group correspond to the description above, while those in the second group show a series of adaptations related to the extreme nutrient poverty of most of the soils. This means nutrient uptake and accumulation take place in the wet season and are used for the growth of new shoots in the hot period, i.e., in high summer. This strategy is a response to the need to avoid nutrient loss, as the

69 Century plants (*Agave americana*) in flower on the Catalonian Costa Brava (Spain). The century plant comes from the cold deserts of Arizona and Mexico and is naturalized in the Mediterranean Basin. It has become so well acclimatized to coastal conditions that it now forms part of our typical image of the coastal landscape. It only flowers once and this may take years or decades. Flowering occurs more often following adverse climatic conditions.
[Photo: Jordi Bartolomé]

nutrients are released by decomposition precisely when soil moisture is highest, but at temperatures that are not optimal for growth. Anyway, the presence of summer-growing plants is also related to the tropical influences in the origins of the Mediterranean floras of the South African and Australian mediterraneans.

Sprouting and vegetative growth

In general terms, sprouting occurs in the spring, when the high temperatures and the availability of water and nutrients allow active growth. There are, however, two clearly distinct strategies; that of the sclerophyllous plants with evergreen leaves and that of the summer-deciduous malacophyllous plants, a group that is abundant in the driest and hottest areas. Evergreen plants normally have deep roots making water available to the plant from the end of autumn, when the first rains have reached the deeper levels, but they delay shoot production until spring and thus avoid possible damage to the new shoots by winter frosts. Summer-deciduous plants, on the other hand, have surface roots and grow in hotter sites, and start to sprout at the end of autumn and continue growing until mid-spring when water starts to become scarce. Vegetation adapted to living in an irregular climate shows great developmental flexibility and in unusually rainy summers or unusually mild winters, or even out of season, additional growth may be produced to take advantage of favourable conditions. Herbaceous and annual plants may show two phases of active growth, in autumn and spring, making the most of the two periods when the combination of temperature and moisture are most favorable.

Growth in thickness of trunks and roots may take place at other periods of the year as the new tissues formed are less sensitive to low temperatures. It must be borne in mind that photosynthetic activity continues throughout much of the year, supplying the resources necessary for these activities. Thus in general, roots start growing actively in the autumn, while growth in trunk thickness is greatest in midwinter. This generalization is subject to many qualifications and irregularities, and so in woody Mediterranean plants it is very difficult to distinguish growth rings, and if they can be recognized, it is very difficult to know if a ring corresponds to an entire year's growth or just part of it, because of the plasticity of the plant's developmental response. The climatic information accumulated in the growth sequences of woody plants, which is so clear and so useful in areas with highly seasonal climates, is difficult to interpret in the Mediterranean biome, where trees and shrubs write down their life history in a script that is difficult to decipher.

Flowering

Flowering behavior shows even greater irregularity, as it is not a process that requires large quantities of water. Some species flower use resources produced in the same season, while others use reserves accumulated in the previous season's growth, thus separating flower production from periods favorable to growth. Flowering appears to be more closely related to the presence and availability of pollinating agents than to water and nutrients. Thus, the intensity of flowering in some shrub species, such as California's chamise (*Adenostoma fasciculatum*), is closely related to the water availability immediately before flowering, as it uses resources synthesised during this period. A closely related species, *A. glauca*, shows a high correlation between flowering intensity and the water availability during the growing season of the previous year, as its flower primordia develop when it ceases active growth, and flowers several months later, mobilizing the resources of the previous year. This results in flowering being staggered over almost the entire year, with minima in winter and summer.

In the western part of the Mediterranean Basin, for example there are species that begin flowering when daylength shortens, i.e., during the autumn. This is the case of *Inula viscosa*, a winter deciduous shrub typical of ramblas, that colonizes many abandoned fields, covering them with its yellow flowers when the first autumn rains arrive. The strawberry tree, an evergreen shrub typical of maquis and holm oak forests, also flowers in autumn, producing its racemes of white flowers just when the previous year's fruit are ripening.

Midwinter sees the flowering of many herbaceous plants of cultivated fields (weeds), growing under tree crops, such as olive, carob, and orange. The most abundant are rocket (*Diplotaxis erucoides*), *Erucastrum nasturtiifolium*, and the Cape sorrel (*Oxalis pes-caprae*) which cover the space between the trees with their white and yellow flowers.

Other herbaceous plants that do not grow on cultivated ground, such as alyssum (*Alyssum maritimum*), also flower in winter, as do some shrubs, such as gorse (*Ulex parviflorus*), laurustinus (*Viburnum tinus*), rosemary (*Rosmarinus officinalis*), and many

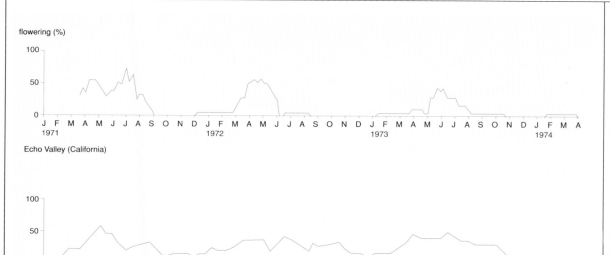

flowering (%)

Echo Valley (California)

Fundo Santa Laura (Chile)

70 **Variation in monthly percentage flowering** in 18 species of woody plant in Echo Valley (San Diego, California) and 20 species in Fundo Santa Laura (Santiago, Chile) over a four-year period. Although the climate and soil nutrient conditions are similar in these two Mediterranean regions, there are phenological differences between their plants. The more uniform distribution of flower production throughout the year in the Chilean region is attributed to its greater floristic diversity and milder climate.
[Drawing: Editrònica]

71 **The composite** *Reichardia picroides* **flowering** on an abandoned agricultural plot in Israel. The flowering of plants in the Mediterranean is spread out at intervals. The most important flowering period is in the spring, as in the case of this olive tree in Palestine, but autumn and winter flowers also decorate the fields with their bright colors.
[Photo: Teresa Franquesa]

species of thyme (*Thymus* spp.). Some introduced trees, such as the almond (*Prunus dulcis*) and mimosa (*Acacia dealbata*) give a touch of color to the landscape during the cold period.

The most important flowering period is when spring arrives, as most Mediterranean species flower between March and June. Some communities, such as cistus scrub (*Cistus*), change greatly in appearance and color as a result of their production of many large flowers, while other communities, such as holm oak forests, only change slightly in colour, when their branches are covered with inconspicuous greenish yellow flowers. Summer is a period when relatively few plants

flower, but there are some. This is the time for succulent plants, some of which are introductions, such as the century plant (*Agave americana*) and the prickly pear (*Opuntia ficus-indica*). Even so, summer flowering is affected by the shortage of water and is limited in comparison with flowering in the winter.

To sum up, in Mediterranean ecosystems the overlapping growth of different species gives rise to an almost continuous production of flowers, which does not stop even in the driest periods in the most arid places, since this is when the succulents flower. This situation is explained by the coexistence of different biological strategies developed to deal with analogous problems.

3. Fauna and animal populations

3.1 The origins of the Mediterranean fauna

Given that the origins of the Mediterranean climate are recent, any attempt to identify phyletic relationships between the fauna of the world's five Mediterranean climatic regions must be centered on past events and, above all, on the relative position of emerged land masses and their greater or lesser proximity in previous geological eras. The history of the relationships between these groups of Mediterranean fauna begins in the Jurassic period of the Mesozoic with the fragmentation of the single supercontinent that formerly contained almost all of the Earth's land masses.

The exchange of common groups between Australia and South America was possible, until the end of the Cretaceous, across Antarctica which had been joined to these continents. However, exchange between Africa and South America only occurred until the beginning of the Cretaceous. In the northern hemisphere exchanges were possible well into the Tertiary as the separation of North America and Eurasia was relatively late. As the Mediterranean climate did not appear until the end of the Tertiary, by which time the continents and the areas that would later become Mediterranean regions had already reached their present positions, the characteristics and adaptations shared by the fauna of these areas cannot be explained as simply being due to biogeographical factors. Rather, they are result of the constraints imposed by the climate of more recent times. Nevertheless, a land connection has existed between Chile and California since the end of the Pliocene which has allowed some animals to migrate.

The first clear signs of a Mediterranean climatic regime appeared during the Pleistocene, after the first glaciation. When summer droughts started to become regular, the fauna responded by adopting various types of survival strategies. Some groups of animals simply died out or only managed to hang on in small refuges where the environmental conditions of the Cenozoic persisted more or less unaltered. Other groups, already adapted to dry summer conditions for non-climatic reasons, survived without problems and would have been favored by selection in an environment that limited the ability of many species to compete. A third group of organisms took refuge in places that were different from those they had formerly lived in, seeking the suitable environmental conditions no longer to be found in their original habitats. Many soil-dwelling creatures simply moved to deeper layers or took to living in caves. Others retreated to areas where climatic oscillations were milder, such as sheltered mountain-sides, gorges, and springs. Some Mediterranean endemic species like the Mallorcan midwife toad (*Alytes muletensis*) probably owe their survival to this strategy. Finally, some groups managed to evolve in this short period of time and in that way adapted to the new conditions.

The effects of the glaciations

The oscillating temperature regime that characterized the late Pliocene and the entire Pleistocene began around three million years ago. Although little is known about the initial effects of these climatic changes, there have been between 16 and 25 oscillations, with their corresponding advances and retreats of the polar icecaps. However, the glacial periods of the last million years during the Pleistocene are better known, and they brought about great changes in the Mediterranean fauna, depending on the area concerned. In Australia and South Africa the changes were not very important, as the animals could escape from the cold into wide open spaces and there were no high mountains to accentuate ice accumulation.

The case of California and, above all, that of Chile is different. These areas contain high mountains that retained ice masses even though they retreated during the interglacial periods. During the glaciations, the fauna was forced to leave mountain areas and move to coastal or interior areas to occupy sheltered valley bottoms and coastal refuges. Some of these valleys and refuges became surrounded by colder areas and this isolation meant the thermophilic animals in the Mediterranean biome then evolved independently for some tens of thousands of years. The situation was exactly the opposite during an interglacial period: heat-loving species

expanded behind the retreating ice while stenothermic species (those adapted to only slight variations in temperature) were left isolated in high mountain areas. Speciation took place in these small high mountain areas, too, and in Chile, for example, there are two distinct types of speciation—that of the valleys and that of the mountains.

The pattern was similar in California. During the glacial periods there were coastal and valley refuges there too, but as the Sierra Nevada is lower than the Andes, less ice accumulated and the degree of isolation and of specialization was less. On the other hand, the advance of the huge North American icecap forced a mass southward retreat of animals despite the effect of the mountains. In the interglacial periods the animals went northwards, although these periods were much shorter than the cold periods. Even so, throughout all these changes most animals could easily move to suitable warmer (or colder) areas and thus avoid extinction. As a result, many species have survived in North America, where the diversity of trees and mammals is greater than in similar regions of Europe.

The situation of the areas around the Mediterranean is, however, quite different. On several occasions an extensive icecap in Europe pushed thermophilic fauna and flora southwards. and there were also secondary ice caps in the Alps and other mountain ranges. As these ranges generally run east-west, they acted as barriers preventing the animals from moving gradually south. This may be why many species became extinct during the glaciations, and even today several European groups of plants and animals show low species diversity.

Yet in some groups this did not happen: at the eastern and western ends of the Pyrenees, the Alps and the Carpathian Mountains corridors with milder climatic conditions continued to exist, where animals could circulate from one side to the other. This is how thermophilic species reached the Mediterranean peninsulas, and they thrived in their gentler, more sheltered climates. As they were partially isolated from their continental relatives, even during the brief interglacial periods, they began to differentiate and acquire special characteristics.

The role of these warm refuges around the Mediterranean in the evolution of local fauna has been well studied. They acted as dispersal centers from which the species and forms developed there spread out, just as in the other dispersal centers in the Northern hemisphere. Southern California, also with a Mediterranean climate is one of these centers, although the largest and best-known of these refugia is the Mediterranean Basin.

72 The Iberian lynx (*Felis pardina*) is the only Iberian mammal included in the Red Data Book of animals in danger of extinction. Today it is recognized as a species that is exclusively Iberian, although it almost certainly once lived in southern France and probably also in central Europe before its habitats were reduced in size. Today it is restricted to a few "mountain islands" in the Iberian Peninsula and many programs have been established to monitor the status of the species. The photo also shows the antenna of a radio-tracking collar designed to keep a track on the animal's activities. It is a solitary, often nocturnal, species although it can occasionally be surprised sunbathing.
[Photo: José Luis Rodríguez]

73 The Ibizan wall lizard (*Podarcis pityusensis*) [Balearic Isles]. The reptiles found today in the Mediterranean Basin began evolving in the second half of the Tertiary Period and replaced almost all the reptiles from previous periods. The movements of the lithospheric microplates of the Earth's crust and successive changes in climate caused seas to retreat and permitted reptiles to spread into new territories. The islands of the Mediterranean have been colonized by a small number of reptile species that arrived in different geological epochs.
[Photo: Javier Andrada]

These warm refugia were divided into smaller ones where different forms and populations were concentrated. For example, within the Mediterranean dispersal center there are nine distinct units which acted as independent refugia. The largest is known as the Atlantic-Mediterranean refuge and it consisted of the area covering the Iberian Peninsula, northern Morocco, and Algeria. Other important subrefugia included the Balkan-Anatolian, the Italian, and the Mauritanian (the area between the Sahara and the Atlas). As the interiors of the Mediterranean peninsulas are mountainous, most thermophilic animals probably gathered in small coastal areas.

Insular speciation

Glacial periods were accompanied by rises and falls in sea levels. Mountain areas and high ground near the coast were converted into islands for a few thousands years by the melting of the icecaps. The opposite happened when the icecaps were advancing: sea levels fell and offshore islands were joined to the coast, forming peninsulas and tombolos (sand bars). As many thermophilic species were restricted to coastal areas, it was easy for them to occupy these new peninsulas, and when the icecaps retreated, islands were formed with populations that were geographically isolated from the rest of the species. This gave rise to a repeated process of differentiation and the formation of chains of local races that is especially evident in the Mediterranean islands. Most of the species concerned were, as might be expected in view of their origins,

terrestrial and thermophilic animals dependent on climatic conditions and unable to move rapidly considerable distances away from adverse environmental conditions. Island microevolution is typical of small Mediterranean reptiles as well as of other groups, for example beetles and terrestrial molluscs, which are also influenced by the same factors.

Most species of small Mediterranean reptiles belong to the Lacertidae, the "true" lizards, and one genus in particular, *Podarcis*, the wall lizard, is perhaps the best example of this process of race and subspecies formation on islands.

The wall lizard *Podarcis muralis* of central and western Europe is close to the original stock which existed in the continent before the Pleistocene. Successive glacial periods forced this species to take refuge in the Mediterranean peninsulas, where it evolved independently, giving rise to different species: *P. hispanica* in the Iberian Peninsula, *P. sicula* in Italy and *P. erhardii* in the Balkans. These species began to diverge in the Pliocene and still continue to diverge today, and this has given rise to new, closely-related, species, both on the peninsulas mentioned (such as *P. bocagei*) and on the Mediterranean islands (such as *P. lilfordi*, *P. tiliguerta* and *P. filfolensis*).

Many of these species occupy a mainland area or an island with many small offshore islands. These small islands have also undergone alternating periods of isolation and communication, but they more often remain far from the mainland. This is because

74 A Lilford's wall lizard (*Podarcis lilfordi*) on the little island of Sa Dragonera, in front of the western coast of Majorca (Balearic Isles). The lizards of the genus *Podarcis*, distributed around the Mediterranean, are small, generally insectivorous reptiles, basically terrestrial and most at home on hard substrata. Yet they live in very varied habitats and this has given rise to an enormous variety of patterning, colors, and shapes caused by the interaction of natural selection and the random factors which placed some individuals and not others in a particular area.
[Photo: Javier Andrada]

the water separating these offshore islets is shallow, and thus smaller climatic changes are enough to modify their connection to the mainland. As a result, Mediterranean archipelagos with many small islets have a wide range of local forms of lizards, generally with distinctive morphological traits and restricted to one or a few islets.

The most notable example of this process is the Ibizan wall lizard (*Podarcis pityusensis*) found on the Balearic Islands of Ibiza, Formentera, and neighboring islets. A total of 42 subspecies of *P. pityusensis* have been described, roughly one for every islet. Notwithstanding the tendency today to reduce this number, this species is still remarkable for its polymorphism. Yet it is by no means unique: 29 subspecies have been described in a closely related wall lizard (*P. lilfordi*) that is native to the Balearic Islands of Majorca, Menorca, and neighboring islets. Other lizard species with many subspecies taxa on Mediterranean islands include *Podarcis sicula*, principally in Italy and neighboring islands; *P. melisellensis* on the Dalmatian coast; *P. tiliguerta* in Corsica and Sardinia; *P. wagleriana* in Sicily; *P. filfolensis* from the Maltese archipelago; *P. peloponnesiaca* in Greece; and *P. erhardii* on the Aegean Islands.

The Mediterranean Basin has long been a cultural and linguistic melting pot and many civilizations have arisen on its shores. It has also acted as a center of animal dispersal and its great historical and geographical diversity has made it the cradle of many new species and subspecies.

The role of humans

The Mediterranean Basin is highly humanized. For thousands of years human civilizations have populated the area and have left such a mark on its landscapes that the Mediterranean Basin's scenery cannot be understood without taking into account the role of humans, even in apparently unaltered areas. Human intervention has taken many forms, the oldest, most persistent and efficient of which is the use of fire. With the added effects of nomadic and permanent agriculture, irrigation, grazing, and tourism, the landscape degradation caused by fire is so ancient that it seems natural. Human use of fire has led to the reduction or even disappearance of the original sclerophyllous forests and their replacement by garrigue and ultimately, by semi-steppe areas such as the thyme scrub (tomillares) in Spain and the renosterveld in South Africa. These changes in the vegetation have led to major and almost irreversible changes in the soils. The brown earths of mature holm oak forests have given way to the typical matorral red earths and rendzinas and even in some areas to virtually bare lithosols.

Just a few thousand years of intensive human intervention has also had great repercussions on the fauna, and its consequences are as important as those of induced by the Pleistocene climatic oscillations or by other processes operating on a much longer time scale. Indirect effects on the fauna are due to the modification of the vegetation and other aspects of natural habitats. The direct consequences are a product of the domestication of wild animals, the introduction of

75 This image of St. Jerome taking a thorn out of a lion's paw in Palestine, drawn by an anonymous artist around 1430, shows the former presence of lions in the Mediterranean Basin. Christian icons, Roman mosaics, and lion hunting scenes in the Atlas mountains from many different periods make it clear that the lion used to be fairly common in the Mediterranean region. The most recent literary references to lions in the Mediterranean Basin are from 19th century North African texts, but many medieval tales also relate the problems facing the inhabitants of areas near the Atlas when the lions' natural habitat was reduced and they took to looking for food around villages.
[Photo: Graphische Sammlung Albertina, Vienna]

species into areas outside their original ranges, and the persecution or elimination of certain species. Large carnivores, such as the lion and leopard, lived on the northern shores of the Mediterranean until the period of the Roman empire, and on its southern shores until the beginning of the last century, but these animals cannot now be considered Mediterranean.

The domestication of animals (except the dog) began in the area near the eastern Mediterranean around 8,000 years ago. One of the first species to be domesticated, the goat, was also a major factor in shaping the Mediterranean's landscape. Changes in the landscape due to the activities of goats, sheep, and livestock in general have had an impact on non-domesticated species at least as great as the direct effects of human intervention. Even though they have been intense everywhere, the effects of livestock on the landscape show variations due to climate relief, and history. As an example, cultural reasons have led many Mediterranean countries to reject the pig, thereby depriving peasants of an animal that can efficiently recycle waste and fend for itself in forests managed as *dehesas* (pasture with oak trees) without major changes to the structure of the pre-existing holm oak forest. The introduction of Mediterranean fauna into other countries has already been discussed. These introductions have sometimes had serious consequences for the ecosystems that receive them, profoundly changing the local vegetation and fauna. The introduction of the rabbit to Australia is a well known case, but, as in other cases of less economic importance, its very complex consequences have only been studied very superficially. Even so, the main effects of human action on Mediterranean fauna are indirect effects due to habitat transformation. Faunal diversity has diminished greatly and has partly cancelled out the effects of topography and climatic change which favor diversification. Repeated fires have eliminated the less mobile animals that take cover in shrubs, and they have favored animals mobile enough to flee and then return to feed on the fresh shoots produced by pyrophytic species after fires. Thus, the tortoises of the Mediterranean coastal countries appear to doomed to extinction, mainly

because of the repeated burning of their refuges. The reduction in tree cover reduces the variety and abundance of insect fauna which in turn affects the communities of insectivorous birds and mammals. The animals living in trees and on their bark are, as a result, forced to find other types of habitats, such as under stones, and they may even become anthropophilous (thriving in close proximity to humans), or even commensal (i.e. living together and sharing food resources) with humans. The same is happening to the plants: many species of forest plants have been eliminated by the use of land for cultivation and this has led to the spread of weeds (plants that invade crops and fields), which are also to some extent commensal with humans, as they take advantage of the concentrations of nutrients around human settlements.

3.2 The Mediterranean fauna

Ecological diversity and the fauna

Mediterranean environments are often described mainly in terms of one or more of their characteristics, such as seasonality, periodic droughts, or physical

76 A genet (*Genetta genetta*) licking its lips after feeding. This member of the Viverridae (civet, genet, and mongoose family) lives throughout the Iberian Peninsula, France, the Balearic Islands, Arabia, and part of Africa. The lack of paleontological remains from the Pleistocene has led to speculation that this species was introduced into Europe by the Romans, Phoenicians or Arabs. The species *Genetta felina*, recently separated from *G. genetta*, reaches as far as South Africa. The genet lives in forests and scrubby areas, climbs trees very agilely and hunts rodents, birds, and reptiles, although it will also eat berries and insects.
[Photo: Jane Burton / Bruce Coleman Limited]

features, depending on the point of view of the person describing them. Nevertheless, the animals inhabiting Mediterranean areas are probably most influenced by the high variability of their habitats in time and space. The Mediterranean environment does not have a uniform climate, unlike tropical jungles where seasonal differences are minimal, nor does it show a regular alternation of seasons of activity and rest as occurs in boreal conifer forests. Mediterranean systems generally show a certain regularity in the succession of their three basic seasons. Spring is short, and in a few days nature wakes from its winter lethargy and transforms the landscape. The summer-autumn period is long, during which the majority of animals either complete their yearly cycle or enter a phase of summer torpor (aestivation). Winter is also short and the animals use up the reserves accumulated in the summer and get ready for reproduction. Nevertheless, large and often unpredictable oscillations in humidity, temperature, and sunshine will often be superimposed on the otherwise regular physical conditions. The geographical complexity of the coastline (especially in the Mediterranean Basin and California), the relief with high peaks in certain areas, a varied mosaic of rocky substrates, soils, vegetation, and differing degrees of human intervention, all combine to make Mediterranean landscapes very difficult to describe concisely. This variablity also offers the fauna a large number of ecological niches. It is, therefore, impossible to describe the composition of Mediterranean fauna in detail. The similarities between the five great Mediterranean areas (the countries of the Mediterranean Basin, southern California, central Chile, the extreme southern tip of Africa [the Cape], and the southern coast of Australia), are comparatively recent and superficial, as climatic changes have acted upon different floras and faunas in each region. Before discussing the animal populations of these areas, we need to make a general comparison of them and outline the main features of their fauna.

Geographical and historical contexts

Southern Australia is the least typical of the Mediterranean areas. It lacks mountains and has been separate from the other continents for a long time. Furthermore, its climate is not typically Mediterranean, since there is some summer rainfall and the vegetation is derived from *Eucalyptus*. The undergrowth is a mallee scrub formation whose composition is very different from the scrub formations of the other mediterraneans. The mallee's fauna is largely native to it or has immigrated from the northern deserts. The role of humans in shaping this landscape was until recently very limited but has subsequently been very marked. Human intervention, which led to forest clearance and the spread of matorral, also included the introduction of non-native animals, both wild, such as foxes and rabbits, and domestic fauna, such as cats and mice, and they have become naturalized in many areas. The rabbit is a native of the Mediterranean Basin, and this explains why it has adapted so well to Australian conditions.

The mediterranean region at the southern tip of the African continent shares some faunal affinities with the other mediterraneans, especially Australia. The scrub communities of the Cape region (fynbos) are very poor and unstructured since the best soils are used to grow crops. Many elements of Australia's and South Africa's flora and fauna originated to the north in tropical or subtropical regions, as the higher latitudes are occupied by the sea. The Cape's landscape is much more heterogeneous than Australia's, partly because there are mountain ranges running parallel to the coast. The Cape fauna, and especially its soil-dwelling animals, shows similarities with that of Australia.

The two American mediterraneans, California and Chile, are the most similar in terms of geography, climate and ecology. Their situation between the sea and a high mountain range running parallel to the coast, the presence of cold currents in the neighbouring seas, and the relatively recent connection of both regions through the Panama isthmus, have ensured similarities between their landscapes despite the large differences between their flora and fauna. In both regions the fauna is derived from tropical areas as well as from nearby temperate areas. Human impact has been similar in the two regions, and is more recent than in the Mediterranean Basin but older than in Australia and South Africa. The fauna of the two American mediterraneans have different origins and very distinct biogeographical histories.

The Mediterranean Basin is larger than all the other mediterraneans put together and has given its name to these biomes. Here, human action first began to modify landscapes, and its colonists and organisms subsequently spread to other mediterraneans. From a zoogeographical point of view, the Mediterranean Basin consists of a group of provinces within the Palaearctic region whose faunal elements are largely derived from temperate European biomes and to a lesser extent from subtropical African biomes. The history of the region is unusual: the connection between Europe and Africa, intermittent in the west and permanent in the east, permitted species to move from one continent to another, thus making the Mediterranean Sea more a means of connection than an insurmountable barrier.

77 Two rodents: *Octodon degus* from Chile (above) and the Mediterranean pine vole (*Microtus* [=*Pitymys*] *duodecimcostatus*) from the Iberian Peninsula (below). Species of burrowing rodents can be found in the Mediterranean regions of Chile, California, and the Mediterranean Basin. They are essentially herbivores and excavate galleries. In the Mediterranean Basin they feed on roots and bulbs and can feed underground from within their burrows.
[Photos: E.A. Janes / NHPA and Jacana]

Naturally, all the regions with Mediterranean climates have different species which carry out similar ecological roles. Thus, the burrowing rodents of the genus *Octodon* from Chile are comparable to the genera *Neotoma* and *Thomomys* in California, and to the Mediterranean genus *Microtus*. Similarly, the iguana communities of the Chilean matorral (*Liolaemus*) find their homologues in those made up of *Cnemidophorus*, *Sceloporus*, and *Uta* in the Californian scrub, or in the *Podarcis* lizards of the Mediterranean garrigue.

The case of the vertebrates

All the mediterraneans contain the five classes of vertebrates found on land—continental fish, including the Agnatha (lampreys and hagfish), Chondrichthyes (sharks, rays and ghostfish) and Osteichthyes (bony fish), together with amphibians, reptiles, birds and mammals. However, the fish found in the mallee are either introductions or members of widely distributed marine groups. Moving down the taxonomic hierarchy, there are 25 orders of non-cosmopolitan land vertebrates, and the Mediterranean Basin is richest with 15 orders. The Mediterranean Basin has close affinities with the Cape mediterranean as they share nine orders. The affinities between the Cape and Australia are minimal, sharing only one non-cosmopolitan order, the pheasants (or perhaps two, if the dingo [*Canis familiaris dingo*, a carnivore introduced into Australia in prehistoric times], is considered as native to Australia). None of these shared orders is endemic to any Mediterranean area. The most widely distributed orders in Mediterranean regions are the Artiodactyla (the even-toed ungulates) and the Cypriniformes (carp-like fish)—and possibly also the carnivores—which are only absent from Australia, and the Phasianidae (the junglefowl and pheasant family), which are absent only from central Chile. If we descend to the family level, distribution patterns are more interesting in that the percentage of cosmopolitan groups obviously decreases, whilst that of

endemic groups increases. The mediterraneans as a whole contain 179 families of non-cosmopolitan land vertebrates. In terms of continental fish, Chile and the Mediterranean Basin are the richest areas with seven families, whilst South Africa has only three families. Chile and the Mediterranean Basin share two families of fish, the Cyprinodontidae (killifish, toothcarp) and Atherinidae (silversides, sand smelts), while Chile and South Africa have no fish families in common. Some families are almost endemic; the Diplomistidae (catfishes) and Pigididae are confined to central Chile.

There is no cosmopolitan family of amphibians (the Bufonidae, or toads, have only been recently introduced into Australia) and as such they are ideal for use in the comparison of different faunas. The batrachians found in the mediterraneans belong to a total of 13 families, seven of which are found in the Mediterranean Basin, the mediterranean with the most diverse fauna. On the other hand. in the Australian mallee, there are only two families of amphibians, the Leptodactylidae (also found in Chile) and the Hylididae, or tree frogs. Other families with restricted distributions include the Heleofrinidae, nearly endemic to the Cape, although extending slightly northwards. All the regions, except Australia and South Africa, have families in common: the Mediterranean Basin and California share the greatest number of families, namely the Bufonidae (toads), Hylidae (tree frogs), Salamandridae (newts and "true" salamanders) and Plethodontidae (lungless salamanders). The lungless salamanders have an unusual distribution as they are found throughout the western United States and northern Mexico, as well as in small areas of Italy and the islands off Italy's western shores, where they are cavedwelling creatures. Their inability to migrate makes this family one of the best indicator of the faunal links between the mediterraneans of both continents.

Reptiles, like amphibians, depend on suitable environmental conditions and are virtually unable to migrate,

78 *Speleomantes* [=*Hydromantes*] *ambrosii ambrosii* is a member of the Plethodontidae, the lungless salamanders, that is found in Liguria in northwest Italy. The European members of the family have adapted partially or totally to life in caves in an attempt to avoid the consequences of the dry Mediterranean climate that arose at the end of the Cenozoic.
[Photo: César L. Barrio Amorós]

79 The mallee ringneck (***Platycercus** [=**Barnardius**] **barnardi***) in the Flinders Ranges in South Australia. The Psittacidae are one of the most typical families of Australian birds. This group's distribution is basically tropical, although it is more diverse and reaches higher latitudes in the Southern Hemisphere. It is most diverse in the Australasian region, and in the Australian continent has differentiated into numerous species that exploit different niches. Some eat seeds, while others eat pollen or take nectar; there are omnivorous and even carnivorous species. Most members of the Psittacidae are gregarious and budgerigar colonies may contain up to a million individuals.
[Photo: Oriol Alamany]

and this makes them favorites for study by biogeographers. None of the 27 families of reptile found in the mediterraneans is cosmopolitan, although one family, the Scincidae (the skinks), is widely distributed and has members in all the mediterraneans. Three more families, the Gekkonidae (the geckos), Colubridae (colubrid snakes) and Elapidae (the cobras, mambas and coral snakes), are represented in four of the five mediterraneans. Once again the Mediterranean Basin has the greatest variety (15 families) and central Chile the least (seven families). Links between Old World regions, especially those between the Mediterranean Basin and South Africa, seem to be closer. These two regions share 11 reptile families and each also shares six families with the Australian mediterranean. In contrast, the Australian region only shares two families with central Chile (Gekkonidae and Scincidae), and two with southern California (Scincidae and Elapidae). In this group also there are no reptile families endemic to a single region, although the Anniellidae (shovel-snouted legless lizards), a family of only two species of limbless, burrowing lizards, are near-endemic to southern California in that they are found in the area of mediterranean climate but also in the nearby arid areas.

80 Majorcan midwife toad (*Alytes muletensis*) on the island of Majorca. At the end of the 19th century it was considered a form of the midwife toad (*Alytes obstetricans*), but when living populations were rediscovered in the 1980s it was reclassified as a separate species. Today it survives in an area of only 77 mi² (200 km²), but fossil remains indicate that it was once more widespread. Life in a rocky biotope has obliged this toad to become an agile climber, quite unlike any other species of the genus.
[Photo: Jordi Muntaner]

Unlike reptiles, birds can travel long distances, and so they are far more difficult to link to a single area. It is therefore not surprising that many bird families found in the mediterraneans are cosmopolitan while others have a very wide distribution. The remaining 64 families show a similar pattern to that found in the orders discussed previously, with the two American mediterraneans standing slightly apart from the other three which, in turn, form a coherent group sharing a large number of families. Of the 24 families living in the Mediterranean Basin, 19 are shared with South Africa where a further 13 families also occur. On the other hand, Australia and Chile only share one family, the Psittacidae (cockatoos, parrots etc.). Logically, endemic families of birds are to be found among birds less able to move long distances, i.e., those that cannot fly. However, there is no bird family limited strictly to a single Mediterranean region.

The situation of the mammals is somewhat different from that of the other vertebrates. Placental mammals arose in the Euro-Asiatic block at the beginning of the Tertiary when Australia was already isolated and had its own ancient mammals. Therefore, the greatest affinities are between the mammals of the Mediterranean Basin and those of South Africa, and also with those of North America. These three regions form a single group, with the greatest connections between the Mediterranean Basin and California, which share 16 out of the 53 families of mammals that are neither cosmopolitan nor marginal for these areas. Central Chile is different and only shares three families with each of the regions mentioned above. Lastly, Australia exhibits its high degree of isolation in the meagre number of shared families: none with Chile, and only one, the Muridae (rats and mice), with the Mediterranean Basin and South Africa, which share over 20 non-cosmopolitan families whereas Australia,

due to the reasons mentioned above, only has seven cosmopolitan families.

This analysis could continue to the genus or lower taxonomic levels, but would get more and more complex. The number of cosmopolitan groups would decrease, even amongst birds, and the number of endemic species would rise. There are many genera and species that are restricted to a single mediterranean region. The clearest examples of this are in the reptiles and amphibians as they are so dependent on their environments. Species such as the Majorcan midwife toad (*Alytes muletensis*), confined to a tiny area of the island of Mallorca, and *Discoglossus nigriventer* of the same family, from Israel (unless it is extinct, as it has not been sighted for decades), are extreme cases of Mediterranean endemic species. Examples of genera include *Pleurodeles* and *Speleomantes*, members of the Urodela (newts and salamanders) restricted to the western Mediterranean Basin. Nevertheless, other genera illustrate the biogeographic links between the different mediterraneans of the world: the porcupines (*Hystrix*) and genets (*Genetta*) are found in the Mediterranean Basin and South Africa and indicate the affinities between the two regions. The inclusion of invertebrates in this analysis would, of course, serve to reinforce these conclusions. Likewise, an analysis of soil-dwelling creatures also demonstrates the degrees of affinity between the five mediterraneans and repeats the pattern shown for vertebrates: the Mediterranean Basin and the Cape region have the most in common along with, to a lesser extent, southern California. Central Chile has a certain affinity with California; and all analyses show that southern Australia is the most unusual of all the regions. This agreement reflects the history of continental movements and the changing patterns of accessibility of each region with respect to the others.

3.3 Mediterranean zoogeographical affinities

Regions with Mediterranean climates have more than just climate in common: their plant communities have a similar physiognomy and structure and their landscapes and land-use patterns are similar. This is all the more remarkable considering that the world's mediterraneans are discontinuous and separated by large distances, and so their biotas have evolved in isolation, starting from very different phylogenetic lineages. No other physically discontinuous region shows as many similarities as the world's mediterraneans do.

From a zoological point of view, as is the case with the flora, the similarities are not due primarily to a shared fauna. There are no animal species common to all five regions other than a few, largely invertebrates, which have dispersed over large distances, often helped inadvertently by humans. Even in these cases, these species tend to be very widely distributed, or cosmopolitan, or even ubiquitous, and not characteristic as such of Mediterranean regions. At higher taxonomic levels (genus, family, etc.) similarities between the fauna of two or more regions gradually increase, especially at family level. Moreover, according to the level under consideration, affinities vary. For example, in terms of the number of shared bird genera, California and Chile are the most similar regions, whereas South Africa has no genus in common with either of these two regions. However, in terms of shared families, the most similar regions are California and the Mediterranean Basin, and the most different are the Mediterranean Basin and Chile.

On the other hand, the similarities between the five regions are reflected more clearly in some cases of ecological and morphological convergence at the level of species and of animal communities.

The faunistic groups of the Mediterranean Basin

The Mediterranean Basin's location at a biogeographical crossroads, especially during glacial and interglacial periods when plant species were moving large distances and there were successive animal invasions, makes it difficult to define a typical Mediterranean Basin species. A Mediterranean Basin species may come from a variety of backgrounds. It may have evolved under a Mediterranean climate or, on the other hand, it may have taken advantage of adaptations that had evolved before the appearance of the Mediterranean climate. It might be an immigrant, fleeing from very arid or cold conditions. Finally, it might even be a newly-arrived invasive species. The Mediterranean Basin is, as a result, a mosaic of forms with different biogeographical origins.

The Mediterranean Basin is a bird-rich area. The dominant element amongst ground-nesting birds are species whose ranges extend beyond the typical sclerophyllous vegetation of the Mediterranean Basin

81 A porcupine (Hystrix cristata) in Italy. The porcupine is an African species only found in Sicily and parts of mainland Italy. The porcupine is one of the world's largest rodents, and its very effective elongated black and white defensive spines make it unmistakable. When attacked, it backs into the animal attacking it, often causing serious injuries. Porcupines live in open woodland.
[Photo: Josep Maria Barres]

Mediterranean tropical plumages

The fish of the world's coral reefs and the birds of the tropics are with reason generally regarded as the most eye-catching of all living creatures. Rich and strident colors are common: the exaggerated technicolor array of macaws, toucans, and birds of paradise, for example, a whole spectrum of colors, is set off by delicate plumage textures or by plain but glossy, outsize bills. Although if you look closely enough, you can often discover beautiful shades of color in many of the commonest Mediterranean birds, such as goldfinches and tits, none of them can compare with the brilliance of tropical birds. Yet there is one exception—the birds of the order Coraciiformes, which include the kingfishers and bee-eaters, avian ambassadors from the luxuriant tropics to temperate Mediterranean latitudes.

For all intents and purposes, the Coraciiformes are tropical birds. It is a diverse order, made up of roughly two hundred species, which can be grouped into ten different families. Four of these families are to be found in Mediterranean climates: two, the Alcedinidae, or kingfishers, and the Meropidae, or bee-eaters, are closely allied to the Todidae and the Momotidae of the West Indies and tropical America. The third family is the Coraciidae, or rollers, which are related to the Brachypteraciidae and Leptosomatidae, two families endemic to Madagascar. The fourth and last family is the Upupidae or hoopoes, distantly related to the Phoeniculidae, or wood-hoopoes, of tropical Africa and the Bucerotidae, or hornbills, of tropical Africa and Asia. Half a dozen of the more than one hundred species of kingfishers, bee-eaters, rollers and hoopoes that live in the globe's temperate and hot zones are found in the Mediterranean Basin, and a more attractive half-dozen would be hard to find.

The kingfishers are the fishermen of the group, as their common names in many languages illustrate (martín pescador, martin-pêcheur, etc.). Their crafty look and disproportionately long bills and short tails give them a characteristically waggish and colorful air as they sit watchful, alert and erect on any convenient branch. Nevertheless, in the blink of an eye, they dive down like an arrow at their prey, normally an unwary freshwater fish swimming too close to the surface. Without worrying about getting a good soaking, the kingfisher momentarily submerges itself, with all its plumage, in the water. All kingfishers have beautiful plumages: the blues and reds of the common kingfisher (*Alcedo atthis*), widespread throughout the Mediterranean Basin (as well as Central and Eastern Europe, India and southeast Asia, China and Japan) contrast with the brown and bluish white-breasted kingfisher (*Halcyon smyrnensis*), found on the coastline of Turkey and the near East (as well as India and Southeast Asia). Then there is the more simplified black and white patterning of the pied kingfisher (*Ceryle rudis*) which overlaps in range with the Smyrna kingfisher but also extends throughout almost the whole of sub-Saharan Africa, as far as the Cape mediterranean. In the Californian mediterranean, the belted kingfisher (*Ceryle alcyon*) is sometimes seen, a vagrant from more northerly latitudes which occasionally even strays as far as Europe.

Bee-eater (*Merops apiaster*) [Felix Labhardt / Bruce Coleman Limited]

Kingfisher (*Alcedo atthis*) [NHPA / Silvestris Fotoservice]

The bee-eaters are the very essence of greenness. Strictly speaking, in the Mediterranean Basin there is only one species of bee-eater (*Merops apiaster*). This bird, with its forked tail, a myriad of different colors and graceful lines, is arguably the most attractive of the world's 24 species of bee-eater, all of which are essentially green. All have the same taste for the bees and other members of the Hymenoptera that form their staple diet. Two African and eastern species, the blue-cheeked bee-eater (*M. superciliosus*) and the little green bee-eater (*M. orientalis*) are present in the Nile delta. The common bee-eater of southern Europe and North Africa is also found in the Middle East and the southern tip of Africa, where the familiar environmental conditions of the Mediterranean Basin reappear.

Lastly, we have the hoopoe (*Upupa epops*) with its predominantly earthy tones and outlandish call and appearance. Its raised crest and the sound of its onomatopoeic song are unforgettable. It is a confiding and friendly bird which nests in cracks and holes in tree-trunks, rocks, or walls and is distributed throughout practically the whole of Europe, Asia and Africa, only lacking from colder and desert areas.

Kingfishers, bee-eaters, rollers and hoopoes bring the colors of the tropics to the Mediterranean world.

The roller (*Coracias garrulus*) which, apart from South Africa, has the same range as the European bee-eater, is a bird of lightly wooded areas. Its robust, sturdy form and unmistakable brown, blue, and black plumage are similar in all the various African, Indian and southeast Asian species of roller (*C. abyssinicus*, *C. benghalensis* and *Eurystomus glaucurus*).

Sylvia communis

Sylvia conspicillata

Sylvia cantillans

Sylvia undata

Sylvia sarda

Sylvia hortensis

Sylvia melanocephala

Bendala

into the broad-leaved Palaearctic forests. Another group of birds stretches across the plains of Eurasia and the steppes of the south and southeast. The endemic Mediterranean species form only a small group. On the other hand, a series of tropical groups such as the Cracidae (curassows and guans), Psittacidae (lories, cockatoos and parrots) and Trogonidae (trogons), amongst others, settled in Eurasia during the Miocene (the Vulturidae also arrived in the Pliocene) but died out in the Mediterranean Basin before the Pleistocene. Thus, faunal relationships between this area and other areas outside Eurasia and North Africa are now rare. On numerous occasions during the Pleistocene, all the European faunal types congregated in the Mediterranean Basin and expanded northwards whenever glaciers began their gradual retreats, without, however, ever totally abandoning the Mediterranean Basin, so that this area has always been connected faunistically to Central Europe, without giving rise to isolation or important episodes of speciation. The Mediterranean Basin's bird fauna is essentially Palaearctic but rather impoverished because its peninsular nature with respect to the Eurasian landmass, with relatively few endemic species. The Mediterranean Basin is an important wintering area for the Eurasian bird fauna.

There are at least 150 species of mammals in the Mediterranean, as well as another 50 whose ranges are restricted to the region. Exchanges of fauna between Africa and Europe have always been infrequent and during most of the Quaternary period the mammals on the Mediterranean's southern shores evolved independently from those on its northern shores. Despite the land bridges formed when the Mediterranean dried up during the Messinian crisis, paleontological research has shown that only seven species of rodent and one species of lagomorph (rabbits, hares, etc.) had previously spread from Europe to Africa via Spain, some surviving until the Pliocene, while five genera of African rodents successfully colonized the Iberian Peninsula. Later on, during the Pliocene and the Pleistocene, the existence of the Straits of Gibraltar prevented any further faunal exchanges. As a result, most of the fauna of southern Europe has Euro-Siberian and Mediterranean affinities, whereas North Africa's fauna is essentially Ethiopian, that is, of trans-Saharan, origin. Secondary elements of the Mediterranean mammalian fauna have originated from three other regions: the Irano-Turanian (including the cold deserts of central Asia), the Saharo-Sindian (the desert belt stretching from Morocco to the Sind in India), and the Oriental, comprising the part of Asia south of the Himalayas. The genuinely endemic Mediterranean species of mammal basically evolved, over periods of isolation measured in hundreds of thousands of years,

as a result of divergence in impoverished island faunas. The last ancient species of tropical origin, the crested porcupine (*Hystrix cristata*), does not differ essentially from African populations; other ancient relict species died out before completing their differentiation into new, endemic species.

The fauna of the other mediterranean regions

Of the biome's five regions, the Mediterranean Basin has suffered the least amount of biogeographical and evolutionary isolation, followed by California, Chile (trapped between the Atacama Desert and the Andes), South Africa, and finally Australia.

Some of the Californian mediterranean's bird species are restricted to the chaparral, but a greater proportion is linked to the boreal zones of North America, while the remaining species are widely distributed throughout the southeastern United States and Mexico. In Chile there are very few species restricted to matorral; most species either extend into the temperate *Nothofagus* forests to the south and in the Andes, or are widely dispersed throughout South America and the New World. Chilean birds, therefore, must have been derived, in equal measures, from the tropical Andes and from the temperate forests to the south. A small number of cosmopolitan species completes the picture. Most bird species found in the South African fynbos are also present in the semi-desert vegetation of the Karoo region. Others fynbos species are shared with the plains of the Orange region and Botswana, and a group is shared with the relict coastal forests.

Chile shares its only two lizard families with California, which possesses a further two families of

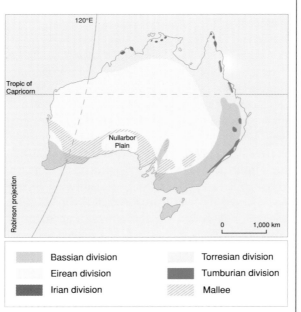

83 Distribution of the mallee bird communities in relation to the zoogeographical divisions of the Australian avifauna. The division coincides to some extent with the differing types of Australian vegetation.
[Drawing: Editrònica, from several sources]

Legend:
- Bassian division
- Eirean division
- Irian division
- Torresian division
- Tumburian division
- Mallee

82 *Sylvia* warblers are small and generally with dowdy plumage. Half of the species in the genus are exclusively Mediterranean and represent one of the few groups of birds which originated in the Mediterranean Basin. They are probably the best example of adaptive radiation in the birds of the region, especially in the western Mediterranean where they are most diverse. The process of species differentiation must have been favored by the fragmentation of the vegetation and the resulting geographical isolation during the Pleistocene glacial and interglacial periods. The drawing shows a whitethroat (*S. communis*), a species which is not restricted to the Mediterranean and extends throughout Europe as far as western Siberia, a spectacled warbler (*S. conspicillata*), a subalpine warbler (*S. cantillans*), a Marmora's warbler (*S. sarda*), a Dartford warbler (*S. undata*), a Orphean warbler (*S. hortensis*) and a Sardinian warbler (*S. melanocephala*).
[Drawing: Marisa Bendala / ECSA]

its own. The Mediterranean Basin has no lizard family in common with either of these two regions.

The affinities of the vertebrate fauna of the Australian mallee are restricted to within the continent, a consequence of its extreme isolation. Only four species of passerine bird and a single lizard, all highly specialized, are restricted to and characteristic of the mallee. In comparison with the neighboring forest and heathlands, there is a greater diversity of lizards in the mallee, although mammal diversity is similar and bird and amphibian diversity is less. Frogs and toads are poorly represented in the mallee and there are no endemic species. The two areas of this region are separated by the plains of the Nullarbor Desert, a very effective barrier to dispersal, and do not have a single species of frog or toad in common. Some genera originating in the southeast, however, did manage to establish themselves in the southwest during interglacial periods. Almost all the reptile families of the Australian continent are found in the mallee, most of them penetrating into the arid Lake Eyre region and the rest into the more humid Bass region (southeast and southwest Australia, as well as Tasmania). Of the 163 species of birds found permanently or seasonally in the mallee, some are endemic, although the majority belong to these two regions (Lake Eyre and Bass Strait) zones. A further six species are related to the Torres region in the north and northeast of the continent. Finally, seven families of mammals, around 30 species, are characteristic of the Australian mediterranean although none shows special adaptations. The low number of species is partially due to the extinctions provoked by humans over the last centuries.

The Mediterranean soil fauna

The fauna of Mediterranean soils is particularly interesting and diverse. Some groups, such as the pseudoscorpions, are found in all five regions and their diversity is comparable to that found in tropical rainforests. On the other hand, other groups are not so well represented such as the termites, whose importance diminishes from South Africa down through Australia and California to the Mediterranean Basin and Chile, in parallel with each region's degree of isolation from the influence of tropical rainforest climates. Hygrophilic arthropods from very different groups have solved the problems associated with extreme changes in daily and seasonal climatic conditions by adapting to life in the soil, as their limited dispersal ability means they cannot seasonally migrate to more humid regions. This adaptation is found in the ancient Collembola (the springtails) and proturan groups, and also in more modern groups such as Coleoptera (bee-

tles) and pseudoscorpions. These groups have evolved from epigeal (living near the ground) ancestors into soil-dwelling creatures by reducing their size, and the development of wingless and eyeless forms and other adaptations to soil life. Yet their main adaptations are not specific to Mediterranean soils, as they developed in soil-dwelling fauna before the appearance of the Mediterranean climate and are also present in other regions. There are close phylogenetic affinities between the soil-dwelling faunas of the different mediterraneans. Geological and palaeoclimatic changes have led to the accumulation of phylogenetic lineages from very different origins. These hybrid origins can be illustrated by looking at Chilean soils in which a significant Palaeantarctic or austral (South America, Australia, New Zealand, South Africa and the Antarctic islands) biogeographical element is found in many groups, such as the gammarids, spring tails, beetles, turbellarians, oligochaetes, isopodsa, pseudoscorpions, Procoptera and others. A second group of xerophilous epigeal species of arid areas, such as windscorpions, spiders, nocturnal ground beetles, pseudoscorpions, some ants and thrombidiform mites, whose distributions are closely linked to the Andes, forms a neotropical element that probably originated when neotropical and Arctic phlyogenetic lineages entered Chile along the Andes from the north. A third group, consisting of the pantropical elements distributed throughout the world's tropical regions, is rare in Chilean soils and represented by a few pseudoscorpions, some millipedes and a number of parasitic nematodes. Most archaic lines must have had a more or less continuous distribution in the Mesozoic, although they are now scattered irregularly throughout the world. These biogeographical anachronisms are mostly strictly hygrophilous and thus limited to the deepest layers of Mediterranean soils. This is the case with most proturans, pauropods, symphylans and microwhipscorpions. The fourth and final group is the cosmopolitan species and consists, on the one hand, of species with a considerable capacity for passive dispersal—helped by their small size and production of resistant forms (in the cases of the rotifers, tardigrades and many protozoans)—and on the other hand, of species accidentally introduced by humans during the expansion of agriculture. These introduced species are great colonizers and are the main component of agricultural and, above all, irrigated land and include acarine mites, springtails, nematodes, earthworms, and a few isopods.

Introductions and invasions

All distribution areas are dynamic and so invasions should not be considered as accidents, but rather as

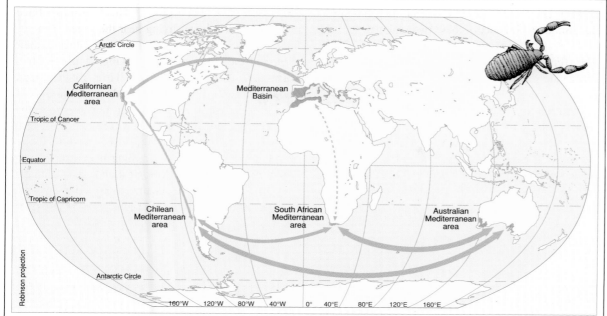

84 **Degrees of affinity between pseudoscorpions** at family and generic level represented by the thickness of the arrows. Although the regions with Mediterranean climate do not share a common fauna, there are some affinities between them. This ancient group is most diverse in the soils and caves of the mediterraneans. The diagram shows the greater importance of "horizontal" affinities (within the southern hemisphere, and within the northern hemisphere) as opposed to "vertical" (between north and south) links, and is evidence of an ancient common geological ancestry in which archaic northern-hemisphere forms evolved in isolation from their southern relatives on the continents of Laurasia and Gondwana. In contrast, vertical affinities are closer and more recent between California and Chile than between the Mediterranean Basin and South Africa. The Sahara, situated between these latter two regions, has acted as a barrier for the fauna of moist soils, whereas the Andes in the New World has tended to act as a bridge for certain latitudinal migrations. Apart from a few cosmopolitan species probably introduced by humans, no species is common to all five regions. This pattern of affinity is repeated fairly closely in other groups of soil-dwelling invertebrates but not in most invertebrate groups.
[Drawing: Editrònica, based on Di Castri, 1981]

processes that take place over a period of time. Potentially invasive species tend to be generalists, with broader ecological tolerances. It is normal in nature for distributions to expand but this process has been accelerated or forced during historical time by the possibility of long-distance dispersal through human activities. Some authors consider that those ecosystems and regions disturbed in the geological and historical past by climatic change or human presence are, in fact, more resistant to invasion by non-native forms, and that they may contain more species capable of invading other regions. The Mediterranean Basin is one such region.

Humans have introduced, deliberately or accidentally, a series of generalist species with wide ranges, some of them ubiquitous cosmopolitan species, into ecosystems originally consisting of species with more restricted distributions and requirements. Over the last two centuries around fifty species of mammals have been introduced with varying degrees of success into southern Australia. In California, some introductions into the chaparral have had limited success, while others are human commensals or live in suburban areas and human settlements, although they show some degree of naturalization. Some domesticated species have returned to the wild, such as the horse (*Equus caballus*), donkey (*E. asinus*), goat (*Capra hircus*), cow (*Bos taurus*), sheep (*Ovis aries*), cat (*Felis catus*) and dog (*Canis familiaris*). In South Africa, the great impact of introduced plants contrasts with the limited impact of introduced mammals. The black rat (*Rattus rattus*), brown rat (*R. norvegicus*), house mouse (*Mus spicilegus* [=*M. musculus*]) and the grey squirrel

(*Sciurus carolinensis*) and the red squirrel (*S. vulgaris*) are all commensals, that is living close to human habitation only without spreading into the fynbos. The rabbit (*Oryctolagus cuniculus*) has only been introduced onto a few islands without fynbos, while neither the wild boar (*Sus scrofa*) nor the fallow deer (*Dama dama*) have become naturalized. Only the tahr (*Hemitragus jemlahicus*, a goat-like ruminant) has been genuinely invasive, and even then only on a local scale.

Most invasive bird species in all the mediterraneans are human commensals and depend on human activity. It is estimated that less than 5% of the Mediterranean bird fauna of California, Chile, South Africa, and Australia is potentially invasive. Unlike other groups such as plants, mammals, parasitic insects and other disease-causing organisms, in the case of birds human-induced dispersal has not been an important factor in invasions in mediterraneans. Once they had arrived, few introductions were successful, and even fewer have had an ecological impact within the Mediterranean fauna. On the other hand, most of the native bird species have had their ranges affected by human modification of landscapes.

The insular nature of the Chilean bird fauna, isolated by the Andes from contacts with other faunas, has meant that this region is less species-rich than California. Only two of the introduced species are well established in Chile and have had an ecological impact on the bird fauna, while in California only the starling (*Sturnus vulgaris*) has had a similar impact. Three introductions have been successful in both regions: the cattle egret (*Bubulcus ibis*), the rock

85 Current world distribution of the rabbit (*Oryctolagus cuniculus*) and wild boar (*Sus scrofa*) indicating natural and naturalized distributions. Where they have naturalized, these animals live and breed as if they were their in natural areas of distribution.

[Drawing: Editrònica, based on Naturalized mammals of the world, *Lever, 1985*]

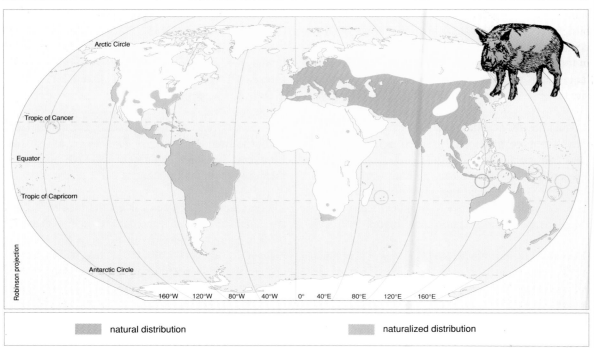

natural distribution   naturalized distribution

dove (*Columba livia*) and the house sparrow (*Passer domesticus*). All three species are widely distributed throughout the world, but because they are ubiquitous and cosmopolitan species they should not be considered to be typical of the mediterraneans.

The large number of unsuccessful bird introductions into South Africa is remarkable, and no bird species native to another continent is found regularly in unaltered areas of fynbos. However, in fynbos degraded by human action, only a couple of African species and the starling (*Sturnus vulgaris*) from Europe have had an ecological impact. Finally, in the Australian mallee,

of the hundred or so introduced bird species, only a dozen have been successful. Almost all are generalist species in their areas of origin, widely distributed throughout Europe or Asia and phylogenetically of recent origin.

Convergence phenomena

When a group of organisms belonging to the same taxon are either genetically or geographically isolated, there is a tendency to progressive divergence in one or more characters while they are undergoing independent evolution. Nevertheless, not all the features that undergo change in isolated taxa necessarily diverge. This is why unrelated species that live in geographically distant areas with similar environmental conditions may develop similar morphologies and ecological strategies in order to adapt to similar environments. This type of evolution is known as convergent or parallel evolution (taxa showing this type of evolution are referred to as ecological equivalents) and the concept applies not only to the phenotype (morphology, behavior, etc.) but also to more subtle aspects such as population distribution and density, diversity and resource-use, and community organization.

When comparisons are made within the same trophic level, the number of species and the density of individuals per hectare for different mediterraneans show similarities in community organization. Two regions may differ clearly in the number of species in the same trophic group (for example, ground-feeding birds), yet this difference may be much smaller when considered in terms of the total density of individuals in the group, due to the fact that the region with the larger number of species has an average species-density that is lower than the region with the lower num-

ber of species. Thus, the excess of species in one group is partially compensated for by the fact that each species has a lower density of individuals.

In general in birds, as well as in other groups such as lizards, convergence is more noticeable at community than species level. This is because a species does not necessarily have a morphological or ecological equivalent in other regions, although there may well be similarities in distribution at the level of trophic group or in feeding methods. Of course, each region has its own characteristics; for example, it seems that birds have divided the Californian region up by habitats, while the Chilean region's greater geographical isolation means they divide it up by geographical areas.

All these convergent adaptations cited are recent and arose after the appearance of the Mediterranean climate. Convergence and its causes are still not well understood. There are some discrepancies, that is to say cases in which, for example, the morphometric space describing the bird fauna of a non-Mediterranean region, taken as a control, can be completely superimposed on the corresponding

86 Ecological convergence of pairs of insectivorous birds in the sclerophyllous formations of California and Chile. The bar diagrams show the distribution frequency of each species at the height (in meters) at which they feed. Species from different genera and families are ecologically equivalent, not only because they feed at similar levels in the vegetation strata, but also because they belong to the same trophic categories: the first four pairs in the diagram seek insects in the foliage, whereas the other two catch insects in flight. Furthermore, the birds of each pair are similarly sized (the silhouettes are all to the same scale).
[Drawing: Jordi Corbera, from several sources]

CALIFORNIAN CHAPARRAL

Chamaea fasciata　*Psaltriparus minimus*
Thryomanes bewickii　*Parus inornatus*　*Empidonax difficilis*
Myiarchus cinerascens

CHILEAN MATORRAL

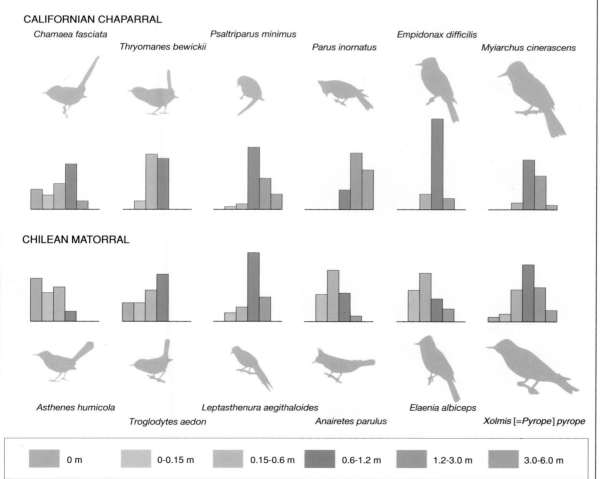

Asthenes humicola　*Leptasthenura aegithaloides*　*Elaenia albiceps*
Troglodytes aedon　*Anairetes parulus*　*Xolmis [=Pyrope] pyrope*

| | 0 m | | 0-0.15 m | | 0.15-0.6 m | | 0.6-1.2 m | | 1.2-3.0 m | | 3.0-6.0 m |

87 A Malachite sunbird (*Nectarinia famosa*) taking nectar while perching on an aloe (*Aloe*) in the Cape Province. In open areas in the extreme south of the African continent there are birds that feed on the nectar produced by aloes and proteas. These species take nectar from flowers and are used by the plants to transfer their pollen. Nectar-taking species are present in all the mediterraneans except for the Mediterranean Basin.
[Photo: Jen & Des Bartlett / Bruce Coleman Limited]

space in the mediterraneans. This fact cannot be interpreted as convergence between faunas as the two faunas in question belong to environments with different climates and vegetation.

Lastly, there are some ecological types that, for historical reasons that transcend the limits of ecological adaptation of the species, are not represented in all the mediterranean regions. For instance, the Mediterranean Basin is the only one of the biome's five regions without a single species of nectar-feeding bird, as the Palaearctic region's flowers have evolved mechanisms for pollination by insects, not by nectar-feeding birds. The ancient co-evolution between insects and plants and the restrictions imposed by the developmental scheme of the animals (in this case birds) has left the flowers in the Mediterranean Basin beyond the scope of the adaptive plasticity of its birds.

3.4 More room for the small than for the big

A mosaic of ecological niches

The complexity of the Mediterranean biome, compared with the uniformity of other biomes, such as the taiga or deserts, has already been pointed out. This complexity is present on many levels, from large geographical units to the microniches only recognizable after a detailed analysis of the ecosystem. The main division is geographical and distinguishes the five regions with Mediterranean climates, whose faunas have different biogeographical origins and vary in composition. Some of these regions are subdivided further into relatively independent areas with similar features: Australia has two mediterraneans separated by the Nullarbor Desert, while the Mediterranean Basin can be divided into the Iberian, Italian, Balkan, Pontic, Anatolian and African areas. At a finer level, the main divisions are climatic and depend more on latitude and altitude. For example, xeromediterranean, thermomediterranean, meso-Mediterranean or sub-Mediterranean communities can be distinguished according to the length of their respective dry seasons.

The regions can be further subdivided on the basis of essentially topological criteria: the main mountain ranges, the basins of the large rivers and the main islands are all types of units whose fauna differ to a greater or lesser extent. In turn, these subdivisions can be further subdivided into even smaller areas a valley, a single mountain ridge, or even to one side of a mountain or its peak. The level below this consists of the features of the landscape, such as a particular forest, and so the process of subdivision can continue down to spaces measurable in millimetres or even smaller units. Ecological complexity is built in this way, with smaller structures inside larger ones, just as Russian dolls fit one inside another. However, the situation is in reality far more complex since the different levels are not clearly separate, boundaries are vague, and a single level includes units of differing types.

Some examples of niches

The most common way of categorizing niches is on the basis of feeding habits, and other aspects of the niche tend to be considered secondary. In the analysis of an ecosystem, the animals present are generally classified according to their type of feeding habits, as well as where and when they perform them. It is taken for granted that herbivores do not compete for food with carnivores, that those animals that feed in trees do not compete with ground feeders, and that nocturnal feeders do not interfere with daytime feeders. Niches are formed on this basis and—in accordance with Gause's principle—are usually only occupied by a single species. Moreover, it is accepted that two very different animals, such as a bird and an amphibian, do not compete, even if they have similar feeding habits, because they use very different methods to obtain their food. Thus, ecological niches for birds, mammals, etc., can be given the same designation.

Let us consider some ecological niches in Mediterranean biomes occupied by similar animals. Only a brief account will be given focusing on birds that are the most studied group. We will begin with the example that led in 1915 to the term ecological niche being coined as a result of a study of the California thrasher *Toxostoma redivivum*. This drably-plumaged omnivore feeds on the ground in densely vegetated areas, and is often overlooked as it is found predominantly in clumps of catclaw (*Acacia greggi*), although it also feeds in other plant communities. In Chilean scrub, another member of the Mimidae (mockingbirds and thrashers), the Chilean mockingbird (*Mimus thenca*), has similar attributes and occupies the same niche. The California thrasher's counterpart in the Mediterranean Basin is the blackbird (*Turdus merula*). In Australia the mallee heathwren (*Serinus cautus*) occupies a similar but not identical niche, and their counterpart in South Africa is a member of the Turdidae (blackbirds, chats, nightingales, etc.), the Cape robin chat (*Cossypha caffra*).

Other birds join together in guilds that occupy similar niches. For example, one can define niches for insectivores that feed amongst the leaf litter, for nectar-feeding birds, for birds that search under tree bark for their food, etc. The most interesting aspect of this distribution into guilds is that each niche or group of niches has approximately the same number of species and the same density of individuals, even in areas separated by large distances. Thus, there is to be found a crepuscular insectivorous species of bird belonging to the family Caprimulgidae, that hunts on the wing, not only in the Mediterranean Basin, but in California, central Chile, and South Africa: respectively, the nightjar (*Caprimulgus europaeus*), the poorwill (*Phalaenoptilus nuttallii*), the band-winged nightjar (*C. longirostris*) and the dusky nightjar (*C. pectoralis*). Likewise, the group of fruit- and seed-eating birds contains between two and four species in each region. All of them are birds of similar size, and have a similar population density in each region—about one bird per hectare (1 hectare=2.47 acres).

Mammals, obviously, also occupy niches but they do not enter into competition with birds because of their predominantly nocturnal nature. In each of the five Mediterranean areas there are homologous niches for mammals: wood mice are represented in Europe by the yellow-necked mouse (*Apodemus flavicollis*), in North America by the California mouse (*Peromyscus californicus*), in South Africa by *Myomys* [=*Myomiscus*] *verreauxii*, in Chile by *Akodon longipilis* and in Australia by *Pseudomys albocinereus*. Analogous lists could be drawn up for burrowing insects, small carnivores, small insectivorous bats, etc. In each case, the niche is shared partially with similar species. The similarity between some niches in certain Mediterranean regions is highlighted by the success of some introductions. For example, the fallow deer (*Dama dama*), a deer native to southern Europe, has been introduced successfully in California, Chile, South Africa and Australia. It has entered local ecosystems throughout the world and has assumed the same role as in its place of origin.

Reptiles are a suitable group for analyzing the division of the habitat into separate niches. The first studies of this type were carried out in the Australian sub-desert mallee and have subsequently been extended to other similar areas of the world. Although the dominant families of small lizards vary from one region to another, analogies can be found in all the mediterraneans. Small European lizards are members of the Lacertidae, whereas American species belong to the Teiidae and Iguanidae, South African lizard species

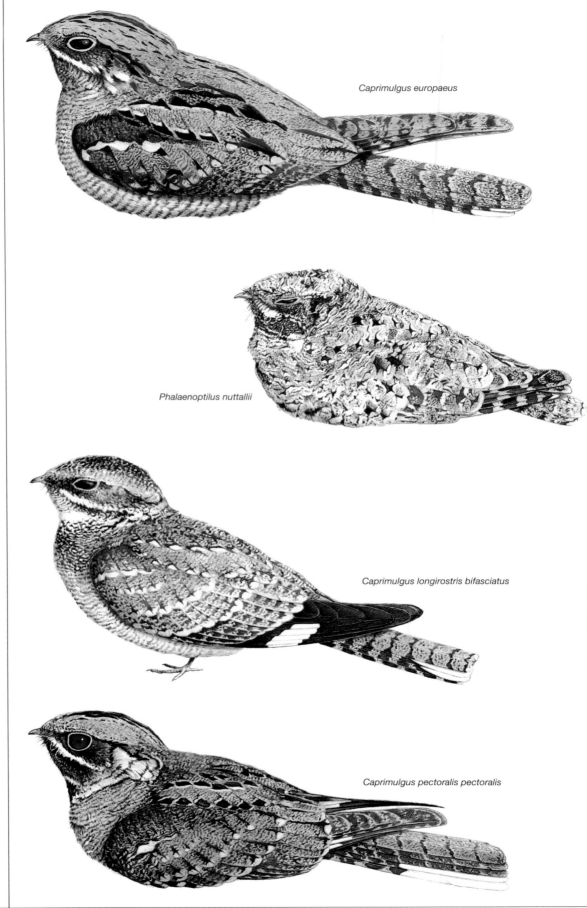

88 **Typical members of the Caprimulgidae (the nighthawks and nightjars) from four of the world's mediterraneans:** the nightjar (*Caprimulgus europaeus*) from the Mediterranean Basin, the poor-will (*Phalaenoptilus nuttallii*) from California, the band-winged nightjar (*C. longirostris*) from central Chile, and the dusky nightjar (*C. pectoralis*) from the Cape. None of these species is exclusive to their mediterranean, but each one has the same ecological role as a predator of crepuscular insects.
[Drawing: Lluís Sanz]

Caprimulgus europaeus

Phalaenoptilus nuttallii

Caprimulgus longirostris bifasciatus

Caprimulgus pectoralis pectoralis

belong to the Scincidae, and Australian species are mainly members of the Agamidae. Yet they are all quick-moving, sun-loving creatures, that are active only during the hottest part of the day and feed almost exclusively on insects. In many areas two species of the same genus live side-by-side but share resources effectively and hardly ever come into contact. Lizards are one of the favorite study tools of today's ecologists and they have played a leading role in research into questions such as faunal convergence and substitution, density compensation, habitat structure, and niche-concealment.

Insects show great variety in Mediterranean biomes and so it is not surprising that this group has developed many extreme cases of specialization in a narrow niche. There are insects that live happily in slightly saline or even salt water, in the nasal cavity of amphibians, that enter and leave slugs through their respiratory orifices, that spend their whole lives on the flowers they were born on, or that feed exclusively on the dung balls rolled and buried by dung beetles. Evidently, in such a complex group no rigorous analysis of niche segregation has been attempted; but there is no doubt that it is much easier to establish functional distinctions between groups of insects than groups of vertebrates.

3.5 Who eats whom?

Animal diets

Animals choose their food instinctively, but in an efficient and economical way. When there is little food available, animals are forced to eat what they can and, as a result, many Mediterranean animals have responded to environmental deterioration by changing their diets. Kites, for example, today depend more on scavenging than on hunting, a response to the increasing shortage of live prey. On small Mediterranean islands with poor insect faunas, lizards, which are naturally insectivorous, have taken to feeding on plants to survive. However, when there is sufficient food, each species employs an optimum feeding strategy consisting of maximizing the amount of energy obtained per unit of effort invested in obtaining food. This type of strategy, which puts efficiency before gross energy yield, is typical of ecosystems with long histories where animals have become adapted to local conditions.

Food choice is very closely related to the anatomical and physiological characteristics of an ani-

mal's digestive system. The specialized stomachs of ruminants and the gizzards (capable of grinding down seeds) of granivorous birds are both the cause and consequence of a specialized diet. Even so, these animals show some flexibility: ruminants prefer soft juicy spring shoots to the tough, hardened stalks of summer pastures. Mediterranean ungulates browse close to the ground searching for new shoots and avoiding old grass. Naturally, when there is no alternative, they will eat tougher plants or migrate to better pastures. Even those animals whose digestive systems are unspecialized have adaptations, especially in mouth size and shape, for selecting food. Insectivores, especially those lacking any chewing apparatus (like amphibians and saurians), will generally only take prey of an optimum size. Therefore, trophic webs are not established by relating one species to another, but rather by relating predators to a certain class of prey size. Consequently, most animals as they grow they have to change their feeding habits.

Food-webs

Other than in highly degraded areas, all the basic trophic levels exist in the Mediterranean sclerophyll forest biome. Producers are normally woody or herbaceous plants and, to a lesser extent, soil algae. The main herbivores are insects, rodents, and ungulates. Birds tend to be rather specialized herbivores, feeding on fruits and seeds. Arachnids, many insects, amphibians, reptiles, insectivorous birds and waders, small birds of prey and the small carnivorous mammals are mainly secondary consumers. The few tertiary consumers are almost exclusively large birds of prey. The decomposers are a very varied groups, ranging from carrion-eating mammals and birds of prey to minute insects and mites living in the soil. However, ultimately, it is the fungi and bacteria that produce the final stage in the decomposition of dead matter.

The organization and structure of the food-web
A more detailed example of the workings of a food-web in a Mediterranean environment will illustrate some of the ideas mentioned above. The example is of an area on a Mediterranean island occupied by a much altered holm oak forest: human action has led to the replacement of the holm oaks by Aleppo pines (*Pinus halepensis*) and a clearing of the undergrowth. The system's producer level consists of the trees and shrubs (*Erica arborea* and various *Cistus* species) along with the

89 The crossbill (*Loxia curvirostra*) is a highly specialist feeder that only eats conifer seeds. It has evolved a stout beak to open fir, pine or larch cones to extract their seeds. Due to differences in cone ripeness and seed quality from one area to another, the crossbill is an itinerant bird that settles where it finds food. When these food supplies are exhausted, it renews its wanderings until it finds another site with enough cones for it to be worth a stop.
[Photo: Hans Reinhardt / Bruce Coleman Limited]

sparse herbaceous layer of grasses. This level provides biomass for other compartments of the web: for frugivorous and graniverous birds such as turtle doves (*Streptopelia turtur*) and crossbills (*Loxia curvirostra*), for the few herbivorous mammals (mainly field mice) and tortoises, and leaf-eating insects and the insects that live in the flowers of *Cistus*. Eventually, the main outputs from this level end up on the ground where they may be consumed by decomposers or used by humans for firewood or, occasionally, for grazing.

Secondary consumers are represented chiefly by insectivorous birds, small reptiles (two species of gekkonid lizards), insectivorous mammals (the most important of which is the hedgehog [*Erinaceus europaeus*]), and carnivores such as the small-spotted genet (*Genetta genetta*). In addition to nocturnal birds of prey that feed on herbivorous rodents and insectivorous reptiles, some larger birds can also be partially considered as secondary consumers, such as crows that eat fruit, seeds, insects, and small tortoises. Outputs from this level go to the compartment of the tertiary consumers, and of course to the decomposers. Humans form part of all levels of consumption, but extract little from this level, and what they do extract is almost always in the form of game. The only tertiary consumers are some birds of prey and

snakes, such as the ladder snake (*Elaphe scalaris*), that feed largely on small birds.

Various smaller scale food-webs superimposed on this framework are located in separate parts of the overall biotope. For example, tree trunks form a production level in their own right. Green algae and lichens obtain energy from sunlight and their nutrients from substances dissolved in the rainwater running down the branches and trunks. Aphids, other insects and mites all live on the sap exuded by trees and provide food for ants and spiders living on trunks. Many birds such as tree-creepers, nuthatches, and woodpeckers search the cracks in the tree bark for the insects and beetle larvae hidden below the bark.

Another important trophic web is located near the ground. The harshness of the Mediterranean summer obliges many animals to seek shelter, as well as food, in cracks in the soil or under rocks. These places have a ground-dwelling fauna consisting of worms, nematodes, springtails and mites, wood lice, myriapods, earwigs, ants, and snails. The largest animals of this community are small reptiles, mainly geckos in the Mediterranean Basin. These creatures, despite being insectivores, tend to feed in the open and use loose rocks or pieces of bark shed by trees only as places for shelter. On the other hand, shrews actively hunt insects and snails under rocks and tend not to stay too long in any one place of shelter.

Trophic relationships

It has already been commented that certain general principles seem to derive from the study of food-webs. For example: trophic chains must be short and, given the energetic inefficiency of food digestion, with a maximum of three or four levels; relatively few animals feed on more than one trophic level (omnivores, in the most general sense of the word); the number of prey species is relatively constant and independent of the complexity of the food-web involved; there are relatively few trophic links in complex communities. All these principles are generalizations, or even simply conjectures that have never been rigorously proven. Some of these conjectures relate to habitat structure and might be studied in the Mediterranean biome. For example, it is assumed that the features of the food-webs are influenced by biomass production levels in the ecosystem, by the degree of vertical structuring, and by the constancy or variability of environmental conditions. Mediterranean communities have intermediate production levels concentrated in spring; intermediate and low vertical structuring in forests and in matorral areas, respectively; and regular, intense, fluctuations in environmental conditions, above all in the amount of water available. Thus, trophic chains should be short, the proportion of species feeding on more than one level should be low, and each species should have few connections with other species. Each species, therefore, should specialize, tending to feed on only a few other species.

These conditions are difficult to prove. Scientific literature has described almost 150 food-webs, although almost always based on insufficient data of varying value. Nevertheless, there are no specific studies that compare the general structure of Mediterranean food-webs with that of any other biome. In the island community of sclerophyll forest described above, ten different trophic chains can be isolated with an average of 2.8 levels per chain. In the Mediterranean matorral community in the Coto de Doñana in the southwestern Iberian Peninsula 250 trophic chains have been identified with an average length of 3 levels. Data from other mediterraneans is much scarcer: in central Chile, the average length of the 43 chains identified is only 2.1 levels, perhaps due to the limited number of species in the area.

Another general quality of an ecosystem's trophic chains (and which presumably changes with the seasons) is their level of *connectancy*; or, in other words, the number of existing relationships in relation to the number all mathematically possible relationships. In a food-web with N members, the total number of possible relationships is $(N^2-N)/2$; although, in fact, the figure obtained is much smaller, and tends to get smaller as the stability of the ecosystem decreases. In areas of Mediterranean climate, therefore, low connectancy values should be expected for the food-webs, and the ones analyzed have given connectancy values between 0.17 and 0.36.

Average length and connectancy values for food-webs in stable ecosystems such as temperate forests or freshwater lakes are, in fact, higher. It is difficult to compare published values as different criteria are used to establish trophic chains. However, we can refer to communities near to those cited above and studied by the same authors. For example, the sheltered gullies found in the garrigue on the Mediterranean island of Minorca have more stable conditions, denser vegetation cover, and are moister since they channel the flow of the rainfall. In these gullies, the average length of trophic chains is clearly greater, with 3.9 levels, and the even more stable freshwater communities on the island, have values as high as 5.1 levels. Connectancy values, between 0.19 and 0.49, are also higher.

These results are not conclusive as the general characteristics of food-webs change in similar ways when complexity increases, and stable communities tend to be more complex. Nevertheless, these results help explain possible differences. It is obvious that in short chains energy circulation is more efficient since most of the energy is lost in each level. These conclusions are all consistent with the fact that in a Mediterranean climate, where resources become scarce in the summer, it is best to feed efficiently and to create short trophic chains. Animals that live in this biome are able, in fact, to feed efficiently by regularly changing their diet. They always seek the foods that are at their peak of production.

A general characteristic of food-webs in Mediterranean biomes is that reptiles play a very important role. Some, especially snakes, prey on mammals and birds and are, at the same time, preyed upon by birds and mammals. Not surprisingly, they form part of long trophic chains. In these chains, the highest level is usually occupied by a predator with a specialized diet,

90 The ladder snake (*Elaphe scalaris*) in the undergrowth of a holm oak forest eating a bird, and a short-toed eagle (*Circaetus gallicus*) perched on an Aleppo pine (*Pinus halepensis*) devouring a snake. Trophic networks in the Mediterranean Basin are very complicated, and while the short-toed eagle feeds almost exclusively on reptiles and amphibians, the ladder snake preys on small rodents and birds. So Mediterranean food-webs include reptiles that eat birds and birds that eat reptiles.
[Photo: José Luis González Grande and Bruce Coleman Limited]

generally a bird of prey with a diet consisting almost exclusively of reptiles. This is the case of the short-toed eagle (*Circaetus gallicus*) in the Mediterranean Basin, the secretary bird (*Sagittarius serpentarius*) in South Africa, the wedge-tailed eagle (*Aquila audax*) in Australia, the red-tailed hawk (*Buteo jamaicensis*) in California, and the grey eagle-buzzard (*Geranoaetus melanoleucus*) in Chile (although the last two species are not specialized reptile hunters). Plants are naturally the first level; although epigeal trophic chains can sometimes be added on to the chains formed by the decomposers made up of the soil fauna. In this way, exceptionally long chains can be formed beginning with bacteria all the way through to nematodes, mites, springtails, pseudoscorpions, ground beetles, insectivorous birds, lizards, snakes and birds of prey.

3.6 Migrants and residents

The animals and plants of the Mediterranean scrublands have to cope with water shortages in summer. If droughts were permanent, these species would have to resort to extreme mechanisms to survive (like desert species) or be eliminated through inability to adapt. The regular nature of the dry period permits animals and plants to adopt regular rhythms so that the principal feeding and reproductive seasons coincide with the end of the winter rest, when water availability is adequate.

Many animals have coupled their rhythms to that of nature in the Mediterranean, whilst others have adopted the alternative strategy of leaving at the

91 The migrations of the white stork (*Ciconia ciconia*) link the shores of the Mediterranean Basin and the Cape mediterranean, although many storks do not in fact migrate beyond tropical Africa. In the Mediterranean Basin, storks have a tendency to breed in or near human settlements—both species live in harmony—and often build nests on church belltowers or other high buildings. Cartwheels are sometimes placed on belltowers to encourage storks to nest. The young birds learn to fly before starting their long journey southwards to spend the winter in warmer regions. During migration, they need to rest in large flocks in wetlands in order to be able to complete their long yearly journeys. The number of storks in Europe has decreased over the last few years, largely because the grave threat posed by the side effects of human civilization, such as electric power lines.
[Photo: Oriol Alamany]

right moment and migrating to find a better environment. The animals of the Mediterranean biome can thus be divided into migrants and residents. Nevertheless, this difference shows many gradations: all animals move from one place to another to look for food or protection, and many do so periodically. However, true migratory animals are, properly speaking, those whose movements occur in more-or-less regularly repeated cycles and that undertake major displacements between far-distant places.

The great bird migrations

Most long distance migrants are birds. Their ability to fly enables them to cover vast distances, even, in extreme cases, between the Arctic and the Antarctic. While not comparable with the great distances travelled by oceanic birds, many land birds take advantage of the contrast of the seasons in the northern and southern hemispheres to move to the places where food is abundant. During the northern summer they move to the north for breeding and then in the winter they move back to the other hemisphere to take advantage of the southern summer. Migrations do not only occur in temperate and cold countries since some tropical and subtropical birds also undertake migrations, far from the areas with Mediterranean climates.

Migration strategies
Many birds migrate over relatively long distances without actually changing hemispheres, merely flee-

92 A flock of snow geese (Anser caerulescens) flying over rice fields in Central Valley, California. The Californian mediterranean is more of a migration flyway than a destination for the many birds that migrate from the north of the American continent to the more arid areas of Mexico or Central America.
[Photo: Jim Brandenburg / Minden Pictures]

ing from the rigors of the Arctic winter to southern areas with milder winters. The Mediterranean Basin is one area with tolerably warm winters where many northern birds choose to overwinter. During the winter, the permanent sedentary species mix with wintering birds from northern Europe and other partially migratory species, in which only some members of the population head for the warmth of Africa while the rest only go as far as the Mediterranean Basin. During the migration periods there are also birds in transit, flying towards the south in autumn and towards to north in spring, that also spend a few days in the Mediterranean Basin before continuing their journeys. In the spring, the return northwards to breeding areas also causes a mixture of species with different migratory patterns to congregate in the Mediterranean Basin.

Outside the Mediterranean Basin such large-scale migrations are less common. In Australia and South Africa migrations are relatively insignificant as there are no land masses to the south for them to migrate to in the spring (except Antarctica, which is of no use for this purpose). Australia might serve as a destination for land birds from northern Asia (and in fact many oceanic birds winter in and around the continent) but most tend to stop to spend the winter in Indochina. A part of the Eurasian bird fauna spends the northern winter in South Africa (for example, many eastern European storks), although only a dozen or so species that breed in the Mediterranean Basin, along with a few that do not, in fact, reach the Cape. It is in this

limited sense that the Cape mediterranean forms part of the Palaearctic migratory system.

There is little opportunity for migration in the Chilean mediterranean. In addition to the problem already mentioned in the case of Australia and South Africe of the lack of land to the south, the proximity of the Andes means that birds need only undertake an altitudinal migration (which is much shorter and less dangerous) rather than a latitudinal migration. Seabirds from the southern archipelagos use the Chilean mediterranean as a stopover area. For example, this is one of the stops of the Andean gull (*Larus serranus*) during the southern spring. The Californian mediterranean is, on the other hand, on one of the six migratory routes of the Nearctic faunal region, but it is more of a stopover point than a final destination. Many species from the northern Pacific, Alaska and even from the other side of the Bering Strait find land here, not in the matorral but in the coastal lagoons and neighboring saltmarshes. However, most continue south to Mexico and Central America.

The Mediterranean Basin is different and its characteristics ensure a great seasonal variety of birds. In addition to the attractions of its mild winters, it also has a series of peninsulas projecting southwards nearly as far as the African shore. These peninsulas funnel vast numbers of migratory birds and reduce the sea crossing to a minimum. Migration is made even easier by the numerous islands in the region which act as temporary resting places. Furthermore, the region has eight large wetlands, including the marshlands of the Guadalquivir delta (Doñana) in Spain and the Nile delta in Egypt, as well as hundreds of other smaller wetlands with high levels of food production. These areas allow migratory birds, above all waders, anatids (ducks, geese, swans, etc.), and ardeids (egrets, bitterns, herons) to rest and feed during their journeys, and they are, in fact, a final destination for many species. For these reasons, nearly half of the entire European bird fauna migrates, either totally or partially, across the Mediterranean.

Migratory routes

Thus, the great migratory routes of the Mediterranean Basin follow the main peninsulas. An important contingent passes over the Strait of Gibraltar, although a second and even larger migration occurs along the Black Sea, the coasts of Anatolia and Syria, Lebanon, Israel, the Sinai peninsula and enters Africa through the Nile delta. The third and least important bird route is along the Italian peninsula and crosses the Mediterranean Sea making use of the area's islands: Sicily, Corsica, Sardinia, and the smaller islands. The first and third corridors are used by European birds heading for West Africa but which generally do not cross into the Southern hemisphere, although more than a few spend the winter on one or other side of the equator. The second (Black Sea) route is the most important and is used not only by birds from Europe, but also by birds from large areas of Asia and goes down to southernmost tip of Africa. These routes are very diffuse and are combined with migrations with shorter routes (partial migrations) that follow broad flight paths with lateral extensions.

Bird mortality during migration is extremely high. Perhaps as many as a third of all birds that undertake a migration die, either on the way out or on the way back. The total reaches many million every year. The principal danger is the weather. A storm or hurricane can kill millions of birds in flight and this may be why the eastern Asian route stops in Indochina. Many cyclones form in the South China Sea, making it difficult for birds to reach Australia. In fact, the main flux of migratory birds in Asia is east-west, probably as a result of the existence of barriers such as the Central Asia mountain ranges and the cyclones of the western Pacific. In addition to deaths due to weather conditions, many animals, including human beings, lie in wait for migrating species.

One may wonder why birds have developed such a costly system, both in terms of energy and survival, for obtaining food. Migratory behavior is evidently a recent acquisition. It is probable that migration in its current form came into existence only in the Holocene. The presence of the glacial icecap must have altered or cancelled any previous migratory behavior. The time needed to develop the migratory habit—in this case 12,000 years—may seem insufficient to implant such a complex and generalized system of behavior. Yet, there is no doubt that it has been sufficient. Moreover, it has allowed other birds, such as a number species of bird of prey that feed on migrating birds, to modify their own migration strategies to take advantage of the strategies of other species.

The arrival of passage migrants

Throughout the Mediterranean Basin, the arrival of migrants in the spring and autumn is one of the highlights of the annual natural cycle. The birds that arrive in the spring are a prelude to good weather and those arriving in autumn signal the beginning of the hunting season. In mid-January, sometimes earlier, the first swallows (*Hirundo rustica*) arrive on the shores of the Mediterranean Sea and they continue their staggered return until they reach central Europe in mid-April. The well-known swallow is so eagerly awaited that it is often used by meteorologists as a phenological observation, i.e. a periodical biological event or indicator. Most other insectivorous species wait until Mediterranean insect life is fully active, and last of all come the storks which arrive when they can find food in ponds and wet fields in the Mediterranean Basin, although they may well continue further north. In the autumn the largest flocks are those of the insectivorous species such as thrushes (*Turdus*), robins (*Erithacus rubecula*), and meadow pipits (*Anthus pratensis*), many of which are partial migrants. Humans wait for the arrival of the ducks, which normally fly high above the reach of guns, but which are hunted as they congregate in the wetlands. They provide abundant game for local inhabitants. Well into autumn, massive flocks of starlings (*Sturnus*), capable of completely carpeting their roosting sites, invade the skies of southern Europe. Normally, birds tend to avoid large mountain ranges and bypass the Alps to the east or the west. If they do have to cross mountain ranges, birds will generally choose mountain passes situated at lower altitude. For time immemorial humans in the Mediterranean Basin have taken advantage of the funnelling of birds through narrow mountain passes to hunt migrants. The hunting of the wood pigeon (*Columba palumbus*), for example, has become highly organized, the same shooting positions being used year after year and passed on from father to son.

Migrants do not only have to run the gauntlet of humans hunters; as mentioned before, other preda-

93 A booted eagle (*Hiera-aetus pennatus*) attacking starlings (*Sturnus vulgaris*) over the city of Barcelona. Some birds of prey in Mediterranean regions have adapted their breeding cycles to coincide with the migration of the birds that form a substantial part of their prey. The Eleanora's falcon is a good example of this in the western Mediterranean. This bird of prey breeds in the autumn in order to feed its young on the northern species of bird heading southwards. This same niche is occupied by the sooty falcon (*Falco concolor*) in the eastern Mediterranean.
[Photo: Joaquim Reberté & Montserrat Guillamon]

tors lie in wait. One of the most important is a bird of prey endemic to the Mediterranean Basin, Eleanora's falcon (*Falco eleonorae*). This falcon remains in the Mediterranean throughout the autumn and is one of the few birds to breed in this period of the year. It feeds its chicks on the abundant prey arriving from northern Europe and at the end of autumn migrates south and spends the winter in southern Africa. Eleanora's falcon commands the western Mediterranean flyway, while its close relative, the sooty falcon (*Falco concolor*), watches over the eastern flyway.

Other migratory movements

The word migration is usually associated with birds, and long-distance journeys of a regular and seasonal nature. This would exclude, logically, movements which follow no fixed rhythm, regular movements over short distances, and migrations carried out by animals other than birds. Nevertheless, these movements are just as important as, and play the same ecological role as, what is normally considered as migration. Their main consequence is to increase the trophic resources available to the animals, and they thus help stabilize populations. Ecological theory suggests that populations which change their spatial location obtain a certain degree of security in the event of a demographic catastrophe. Given the environmental heterogeneity resulting from the mosaic of landscapes in Mediterranean areas, these

regions are, therefore, ideal for a species to diversify its place of residence through regular or irregular movements. These may consist of journeys of thousands of kilometers in the case of migratory birds, or may well amount to a journey of just a few millimeters in the case of the soil microfauna.

Many non-sedentary Mediterranean animals carry out short seasonal movements. A large part of the bird fauna of the Australian mallee, consisting predominantly of small members of the Psittacidae, moves gradually between the north and south of the continent following the seasonal changes in the vegetation and, in particular, seed production. In other mediterraneans, the proximity of relatively high mountain ranges allows animals to take advantage of the variations in climate that accompany changes in altitude. Some animals carry out seasonal migrations in search of fresh pastures between high and lowland areas, or between neighbouring areas whose periods of maximum production are out of phase. Vertical bird migrants in the Chilean mediterranean include swifts, flycatchers, swallows and various finches, which are in the main insectivorous, partially migratory species that feed in the mountains or on the plain (as they also perform horizontal migrations), depending on when their prey becomes active.

The migrations of winged insects
As mentioned above, migrations are often thought to be exclusively performed by birds. Nevertheless, animals of other groups carry out large-scale move-

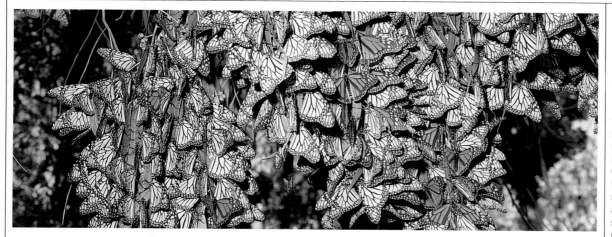

94 The monarch (*Danaus plexippus*) (above) and the plain tiger butterfly or African monarch (*D. chrysippus*) (below), are both migratory species of butterfly. The monarch performs one of the most spectacular of all migrations by winged insects. It migrates up to 3,725 mi (6,000 km) south from the northern United States and Canada to the southern United States and Mexico in search of a warmer wintering area. However, the most spectacular aspect of the migration is that the same individuals migrate north again in the spring, a phenomenon unknown in any other species of butterfly or moth. In Europe, the plain tiger butterfly or African monarch, an Asian and African species, also undertakes migrations of up to 1,240 mi (2,000 km) from Africa to southern Europe. During the 1980s and 1990s, large migrations have been recorded on the east coast of the Iberian Peninsula and groups made stops, above all in wetland areas. However, unlike the monarch, with the plain tiger it is not the same individuals that undertake the return migration.
[Photos: Jeff Foott Productions / Bruce Coleman Limited and Josep Maria Barres]

ments. The best known case among the land animals of the mediterraneans is the monarch butterfly (*Danaus plexippus*). This species has been well studied because of its seasonal journeys over thousands of kilometres between Canada and Mexico. Although the main migratory routes run inland down the United States and then to Mexico, many of the butterflies from the Pacific coast of North America spend the winter on the warm coasts of California and Baja California where they choose sheltered and wooded areas in preference to the matorral. Very often, numerous individuals are carried by the wind and appear in Europe (they breed in the Canary Islands), and by using the islands of the Pacific as stepping stones the species has spread as far as Australia. The monarch has a little studied European counterpart, a species of the same genus, the plain tiger butterfly or African monarch (*Danaus chrysippus*), native to the eastern Mediterranean, which migrates from Africa to Europe. Other species of butterfly carry out smaller-scale migrations between the Mediterranean Basin and central Europe.

The coasts of the Mediterranean Sea, home to this biome's typical forms, are the scene of all types of seasonal movements. Mountain ranges, generally parallel to the coast, encourage vertical migrations. Many of the animal groups in the Mediterranean Basin regularly migrate vertically and in many other ways. Amongst the insects, the plain tiger is not the only migratory species: other butterflies such as the painted lady (*Vanessa carduii*) travel to higher areas in spring and summer, as does the 7-spot ladybird (*Coccinella septempunctata*).

In southern California conditions are very similar to those in the Mediterranean Basin. The presence of nearby mountain ranges and coastal archipelagos has encouraged the formation of endemic insect species. These are, logically, less mobile, although the more widely distributed species carry out long migrations. The 3,725 mi (6,000 km) migration of the monarch butterfly has already been discussed. Ladybirds also carry out long-distance seasonal movements, just like many other ladybird species in mountain systems in the Northern hemisphere: they gather on mountain summits at the beginning of summer and move down in autumn. One Californian species, *Hippodamis convergens*, congregates by the million just before and after its annual journey.

The migration of fish and mammals

Long migrations are more difficult for animals that cannot fly. Nevertheless, large herbivores respond to the phenological changes in the vegetation by mov-

ing to areas with fresh shoots. These movements are most noticeable in mountain areas, where they consist generally of changes in altitude. In Mediterranean mountains, various species of wild sheep and goat such as the Spanish ibex (*Capra pyrenaica*), mouflon (*Ovis musimon*) and pasang (*Capra aegagrus*) move up to summer pastures as the snow retreats and return to the valleys with the first snow falls of autumn. The animals that perform vertical migrations also include non-mountain species more typical of Mediterranean forests, such as the wild boar (*Sus scrofa*), fallow deer (*Dama dama*) and, in California, the Californian mule deer (*Odocoileus hemionus*), a forest species that often lives in matorral. Transhumance (human movement of domesticated animals to different regions in search of seasonal pasture) probably originated as a result of observations in prehistory of the movements of wild animals.

In South Africa many bovids show migrations of this kind. One of the symbols of South Africa, the springbok (*Antidorcas marsupialis*), an antelope, in the past formed herds of millions of individuals that migrated seasonally in search of pasture in the karoo and nearby regions. Nevertheless, legal as well as illegal hunting has drastically reduced the species, although today it is protected in parks and reserves. The story of the springbok is very similar to the well-known story of the American bison (*Bison bison*).

Other flying animals also undertake long-distance migrations. For example, various species of bat take refuge in winter in sheltered caves and disperse in spring when their insects prey is once again plentiful. Bat migration has been studied by ringing individuals, as in birds, and migrations of hundreds of kilometers have been recorded. Some bat species in the Chilean mediterranean show a similar behavior, gathering in caves, while others (above all *Lasiurus borealis* and *L. cinereus*) carry out true migrations to the north. Non-flying mammals are less able to move large distances and tend not to be so dependent on insects or fruit. Many large herbivores have been introduced from the European mediterranean and often migrate seasonally in search of pasture. The red deer (*Cervus elaphus*), roe deer (*Dama dama*), Spanish ibex (*Capra pyrenaica*) and above all the wild boar (*Sus scrofa*) move up or down mountains with the seasons.

Possibly the most well-known wingless vertebrate migrant is the eel (*Anguilla anguilla*). The juvenile stage of this eel does not form a part of the Mediterranean biome, but the adult stage does when it enters the biome's rivers at the beginning of its upstream journey. Human interference has dramatically affected this type of migration as the construction of dams and reservoirs has greatly reduced the eels' chance of reaching the higher stretches of the watercourses. This process has occurred throughout the whole of the Mediterranean Basin and has led to the disappearance of the eel in the interior of many countries, including Spain.

Sedentary animals

Sedentary animals are defined here as those animals which move only short distances. In general, sedentary species tend to stay within their territories, if they have them, or to make irregular and erratic movements. Many show some regularity in their movements, and this may be linked with the cycle of the seasons. Many flying Coleoptera perform make synchronized daily flights between feeding and breeding areas. The small snails so abundant in Mediterranean matorral tend in the summer to climb plant stems and hundreds may gather on the upper branches and on dried-out inflorescences. Flies and mosquitos head for more humid areas and congregate in thousands on cool, shady walls, only becoming active again at dusk when the heat of the day has passed. Ground-dwelling fauna make short upwards and downwards movements in accordance with daily fluctuations in temperature. In general, these short movements are a response to the daily cycle of night and day (nyctohemeral). On the other hand, seasonal movements are linked to annual phases of activity and rest, determined by the alternating wet and dry seasons characteristic of Mediterranean environments. Animals journey to find a suitable place to shelter during the winter, or to spend the summer.

3.7 The rhythm of animal life

In nature, animals usually adapt to changes by anticipating them, if possible; but this is only possible if the changes occur with a certain regularity. For the fauna of Mediterranean and other regions, the most important regular changes are related to the alternation of day and night that produces circadian rhythms. Also very important, and even more complex, are the rhythms derived from the changing of the seasons.

On a time-scale somewhere between short annual cycles and the great Quaternary climatic oscillations, irregular changes occur in the Mediterranean biomes that are certain to repeat but lack periodicity. Generally, these changes are catastrophic and, unlike climatic changes, their effects are local. An area may remain unaffected for many years, but sooner or later these changes will occur after a period that is never more than a few centuries. Bearing in mind the inherent variability in the frequency of occurrence, the flora and fauna have been unable to predict these catastrophes by acquiring endogenous rhythms permitting them to take action in advance. Yet the certainty that they will eventually happen at some time or other means that there has been selection pressure that has favored the dominance of Mediterranean environments by organisms preadapted to these catastrophic accidents.

95 **The Moorish gecko (*Tarentola mauritanica*)** has become a species of human-influenced areas, living near humans and even in large cities. Under normal conditions, in garrigue, scrub, and predominantly open areas, this Mediterranean member of the Gekkonidae (the geckoes) is largely diurnal and most active in the morning. However, its contact with humans has made it nocturnal and it now specializes in capturing insects which fly around artificial light sources, which makes it an especially beneficial species for humans. [Photo: Javier Andrada]

Daily rhythms

Adopting a nocturnal life represents, for ground-dwelling creatures, a way of avoiding two types of dangers. First, they escape predation by the animals, such as birds, that hunt by sight. Second, they avoid desiccation caused by the heat of the sun. Competitive exclusion is another reason why many animals are active at night, since nocturnal activity allows them to occupy niches that are inaccessible to diurnal species.

All these factors play their part in mediterraneans. The main diurnal creatures are birds, reptiles and flower-inhabiting insects. Nocturnal animals are those whose sensory environment is based on their olfactory sense (mammals) or those that need more humidity to survive (epigeal insects and amphibians). Molluscs show no preferences, and are active whenever there is sufficient humidity. However, the need for water during the summer dry period induces many animals to change their behavior patterns, and so their nocturnal or diurnal character will depend on the season of the year.

Moreover, the daily activity peak may vary within the day or night, depending on daylength, temperature, and humidity levels. In the most northerly latitudes, the internal clocks of animals are relatively fixed, and since prehistoric times humans have used them to complement the observation of the sun to estimate the time of day. In the mediterraneans, biological rhythms are less stable, although they do still show definite patterns.

At dawn, the first animals to stir are generally small passerine birds. They roost communally in trees and begin the day with socializing calls and cover the branches with song. In America, hummingbirds are obliged by metabolic necessity to begin to search for nectar as soon as the day begins to break. In general, small birds have a much higher metabolic rate and need to feed with a greater frequency. Other diurnal birds become active only once the sun has actually risen.

Reptiles leave their nighttime shelters once the sun begins to warm the day, tending to remain inactive on cloudy and cold days. Nevertheless, during the hottest summer months, they return to shelter after a quick spell of hunting insects before the heat of the sun becomes too great, although they may hunt again in the evening or even at night. Many epigeal insects begin to move just before the reptiles stir and are followed by flower-inhabiting and flying insects once the day has got underway.

The activity of small mammals is discontinuous, in short bursts, alternating with resting in the shelter where they pass the night. Naturally, if the day is too hot or too cold, they will remain hidden. Shrews, on the other hand, are always on the move, almost constantly searching for food, and they tend to be most active at night. Large herbivores prefer feeding early in the morning once the sun has melted the frost, and later retire to the shade of the bushes or rocks to chew the cud. They feed again before the evening and generally rest at night.

During the afternoon and most noticeably at dusk as the sun begins to set, many insects vanish, generally to a safe shelter under a stone or, while flying species remain immobile on plants. Other groups, mainly dipterans (the "true" flies) and small lepidopterans that do not take nectar and need a moist environment, are mainly active in the evening or at night and only come out as the sun sets. At the same time, birds such as the nightjars, specialists in the capture of these insects, and insectivorous bats, the best known of all crepuscular animals, take to the wing. Some species that are now human commensals, such as Mediterranean geckoes, hunt mosquitos and moths, and thus their pattern of activity shows a peak in the first hour after sunset when flying insects gather around street lights and buildings.

Many mammals show most activity at night. This is true for shrews and mice, although they are also active by day, and for some bats that continue their crepuscular activity throughout the night. An record of activity patterns for various species of rodent in the Cape region shows very diverse patterns; most are nocturnal, even when they live below ground. It was thought that small carnivores (weasels, civets, mongooses, etc.) generally hunted by night. However, recent experiments carried out with radio-tracking devices have shown that much of their activity takes place during the daytime.

The members of the owl family (Strigidae) are highly specialized nocturnal birds of prey and only very rarely fly in daylight. However, they are not alone: other Mediterranean species such as the stone-curlew (*Burhinus oedicnemus*) also tend to be active at night. Nocturnal reptiles are scarce because of their heat requirements, while terrestrial amphibians (Bufonidae, Discoglossidae, Leptodactylidae, etc.) need humidity and are, therefore, forced to hunt and reproduce at night. Among the invertebrates, nocturnal groups include insects that live under stones such as nocturnal ground beetles (Tenebrionidae) and ground beetles (Carabidae), worms, and most spiders and snails.

There is least animal activity just before the sun rises. Some groups do remain active, but in general animals seems to rest before a new day starts. At this hour, large carnivores such as the lynx (*Felis lynx*), wolf (*Canis lupus*), and the puma (*Felis concolor*) begin to patrol. A little later on, as the sun rises, the daily cycle begins all over again.

Seasonal rhythms

It has already been commented in the section on migratory movements that the Mediterranean fauna undergoes regular annual changes. In fact, in the Mediterranean biome the two basic climatic resources, humidity and temperature, fluctuate throughout the year, but in opposing ways: while temperatures peak in the summer and are lowest in winter, changes in humidity levels are exactly the opposite. Thus, around the winter and summer solstices, climatic conditions are not very favorable for the fauna. The best times of year for wildlife are the equinoxes when temperatures are high and conditions are not excessively dry. This situation modifies annual patterns of activity and many species show an annual activity pattern that is bimodal, with two peaks instead of the single peak found in temperate and cold forests.

However, this pattern is not applicable to every species, rather it corresponds to the fauna as a whole. For species particularly, the period of maximum activity will depend on the resources needed and their degree of adaptation to the appearance of these resources, although many species have endogenous rhythms that are independent of environmental conditions. In Mediterranean animals, annual cycles differ mainly in the intensity and the timing of the rest periods. Hibernation is rarely total and may consist of no more than a few weeks' rest in a sheltered place, whereas aestivation (torpor during summer heat), absent in species at higher altitudes, is frequent and is usually longer and more intense than winter dormancy. Nevertheless, total activity is not the only element in the annual cycle: gamete formation, the different phases of the reproductive cycle, feeding, building-up of reserves, and in some cases, aestivation, molting and migration, are all factors in the annual cycle. Each of the factors has its own pattern in time, and this provides numerous opportunities for resource separation and use.

Obviously, differences also exist between the annual patterns of activity in the different Mediterranean

96 A barn owl (*Tyto alba*) catching a mouse. The members of the Strigiformes (the order consisting of the Strigidae, the owls, and the Tytonidae, the barn owls) hunt nocturnal animals and have developed an extraordinary sense of hearing capable of detecting the rustling of small rodents. Their vision is also adapted to darkness and their eyes are located on the front of the face to give stereoscopic vision. Nocturnal birds of prey, as well as other birds, regurgitate the indigestible parts of their prey in the form of pellets.
[Photo: Kim Taylor / Bruce Coleman Limited]

97 A group of emus (*Dromaius novaehollandiae*) migrating across stubble fields invaded by purple viper's bugloss (*Echium plantagineum*), an introduced European plant. The southern Australian fauna's activity is affected by the pattern of rainfall; in the winter, the emus head for the rainier parts of the southwest Australian mediterranean, where they compete with farmers. In their search for food in the winter they may destroy fields of wheat, bringing them into direct conflict with humans. To avoid killing the emus, specialists have suggested that some areas should be left unplanted so that the emus can find other types of food and to serve as a refuge.

[Photo: Jean-Paul Ferrero / Auscape International]

regions. In the Mediterranean Basin and in California, activity is at its peak in spring, in central Chile it continues until the beginning of summer, and in Australia, well into the summer. In fact, the flora of the Australian mallee, tropical in origin, seems to have conserved the patterns of annual development its Tertiary ancestors followed, and many species sprout in midsummer. This obliges animals to feed when it is very hot. The activity of the fauna of southern Australia is thus regulated to a large extent by humidity levels. In contrast, in the two mediterraneans of the Northern hemisphere, the secondary winter rainfall peaks are much smaller than those in the Southern hemisphere.

Spring activity

The cycle begins every year when hibernating animals wake from their winter lethargy. A number of plants, many of which produce shoots and flowers during the winter, are the first organisms to become active again when their endogenous rhythm has been stimulated into action by the increasing daylength. In animals, on other hand, awakening is caused by external conditions, mainly temperature. Nevertheless, many animals regulate their activity, especially their reproductive period, through the photoperiod, and so changing the photoperiod can advance or delay sexual activity. By means of this synchronization mechanism, mammals, for example, can initiate gestation so that birth occurs in spring, when there is an abundance of food for the offspring. Aphids and hymenoptera, amongst other animals, lay their diapause eggs before the winter so that the resulting parthenogenetic females will appear in spring and start the reproductive cycle anew. The most stable and fixed cycles are those of butterflies and moths since their caterpillars depend on specific plants, and their mating and egg-laying has to

adapt to the plant's coming into growth. The number of annual generations depends on the species and area and is always related to the availability of food.

Apart from winter-flowering species, the first flowers appear in January (in July in the southern hemisphere) and within a month the first insects, above all nocturnal species which need more moisture, begin to emerge. The epigeal fauna of myriapods and nocturnal ground beetles rouses itself, and, before spring arrives, many coleoptera are out and about eating fresh plant leaves. Flower-living beetles, nectar-gathering hymenoptera (bees and wasps, etc.) and butterflies appear when flowers are most abundant, in March and early April (September and early October in the southern hemisphere). By this time migratory birds have arrived and insectivorous species begin their breeding seasons by feeding intensely and preparing their nests.

In March most small mammals, which are at least partly active throughout the winter, renew their activity. Sufficient food is available for the insectivores and rodents to begin their breeding cycles. The secondary rainfall peak in spring also allows the continuation of the snails' winter activity and the start of the amphibians' activity, allowing them to lay their eggs one or two months before the summer begins. Large herbivores make use of the abundant spring biomass to suckle the young born after the winter gestation period.

Once spring is well underway, the animals towards the top of the trophic chain appear: snakes, birds of prey, and coprophagous species are all most active during the summer. The presence of recently-born litters of mammals and fledgling birds at the end of spring also leads to an increase in the presence and activity of small carnivorous mammals.

98 A beetle of the genus *Trichostetha* on a member of the protea family, a species of the genus *Leucospermum* in South Africa. When the spring flowering period arrives, the flower-inhabiting insects become active. As they visit flowers, they brush against the plant's stamens and stigmas and a certain amount of pollen sticks to the insect's body. This will then be left on the stigmas of other flowers the insect visits. This is why animal-pollinated plants have sticky pollen grains and colorful flowers.
[Photo: Colin Paterson-Jones]

Summer activity

By the beginning of summer the most intense period of vegetative growth is over, although many species are still in flower and others will continue into the autumn. This leads to a drop in the number of flower-inhabiting and nectar-feeding beetles, although many butterflies and moths remain active throughout summer after the May-June (November-December in the Southern hemisphere) peak, and parasitic and granivorous hymenoptera such as ants increase their activity. The abundance of dung left by large herbivores leads to an increase in the activity of coprophagous (dung-eating) beetles, as they accumulate food reserves. Defoliating beetles (*Melolontha, Polyphylla*) also tend to be at their most active towards the summer solstice. Small orthopterans (crickets, grasshoppers, etc.), capable of feeding on tough foliage, and many sap-sucking Hemiptera (Cicacidae, Pentatomidae, i.e., the cicadas and shieldbugs, etc.) are also most active in the sum-

mer. Carnivorous arthropods such as the ground beetles (which were preceded by herbivorous members of the same family), tiger beetles (Cicindellidae), and wolf spiders (Lycosidae) are common in summer when the Mediterranean insect fauna is still very varied and abundant. In stark contrast, snails, faced with the beginning of the summer drought, seal off their shells and take shelter under stones or at the top of the stems usually of herbaceous plants.

Turning to the vertebrates, amphibians are extremely limited by the lack of water. In summer, they only hunt at night and even then will not leave their cool shelters if the night is too warm. Like the snails mentioned above, amphibians aestivate in summer, and some species stay buried for a number of weeks. In Australia, two species of the Leptodactylidae, *Limnodynastes dumerili* and *Neobatrachus pictus*, spend the summer in mud-sealed underground cavities

99 The cicada *Lyristes plebejus* is common in the Iberian Peninsula. Chirring, or churring, the characteristic shrill sound made by cicadas is a typical noise of the Mediterranean summer. It is the male of the cicada family, the Cicadidae, that possesses the two stridulous organs (one on either side of the abdomen) that produce their characteristic song. The purpose of the male cicada's summer call is to attract females, but the "song" varies from species to species.
[Photo: J.H. Brackenbury / Bruce Coleman Limited]

to maintain their moisture. In the Mediterranean Basin other species adopt similar survival strategies: the sharp-ribbed salamander (*Pleurodeles waltl*) generally spends the summer in cracks in the dry mud at the bottom of ponds. Many animals will emerge briefly from these types of shelter with the typical late-summer storms, giving rise to the popular belief in it "raining toads." It has often been said to rain spadefoot toads (*Pelobates cultripes*), a Mediterranean species which sometimes suddenly appears by the million in areas wetted by a summer storm.

Reptiles play a very important role in Mediterranean communities in the summer. There are still sufficient insects around to feed on and their heat requirements are covered. They breed late on in the season and most species lay their eggs at the end of spring or at the beginning of summer, allowing them to produce a second batch in early October. They are active throughout the summer, perhaps taking shelter for a few hours during the hottest part of the day. During the long summer period there is great competition for food and various species specialize in order to overcome the problem. Each mediterranean, for example, possesses one or more species of ant-eating lizard that feed on the abundant supply of granivorous ants (the worm lizard [*Blanus cinereus*] in the Mediterranean Basin, or *Phrynosoma coronatum* in California). Snakes have to cope with the reduction in the populations of small mammals and young birds, so they change their feeding strategy and take small lizards, snails, and large

insects, and they remain active throughout the summer.

Insectivorous species form only a small percentage of the total bird population, which contains more frugivores and granivores. For example, pigeons (*Columba* sp.) and the turtle dove (*Streptopelia turtur*) reproduce relatively late, at the beginning of the summer when fruits and seeds are abundant, although like all other animals they have to search ever harder for food and water as the summer progresses. The concentration of animals near irrigation ponds and permanent springs makes them easier to hunt and, therefore, represents a risk for them. Birds of prey also modify their diets and come to depend basically on reptiles and not on the small mammals they mainly hunted in spring. Small mammals become less and less common until their populations reach their smallest size in the autumn. This also forces the small Mediterranean carnivores to change their diets, and towards the end of the summer and autumn even the most carnivorous species, such as the genet (*Genetta genetta*), have to include a substantial amount of fruit in their diet. Clearly birds of prey may frequently resort to feeding on animals run over by vehicles on roads. In general, all Mediterranean vertebrates except for reptiles, decrease or interrupt completely their activity during the hottest days of summer, the dog days.

Autumn activity
Autumn is the rainy season in Mediterranean regions; it begins with heavy storms in mid-summer. As

100 Common frogs (*Rana temporaria*) in the Montseny mountains (Catalonia). The western Mediterranean Basin is the southern limit of the range of the common frog, as the temperature and summer droughts prevent this species from colonizing further south. In the Mediterranean Basin the common frog hibernates from October or November until the end of February. It probably hides at the bottoms of a pond or under the same stones it uses in summer to shelter from the heat. Pairing takes places when the animals finish hibernating, from the end of winter until spring.
[Photo: Josep Maria Barres]

autumn begins warm air over the sea causes the Mediterranean region's typical rainfall regime of short, intense, irregular rain. Although most animals and plants have already finished their annual cycle, the renewed availability of water often encourages plants to shoot again and animals to finish their aestivation.

For some animals, such as small reptiles, the renewed autumnal activity of the environment provides a second opportunity to reproduce and increase populations. They and other vertebrates make use of this opportunity to accumulate food reserves and to prepare for the winter cold or, in the case of birds, to prepare for migration. Activity during this period seems not to depend on available resources, but on endogenous rhythms synchronized with the number of daylight hours and the organism's hormonal status. Even if temperatures are still high and food plentiful, when animals have accumulated sufficient reserves, they find a shelter where they will spend their winter resting period. It has been shown that in autumn Mediterranean reptiles become less active and enter into deep hibernation at temperatures higher than those they tolerate in spring when they begin their breeding cycles.

Winter activity

The first cold days of winter interrupt the short-lived relief for aestivating animals provided by autumn rains. The weather is still good in the Mediterranean Basin in early October, when far to

the north the first snows have already fallen, forcing the very last migratory birds to leave. The arrival of these birds in the Mediterranean Basin or in California, whether as temporary visitors or for the whole winter, is a signal to the local resident fauna that winter has begun.

Poikilothermic animals, those whose body temperatures vary with that of their surrounding medium, are obviously more sensitive to low temperatures and are forced to hibernate, especially if they are relatively large. Large snakes normally hibernate, whereas small lizards show a reduced dormancy and emerge from their shelters on sunny days, even in mid-winter. Adult insects act similarly, although their populations are very reduced, since most species spend the winter in the form of eggs or in other resistant phases. On the other hand, winter is a gentle season for amphibians, except in sub-Mediterranean mountain areas or at high latitudes. Thus certain species modify their annual biological cycles and breed in mid-winter rather than in spring, as is normal in other members of their group. This behavior occurs not only in typically Mediterranean families such as the discoglossid toads, but also in species from higher latitudes at the southern limit of their distributions in the Mediterranean Basin. For example, the common frog (*Rana temporaria*), the most northerly distributed amphibian in Europe, spawns in spring and summer over most of its range, but in some areas of the Iberian Peninsula it spawns in December.

4. Life in rivers and lakes

4.1. Changing aquatic systems, adaptable flora and fauna

The main characteristic of Mediterranean aquatic systems, which results from the fact that the hot season coincides with the dry season, is the irregularity of the flow of water and the consequent variations in the levels of lakes and natural pools. The water regime shows peak flow in the autumn, associated with the main period of rainfall in both the northern and southern hemisphere, except for Australia where the rainfall peak is in winter. Wetter areas may have two peaks, one in spring and another in autumn, as occurs in Mediterranean watersheds with headwaters high in the mountains (such as the Pyrenees, the North American Sierra Nevada and the Andes), where the springtime increase in flow is a direct consequence of the snow melting in the headwaters.

Watersheds with a strictly Mediterranean water regime show a series of common general characteristics. They are short (except for the partially Mediterranean watershed of the Murray River in Australia), show very sharp changes in flow and have a clear tendency to be seasonal. In the hot period the bed is totally or partially dry, or flow may be intermittent. The most seasonal rivers are the ramblas of the Iberian Peninsula and the African wadis, which are dry for most of the year. The intense dry periods are usually accompanied by potentially catastrophic seasonal surges or floods during the period of maximum rainfall, which may be autumn or winter, depending on the continent. However, floods may also occur at other times of year, especially in summer. These river systems consist of two distinct zones. The upper zone, which is known as the *rithron*, has a steep slope, a substrate of pebbles and turbulent flow, and corresponds to the stretches in the headwaters of the watersheds that drain mountainous systems, whether low or high. The rest of the river, which is known as the *potamon*, with a gentle slope or no slope at all, shows laminar flow over a gravel and sand substrate, and consists of the part of the river that flows along the flat land to the sea or a landlocked basin. The relative importance of the *rithron* and the *potamon* depends on the basin's geomorphological structure. The entire range of intermediate situations is found in the Mediterranean bioclimatic area.

It is usually the stretches of the potamon, with their alluvial plain, that suffer and record the effects of surges and floods most clearly. Even the rivers that have large drainage areas high in the mountains, ensuring flow during the dry season, are affected by the seasonal regime, which means that the lower reaches are subject to regular floods that fertilize riverside cultivated land or wash it away, depending on the magnitude of the surge. Complex alluvial plains, like that of Australia's Murray River, may show formation of "billabongs" isolated stagnant pools, which may be abandoned meanders, pools that have flooded as a result of the freshet, or pools derived from the surface water table, etc. and which in the rainy season receive the force of the water, of unpredictable strength and flow.

The natural composition of water depends on its mineral content, and this in turn depends on the composition of the rainfall (which varies from area to area), on the transport of salts in the air, on the flow of underground water to the surface, and on the solubility of the rocks forming the drainage basin's substrate. If the substrate is calcareous or evaporitic, salt levels are higher. The Australian mediterranean's basins have the highest salinity values (500–3,000 mg of salts per liter of water), except for the headwaters of the Murray River, whose water is less salty than that of the other rivers of the Australian mediterraneans. The salinity values found in the very short rivers of the Australian province of Victoria depend on the composition of the rainfall, but in much of the Murray's basin the salinity is brought from salty rivers from the mallee zone. In the other mediterraneans the average salinity of the water is 120 mg/l, much less than in Australia. The degree of salinization shown by enclosed lake basins depends to a great extent on the water's rate of renewal. Thus, where the geological development of the landscape and climatic conditions lead to the formation of large areas lacking any drainage system, the runoff cannot follow a clear route to the sea and thus accumulates, forming puddles and shallow pools of *athalassohaline* water (salty water that has never been in contact with the sea). If evaporation is not greater than precipitation, the flooding lasts all year round, but often these endorheic areas (areas that do not reach the sea) are subject to annual cycles of flooding and desic-

cation, or even cycles lasting more than one year (for example, the Laguna de Gallocanta lake, in Spain).

In the areas with karstic relief in the Mediterranean Basin, the calcareous rock dissolves leading to the formation of cavities that are continuously widened and extended by the flowing water, which as a result contains many dissolved salts. Lakes may form in these systems when karstic depressions sink, and as their water is continuously being renewed they do not become very saline. Tectonic movements may occur in karstic lakes, and in some cases they cause the bottom to collapse periodically, as happens in the large, deep lakes of the Mediterranean Basin, such as the Dead Sea, which is very highly saline because of its evaporation-precipitation balance and the composition of its substrate. Karstic aquifers are very important throughout the Mediterranean Basin and in many cases form rivers with springs called resurgences that often reach the surface with a high artesian pressure. Their flows are, to a greater or lesser extent, independent of the rainfall, and depend more on the depth and the surface area of the associated karst (for example, the headwaters of the River Jordan or those of the River Llobregat just south of Barcelona, Spain). These waters are inter-

esting because they are the refuge of the last survivors of the biota that originated in the waters of the ancient Tethys Sea. The quantity of water and its distribution in time are largely the consequence of the Mediterranean climate, but they also depend on the terrestrial systems they are linked to. All of the water that the rivers receive, except for direct rainfall, has come from the terrestrial systems in its watershed through which it has flowed previously. The distribution and area covered by forests, their floristic composition, the pattern of organic material input from the watershed, and the presence and development of the riverbank vegetation, all affect the water's composition and dynamics, and the biology of its ecosystems. The Mediterranean forests of *Eucalyptus* affect the colour and acidity of Australia's mainland waters, and this means that the limnological characteristics of Australia are distinct from those of the other Mediterraneans. For example, the *Eucalyptus* forests are responsible for the water's dark, earthy color that is typical of the phenolic compounds produced in leaf, bark, and fruit decomposition, as well as the water's greater acidity (pH between 5.3 and 6.5). In the other Mediterranean areas most waters are basic (pH values above 7.8) and clear.

101 Adaptation to the highly seasonal water regime of the Mediterranean climates is a common characteristic of species adapted to aquatic environments. Temporary flooding, due to surges, forms ephemeral pools. But this does not mean they lack communities of aquatic organisms, or that they lack importance in the site's annual cycle, especially since they fertilize the system. The photo shows a temporary wetland in Western Australia with riverside eucalyptus (*Eucalyptus largiflorens*), helophytes, such as *Eleocharis*, that are well-adapted to seasonal changes in water supply, and fully aquatic plants, such as *Marsilea drummondii*. The latter is a fern that roots in the mud by means of its creeping, highly branched rhizome and grows on the surface of the water in times of flooding.
[Photo: Wayne Lawler / Auscape International]

On the other hand, the vegetation of the drainage basin not only affects whether the river systems are autotrophic or heterotrophic in nature, but also influences the input of organic material from terrestrial systems. In general, rivers flowing through forests are basically heterotrophic, that is to say, they are dependent on the input of organic material from the forests, not only because of the quantity of the input, but also because the riverside vegetation totally or partially covers the river's bed, thus reducing the light supply. If the river is wide, this sunshade effect disappears and the production of algae and larger aquatic plants may equal or exceed the input of organic material of terrestrial origin, and in this case the river behaves as an autotrophic system. In Mediterranean systems, the *rithron* is heterotrophic, while the *potamon* is autotrophic, since the forested stretches were originally found in the mountain headwaters. The Chilean mediterranean's rivers are an exception to this, as their headwaters are above 8,200 ft (2,500 m) and lack forests. Another exception is the basin of Australia's Murray River, as it has forested areas in the center of the alluvial plain. In Mediterranean systems the rivers even export primary production to terrestrial systems. Sometimes, masses of aquatic vegetation are stranded and used by different groups of organisms in rivers, lakes and seasonal billabongs, etc., during the dry period. On other occasions, the water plant masses are dispersed by flooding and deposited outside the aquatic system. Except in the case of Australia, the Mediterraneans show a peak input of organic material from deciduous forests in the autumn, coinciding with greatest water flow. This high flow permits rapid transport and distribution of the material introduced into the system's and the rapid decomposition of the main component (the leaves, which represent more than 60% of the material) in a short period (2 to 3 weeks in systems in the Iberian Peninsula). This prevents the growth and permanent settlement of strictly triturating organisms in these systems. However, the Australian mediterraneans show peak input of materials from the *Eucalyptus* forests in the summer, when river flow is low and when seasonal watercourses may even be dry. Both these situations, caused by different processes, postpone the availability of organic material for consumption by triturating organisms.

The biota of the continental waters of the different mediterraneans has the same adaptive strategies, as they all live in similar conditions in areas that are continents apart. Yet, as happens with the terrestrial flora and fauna, these biotas do not have species in common, and often not even the same taxonomic groups. For example, major groups of organisms that are abundant in the northern hemisphere, such as the Salmonidae (the salmons) or the species of the genus *Perla*, and those of the Perlidae (stoneflies) and Perlodidae are replaced by other families in the southern hemisphere, or in the southern region.

The effect of human activity

Human intervention in natural aquatic systems goes back to antiquity, and in the mediterraneans it is very widespread. Many ancient civilizations arose in the arid and semi-arid Mediterranean valleys suitable for agricultural exploitation. The need to construct reservoirs and irrigation channels to distribute water for urban and agricultural use, complemented by aqueducts in the Roman period, as well as the collection and channelling of sewage in major settlements, and even dams and industrial usage, have as a whole created a water use culture favoring maximum exploitation and the development of large-scale hydraulic engineering projects. The consequence of this today is the extensive system of reservoirs along the Mediterranean's river systems, built to store water and to prevent flooding by controlling water flow. These seasonal flood prevention schemes are most destructive when they turn rivers into straight canals with treeless banks, as they increase erosion and the destructive effects downstream. The large-scale urban settlement of the world's Mediterranean areas, with major population centers on all the continents, is a consequence of their mild climate, and this, together with the conversion of large areas of forest and matorral into irrigated croplands, has so profoundly transformed the Mediterranean regions that it is now difficult to recognise any trace of the original landscape. There have been major erosive processes, and even desertification (especially in the Mediterranean Basin), together with pollution by urban, agricultural, and industrial discharges. The overall result is major transformations in many of the Mediterranean's aquatic systems, e.g. stagnant, polluted rivers, modified river beds, eutrophic lakes and reservoirs, salinization of watercourses and of surface waters due to irrigation, etc. The introduction of non-native fish into the Mediterranean's aquatic environments for sport or for commercial reasons, made easier by the climatic similarities between the world's different mediterraneans, has been one of the reasons for the impoverishment of the river fauna. The carp (*Cyprinus carpio*), from the Near East, has spread throughout the Mediterranean Basin and has been introduced into California, Chile, and Australia. Other species sharing this artificial distribution include the common trout (*Salmo trutta*) and the topminnow *Gambusia holbrookii*. The level of alteration present in the fish community varies greatly from country to country and from basin to basin. In the Iberian

Peninsula as a whole the percentage of non-native species is 36%, but in Catalonia it reaches 47%.

4.2 Mediterranean aquatic environments

Living dangerously in rivers, torrents and ramblas

Mediterranean rivers' water regimes, with their dry seasons and periods of peak flow associated with surges, determine the wide range of conditions the species living there must adapt to. In permanent rivers and watercourses, the insect fauna is basically dominated by species with more than one generation per year. Many members of the caddis-fly order, Trichoptera, (such as the tiny grazing hydroptilids) produce up to four generations a year; although due to different heat tolerances, some species produce a single generation per year in the winter or in the summer (members of the Baetidae in cold water and of the Ephemerellidae in warm waters). Two well-defined insect communities alternate over the year. One appears with the floods, in autumn or winter depending on the area, and lasts until the middle or end of spring, when rheophilous species (those that thrive in running water) dominate the benthic animal communities. The second insect community is typical of the summer, and its species typically show life-cycles so synchronized that when the flood begins they are either dormant (as resistant eggs) or in the flying stage of their life cycle and outside the aquatic system. In some groups, including many coleopterans and heteropterans, the adults—the only ones present in the watercourse at the beginning of the flood period—can escape the river systems and seek shelter somewhere else or, like the coleopterans of the family Elmidae and the ephemeropterans of the family Ephemeridae during their first stages of development, they may burrow into the sediments of the river bed. In Mediterranean rivers, the new water cycle begins with the flooding of the river system and a rise in water levels, resulting in the appearance of the winter communities of insects. In rivers in the Iberian Peninsula, within only 20 to 25 days most of the species forming the winter community may be found. The fish communities also show the ability to adapt to the environmental variability of the rivers and to persist after surges, floods, and droughts. In short, with seasonal rivers the composition of the fauna is very similar along the entire course, except at the mouth, and sometimes, in the headwaters. In the permanent, more structured basins a larger number of species occurs, and a certain kind of distribution may be observed, determined by biological, biogeographical, and physical factors (the river's current, water temperature, and arrangement). During the dry period, the

102 **Fishing "net" set by** **_Hydropsiche_**. The larvae of some species of the Trichoptera use silk-fibers to weavefishing "nets" to trap food borne downstream on the current. The nets of _Hydropsiche_ are well known and as the larvae grow older they weave nets with larger and larger mesh-size, so that they abandon the microphagous eating habits of their early larval stages and develop into carnivores. Freshwater trichopterans are common and important especially in flowing water, where they form the largest fraction of the biomass. The silk they produce allows some species to build hiding places and mobile sheaths.
[Photo: Freider Sauer / Bruce Coleman Limited]

animals concentrate in pools where they remain until the river bed is once more flooded, allowing them to expand. The two regions with Mediterranean climates in the northern hemisphere have similar faunistic models. The dominant species in the Mediterranean Basin, except in its mountain areas, are basically members of the Cyprinidae (80%)—_Barbus_, _Leuciscus_, _Chondrostoma_, _Rutilus_, _Anaecypris_, _Iberocypris_, _Pseudophoxinus_, etc.—together with other groups, such as the cobitids (loaches). The rivers of the California mediterranean are also dominated (30%) by members of the cyprinids (carps and minnows; _Gila_, _Rhinichthys_, _Lavinia_, etc.) and other groups, such as the cottids (sculpins; _Cottus_) (28%). In the southern hemisphere, in Chile between the basins of the Limari and the Bío-Bío, for historical but also ecological reasons, the number of species is very low, and the most important typical freshwater species are members of the Pygididae (_Pygidium_), Characidae (characins, such as _Cheirodon_), Atherinidae (silversides, such as _Basilichthys_), and Trichomycteridae (parasitic catfish, such as _Trichomycterus_). In Australia there are many species of galaxias (Galaxiidae), also found in South Africa and at the southern tip of South America, Percichthyidae (striped bass, such as _Maccullochella_, _Macquaria_) and Kuhliidae (flagtails, such as _Nannoperca_, _Nannotherina_). The galaxiids' distribution is basically the result of continental drift. Linked to these environments are some vertebrates that are not strictly aquatic. For example, the frogs are represented by _Rana perezi_ and _R. esculenta_ in the Mediterranean

103 *Lepidogalaxias sala-mandroides* **emerging from the hole it had burrowed,** in Western Australia. Some Australian galaxiids bury themselves in the moist mud of temporary rivers in order to survive the dry period. It is said that their mud-digging nature allows Australian aborigines to gather them directly from the mud, without even having to fish for them. The galaxids are the most characteristic fish of Australia's temperate rivers and lakes, and show an interesting biogeographical distribution. They are restricted to Australia, Tasmania, New Zealand, South America, South Africa, and New Caledonia. They are small fish, measuring less than 6 in (15 cm) long. Many species are migratory and almost all are found in freshwater. *Lepidogalaxias sala-mandroides* has been largely displaced by the topminnow (*Gambusia affinis*), an American species that has spread over the entire world. [Photo: Reg Morrison / Auscape International]

Basin, *R. aurora* and *R. catesbiana* in California, and *Heleioporus eyrei* in Australia. The reptiles are represented by water snakes, such as the genus *Natrix* in the Mediterranean, *Diadophis* in Australia, and *Austrelaps* in Australia. There are also freshwater turtles, also called terrapins, such as *Mauremys caspica* and *Emys orbicularis* in the Mediterranean Basin. Some mammals also exploit these aquatic systems, for example, the otter (*Lutra lutra*).

Seasonal rivers follow a simpler model than permanent ones. Those that remain dry for part of the year (4 or 5 months) are colonized by communities similar or equivalent to the winter communities in permanent rivers, with a few different species. The greater the seasonality, the greater the domination by more cosmopolitan species with short lifespans, such as the dipteran midges and flies (chironomids, simuliids and ceratopogonids). In extremely seasonal rivers (dry for ten months a year or more), and in the pools left after the flooding typical of ramblas and wadisin arid areas (the southeast of the Iberian Peninsula, North Africa, South Africa, and the transition zone between California, Arizona, and Mexico), the community that is established corresponds to a seasonal lentic (ponds and lakes) system, with microcrustaceans, mosquitos, water bugs, etc. This seasonality also means that there are fewer fish species in moderately seasonal rivers and even fewer in ramblas where, at most, fish survive in the pools near the mouth (Ciprinodontidae and *Gasterosteus*). Thus, in Mediterranean regions, the longer period the river bed is dry, the greater the decrease in species diversity. Yet there are no major differences between permanent rivers and moderately seasonal ones that retain a system of more or less persistant puddles and deep pools throughout the summer period.

Resisting the current in mountain zones

The varied community microalgae of the *rithron* that develops on the substrate of rocks, blocks, and pebbles is basically dominated by diatoms and mosses capable of resisting partial desiccation. The organic material input from the forests is rapidly decomposed and transported as small particles in suspension or retained in the areas of sedimentation in the deep pools that alternate with waterfalls. These areas are dominated by filter feeders, such as simuliids and trichopterans (members of the families Hydropsychidae, Helicopsychidae and Leptoceridae) able to catch these small particles by making nets or using appendages with combs of bristles (Simuliidae and Leptoceridae). Plant-detritus eaters are also abundant, alternately consuming algae and particles trapped by the sediment or that have settled on it; one example is the mayflies of the families Caenidae and Ephemerellidae, the coleopterans of the Helodidae and most chironomids. Other organisms typical of river stretches in mountain areas are the grazers adapted to currents, such as the Ephemoptera with flattened bodies (Heptagenidae in the northern hemisphere and in South Africa, and Leptophlebiidae in Australia and Chile), molluscs (*Ancylastrum*), and some species of small coleopterans of the Elmidae family, trichopterans of the Glossosomatidae and those of the genus *Baetis*, which can withstand strong currents. The dominant predators are the walking invertebrates, such as the large plecopterans (*Marthamea* and *Eoperla* in the Mediterranean Basin, and different genera of Grypoptrigydiae and Eustenidae in South America and Australia), or the trichopteran Rhiacophylidae. In these mountain stretches, the fish are northern hemisphere members of the Salmonidae; the common trout (*Salmo trutta*), in the Mediterranean Basin and the rainbow trout (*Salmo gairdneri*), *Oncorhynchus clarkii* and *Cottus* in California.

Steady flow in the plains at lower levels

In the stretches of *potamon*, the river flows along a very low gradient and its bed consists totally or partially of a soft substrate. The water circulates without turbulence, with a laminar flow that allows the development of submerged meadows of macrophytes (*Potamogeton*, *Myriophyllum*) or, in areas with substrates of a larger grain size, large masses of filamentous algae (*Cladophora*), which act as a support for microscopic algae (diatoms) and a refuge for much of the mobile macrofauna (cladocerans, copepods, ostracods, acarids) to hide from predators such as odonates (Agrionidae, Hemiphlebiidae), heteropterans (Notonecnidae, Gerridae) and fish. In these zones, fine particulate materials tend to settle, allowing detritus eaters to dominate, mainly the mayflies (Ephemeridae, Polymitarcidae) which tunnel into the soft substrates.

When there are forested areas near a river, as in Australia, there are many grinding plecopterans (Nemouridae, Notonemuridae), and some species even take part in the decomposition of the remains of trunks and branches (Austroperlidae). In these parts of the river cyprinids are prolific, such as *Barbus*, *Rutilus*, *Chondrostoma*, and *Leuciscus* in the Iberian Peninsula. They have specialized feeding mechanisms that consisting of a protractile jaw with pharyngeal teeth; most eat small aquatic insects, crustaceans, grubs, and detritus in variable proportions.

Surges and floods introduce moments of peak flow into this apparently complex and highly structured community organization, mobilizing the substrates and uprooting some of the submerged plants. The effects of freshets on the fauna of these systems are only partially known, but they appear to be related to the predictability of the different events, so that annually-occurring surges do not destroy the structure of the community. However, insect communities take from three months to a year to recover from the effects of a surge, while fish communities take even longer to restructure.

Residual pools, refuges in the dry season

The summer dry period typical of the Mediterranean climate conditions the persistence of most watercourses to such an extent that most have some seasonally intermittent stretches. In these stretches, which are present mainly in the lower stretches of short rivers, the river turns into a chain of more-or-less residual pools.

Organisms lacking any other strategy to get through the dry period take refuge here, such as fish (which may reach densities of 25 individuals per cubic meter) or species with a summer cycle that finishes in these pools. The pools are sporadically visited by flying animals (adult coleopterans, heteropterans) from lentic systems, artificial pools, pools on alluvial plains (billabongs), lakes, and reservoirs.

104 **Viperine snake** (***Natrix maura***), a species from the Mediterranean Basin. Many of the predators of Mediterranean lakes and rivers are birds, mammals, and reptiles that alternate between terrestrial and aquatic life. The viperine snake eats, mainly within the water, prey such as fish, amphibians, water snails, and leeches, but it also leaves the water to catch oligochaete worms and even small mammals. It is thought that the Romans, because of their religious customs in which snakes were considered benefactors, helped to introduce it to the islands. It lives in temporary waters, preferably still ones, and also in the brackish waters of low-lying areas.
[Photo: Javier Andrada]

105 Salt lake in Western Australia. In Western Australia salt lakes are very abundant, both on the coastline and in the interior. They are almost all seasonal and shallow. Their salinity varies greatly, from 3 parts per thousand to 350 parts per thousand, i.e., ten times saltier than seawater. The fauna of these waters consists of halobionts, typical of salty water (more than 50 parts per thousand), and halophilic organisms (organisms of relatively salty water, between 10 parts per thousand and 60 parts per thousand) and others that tolerate salinity (less than 20 parts per thousand). In general, the higher the salinity, the fewer the species that live in the water. In the saltiest water, the larvae of ephydrids (*Ephydra*) and other salt pan flies dominate in the absence of competition, and they have even been used as food by humans.
[Photo: Jaume Altadill]

Billabongs, lakes on alluvial plains

Australia's billabongs are more-or-less permanent pools of water associated with a river course, but they only make contact with the main river during a surge in the rainy season. They are very rich in aquatic vegetation, and especially floating plants (such as the liverwort *Riccia*, the duckweed *Lemna*), some of which (for example, the fern *Azolla*) are favored when the dissolved nutrients have been exhausted. Their animal communities are very varied and differ greatly from those in the nearby main watercourse (280 species live in the pools but only 112 in the river). When the river's level has gone down once more, some billabong species will by then have colonized the main watercourse, displacing the species that were originally there, such as as the bivalve *Velesunio ambiguus* for example which has replaced *Alathyria jacksoni*. Australia's billabongs shelter many introduced species, which may turn into invasives, such as the liverwort *Riccia* and the water hyacinth (*Eichhornia*). Non-native fish species (*Salmo*, *Gambusia*, etc.) may even eliminate the native species, either by predation or through competition.

Variable salinity and heat in shallow inland pools

Salt lakes have their own special annual dynamic, because on to the hydrological cycle of water shortage followed by plenty due to abundant rainfall, a very marked cycle of salt concentration and dilution is superimposed; in the summer low-water period, the high rate of evaporation leads to an increase in the water's salt concentration, which may reach two or three times that of sea water (34-35 parts per thousand). The intense sunshine on the pool has a further effect on its organisms—an in increase temperature. This is because the water, which forms a thin layer over the sediment, gradually warms up and may reach 104°F (40°C) or more. So, obviously, conditions in these pools are very special. In the summer these pools may easily dry out completely, turning the formerly water-filled basin into a broad plain covered with a crust of salts. This type of temporarily flooded *athalassohaline* pools are abundant in the Mediterranean area, while in other parts of Australia permanently flooded ones are more common.

The salinity of athalassohaline pools may be due to the presence of greater quantities of some ions than others (the carbonates, chlorides and sulfates of sodium, potassium, calcium, and magnesium and combinations of these), and the dominant ions show a regional distribution: Australian salty lakes tend to be dominated by chloride ions, while those in the Mediterranean Basin (such as the Jordan Basin) are mainly carbonate-rich and strongly alkaline. These waters usually have high pH values, which may vary between 7.5 and 11. In summer the salt content of the water leaves a record in the sediment where different salts precipitate depending on the different proportions of ions in the water (aragonite, gypsum, dolomite, mirabilite, epsomite, halites, etc.). Evaporitic materials are dissolved at regular intervals and then redeposited, thus forming a concentric zonation of minerals from the least soluble to the most soluble, from the edge to the center of the basin. The dry sediment is covered by a hard crust of salts and contains the seed bank that will recreate the plant communities.

Life in salty Mediterranean pools forces organisms to adopt strategies to cope with the major changes in salinity and temperature that occur during the yearly cycle. These communities generally have few species, yet over the year they may show a succession of organisms increasingly adapted to the strong salinity and high temperature. A carpet, typically consisting of algae and microorganisms (heterotrophic and autotrophic bacteria), forms over the sediment and especially over the shallow edges of the water, as do extensive meadows of charophytes that completely cover the basin, especially *Lamprothamnium papulosum* and *Chara galioides*, sometimes accompanied by species of the monocotyledon genus *Ruppia*, which can also support salinity higher than that of seawater. When the dry season approaches, the charophytes produce numerous oospores and leave a reserve in the sediment that can recolonize the basin during the next flood, while *Ruppia* has a survival strategy in the form of roots buried in the soil. The most important phytoplankton species are *Nannochloris* and *Dunaliella salina*, while the zooplankton is dominated by the ciliate *Fabrea salina* and the genera *Hexarthra*, *Asplanchna*, *Keratella*, several cladocerans and copepods, but also many species that are often cosmopolitan (*Brachionus plicatilis*, *Cyclops viridis*). The Dead Sea is an extreme case of salinization, with a salinity of 280 parts per thousand (compared to 34-35 parts per thousand for seawater!) and a depth of 1,050 ft (320 m) of highly alkaline waters with a very high concentration of calcium, magnesium, potassium, and bromine. This is the preferred habitat of *Dunaliella salina* (despite its name) and of halophilic bacteria, which are aerobic in the upper layers and anaerobic in the underlying ones. Due to its depth and high salinity, which both affect the water's density, this lake does not form a thermocline every year, as generally happens in temperate lakes; the Dead Sea shows cycles that are longer and the period of stratification (formation of distinct temperature layers) may last for several years.

The salty waters of estuaries and coastal marshes are exploited by the North American topminnow *Gambusia affinis*, which has been introduced into all the continents; in Australia's lakes this fish represents a serious danger for the endemic species of the equivalent habitat, mainly the member of the Galaxiidae *Galaxias maculatus* and *Atherinosoma microstoma*. Both these species can tolerate high salinity and high

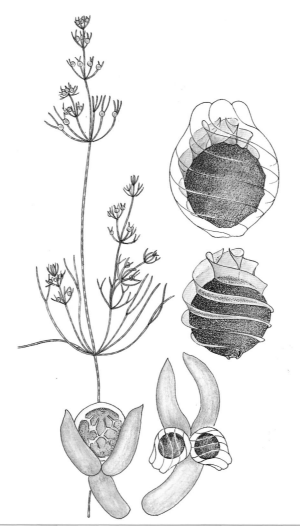

106 **Oospore (above) and vegetative structure (below) of *Nitella opaca*.** Many organisms of temporary waters form resistance organs in the dry season. The charophytes or stoneworts, a group of algae found in fresh and brackish water, form oospores less than 1 mm long that are protected by a thick wall and can survive the Mediterranean summer's long dry period in the dry mud. The oospores of *Nitella* can lie dormant for many years. The vegetative body is organized in nodes and internodes and can reach 3 ft (1 m) in length in deep permanent waters, while in temporary lakes caused by flooding it roots in the mud at the bottom, forming small cushions. It is a widely-distributed species that grows in the winter, completing its entire life-cycle in a few months, leaving behind it resistant oospores to rebuild the population in a few days when the autumn rains arrive. Their sensitivity to high levels of phosphorus in the water, as caused by pollution by detergents, make the charophytes useful indicators of the health of aquatic ecosystems.
[Photo: Montserrat Comelles; drawing: Miquel Alonso / ECSA]

107 **Karstic pool** formed by the Fuentona de Muriel spring in the Sistema Ibérico (NE of the Iberian Peninsula). Karstic aquifers are present in all the mediterraneans and gather and channel underground water by means of interconnected conduits leading to the point of emergence, which may be a fountain, a spring or a pool. The relative permanence of water in karstic systems, with an underground supply, allows aquatic vegetation to develop throughout the basin. However, the form of a karstic flooded basin, usually lacking a bank with a gentle slope, generally prevents the growth of a belt of helophytes on the banks.
[Photo: Rafael Vela]

temperatures, as can the endemic isopod *Haloniscus searlei* and the larva of the chironomid *Tanytarsus barbitarsis*. The Anostraca (fairy and brine shrimps) are typical members of coastal athalassohaline systems: the cosmopolitan species *Artemia salina* is found in all the continents except Australia, where it is replaced by *Parartemia*.

Seasonal pools with moderate levels of mineralization that vary over the year have more species. They easily develop concentric areas of vegetation around the basin. The outermost ring of helophytes (*Phragmites*, *Juncus*) is replaced towards the center by semiaquatic plants that root in the mud: water buttercups *Ranunculus* may flower so abundantly they cover the water's surface with a circle of white flowers. In the first stages of flooding, meadows of pioneer charophytes are usually formed (by the genus *Tolypellia* in highly mineralized waters, *Nitella* in purer waters), which are later replaced by other charophytes or completely aquatic flowering plants, such as *Zannichellia palustris* or *Potamogeton pectinatus*, which can both stand very high levels of salinity.

The deep, still waters of karstic lakes

In the Mediterranean world, karstic lakes are the only natural lake systems where water is present long enough for them to give rise to an ecological organization of their own. Their highly mineralized waters contain high levels of carbonates and bicarbonates due to the dissolved materials from the substrate. The abundance of fine materials in suspension greatly reduces transparency.

The phytoplankton of Mediterranean karstic lakes shows a marked annual succession, as the regime of alternating stratification and mixture of the waters gives rise to a new situation twice every year, requiring colonization by pioneer species that reproduce rapidly. These consist of diatoms (*Cyclotella*), chrysophytes (*Dinobryon*), cryptophytes (*Rhodomonas*, *Cryptomonas*), dinophytes (*Ceratium hirundinella*, *Peridinium*), and some chlorophytes (*Ankistrodesmus*). However, primary production is usually low due to the lake's nutrient deficiency, as the abundance of calcium leads to scarcity of free phosphorus. The zooplankton is dominated by copepods, then rotifers and cladocerans, all of which have longer, yearly, life-cycles and different feeding patterns at different stages of their life-cycle. The deep benthos is poor, because the bottom sediments do not form a stable system, as they are kept in suspension by the entrance of underground water. The benthos is richer at the edges of the lake, where there are chironomids, nematodes, and bivalves (*Unio* in Europe and *Velesunio* in Australia). There may be a well-developed plant community on the shores that is especially rich in charophytes (*Chara*, *Tolypella*, *Nitella*) and rooted flowering plants (*Myriophyllum*, *Potamogeton*). The alkaline waters of karstic systems typically contain isopod crustaceans (*Gammarus* in the northern hemisphere, and *Hyalella* in the southern hemisphere and the phreatoicidean *Metaphreatoicus* in comparable habitats in southern Australia). Swimming crustaceans with planktonic larvae are also important (*Atyaephyra*, *Dugastella*), which live on the shores in the adult state.

Waters in reservoirs, neither still nor flowing

Reservoirs built by human beings have introduced an aquatic habitat into the mediterraneans that would not naturally be found there: large basins permanently flooded by waters from the mountains, and furthermore they are subject to a variable water regime that is a response to the water needs of the human settlements downstream. Yet in spite of these hydrological variables, reservoirs usually behave as temperate lakes when the discharged water is taken from the upper layers, as nothing prevents the depths of the water column from continuing to behave like a natural lake and forming an annual thermocline; the banks are subject to non-climatic variations of water level, preventing the normal development of lake shore systems. The effect of human intervention is much greater on life in the sediment if the sludge at the bottom is regularly emptied, as this governs the biology of the entire basin. This action completely disrupts the biological and chemical activity of the reservoir as well as that of the river downstream, as it is very difficult for the river to assimilate discharge of the highly reduced sludge that accumulates below the thermocline.

The time its takes to renew the reservoir's water greatly affects the composition of the aquatic communities: the faster the water is renewed, the more the system is like a river; and the slower its renewal, the more the system resembles a lake.

In terms of their thermal cycle, the deeper parts of the reservoir, closest to the dam, behave as lacustrine systems. Mediterranean Basin reservoirs stratify during the summer. That is to say, the surface layers of water (the *epilimnion*) become warmer due to heat exchange with the atmosphere, while the lower layers (the *hypolimnion*) stay cold; these two layers are separated by a layer where there is a sharp change in temperature, the *thermocline*, although the depth of thermocline formation is influenced by human use of the reservoir. Dissolved oxygen levels vary over the year, mainly due to biological processes (respiration and photosynthesis) and summer stratification: when the water is mixed, all levels are well oxygenated. On the other hand, in conditions of stratification, the upper layer is well oxygenated as it is in contact with the air and contains most of the primary producers (phytoplanktonic algae) while the lower layer may become oxygen deficient, mainly due to respiration by organisms and the oxidation of organic material; when the water is eutrophic, oxygen may easily be exhausted. The high degree of eutrophication of reservoirs is a consequence of their high nutrient content, and means that one or another type of plankton community may develop. In the phytoplankton, eutrophication changes the basic communities of nutrient-poor waters, causing the increase of other species, and when conditions are highly eutrophic the alteration of the normal proportions of phosphorus and nitrogen leads to domination by cyanobacteria. With regard to the zooplankton, eutrophication leads to the dominance of cladocerans and rotifers with short life cycles (sometimes, only a few days) over the copepods: both cladocerans and rotifers show clear cyclomorphosis (seasonal variation in body shape) in relation to the summer temperature of the water. In highly eutrophic conditions, the rotifers are totally dominant. At the bottom of the basin, benthic mud-dwelling communities build galleries or tubes, or move over the surface in contact with the sediments; the most abundant are the tubificids (*Tubifex*) and chironomid larvae (*Chironomus*). They, and other organisms living here, can tolerate very low oxygen concentrations; the tubifex worms survive by exploiting the detritus that arrives from the shores, while the chironomids basically eat the phytoplankton that is continuously settling on the bottom. Some molluscs and the larvae of the chaoborids (the gnats, such as *Chaoborus*) also seem to be adapted to life in this mud. In the shallow water at the edge, whose level fluctuates greatly, the communities resemble those of freshwater pools. They consist of organisms with short life-cycles that are not particular what they eat, and that also produce many offspring. This variation in water level normally prevents the growth of a belt of helophytes. When the water level is stable enough for them to grow, communities of aquatic flowering plants (*Potamogeton*, *Myriophyllum*) develop in still backwaters. The fish that live there are species found in river systems and especially introduced species.

4.3 The vegetation of riverbanks

The sites that edge Mediterranean rivers are privileged with respect to their abundance of water, as however dry the landscape may be, by the side of the watercourses the drought is less harsh. Bands of vegetation, called *gallery forests*, form along the riverbanks and follow the watercourses, giving rise to azonal landscapes, that is to say, landscapes more determined by local ecological conditions than by general climatic ones. The species composition and physiognomy of these riverside landscapes are highly conditioned by the depth, abundance and maintenance of the water table, and also by the size and frequency of the surges, as these are the factors that determine the type of plants that can live there. The Mediterranean regime shows a whole range, from permanent rivers to watercourses that only flow during the rainy season; the permanent rivers show more regular regimes than temporary watercourses. Large surges force the riverside plants to deal alternately with the shortage of water and its destructive effects.

Forests along permanent watercourses

The Mediterranean rivers with the highest flows have a strip of forest vegetation dominated by the winter deciduous trees typical of other biomes, as these displace the evergreens in the absence of the summer dry period. These communities are pure or mixed forests, such as the groves dominated by the alder (*Alnus glutinosa*), white poplar (*Populus alba*), elm (*Ulmus minor*), willows (*Salix*), and oriental plane trees (*Platanus orientalis*) or walnuts (*Juglans regia*) in the Mediterranean Basin.

In the Californian mediterranean there is another species of plane (*Platanus racemosa*, the Californian plane) and remnants of the formerly abundant, rich, deciduous oak riverside forests of valley white oak (*Quercus lobata*); these oaks are now uncommon because they were felled for timber, the valley bottoms they grew in have been ploughed for intensive cultivation, the water table has lowered due to intense exploitation of the aquifers and the lower areas have

108 **Autumn appearance of a deciduous riverside forest** in the heart of the dry southern Mediterranean (the headwaters of the Turia, in the Autonomous Community of Valencia, in the east of the Iberian Peninsula). These gallery forests consist of poplar (*Populus nigra*), ash (*Fraxinus oxycarpa*), etc., and sharply contrast with the sclerophyllous shrub vegetation of the rocky areas through which the rivers and mountain torrents run.
[Photo: Ernest Costa]

been developed for housing. In the South African mediterranean the treeless shrub communities are dominated by deciduous plants. In the Chilean mediterranean there are small deciduous forests along the watercourses, dominated by the willow (*Salix humboldtiana*) and *Psoralea glandulosa* accompanied by small evergreen trees, such as winter's bark (*Drimys winteri*), a laurel-like plant typical of the sub-Antarctic forests, *Luma chequen* or *Crinodendron patagua*. The Australian mediterranean's evergreen riverside forests consist of eucalypts such as *Eucalyptus camaldulensis*, usually accompanied in south Australia by *E. largiflorens*, while in the southwestern mediterranean it shares this habitat with the yarri (*E. patens*) and *E. rudis*.

Most of these trees and shrubs can sprout from their rootstock, an adaptive response to the possible destruction of the aerial parts by violent surges. In nitrified soils and those relatively rich in chlorides, whether due to the soil composition or to proximity of the sea in rivers with a low flow, the deciduous trees are replaced by small halophytic evergreen trees, such as the tamarisks (*Tamarix gallica* and *T. africana*) that grow in the Mediterranean Basin.

Woodlands along intermittent watercourses

In the most arid parts of the Mediterranean regions, dominated by summer deciduous trees or by very dry evergreen shrub communities, the riverside vegetation is also impoverished as it has to withstand relatively dry conditions in the summer period. This leads to riverside shrub formations dominated by evergreen shrubs or by small trees.

In the Mediterranean Basin, in the middle of the river beds in the deepest zones which always retain some water, shrubby willows such as *Salix elaeagnos* thrive. On the banks or in the middle of the ramblas and in even harsher conditions, there may be oleanders (*Nerium oleander*) and even the chaste tree (*Vitex agnus-castus*) or myrtles (*Myrtus communis*). The fact these plants are evergreen clearly indicates the limited availability of water.

Riverbank nitrophily

When water courses dry out regularly or only flow in the rainy period, the zonal vegetation growing along the banks colonizes the bed of the river. In fact, these river beds are terrestrial systems that are periodically devastated by short, violent, surges, which are too frequent for a well-developed aquatic community to establish. This vegetation is formed by fast-growing pioneer species with short life-cycles that produce many offspring in time for the next surge. Most are herbaceous plants, like *Aster squamatus* and several species of rushes, but there are also some shrubs, such as the elecampane (*Inula viscosa*), a summer-deciduous plant that flowers in the autumn that is typical of the ramblas of the Mediterranean Basin.

All these species show a marked tolerance to the nitrogenated compounds carried by the surges and even show a taste for them. This is why these environments are a dispersal route for many nitrophilous species, such as the elecampane (*Inula viscosa*). Later, these species colonize moist sites nitrified by human activities, such as the edges of paths. Thus ruderal, nitrophilous, and riparian come together.

109 **Oleanders (*Nerium oleander*) flowering in a rambla in Andalusia,** in the south of the Iberian Peninsula. The oleander's typical habitat is the most southerly seasonal ramblas of the Mediterranean Basin with their highly variable water regime. The oleander is an evergreen shrub with dark green leathery leaves and abundant large flowers that form patches of color in the middle of very dry landscapes. The fruits of the oleander are not edible, and in fact almost all parts of it contain substances poisonous to humans. [Photo: Ernest Costa]

5. Dynamics and variability

5.1 Sun and shade, up on the mountains

Even a hypothetical pristine Mediterranean landscape would show variations dependent on relief. In fact, changes in exposure or altitude increase the effects of the ecologically limiting factors that already act to determine the vegetation under normal conditions. This is why the markedly mountainous Mediterranean world shows such highly variable landscapes.

The effects of exposure

Variation due to exposure plays a major role in determining the microclimate, as the amount of sunlight affects temperature and water availability. This variability results in major landscape changes in areas of transition between different vegetation types, as happens in the transition areas between scrub and woodland, or from an evergreen forest to a deciduous one. In these conditions, the effect of exposure gives rise to a landscape mosaic. It is a major factor in most of the mediterraneans, except for the two Australian areas where the smooth, ancient relief has given rise to large extents of essentially homogeneous landscape.

Generally, north-facing slopes (north-facing in the northern hemisphere, and south-facing in the southern hemisphere), that is to say, in the shade, receive less sunshine and are thus cooler and moister; on the other hand, the south-facing slopes (in the northern hemisphere, and north-facing in the southern hemisphere) are in the sun and thus warmer and drier. The sunny side and the shady side of a Mediterranean mountain may be spectacularly different.

The effects of altitude

Altitude is another factor leading to variation. Increases in altitude mean lower temperatures and higher rainfall. Modest increases in altitude do not cause important changes in the landscape, as they are limited to the replacement of one Mediterranean formation by another, such as the replacement of scrub by sclerophyllous forest. When there is a large increase in altitude, the landscape changes more drastically. In fact, only in the Mediterranean Basin and the Chilean mediterranean are there mountain ranges over 9,840 ft (3,000 m) that are able to cause the appearance of extra-Mediterranean landscapes within the Mediterranean biome.

The transition to the mountain biome in the Mediterranean Basin

In the Mediterranean Basin, one can distinguish between the mountain chains in the north, where increasing altitude gives rise to a decrease in temperature and an increase in precipitation, and those in the south, where the dry period is felt even at the highest levels. In the northern mountains, such as the Pyrenees, the Alps, Mount Etna, and Mount Olympus, a different layer appears above the evergreen forests. It is dominated by Mediterranean deciduous trees, such as oak (*Quercus humilis*) and Spanish chestnut (*Castanea sativa*). Further up, the forests consist of typical temperate deciduous trees, such as beech (*Fagus sylvatica*), or extra-Mediterranean conifers, such as silver fir (*Abies alba*), Norway spruce (*Picea abies*), and Scots pine (*Pinus sylvestris*), followed by stands of pine (*Pinus uncinata*), and then at the very top the vegetation consists mainly of meadows. In the southern mountain chains, such as the Atlas Mountains in Morocco and the Sierra Nevada in Spain, the deciduous tree layer is absent as the summer drought occurs even at higher altitudes, and the layer of sclerophyllous forests is replaced by one dominated by Mediterranean conifers, such as Atlantic cedar (*Cedrus atlantica*) or the black pine (*Pinus nigra*), and followed at the peak by a zone with cushion-forming, spiny shrubs and dry meadows.

The transition to the mountain biome in the Chilean mediterranean

In the Chilean mediterranean the Andes may rise to 22,960 ft (7,000 m) within about 60 mi (100 km) of the coast, leading to a very clear altitudinal zonation. To the north, the sclerophyllous vegetation is replaced by shrub communities and a few forests of Chilean incense cedar *Libocedrus* [=*Austrocedrus*] *chilensis*, followed at greater heights by cushion-forming plants, often of a sub-Antarctic nature. To the south, the Mediterranean vegetation is replaced by deciduous forests of southern beech *Nothofagus obliqua*, replaced in turn at greater height by temperate evergreen forests called *Valdivian forests*.

110 **Hedgehog broom** (***Erinacea anthyllis***) **and donkeys** (***Equus asinus***) **in the Atlas mountains** in Morocco. These pulvinate, or cushion, plants are an adaptive response of many plants to environments where the wind is strong. Wind's action on plants has both mechanical and physiological aspects. To defend themselves from mechanical breakage of branches standing above the soil level, these plants have developed aerodynamic forms. Physiological defence against water loss due to wind-caused transpiration is also increased by adopting a cushion form, as this reduces the effect of the wind. The ass is probably of African origin and is and has been widely used as a draft-animal throughout the Mediterranean Basin, although the general use of the internal combustion engine has caused a reduction in their numbers.
[Photo: Jordi Bartolomé]

111 Ponderosa pine (*Pinus ponderosa*) in the Yosemite National Park. The mountainous zone of the Californian mediterranean is dominated by ponderosa pine, which grows amongst an array of other conifers. This pine is a large tree and some specimens may reach a great height.
[Photo: Teresa Franquesa]

The transition to the mountain biome on the Mediterranean slopes of California's sierras

Although they are not as high as the Andes, the sierras that separate the Californian mediterranean from the Great Basin and Mojave deserts reach 13,120 ft (4,000 m)—14,491 ft (4,418 m) in the case of Mount Whitney. The layer of Mediterranean chaparral vegetation may reach 4,920 ft (1,500 m), the altitude roughly marking the lower limit of winter snows. Above this, up to 7,872 ft (2,400 m), there is a Mediterranean montane layer of conifers, such as ponderosa pine (*Pinus ponderosa*), or white fir (*Abies concolor*), red fir (*Abies magnifica*), lodgepole pine (*P. contorna*) and very locally giant redwood

(*Sequoia-dendron giganteum*). There is no deciduous tree layer like that found in the northern part of the Mediterranean Basin, since in California the deciduous oaks, such as *Q. lobata* and California black oak (*Q. kelloggii*) and other deciduous trees, such as Californian chestnut (*Aesculus californica*) are dispersed and only form groups in particular sites within the chaparral or in conifer forests, but never forming a definite layer. Between 7,872-8,200 ft and 9,840 ft (2,400-2,500 m and 3,000 m), there are sub-Alpine forests in the Californian Sierra Nevada, dominated by white-bark pine (*Pinus albicaulis*). Above 9,840 ft (3,000 m), there are open alpine meadows.

Calcareous or siliceous

Another important factor leading to variations in the landscape in the mediterraneans is the composition of the underlying rock substrate, as this greatly affects the soil's structure and water retention capacity. Areas with calcareous and siliceous (silica-rich) substrates may differ greatly even if they receive comparable rainfall, because the permeability of calcareous substrates gives rise to drier conditions. Calcareous substrates are very common in the Mediterranean Basin, but they are relatively unimportant in the other areas of the biome. In transitional zones between forest and scrub, the change from a siliceous to a calcareous substrate causes a sharp change in the landscape, as formations that clearly correspond to more arid zones start to dominate. The major difference in composition and properties between soils derived from calcareous or siliceous substrates also gives rise to similar formations dominated by different species, as some species have nutritional requirements restricting their distribution to one or other of these substrates. For example, in the west of the Mediterranean Basin, the forests of holm oak (*Quercus ilex*) and those of *Q. ballota* grow on both types of substrate, but forests of cork oak (*Q. suber*) are restricted to siliceous soils. In maquis and garrigue environments, cork oaks tend to be located on calcareous substrates; matorral environments have communities with different species on the two substrates, and these are called calcicolous (lime-loving) and siliccolous (growing in silica-rich soil) matorrals. Silicolous matorrals are more dense and luxuriant. Although they share some species, such as rosemary (*Rosmarinus officinalis*), silicicolous matorrals are dominated by cistus (several species *Cistus*) and heathers (*Erica*), although Western Mediterranean heath (*Erica multiflora*) grows on calcium-rich soils. The dichotomy between calcareous and siliceous soils, thus not only affects water availability for the plants but also the species composition of the vegetation.

112 **Garrigue that has been grazed and heavily degraded in Albania.** The secondary effects of human presence degrade the landscape in several different ways. Grazing increases aridity, because livestock can devastate large areas.
[Photo: Lluís Ferrés]

5.2 From forest to scrub: a one-way route?

The Mediterranean biome is a transition between desert and temperate forests, and this is why both forest and scrub occur spontaneously, that is to say, tree formations and shrub formations alternate, replacing each other in space, depending on each site's characteristics. The presence of forests or scrub depends in principle on the availability of water, and this is in turn closely related to rainfall, but it also depends on soil depth and substrate composition, the two factors that determine the soil's water storage capacity. The balance between forest and scrub in contemporary landscapes now mainly favors scrub, due to the disturbances that limit the stage of development of the vegetation. The main factor affecting the Mediterranean landscape today is intense human action.

Factors causing degradation

Grazing and the use of fire, together with abusive forest management, result in the degradation of many forests into their corresponding scrub formations. Secondary succession processes ought to lead to forest reestablishment if these disturbances cease; but if they occur repeatedly, succession is slowed down and the vegetation does not develop beyond a certain stage. In the case of fire, this recurrence is very common, as the fire itself eventually benefits pyrophytic species, which in turn increases the chance of fires occurring, thus generating a vicious circle. The case of grazing is different, as the destruction of the vegetation is not drastic but selective, thus favoring some species over others. Furthermore, over the last few years many formerly grazed areas have been abandoned, and this has allowed some recovery of the vegetation.

In any event, erosion may prevent the regeneration of forests from degraded communities. Soil loss due to fires and excessive grazing leads to reduced fertility and water retention capacity, and this totally prevents forest regeneration. Fires also cause a loss of nutrients as volatile gases; in the case of the essential nutrient nitrogen, these losses may exceed 10% if temperatures exceed 392°F (200°C).

The pathway from forest to scrub is often a one-way route. In all the areas of the biome, but especially in the Mediterranean Basin, human activity must be borne in mind when considering the vegetation's distribution and condition. The apparently hard, resistant scrub community can itself be degraded, since simply thinning the vegetation may cause irreversible soil changes due to the mineralization of the humus (conversion to an inorganic state through decomposition by soil micoorganisms) and the loss of soil structure, leaving it vulnerable and at the mercy of erosive factors.

Human interference

Humans exploit and modify the ecosystems, introducing changes affecting their function and structure. This occurs in almost all the biomes in the biosphere, but is especially acute and dramatic in

113 Vegetable plots in the east of the Iberian Peninsula, surrounded by an eroded landscape with kermes oak (*Quercus coccifera*) and the western Mediterranean heath (*Erica multiflora*). Landscape degradation cannot be understood without taking human activities into account, especially in the Mediterranean Basin, which was settled by humans in ancient times. In fact, agricultural practices have led to the transformation of many forests into fields and have degraded the surrounding land.
[Photo: Rafael Vela]

the Mediterranean biome, and in the Mediterranean Basin in particular. The intense humanization in the last 200 to 300 years of the Mediterranean areas in California, Central Chile, south and southwest Australia and South Africa has already profoundly changed their landscapes. But human pressure on the Mediterranean Basin goes back 10,000 years and it can be said that the landscape changes associated with the climatic changes at the end of the last glaciation cannot now be distinguished from those caused by human activity. The landscapes of the Mediterranean Basin are the result of intense interaction with the human species, which has been and still is one of the most important ecological factors.

The intense and continuous exploitation of forests, grazing, the use of fire as a tool of change, together with ploughing to introduce crop cultivation, have all left their mark on this fragile biome; its soils are easily eroded and show little or no soil formation. Vegetation now only occupies one third of the Mediterranean Basin as it has been removed from the best soils that are now occupied by crops and grazing land, and relegated to marginal, unproductive land. Furthermore, many of these landscapes are, or have been until recently, used for forestry or livestock purposes. And even when this has not occured, they have not been preserved unaltered, as fortuitous fires occur regularly.

Thus, humans are directly responsible for the distribution patterns of landscapes and ecosystems in the Mediterranean biome, and for their general appearance. The biome's inherently low productivity, poor soils, the limited water retention capacity of its soils, and the role of fire have all been made worse by human intervention. The mediterraneans have never been covered by dense, productive forests, but they would have been, or might have been, biologically richer without such intense human pressure. The existence of slow-functioning but luxuriant forests and matorrals on the biome's richest soils, would force us to rethink, in part, our concept of Mediterranean landscapes.

3
Humans in sclerophyllous formations

1. The human settlement of sclerophyllous formations

1.1 A place to settle: the mediterraneans in the Paleolithic

Human settlement of sclerophyllous formations is so old that, at least in the Mediterranean Basin itself, human settlement was before, not after, the development of the climate needed for the Mediterranean-type vegetation. Not only is the Mediterranean Basin an area with a long history of civilization, but it is also a densely populated area with cities with powerful infrastructures, and has a fine network of penetration of settlements reaching the furthest corners of the remotest mountain valleys. It includes many islands, some very small and some large, and they too have been settled, except for the smallest and most inhospitable ones. The other mediterraneans were originally less densely populated, but they also have developed settlement systems based on large cities with a large hinterland created by "capillary" spread.

The Mediterranean Basin's relief has also divided the territory into separate compartments, allowing the appearance of cultural differences, if not in every valley at least in every part of the territory where a group of humans has settled for long enough to be considered a settlement. Yet since antiquity the relative ease of communication along sea routes has allowed the creation of a common cultural background, leading some to speak of a "Mediterranean" culture. Rather than a single culture, it is better to talk of a place of contact between diverse cultures, all dealing with an environment harsher than many northern Europeans imagine, but which can be used in many different ways that favor interchange. These cultures adopted similar behaviors with respect to natural resources by incorporating elements of one and other, both in the material and symbolic realms.

The situation in the other mediterraneans would appear to be very different in this respect. Yet to some extent it reproduces, centuries or even millennia later, models already tried in the Mediterranean Basin. At least in California and South Africa, the mixture of peoples and cultures (until recently prevented by South Africa's former policy of apartheid) appears to be following similar models.

Neanderthals and modern humans in the basin

Humans reached the Mediterranean very early. There are traces of *Homo erectus* in places almost on the coastline, such as at Tighenif (the former Ternifine), which is in the Tell Atlas 31 mi (50 km) southeast of the city of Oran (Algeria), where three jaws and a parietal were found in materials dated at 700,000 years old; on the terraces of the rivers Aglí, Tet, and Tec, three rivers in the French region of Rousillon, where cut stone remains 600,000-900,000 years old were found; and at Lo Vallonet cave in Roquebrune-Cap-Martin on the Provence coastline where even older cut stones, about a million years old, were found.

Yet at that time the Mediterranean was not as we now know it. Studies of pollen in the strata containing remains of humans, or of human work, show that in the middle of the ice age there was still vegetation more like a steppe with scattered pines than a holm oak forest or maquis.

About 400,000 years ago (at approximately the end of the Mindel glaciation), many millennia after the *Homo erectus* of Tighenif, and the settlements on the terraces of the rivers in Roussillon, there were humans living in the Aragó cave in Talteüll, near the Aglí. They lived in climatic conditions colder than today, but major elements of the flora that would later dominate the Mediterranean were already present in sheltered areas that received the midday sun. The pollen of holm oak, cypress, Mediterranean pines, mock privet, plane trees, and vines has been found in the sediments associated with the human remains in the Aragó cave, together with species indicative of a colder climate than today's (Scots pine, birch, and steppe grasses). In fact, as the site's finder Henry de Lumley pointed out, the Aragó cave's permanent or temporary inhabitants—ancient

114 The Neolithic remains in Mnajdra, part of the megalithic complex of *Hagar Quim*, on the southern coast of the island of Malta in the central Mediterranean, show how long ago humans settled the Mediterranean Basin. In fact, archeological remains suggest the island has been inhabited since 7000 B.P. In 5800 B.P. megalithic temples began to be built, but for no apparent reason new constructions ceased after 4500 B.P. The purpose of the grooves, up to 29 in (75 cm) deep, in the stones is also unknown, although they may well be marks left by the as yet unknown system used to transport them.
[Photo: Lluís Ferrés]

humans intermediate between *Homo erectus* and Neanderthal man—could exploit several nearby different ecosystems. Further from the sea and at a higher altitude than today, the cave was probably located in a mountainous environment at the base of a plateau. Those dwelling there could hunt moufflons *Ovis musimon*, goats and ibex on the steep slopes where the cave had its entrance. They could reach the plateau by means of a steep but climbable chimney to hunt the tundra animals, such as musk oxen, reindeer, and wolves. If they went down to the plain the cave overlooked, they could hunt the herds of horses, bison, aurochs, rhinoceros, and elephants. Furthermore, on the wooded slopes where the cave was located they could find deer, fallow deer and panthers and also the rivers where they could hunt beavers and all the animals that went to the water to drink.

There was an unusual situation around the Mediterranean between 130,000 and 40,000 years ago. Two different human populations occupied the northern and southern shores. To the north, the Neanderthals occupied all the habitable areas of the western part of the Eurasian continent, from the Atlantic to the edge of the central Asian highlands. To the south, modern human beings seem to have occupied most of Africa (although no traces confirming their presence on the Mediterranean's African shoreline at the time have yet been found) and apparently part of Southeast Asia. The controversial remains found on Mount Carmel on the Mediterranean's eastern shores (right on the—uncertain—frontier between the areas occupied by these two great human groups) appear to show that Neanderthals and modern humans may have lived together for tens of thousands of years before modern humans came to dominate, as they eventually did all over the world. Electron-spin resonance dating in 1989 of modern human remains from the caves of Qafzeh (near Nazareth) and Skhül (on Mount Carmel) gave ages of 90,000 and 100,000

years (only exceeded in the area by the Neanderthal remains in Tabun, also on Mount Carmel), which are much older than the Neanderthal sites at Kebara, which is also on Mount Carmel (60,000 years old), and at Amud near the Sea of Galilee (55,000 years). For some authors the coexistence of Neanderthals and modern humans in the western Mediterranean over a period of at least 40,000 years shows that they were separate species, and this would explain the sudden disappearance 35,000 years ago of the last European Neanderthals without any observable mixing with the populations of modern human beings. For other authors, this apparent coexistence did not occur, as the presence of Neanderthals and modern humans in an area where their spread coincided might reflect episodes with climatic conditions that favored one or other of the two groups. Neanderthals would have dominated the area during colder periods and modern humans during warmer ones. In any case, since about 35,000 years ago modern humans expanded to occupy the entire Mediterranean space and thus no more Neanderthal remains are to be found on the northern shores of the Mediterranean and even fewer on the eastern shores.

The first modern humans in Mediterranean South Africa

Some of the oldest hominid remains have been found in South Africa, but the oldest human remains in the strictly Mediterranean area, the southeast of the Cape province, are roughly contemporary with the European Neanderthals. These remains consisted of the upper part of a cranium and a fragment of jawbone found in 1953 in Elandsfontein, a farm, near Saldanha Bay on the Atlantic coast of the Cape province. Some authors consider that they might be an archaic form of *Homo sapiens*, but some characters and above all their age (between 200,000 and 500,000 years old) suggest *H. erectus*. These human remains were found with remains of the Pleistocene fauna and artifacts comparable with the Acheulian industries in other parts of Africa, Europe, and western Asia. The acidity of the soils of Africa's southern tip does not favor the conservation of bony remains, but in many points Acheulian artifacts have been found that indicate this area was colonized by modern humans in ancient times.

There are traces of the deliberate use of fire since at least the beginning of the middle Pleistocene, 125,000 years ago, presumably to favor the

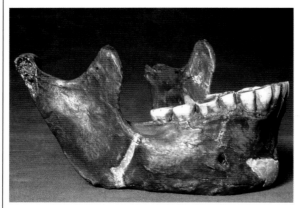

115 Jaw of the Tabun Neanderthal from Israel. The striations left by food on the teeth tell us that Neanderthals mainly ate meat, and this hypothesis is consistent with the information obtained from studies of the glaciations: in the Near East, in the times of the Neanderthals, animal resources were probably much more abundant than plant resources.
[Photo: Israel Antiquities]

appearance of the bulbous plants that probably formed one of the basic foodstuffs of the humans in South Africa at the time. Thanks to fortunate circumstance of its receiving alkaline water from a spring, shells and human bony remains have been preserved in a cave on the coast of Tsitsikamma at Klasies River Mouth, together with traces of stone-working industries, all datable to between 120,000 and 90,000 years ago, during the Riss/Würm interglacial period. They are not only clearly modern humans, but are probably the oldest evidence of the systematic human usage of marine resources, such as bivalve molluscs, seals, and penguins (but there are few traces of fish and flying marine birds, perhaps indicating the lack of the technology necessary to catch them), although the humans of the Klasies River site hunted terrestrial animals, such as African antelopes, Cape buffalo, and wild boar.

In the contemporary South African mediterranean, in the last glaciation (Würm), the climate was not only colder than today, but also drier. Perhaps this is why traces of human settlement are less frequent until the end of the glaciation. Twelve thousand years ago, when the current interglacial period began, these remains started to become more common, although until the arrival of the Khoikhoi (Hottentots) the area's inhabitants maintained a hunter-gatherer lifestyle. The Khoikhoi were nomadic shepherds, a mixture of San (Bushmen) and Bantu who, in the first centuries A.D., moved west and south from the banks of River Zambezi and the north Kalahari desert, through present-day Botswana and Namibia, accompanied by their livestock which consisted first of sheep, then later cattle. They were the inhabitants encountered by the first Portuguese sailors to land on the Cape peninsula in 1488, and

116 Tsitsikamma Cave, South Africa. The name comes from the Khoikhoi word *tsitsi*, which means rain; *tsitsikamma* means running water. The caves of the southern coast of the Cape Province have many human remains associated with Paleolithic and Mesolithic workings, but the state of the fossils does not allow significant conclusions and the stratigraphy is often doubtful. Yet the remains are apparently from the end of the Pleistocene and beginning of the Holocene, and they are of individuals of the San (Bushmen) in the late Pleistocene and Khoikhoi (Hottentots) in the early Holocene.
[Photo: H.J. Deacon]

by the settlers from the Dutch East India Company who established the colony that would later become Cape Town.

The first humans in the contemporary Chilean mediterranean

The date of the arrival of the first humans in the Americas and the dates of the successive waves of settlers are still under debate, but few traces of human presence or activity are older than 12,000 years.

However, one of the oldest of these rare traces is in the locality of Monteverde, on the terraces of the Chinchiuapi torrent, a tributary on the lefthand side of the river Maullin, in the Chilean province of Los Lagos, more than 248 mi (400 km) south of the River Bío-Bío, the approximate southern limit of the Chilean Mediterranean bioclimatic area. The finds in Monteverde include remains of stone-working industries together with charred logs that were carbon-14 dated to 33,000 years ago. No human remains have been found at this level, nor remains of the animals they might have eaten, and most authors do not accept this date. Yet remains found on a more recent level at the same site (13,000 years old) include not very different worked stone, together with the remains of mastodons and edible or med-

icinal plants, and what were possibly the lower parts of the poles to hold cabins or tents up.

There is also good documentation in Chile of the retreat of the glaciers from the Andes between 16,000 and 10,000 years ago, and of the gradual establishment of the current distribution of the layers of vegetation, as well as of the existence of a period with a warmer, drier climate 7,000 years ago (coinciding with the climatic optimum known from the Old World). There are however no traces of the beginning of the extension of the agricultural system that developed in Peru in the centuries after this climatic optimum. It seems that by the time of the Spanish Conquest, this system had already spread to Chiloe Island.

The first human settlement of modern-day California

Although it is generally accepted that the settlement of the Americas started from the north, along the land bridge that existed during the glaciations between northeast Siberia and Alaska, there are no remains of human occupation in California as old as those of Chile. Human remains found at the Rancho La Brea, now a park within the Los Angeles conurbation, have been carbon-14 dated as 9,000 years

117 Near this small pond on the Rancho La Brea in the City of Los Angeles (seen here in a photograph taken at the beginning of the century), many finds have been made, including a large number of bones of prehistoric animals, a human cranium, and some fragments of postcranial bones belonging to Amerindians, dated at more than 10,000 years old. The remains of stone tools, found at archeological sites from the Arctic to Argentina, suggest that humans had reached the American continents (through the Bering Strait), well before this. [Photo: The Huntington Library (California)]

old. The same site has also yielded bones of *Smilodon*, an American feline about the size of the contemporary lion and similar to the Eurasian sabretoothed tiger (*Machairodus*), that are more than 15,000 years old and bear marks that might be due to stone instruments used by humans.

There are few traces in California of the Clovis culture, that left the oldest traces of human presence in North America (12,000 to 11,000 years old) from Alaska to Mexico. The only traces attributed to this culture have been a few arrowheads found in places like Lakes Borax, Tulare and Mohave, but they are more recent, about 10,000 years old. The traces found between then and the arrival of the Europeans seem to indicate a notable continuity in settlement and material culture. This suggests settlement by different hunter-gatherer peoples, most of them sedentary and with some social stratification, whose economic base was the collection of the acorns from the chaparral and, on the coast, fishing and hunting marine mammals.

Probably, the same Hoka-speaking (Diegueño, Chumash, Pomo), Uto-Aztec-speaking (Cahuila, Paiute, Shoshon) and Penutia-speaking (Maidu, Miwok, Yokuy) peoples found by the first European colonizers were direct descendents of the first inhabitants and partly kept to their life-style.

The first Australian mediterranean humans

Humans reached Australia at least 36,000 years ago (the carbon-14 dated age of charred wood from a fire found near Arumpo, in southwest New South Wales), and most authors accept that they could not have arrived much before this. Indirect indications, such as the extinction of many species of Australian animals for no apparent reason between about 50,000 and 30,000 years ago, and the discovery near Beachport (on the southeastern edges of the South Australian mediterranean climatic zone) of a stone axe that is apparently at least 50,000 years old, seem to indicate a human presence for which there is no other or direct evidence.

In southwest Australia, outside the strictly Mediterranean area, on the coast between Cape Leeuwin and Cape Naturaliste, there are sites like the Devil's Lair, with human remains dating back 30,000 years and animals remains accumulated by the humans that settled there. These animal remains consist of the bones of a large variety of

marsupials and monotremes, most now extinct. In Seton Cave, on Kangaroo Island near Adelaide, human remains have also been found together with those of extinct members of the Quaternary fauna, specifically teeth of *Sthenurus*, a now extinct genus of large kangaroos with a short tail and a flattened snout.

1.2 The Mediterranean cradle: from the Neolithic onwards

The most important neolithic revolution took place in the Fertile Crescent, and started on the eastern shores of the Mediterranean, and the main route for its spread west were the Mediterranean and its coastline. About 10,000 years ago contemporary bioclimatic conditions consolidated in the Mediterranean Basin, and they have lasted with minor oscillations until the present day. This was when what was known as the Natufian culture arose in the regions between the western shore of the Mediterranean to the middle basin of the Euphrates. They were sedentary hunter-gathers that lived in relatively large settlements and made use of a wide range of resources. With respect to supplies of animals, they basically hunted gazelles, but also goats, sheep, wild asses, hares and aquatic birds; they also fished, both on the coast and inland. With respect to foodstuffs of plant origin, they collected wild legumes, nuts, and cereal grains, which they stored and preserved for periods when they were scarce or absent—in effect, a lifestyle that was not unlike that of the Native Americans until California's occupation by Europeans.

The first domestications

The heart of the area where the Natufians lived 10,000 years ago appears to have been one of the several areas pre-adapted to exploitation that appear to have encouraged the birth of agriculture in different points of the world at that time. These were sites mainly occupied by grasslands rich in large-grained grasses, and already intensively exploited by herbivores and thus preadapted to conscious exploitation by human beings. The successors of the Natufians used these grasses and selected varieties of suitable cereals that produced spikes with rachises that were resistant enough not to break when the plant was harvested with a bone or flint sickle. The remains they left have been dated to the second half of the

118 Different animals portrayed in hunting scenes in cave paintings from the Valltorta gorge (in the north of the Autonomous Community of Valencia, in the eastern Iberian peninsula). The left-hand column shows wounded animals, the central column shows animals that have fallen down or are dead, and the column on the right shows animals in other positions. Rock paintings can provide a lot of information about human life in the late Paleolithic and in the Mesolithic. It is now thought that as the main animals represented are edible, the paintings were intended to favor the hunt by magic, i.e. they formed part of an underlying fertility cult. So-called Venuses, sculptures that are clearly women but whose lines are unclear, were later added. Many of the pictures show the outlines of men hunting together, allowing the deduction that they must have had some language that allowed them to communicate and coordinate with each other.

[Drawing: Editrònica]

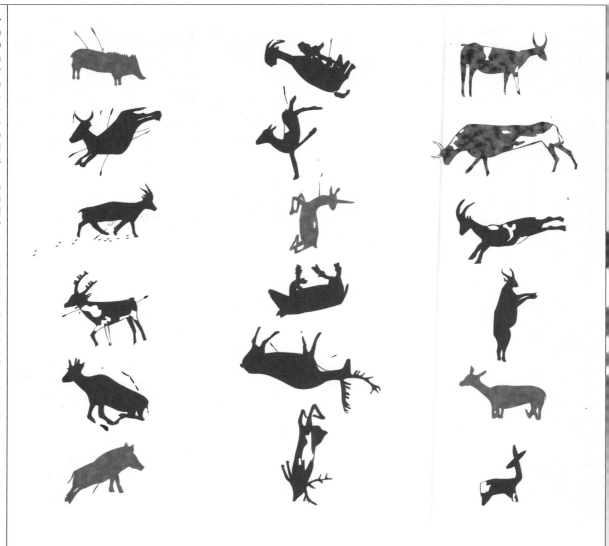

tenth millennium B.P. (9,500 to 9,000 years ago) and contained a dozen domesticated plants, including three cereals, emmer (*Triticum turgidum* subsp. *dicoccum*), einkorn (*T. monococcum*) and barley (*Hordeum vulgare*), four legumes, lentil (*Lens culinaris*), pea (*Pisum sativum*), bitter vetch (*Vicia ervilia*), and chickpea (*Cicer arietinum*), and a fiber-yielding plant, flax (*Linum usitatissimum*). These plants were the basis of this early agricultural complex, which spread from the Fertile Crescent, first to Anatolia and then to Egypt, Greece, and throughout the whole Mediterranean (as well as Europe, southeast Asia, and part of Africa).

In the next millennium, the domestication of plants and animals consolidated as did the associated technologies, especially pottery, in their area of origin and spread through the neighboring areas, mainly Anatolia (where there was a major settlement at Çatal Hüyük) and the northern Aegean sea. The eighth millennium B.P. (between 8,000 and 7,000 years ago) is the Neolithic in southeastern Europe.

Yet the oldest known Neolithic settlements on the southern shores of the Mediterranean are in the second half of the seventh millennium B.P. (Fayum and Merimde, 6400 and 6200 B.P.), the same age as those in the western Mediterranean Basin, although the following development of Egyptian civilization was very different from the distant Neolithic settlers of the Iberian peninsula and Africa Minor.

From the east to the west

In fact, by 6,500 years ago, the Neolithic advances had already reached the Mediterranean coasts of the Iberian peninsula, at the western end of the Mediterranean Basin, as shown by finds of cultivated cereals in the cave at Or, in Beniarrés, near Gandía in the Autonomous Community of Valencia, or in the cave of Los Murciélagos in the Sierra de Cabra, southeast of Cordoba in Andalusia. However, the Iberian agricultural and cattle rearing societies coexisted at the margins of the land they exploited with people of hunter-gatherer lifestyles, who were recorded in the art by those known as the mountain painters who created highly stylized rock paintings, generally with hunting scenes, scattered in many caves and refuges, especially in the Autonomous Community of Valencia.

The cave paintings in this region are all schematic and are usually in a single colour. The human figure is their main character; the figures include men represented with skirts, phallic symbols and ornaments apparently made of feathers, and women with long skirts and bodices. The most frequently-represented scenes are hunts using bows, arrows and other jaculatory weapons, and the most frequently-hunted animals are deer, wild boar, bulls, and goats, presumably the most common species at that time.

The sheltered sites with cave paintings rarely contain archeological remains (meaning the paintings can only rarely be dated by relating their artistic style with the characteristics of the associated industries), but both the animals represented and the clothing of the human figures suggest they were painted in a period when the climate was not harsh, allowing the deduction that they must have corresponded to some time within the post-glacial period.

One of the most interesting finds, because of the large number of paintings it contains, is the Valltorta gorge in the Maestrazgo region (Autonomous Community of Valencia, Spain), between the towns of Albocàsser and Tírig, about 31 mi (50 km) north of Castellón de la Plana. There, along a rambla that only rarely bears water are about fifteen caves with hundreds of cave paintings. The main ones are in the Civil, Cavalls, and Saltadora caves in the farm of Josep.

Hunters and livestock rearers

The Cueva Negra (Black Cave) of Ares in the Maestrazgo, only 12 mi (20 km) from the Valltorta gorge, contains archeological remains between 7,500 and 6,500 years old, showing features typical of the Neolithic. The oldest traces of pottery are in a level more than 7,500 years old, and more recent layers (only 6,500 years old) contain cereal grains and the bones of domesticated animals as well as ceramics. This suggests that, at least toward the end of the second half of the eighth millennium B.P., the Mediterranean shore of the Iberian peninsula was shared by different populations at different levels of cultural development. On the one hand there were groups of hunter-gatherers who drew representa-

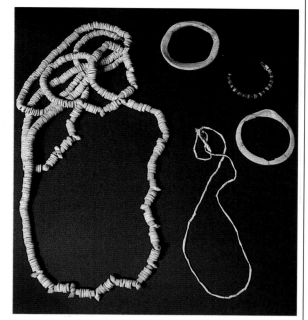

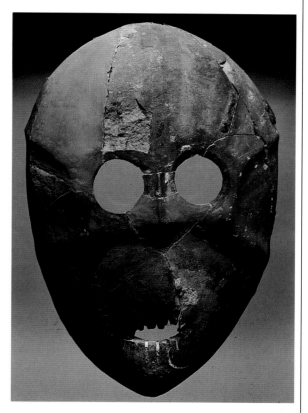

119 Neolithic Iberian jewellery and a Neolithic mask from Israel. The Neolithic saw the domestication of some animals and plants and the use of baked clay objects. The Mediterranean Basin was one of the three places on the planet where the cultivation of plants for human food supply first started. From the Near East, Neolithic culture spread along the entire coastline until it reached the western tip. This had varied consequences: Most agricultural societies became sedentary, a fact with demographic consequences—increase in the population. Human groups also became hierarchical, leading to increasing differences between individuals. The possession of land and of objects—even those of wood or pottery—began to be considered as a social feature. Funeral rites were also developed, as well as rituals intended to please the gods. It seems offerings were made all over the planet in ceremonies of greater or lesser complexity.
[Photos: Jordi Gumi and Zev Radovan]

tions of the animals they hunted, perhaps with the intention of favoring their hunt by magic. On the other hand, there were the groups of humans that knew how to work clay and who, at least since the middle of the seventh millennium B.P. knew how to work the land and rear livestock.

The Mediterranean area offered resources for both life-styles, and in the then remote eastern coastline of

Believing in the Mediterranean

L'École de Platon (1898) [Jean Delville (1863-1953), Musée d'Orsay / © Réunion des Musées Nationaux]

"Be fruitful, and multiply, and replenish the earth, and subdue it." This phrase from the Old Testament is a clear precept. A precept that has been followed to the letter. In fact, many basic patterns of human behavior as regards relationships with the natural environment were foreshadowed, if not explicitly stated, in the moral codes associated with religious beliefs. This is the case of Jehovah's command to the Israelites, so inappropriate on the eve of the twenty-first century, a time characterized by a demographic explosion and excessive human pressure on the planet. On closer inspection, many of the environmental problems faced by the Earth today have their roots in cosmological (and ecological) concepts contained within the religious beliefs that have developed over the centuries in the Mediterranean area.

We do not know how many pupils Plato had in his Academy in Athens. The symbolist painter J. Delville painted twelve pupils in the painting reproduced above, thus committing an undeniable iconographic error. Choosing twelve disciples implies adopting the old Mesopotamian tradition based on the sexagesimal system and shows broadly Mediterranean origins. A large part of humanity professes one of the religions of Mediterranean origin, specifically Judaism, Christianity, and Islam.

In fact, all three religions, and all their different forms, such as the Roman Catholic church, the Anglican and other Protestant churches, the Orthodox churches, etc., within Christianity, the Shia and Sunni divisions of Islam, and Pharisaism, cabbalism, and Hasidism in the case of Judaism, are variants on the fundamental concepts established by Judaism. This must be why the members of these religions are so intolerant of each other: they are too similar to forgive each other their differences. To sum up, Judaic or Judaeo-Christian princi-

ples have dominated the soul of Mediterranean civilizations, including their environmental attitudes, for more than 2,000 years—the soul of the Mediterranean civilizations, the entire Christian western world, and the many Islamic societies in Africa and Asia, and practically every corner of the planet that these civil sectors have reached, in fact everywhere. Even Marxism, with its messianic and salvationist attitude cannot escape being a distant derivative of Judaism. All this has undoubtedly had very important ecological consequences.

Hands of Fatima (North African Amulets) [Jordi Vidal]

Judaism, Christianity, and Islam are monotheistic religions based on revelation (doctrines disclosed directly by a single god, not the result of human reflection or reason), that believe in the existence of an immortal soul (to such an extent that all the actions and responsibilities of the human beings alive today are subordinated to achieving their trasnscendence to the immortality of the soul), and generate communities of believers and churches devoted to communitarian practices (these public rites are considered essential for the wellbeing or the survival of the entire group). This revelation and the human believers' direct and privileged relationship with the divinity stimulate the development of anthropocentric attitudes, a fact with great environmental consequences.

A Christmas scene made of paper and clay (18th century), Museu de Lluc, Majorca [Jaume Gual]

In fact, Mediterranean religions are anthropocentric. Even more, they proclaim humans should rule the world. Humans are considered to be the masters of the universe, and all its resources are at their command. No moral limits are placed on the exploitation of the environment, thus precluding all ethical considerations about people's relationship to nature. This is the reason why the environmental movement has had to base its arguments on the unsuitability in practice of certain human activities, rather than their possible moral inappropriateness. On the other hand, Christian religious art and Jewish and Islamic terminology have also implanted the unquestionable image of a white, male God, a fact with equally important anthropological and environmental consequences (the non-white ethnic groups, and indirectly the area they lived in, necessarily belonged to inferior classes, and could thus be appropriated by the whites).

On the more benign level of the use of symbols, the Mediterranean character is clearly developed, especially in the Judaeo-Christian religions. Some clear examples of this are the paschal lamb, the mystic *vesica piscis* (an almond-shaped aureole surrounding pictures of Christ, the Virgin, etc.), the soul-saving values attributed to the oils used in extreme unction. Bread made from wheat and wine made from grapes both play a very important role in the Christian eucharist, but these Mediterranean products are of doubtful evocative value for Christians beyond the seas. They are benign symbols that should not distract attention from a basic, much less agreeable fact: modern humanity's thoughtless, aggressive attitude

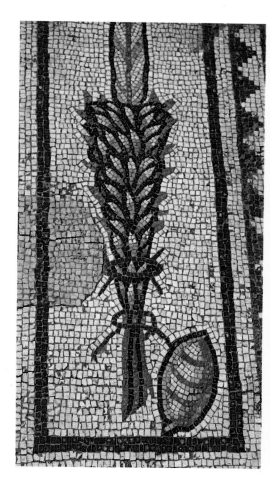

toward the rest of the biosphere is closely associated with the Jewish and Mediterranean roots of the dominant cultural values in the countries of the so-called developed world.

The four species required for the Feast of Tabernacle (citrus fruits, palms, branches from a leafy tree and from a willow growing by a river) [Zav Radovan, Jerusalem]

120 A Phoenician coin showing a warship propelled by oars, dated at 2375 B.P., from the National Archeological Museum of Beirut. Note the oar at the rear that served as the helm, and the mythical beast, the hippocampus (half horse, half dolphin) underneath. The Phoenicians were Semites that settled the Mediterranean coasts and excelled at sea trade. They did not ignore agriculture, but for a long time they held a monopoly on dyed woollen fabrics. They had sought out raw materials, especially purple (they extracted Tyrian purple from the mollusc *Murex*), and this took them to the other end of the Mediterranean, and according to some, even to the Canary Islands. On these voyages they founded a series of settlements on the North African coastline (Carthage) and in the Iberian peninsula (Gades, now Cádiz), although such settlements could not be considered an empire. They not only traded their own products, but also traded on commission and obtained profits from other people's products by distributing them. The Phoenicians improved western civilization's navigation techniques, explored the Mediterranean's coastline and organized commercial exchanges between distant regions and civilizations.
[Photo: Erich Lessing / Archiv für Kunst und Geschichte, Berlin]

the Iberian peninsula and the rest of the Mediterranean Basin while some pre-existing populations adopted the new colonists' culture and way of life, and maybe even their language, others stuck to their hunter-gatherer traditions until the first centuries of the Christian era. Still other groups, although adopting the new arrivals' technologies and culture, rejected their language and kept their ancestral language. This was the case, on the Mediterranean's European shores, of the Etruscans and the Iberians who retained their language until the Roman colonization, and of the Basques until well into the modern era (the area where Basque is still spoken is not Mediterranean, but historical records show it was formerly spoken over a much wider area, including much of the Ebro Basin, the Pyrenees, and perhaps as far as what is now the eastern edge of Catalonia). On the southern shores, the Berbers have kept their own language, in spite of the many occupations and invasions of North Africa over the last millennia.

In any case, by the beginning of the sixth millennia B.P., Neolithic culture had already taken root throughout the Mediterranean, and the small groups of hunter-gatherers that persisted were probably clearly a minority, if they continued to exist at all.

1.3 Native peoples and settlers from outside

Colonization of the Adriatic coastline and the western Mediterranean took place long before the settlement of many regions in northern Europe, showing the importance of sea routes in spreading the Neolithic's cultural advances. The fact that sea navigation was known in the eastern Mediterranean, even before the domestication of plants and animals, is convincingly shown by finds of obsidian from the Aegean island of Milo in several very old Neolithic settlements in Anatolia and the Fertile Crescent. The first Neolithic settlers of Crete and Thessaly arrived by sea from Anatolia. Despite the riskiness of the means of sea travel available to the Aegean's early navigators, they established the first contacts between the different shores of the Mediterranean. The colonization of new lands and the exchange and mixing of populations has been constant throughout the Mediterranean's history—a story of strong identities derived from repeated interbeeding of different peoples, of deep roots grafted on the yearnings of emigrants or the banished. The present distribution of peoples and cultures on the shores of the Mediterranean is the result of all the population movements, occupations, and colonizations that have taken

place over many thousands of years; many starting from within the Mediterranean, although many others started from places far outside it.

The routes of antiquity

The Iron Age is usually considered to have begun about 3,000 years ago, at the beginning of the third millennium B.P., at the same time as the colonial and commercial expansion of the peoples of the eastern Mediterranean (both Phoenicians and Greeks) toward the west. The Phoenician city states (Tyre, Sidon, Tripoli, Byblus, Beirut)—freed from Egyptian domination two centuries previously, due to the crisis in the power of the Pharaoh caused by the invasions of the "people from the sea"—reached as far as the Strait of Gibraltar and beyond with their boats before 3000 B.P. The Greek city states, after the dark age that followed the collapse of Mycenean culture, began to establish colonies on the Anatolian coastline, and after the 28th century B.P. they began to expand. This period 3,000 years ago also saw the consolidation under kings David and Solomon of the first Jewish state in Palestine and the definitive defeat of the Philistines, the last of the "people from the sea." The decadent Egyptian Empire of the 21st dynasty—divided into upper Egypt (governed from Thebes by the high priest of Amon-Ra) and lower Egypt (governed by the legitimate pharaohs from Tanis in the Nile delta)—was the other main figure in the eastern Mediterranean around 3000 B.P.

The eastern sailors basically used three routes to reach the western Mediterranean. The first followed the coastline of Anatolia, the Aegean archipelago, and the southern coasts of Greece down to Corfu. From there, with a favorable wind, in a light sailing boat like those used by Greeks and Phoenicians, the

Strait of Otranto could be crossed in less than one day (unless they chose to follow the Adriatic coastline in search of amber from the north and tin from Bohemia, as Mycenean traders were doing by 3700 B.P.). They could then follow the Italian coastline to the Strait of Messina, where they could choose between the northern route along the coasts of the Tyrrhenian Sea, or the routes to the south or to the west, following the eastern coast or the northern coast of Sicily.

A second, southern, route, followed the coastline of Africa, as far as the Columns of Hercules (the Strait of Gibraltar) and Tartessus. This was the route of the Phoenician navigators who founded Utica around 3100 B.P. at the mouth of a bay that has now been filled by the alluvial deposits of the river Medjerda, to the northeast of the modern city of Tunis. Later they founded settlements on the Atlantic coast, beyond the feared Strait of Gibraltar, such as Gades (now Cádiz, in southern Spain) and Lixus (close to the modern city of Larache, in northeast Morocco). According to tradition they later founded the city of Carthage in 814 B.P. not far from Utica. After the fall of Tyre to Nebuchadnezzar II of Babylon in 574 B.P., Carthage became the center of the Phoenician, or Punic, civilization.

The third route was also opened up by the Phoenicians and ran through the middle of the Mediterranean, based on a chain of island stopovers—from Cyprus to Ibiza—some large distances apart, thus making it necessary to sail far from land. On journeys from Cyprus to Crete, from Crete to Malta or Sicily, from Sicily to Sardinia, from Sardinia to Minorca, it was necessary to navigate for a few days without land in sight. They navigated using the sun and the pole star, which the Greeks knew by the name of Phoinike (that is to say Phoenicia), an appropriate compliment for the marine skills of their competitors, who on their return spread terrifying rumors about the remote coasts of Iberia, Mauritania, Sicily, and the Balearic Islands. These legends were transmitted to the Greeks who incorporated giants like the Geryon, Cyclops and Lestrigons, sorceresses like Circe, fabulous beings like Scylla, Charybdis and the sirens, almost all of which lived in what for the sailors from the eastern Mediterranean was the far west.

The new western cultures

The reality was less fantastic, but the people living in the western Mediterranean were just as dangerous as the beings in the stories, and real riches were to be found there. Far to the west, the Tartessans developed

121 **Frescos from Akrotiri (Santorini, known as Thira today, in Greece) from the fourth millennium B.P.** These important murals were discovered only recently (1967), and it is hoped that remains of a more prosperous civilization earlier than Minoan Crete will be found under the volcanic ashes. The interpretation of archeological remains in the eastern Mediterranean seems to show sea voyages were frequent earlier than had been thought, but this is not too surprising considering that the scattered islands encouraged the development of navigation. Abundant prehistoric remains on the isle of Milo show the arrival of human beings, who are thought to have been searching for obsidian. Tips made of this volcanic mineral have been found on the mainland and on other islands in the Mediterranean where it does not exist spontaneously.
[Photo: Erich Lessing / Archiv für Kunst und Geschichte, Berlin]

from the 8th century B.C. onwards an original civilization clearly influenced by the Phoenicians who had been established for more than two centuries in Gades. Likewise in the Balearic Islands, at least in Majorca and Minorca, the culture known as *Talayotic* was flourishing; its largest monuments appear to date from about 3000 B.P. or a little earlier, but it developed without interruption until the Roman occupation at the end of the 22nd century B.P. On the Mediterranean coastline of the Iberian peninsula a profound transformation took place throughout the first half of the third millennium B.P. Successive waves of Indo-European crossed the Pyrenees about 3000 B.P. and their influence, together with that of the eastern colonists, profoundly changed the local bronze cultures, the predecessors of the Iberian culture that from the 25th century B.P. stretched from what is now southern Andalusia to the coast of the Languedoc and inland to the middle basin of the Ebro (where it came into contact with Basques and Celtiberians), as well as occupying a part of La Mancha. The Etruscans undeniably dominated the Italian peninsula at the beginning of the third millennium B.P. However, at the beginning of the 27th century B.P., an obscure Italic people, the Latins, began to lay the foundation of what would later be Rome. By the early Christian era, and for the next few centuries, Rome was the dominant civilization of the ancient Mediterranian world.

Less is known about the situation at that time in the southern part of the Mediterranean Basin. To the west of Egypt, the Libyans were nomadic shepherds who waged regular wars with their rich neighbors, and in certain periods exercised some power over Egypt. The 22nd and 23rd dynasties (2945 B.P.-2730 B.P.), which corresponded to periods of chaos and disorder in Egypt, were Libyan dynasties. Little is known about the Garamantes, Numidians, and Mauritanians (possibly the ancestors of modern-day Berbers), until the Punic Wars, when they intervened alternately in favor of the Carthaginians and the Romans, and some of their kings entered history by way of Latin sources. Most of these peoples on the Mediterranean's northern and southern shores were descended from the population surplus that had emigrated in all directions from the Fertile Crescent after the Neolithic revolution. In fact, the physical differences between the human populations of the northern and southern sides of the Mediterranean are not very significant, although there is high variability within each population. Probably the invasions after about 3000 B.P. have only in a few cases supplied genetic material distinct from the common gene pool of all the Mediterranean peoples. The Celts that invaded Gaul and much of the Iberian peninsula, the Germanic peoples that destroyed the Roman Empire, and the Slavs that completed the task in the Balkans, were all Indo-European peoples that separated from the common stem 4,000 years ago, and thus with a very brief separate evolutionary history. Other invasions by populations that had been separated for longer periods, such as the North African Arabs' invasion of the Iberian peninsula in the 6th century, or the Turk's invasion of Anatolia, the Near East, and the Balkans in the 15th century, never consisted of large numbers of people. They could conquer territory, and even impose their language and culture, but their genetic heritage was rapidly diluted in the demographically larger population.

From one end of the basin to the other: Greeks and Berbers

We cannot follow the demographic history of all the Mediterranean peoples, so we will limit ourselves to two very different examples from the Mediterranean Basin—the Greeks and the Berbers. Modern Greeks, the inhabitants of the south of the Balkan peninsula and the islands of the Aegean, live in the eastern end of the northern shores of the Mediterranean, while the Berbers live in the Maghreb, at the western tip of the southern bank. These two examples of Mediterranean populations lack significant overall differences, but each shows considerable variability, the result of their different histories.

Both Greeks and Berbers are pale-skinned peoples with relatively dark pigmentation. Most Greeks have high skulls that are short from front to rear (brachycephalic), while most Berbers have skulls that are relatively longer from front to rear (dolychomesocephalic). Both populations are dominated by individuals with curly, dark hair, although both have some fair individuals and there are even some redheads, especially in the Berber populations of the Rif mountains. Both populations have many individuals with oval-shaped or elongated faces, but the latter are more frequent among the Greeks. Their noses are also narrow and elongated with a straight or slightly convex profile, especially the Greeks, who often have noses with a high, protruding, convex bridge. These similarities can be explained by their common origin in the populations of the eastern Mediterranean, but their differences can be explained by the genetic contributions from the other populations with which they have been in contact.

The Balkan Mediterranean peoples

Europe's first farmers and livestock raisers settled in the Balkan peninsula, in what is now Thessaly and perhaps in Macedonia and central Greece too, before 8000 B.P. Remains of the physical types discussed above as dominant in modern Greeks have been found in Anatolian burial sites about 5,000 years old. The Neolithic farmers that reached Greece exploited ecological niches very different from those exploited by the hunter-gatherers resident there since the Paleolithic and Mesolithic, who were displaced or assimilated. The cultivated plants identified in the oldest Neolithic settlements in Greece correspond to the group of plants domesticated in the Fertile Crescent: emmer, einkorn, and barley, as well as legumes, such as the pea. The domesticated animals were mainly sheep and goats, although there are also remains showing they reared cattle and pigs.

The archeological remains left by the inhabitants of ancient Greece, between 8,500 and 5,000 years ago, appear to show an economy based on farming and rearing livestock, with simple lifestyles and a patriarchal social organization. They knew how to navigate, as shown by the fact that to the northeast, in Thrace and eastern Macedonia, Neolithic remains are not as old as those found in Thessaly and southern Macedonia, apparently showing that the first agriculturalists reached the Greek coast by ship.

Bronze also arrived by sea from Asia Minor, where it had been used for about a thousand years. Bronze is first known in Crete about 5000 B.P., and it then spread to the Cyclades and to mainland Greece. The first urban "Greek" culture developed in Crete at the same time as the spread of bronze. The texts found at Knossos, Crete (dated to between 3450 and 3200 B.P.) that have been deciphered were written in an archaic Greek, resembling that used in similar texts found in Mycenean cities on the Greek mainland. This discovery is important in terms of the identity of the populations affected, because, although the term "Greek" was not used until the Roman period, all the Greek populations recognized their "kinship" because they spoke the same language. In fact, these Greek populations divided the known world into two groups: those who spoke their language, and those who spoke a strange tongue, whom they called barbarians.

A long and obscure period of Hellenic history began about 3200 B.P. Traditionally, the classic Greek historians attributed this change to the arrival of successive invasions of iron-working, Indo-European populations (Thessalians, Boeotians, and Dorians) from the north. Yet, the evidence of archaeology, in spite of showing important destructions that represented the

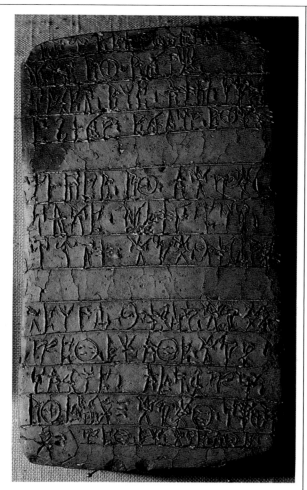

122 **Clay tablet in the Linear B script, from the Pilos site** (now in the British Museum), dated at between the 15th and 12th centuries B.C. The Mycenaean Greeks (or Achaeans) used a syllabic script, called Linear B, that was derived from the not yet deciphered Minoan script —Cretan—that is called Linear A. The documents written in Linear B that have been deciphered are inventories (oarsmen, herds, or quantities of grain), receipts, and working instructions. The Greek language became important later, when it adopted the Phoenician script and developed a more effective system than its former syllabic alphabet. The Phoenicians were inspired by Egyptian hieroglyphics and created a script with a single ideogram for each phonetic value. After the changes by the Greeks *aleph* (bull) became *alpha*; *bet* (house) became *beta*; *gami* (camel) became *gamma*, etc. The interesting things about the Phoenician alphabet is that it is the only one that has survived and that it has given rise to all the major script systems in the world, except those of the Far East. [Photo: The Ancient Art and Architecture Picture Library]

definitive collapse of Mycenean civilization, appears to indicate an absence of major population movements. Rather it seems to show that these movements, where they occurred, were small in size and limited to the Hellenic world, with some maritime migrations to different points of the Mediterranean, mainly the coasts of Asia Minor. The Ionian descendants of these Greek immigrants left Asia Minor when the population reached high densities, going by sea to places as far away as Massalia (now Marseilles) in Provence, or Rhode (now the town of Roses in Catalonia), where colonies were founded between the 8th and 5th centuries B.C.

Greece's golden age was in the 7th and 6th centuries B.P., when the democracy of Athens was established. This golden age was followed by Alexander the Great's Hellenic empire, which some would say laid the basis for western civilization. This period saw the rise of a brilliant philosophical and humanist movement to which we owe our awareness of individual freedom and rationalism in philosophy: This movement also provided the basis for western conceptions of art, literature, and the development of scientific thought.

123 **An Italian coin with a head bearing a laurel wreath.** The goddess Diana next to laurel on the other face. Below, a drawing of a flowering and a fruiting branch of the laurel (*Laurus nobilis*). For the ancient Greeks the laurel was sacred to Apollo. Classical athletes won wreaths of laurel leaves (and were thus called *laureates*) and laurel wreaths were also used to crown the emperors of Rome. The laurel is a typical tree of the Mediterranean Basin that grows in moist but temperate sites (gorges, at the base of rocky slopes). It might not be a native to the area, as it is has been cultivated since antiquity. It is known to have been present in the Cenozoic and thus it is not unthinkable that some populations might have survived in refuges in sheltered streams, although with a distribution restricted by the cold periods of the Quaternary.

[Photo: Jordi Vidal; drawing: Eugeni Sierra]

The Greeks of those times successfully resisted attempts to incorporate them into the Persian Empire, and struck back against Persia when Alexander expanded to the Indus River and the shores of the Aral Sea. The decadence of the Hellenic empires created by Alexander's successors coincided with the rise in the Mediterranean of a new power, Rome, which incorporated Greece into its territories in 2146 B.P., the same year as the destruction of Carthage.

The Mediterranean peoples of the Maghreb

The Berbers appear to have reached the southern shores of the Mediterranean, now known as the Maghreb, about 10,000 years ago as the proto-Berbers, with the culture known as *Capsian*. They settled the Mediterranean, displacing or absorbing the preceding Paleolithic inhabitants who did not know how to cultivate the land. In fact, they were a people consisting of different tribes joined by their (Afro-Asiatic) language whose original and characteristic script spread to cover the area from the Mediterranean to the Niger, and from the Egyptian oasis of Siwa to the Canary Islands, more or less the space now occupied by Berber-speaking populations. In North Africa, the change from hunting and gathering to the domestication of plants and animals was a slow process that lasted from the 7th to the 5th millennia B.P., with the different Maghrebian societies developing at different rates. Some were backwards in adopting pottery, while others did not adopt livestock-rearing until much later. The Neolithic is thought to have entered this part of Africa with groups of fugitive Canaanites who arrived by sea from the eastern Mediterranean, and interbred with the ancient Berbers. The Neolithic economy established in the Maghreb was the same as in the rest of the Mediterranean, although it had some subtropical influences. Some tribes were genuine peasants settled on their own land, such as the populations of the Rif mountains and the Kabylie region of Algeria, where olives, figs, date palms and apricots are still cultivated. Other tribes knew how to work the land, and cultivated cereals, especially the tribes living in southern Tunisia and Morocco's Anti-Atlas mountains. There were also semi-nomadic tribes that cultivated trees such as the date palm and reared livestock, such as cows, goats and sheep, and were always roaming in search of new pastures. The last type was represented by the nomadic desert tribes whose economy was based on trade. The Phoenicians, or perhaps the later Carthaginians who succeeded them and increased the Punic colonization of the Maghreb coastline, introduced the use of metals, new crops and the domesticated horse. The date of the introduction of the camel into the Maghreb area is even less certain. Pliny does not mention it, but Julius Caesar captured 32 camels from Juba I, king of Numidia and Mauritania, when he beat him in 2046 B.P. in the battle of Thapsus. As a result of the Punic Wars the western Berbers entered written history (the Libyans, further east, had already been mentioned due to their repeated confrontations with the Egyptians) when king Masinissa of Numidia and Mauritania formed a unified state. Rome's victory over Carthage and destruction of the city in 146 B.C. (the same year as the Romans occupied Greece) represented the incorporation of all North Africa into the Roman Empire. Yet the process of Romanization appears to have been relatively superficial in North Africa's rural areas, unlike in other regions of the empire, such as Gaul and Hispania. More than 500 years later when the Roman Empire was in decline, Augustine, the African bishop of Hippo, on what is now the Algerian coast, wrote between A.D. 395 and A.D. 430 that the peasants in his diocese still spoke the Punic language and considered themselves Canaanites.

The Roman-German connection

The period of the Roman Empire, at least from Augustus to Constantine (27 B.C.-A.D. 337) was the only period in history when the Mediterranean Basin has ever been under the control of a single power. A power that reached far to the north and to the east of the Mediterranean, and which on the one hand caused the extinction of many pre-existing cultures and assimilated under its influence the populations of many provinces (especially the most western ones), but on the other hand incorporated into the common heritage many of the common practices, traditions and rites of several of these cultures. The splendor of the empire, felt not only in Rome but in all its provinces, at least in their capitals and in many of the lesser cities, attracted foreign settlers (leaving to one side the slaves brought from military campaigns on the borders of the empire). Celts, Germanic peoples, and Sarmatians advanced from the north, while Armenians, Persians, and Babylonians entered from the east. Ethiopians and Garamantes entered from the south. They all ended up creating a group of human populations and types with enough features in common to have been called a Mediterranean race. Later, on a more or less local scale, there would be the fruits of successive waves of immigration, not very important in their demographic effects, but sufficient to add additional variation to some populations. Examples include: the Germanic barbarians who spread from the north into the Iberian peninsula, into Occitania, into Italy (especially in the north, but even into Sicily, which was invaded by the Vandals in the 5th century and by the Normans in the

124 **A milestone on the former Via Augusta** found in 1967 and now in Bisbal del Penedès (Catalonia). Milestones were columns that the Romans placed along major communication routes to mark how far the place was from Rome, measured in Roman miles, which consisted of a thousand (mil) paces. The Roman empire built more than 49,700 mi (80,000 km) of roads across its territories, some more than 30 ft (9 m) wide. This road network allowed the transport of goods, administrators, and legionnaires, meaning limited military forces could control large areas. There are still many roads following the Roman routes, and while some are now highways, others have changed little since when they were built.
[Photo: Lourdes Sogas]

11th century) and even into North Africa; Arabs from the Near East who invaded North Africa and the Iberian peninsula; black Africans who also invaded North Africa; Turks from central Asia who have also left their mark on the eastern Mediterranean.

The Romans were more culturally influenced by the Greeks than the Greeks by the Romans. Under the Byzantine emperors, from the 4th century to the 15th century, the Greeks lived a hectic period of invasions, wars and occupations, none of them long-lasting or demographically important. The barbarian invasions of the 4th century did lead to an increase in population, basically because the entire interior of Greece was depopulated and most of the native population concentrated in the cities and on the coast, while the interior was successively invaded by Visigoths, Ostrogoths, Huns, and Slavs, although only the Slavs settled, as shepherds, in the interior part of the Peloponnese. The Arabs (in Crete), Bulgars, Normans, Venetians, Burgundians, Franks, Catalans, Serbs, Albanians and finally Turks, invaded and conquered the cities and regions of Greece during this period until in 1460 the Turks completed their occupation which lasted until 1830, the year of the establishment of the Kingdom of Greece, whose territory was limited to the Pelopennese, Attica, Levádhia, the Northern Sporades, Euboea, and the Cyclades. The Turkish tribes established in central Asia adopted Islam after the 7th cen-

125 An astrolabe made in Toledo in 1068 by the Al-Andalus craftsman Ibn Said al-Sahli. On the back it has the typical zodiac calender with the names of the months of the Julian calender in Arabic and Latin, with a calender diagram around the center. Islam, with its music and language, unified the entire area from the Maghreb to central Asia and brought the civilizations of the Mediterranean Basin into contact with those of the east. The results of the Islamic influence include the arrival in the west of high-quality steel, the stirrup (invented by the Mongols), silk, paper, porcelain, and technical improvements, such as what we still call Arabic numerals (in fact of Hindustani origin, although they then lacked a symbol to represent the zero), and others in metalwork and agriculture (especially the oasis) and the astrolabe which was a Chinese invention improved by the Arabs.

[Photo: Museum of the History of Science, University of Oxford]

tury A.D., and from the early 8th century onwards they were pushed by the Mongols from Turkestan toward Anatolia. They in turn displaced much of the Byzantine population of Asia Minor to Greece. The conquest of Gallipoli in 1354 was the start of the Turkish settlement of Europe. The Turkish conquest of Constantinople a century later marked the definitive end of the Byzantine Empire. In spite of four centuries of Turkish occupation, the religious and social barriers between Greeks and Turks coupled with the migrations in both directions after Greek independence explain the fact that the Turkish contribution to the current population of Greece was very small. Nor was the population of North Africa, at least the peasantry, much influenced by the basically city-dwelling Romans, even though it was one of the main areas supplying cereals to Rome. The Vandals, one of the Germanic peoples whose invasions from the 4th century A.D. onwards finished off the Roman empire, settled in North Africa in A.D. 429 and gradually disrupted the entire area's Roman agricultural infrastructure and exchanges with Rome. From their African bases they conquered all the islands in the Mediterranean, until they were defeated in A.D. 534 by the Byzantines, who incorporated the kingdom into the Eastern Roman Empire, the only remaining part of the Empire after the fall of Rome to the Barbarian Odoacer in A.D. 476.

The expansion of Islam

In the 7th century, the Arabs were moved by their new faith, Islam, to conquer the world. Little more than a century after the Byzantine occupation of North Africa, the first conquering Arabic tribes reached what is now Tunis, and in 670 they founded the city of Kairouan. By the early 8th century they had already reached the Atlantic coastline of

Morocco, and shortly afterwards they crossed the Strait of Gibraltar and settled in the Iberian peninsula. The Arabs rapidly converted the Berbers to Islam, although in rural areas there were (and still are in spite of "Arabization" campaigns by some governments in the Maghreb) centers of resistance to this acculturation and these areas retained the language and customs that preceded Islam.

From the second half of the eleventh century onwards, confederations of Berber tribes created large Berber states in the Maghreb and in Al-Andalus (the zone of the Iberian Peninsula occupied by the Muslims). The Zenága tribe established the Almoravid empire, which between the middle of the 11th century to the middle of the 12th century stretched from Senegal and the Niger to the Ebro, with its capital in Marrakesh. The Masmuda tribe gave rise to the Almohad empire, which succeeded the Almoravid empire in the Maghrib and Al-Andalus after they captured Marrakesh in 1146 until they lost it to the Benimerines in 1269. Coinciding with the decadence of the Almohad empire, several Berber states formed in the Maghreb from the middle of the 13th century and consolidated in the 14th century. These included the Hafsites from Ifriquiya from 1229 onwards, the Abdalwadis in Tlemcen from 1236 onwards, the Benimerines who arose from the Zenata confederation, the traditional rival of the Zenága, the successors of the Almohads in modern-day Morocco from 1269 onwards. The expansion of the Ottoman empire in the mid 16th century ended their independence, except for Moroccco, where the last Berber monarch was displaced in 1542 by the Arabic Sadi dynasty, the origin of the current Sharifian monarchy. Today, the presence of non-Arabized Berbers is basically concentrated in Morocco and in some regions of Algeria (Kabylia). In general, the Berbers living in the mountainous areas from the Rif to the Sahara have kept their language and customs, while those that live in the plains are completely Arabized.

The case of the Iberian peninsula

The Iberian peninsula may serve as an example of the genetic reconstruction of the history of the human population of a peripheral territory of the Mediterranean Basin, recently studied by applying cartography to a principal component analysis of gene frequencies. The analysis, based on a study of 635 population samples for 20 gene systems with a total of 54 different alleles, identified three main components, which correspond to the basic demographic contributions to the population of the territory in question.

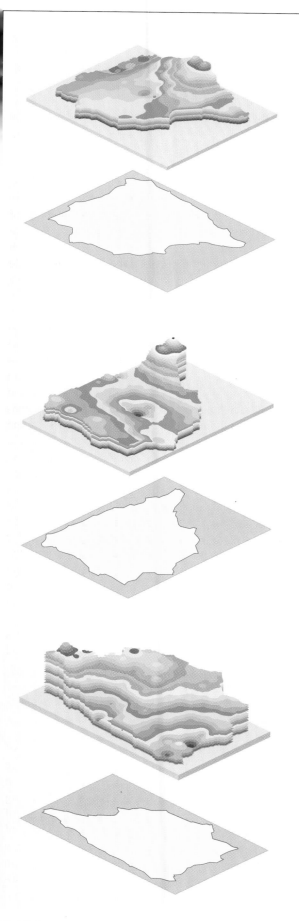

The first main component accounts for 21.7% of the variation in the populations studied, and shows the difference between the Basque population, with a genetic inheritance mainly derived from the Paleolithic or Mesolithic populations present before the waves of populations that took Neolithic culture to western Europe, and the rest of the peninsula's population which is dominated by genetic elements derived from these later arrivals. The second main component accounts for 14.5% of the variation and shows the distinctiveness of the descendants of the first infiltrations of the Neolithic populations to reach the eastern tip of the Pyrenees; that is to say, it distinguishes the populations of the northeast of the peninsula (Catalonia, Andorra, the eastern half of Aragon and the north of the Autonomous Community of Valencia) from those of the rest of the peninsula. Although this area, because of its geographical location, has been crossed by all the invaders and has received many waves of immigration in the course of its history, it appears that none of them has had such great demographic effects as the first. The third main component accounts for 12.3% of the variation and shows the differences between the east and south of the peninsula and the populations in the north and west, that is to say, the Mediterranean regions from the Atlantic ones. This corresponds to the distinction established during the course of the third millennium B.P., between populations with Iberian culture, of non-Indo-European language, and Celtic or Celtiberian cultures, of Indo-European languages, that were introduced with the migrations of the first Iron Age.

None of the other historically known immigrations, not even the long Muslim occupation of Al-Andalus, from the beginning of the 8th century to the end of the 15th century, accounts for such a large percentage of the genetic variability within the population of the Iberian peninsula. This should probably be attributed to the great difference in numbers between the large population of the 8th century Iberian peninsula and the limited numbers of the invaders, many of whom were Berbers. The Islamization and Arabization of the population must have been mainly a cultural phenomenon, without important changes in the population, except in very localized cases.

Exporting Mediterranean culture overseas

The colonial tradition of the ancient Greek and Phoenician navigators was revived with the 15th century's voyages of discovery, when sailors from Portugal and Castile went far into the Atlantic, leading to the first crossing of the Cape of Good Hope in 1487 and the discovery of America in 1492. The overseas expansion of Europe started 500 years ago and was led

126 Genetic maps of the Iberian peninsula. The geographical distribution of gene frequencies makes it possible to explain the histories of different populations. Using the large quantity of data available for gene frequencies (such as blood groups) in the Iberian peninsula, it is possible to draw a distribution map for each allele. The information contained in dozens of these maps can be summarized by means of a statistical method called principal component analysis. The drawings show the distribution of the values of the three principal components, which explain 48.5% of the variability (21.7% in the first, 14.5% in the second, and 12.3% in the third). Colors (from warm to cold) are assigned to each interval in the values of the three components, and in each case the map of the peninsula has been turned to show most clearly the rises and falls in the distribution of frequencies. The results show that the first principal component —with a peak in the western area of the Pirineos and a minimum in the northwest— reflects the difference between the Basque population in the Mesolithic or perhaps in the upper Paleolithic (top diagram). Geographical, ecological, and cultural factors contributed to maintaining this differentiation. The second principal component—with a peak in the northeast and a minimum in the center of the peninsula—reflects the arrival of populations bearing the Neolithic economy, which spread from the east of the Pyrenees first along the Mediterranean coastline, and then toward the Meseta (central diagram). The third principal component shows the separation between the Atlantic and Mediterranean basins that was expressed culturally in the Celtic and Iberian traditions.
[Cartography: Editrònica, based on Bertranpetit and Cavalli-Sforza, 1991]

Legionnaires and feudalists, traders and manufacturers

Roman glass vase [Index]

Barcelona is now a city of two million inhabitants, with a million more in its metropolitan region. Yet two millennia ago when Roman legions reached the flat area the city now occupies, they found a swampy landscape with just a few scattered settlements of an Iberian people, the Laietani on the hills that stood out, such as Laie and Barcino. Inland, the area was covered by forests of deciduous and evergreen oaks, with a few scattered clearings of cultivated land. They decided to stay there.

Barcelona (1740) [Barcelona Council / BIMA]

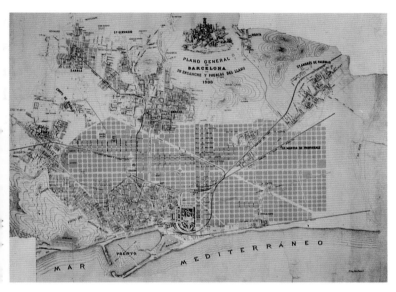

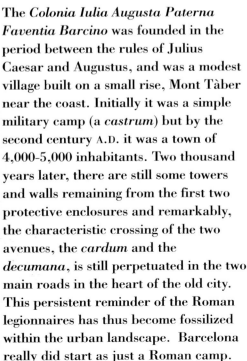

The *Colonia Iulia Augusta Paterna Faventia Barcino* was founded in the period between the rules of Julius Caesar and Augustus, and was a modest village built on a small rise, Mont Tàber near the coast. Initially it was a simple military camp (a *castrum*) but by the second century A.D. it was a town of 4,000-5,000 inhabitants. Two thousand years later, there are still some towers and walls remaining from the first two protective enclosures and remarkably, the characteristic crossing of the two avenues, the *cardum* and the *decumana*, is still perpetuated in the two main roads in the heart of the old city. This persistent reminder of the Roman legionnaires has thus become fossilized within the urban landscape. Barcelona really did start as just a Roman camp.

Later, however, it became a center of feudal rule. After the dark centuries of the barbarian invasions and the early Middle Ages, during which it knew brief periods of splendor (as the capital of Ataulf's Gothic kingdom in 415) and ruin (when the city was taken and destroyed by Al-Mansur in 985), Barcelona consolidated itself as the capital of a county, bordering Al-Andalus. At first it was a feudal territory of the sovereigns of the house of Charlemagne, then later it attained independence, around which other areas became associated, forming a territory that some already called *Catalunya* (Catalonia).In 1162 the Counts of Barcelona became kings of Aragon and this was followed by the resettlement of the Balearic Islands and Valencia, then becoming lords of Provence and Corsica, occupying Sardinia and Sicily and establishing the bases for a Mediterranean seapower to rival Genoa, Pisa, and Venice. Barcelona became so rich that its city walls became too small. By the middle of the 14th century new city walls had to be built to incorporate the new buildings outside the walls to the west, almost doubling the size of the city. The military camp and Roman town had become a medieval city.

still numerous open spaces in the then new Raval district.

Barcelona had become a city of traders and artisans, a more-or-less enlightened commercial center.

And this medieval city dominated a large area around it. In Roman times some of the wetlands had been drained and almost the entire plain of Barcelona was under cultivation. The forests were limited to the highest and steepest parts of Collserola, the hills to the north and west of the city. In the early Middle Ages, despite the ups and downs of population growth, land use had changed little from Roman times. However, when the city started to flourish in the late Middle Ages, the urban space grew and at the same time agriculture became more intensive. It became necessary to bring irrigation water from the River Besòs and to dig many channels in the slopes of the Collserola hills for water catchment. Yet wheat production was still insufficient, and only control of overseas wheat markets, such as those in Sardinia and Sicily, kept the people of Barcelona from going hungry.

Before the new city walls were finished, the plagues of the 14th century decimated the population of the city and the whole of Catalonia. Foreign wars, new epidemics, famines, and a destructive civil war brought the city to its knees at the end of the 15th century. Barcelona's population did not recover to the level of the mid 14th century (35,000 people) until the late 18th century. That century began disastrously with the War of the Spanish Succession (1705-1714), when Barcelona fell after a long siege, (a defeat that was the end of the freedoms of the Catalans, and was followed by the destruction of an entire district of the city, the Ribera, in order to construct a fortress). In spite of this the 18th century saw the beginning of the economic and demographic recovery of both Barcelona and Catalonia. The population and economy grew, and industrialization began, at the time unique in the Mediterranean. Buildings invaded the

Rose window in the Palace of the Generalitat
of Catalonia (15th century), Barcelona
[Ramon Manent]

It was also becoming an increasingly powerful urban area by enclosing more and more territories. Even though the population was declining, the late-medieval food production methods were still insufficient, because there were not enough manual laborers in the countryside and some marginal sites first cultivated in the centuries of expansion were abandoned. Resorting to external supplies became increasingly problematic as the Mediterranean was very unsafe. But all this changed at the beginning of the 18th century, when population growth, the intensification of agriculture, and the needs of artisans seemed to conspire to ensure the elimination of the remaining forests on the heights of the Collserola hills and beyond. The price of firewood rose faster in the second half of the 18th century than the price of bread. In addition, the drying out of the last wetland areas to the east of the city began. The city spread unceasingly into the entire territory.

In the 19th century, industry continued to grow, mainly due to the thriving textile industry. The city's population increased with the arrival of many agricultural workers and peasants in search of work in industry. This was the time to tear down the walls, not to build larger ones. Thus, in 1854 the city got rid of its former protective belt, and the eighteenth century urban center began to grow over the surrounding plain. The design of Barcelona's "Eixample" district (*eixample* in Catalan and *ensanche* in Spanish both mean an extension) was a return to the classic grid pattern of Roman origin. Twenty centuries after the Romans built their camp, a new version of the *castrum* was built, but to fulfill the requirements of a very different urban system. After absorbing all the surrounding populations in its accelerated growth, Barcelona began to be a modern industrial city.

The ceramic dragon in the Güell Park, built between 1900 and 1914 by Antoni Gaudí, Barcelona [Barcelona Council / BIMA]

Legionnaires and feudalists, traders and manufacturers: the forerunners of a quaternary sector that may well inspire the Barcelona of the 21st century.

127 The interior of a colonial house in Mediterranean Chile. The colonization of new territories led to the export of the customs of the Mediterranean Basin and their mixture with local customs, creating a characteristic style. Architecture is just one example. Especially in sites with climates similar to the Mediterranean Basin, the arrangements typical of the Mediterranean Basin were used in the houses. In colonial construction it is possible to recognize, among the characteristics typical of the local people, the floorplan of a Roman villa with the *impluvium* turned into a central patio around which the rooms are arranged. This arrangement still exists in some places in the Mediterranean (the typical Majorcan *patio*). Another typically Hispanic detail is the abundant use of terra cotta floor tiling.
[Photo: Jaime Álvarez / Fotobanco]

by people from the Mediterranean. Thus, the first cultural features and the first domesticated plants and animals introduced into the Americas (and into the Portuguese colonies in Africa and Asia) were also Mediterranean. These introductions, spread either by force of conquest or by the invasive nature of some organisms, were not always welcomed. The first trials were in the Azores in Madeira and in the Canary Islands. The Azores were uninhabited until the first half of the 15th century, when they were colonized by the Portuguese, who introduced their crops and domesticated animals. The Canary Islands were inhabited by the Guanches, a population derived from the Berbers, who did not know how to work metal but who had introduced agriculture and livestock rearing to the islands at some unknown time, possibly in the first centuries of the present era. They were colonized by Castile. In Porto Santo, a small island close to Madeira, rabbits introduced by colonists destroyed the island's entire vegetation. In Madeira a devastating fire had virtually the same effect. In the Canary Islands, at the end of the 16th century there was not a single Guanche left, although the islands had a large population (about 20,000 inhabitants) according to all the records dating from the period of the conquest. By the middle of the 16th century all these islands had populations that were basically of Mediterranean origin. The arrival of the first European colonists in the Americas, starting with the voyages of Columbus, opened up a new route for the export of Mediterranean culture. The first

European settlers of the Americas came from the Iberian peninsula, and to a lesser extent from the other European dominions of the kingdoms of Castile and Portugal. They took their languages with them, which were derived from the Latin—the language of communication throughout the Mediterranean in the first four centuries A.D. They also took their religion, Christianity, born in the Mediterranean where it was most deeply rooted. The imposition of all this on the native populations has left a deep and lasting mark on most of the Americas. The area now known as Latin America, or sometimes as Ibero-America, is the area with the greatest number of people speaking neo-Latin languages and belonging to the Roman Catholic Church. Many features of the folklore, popular music, and poetry, as well as the religion, social relations, and social stratification of many populations in Latin America reveal that their traditions came from the old Mediterranean. This is true to a lesser extent of the other areas colonized by the Spanish and Portuguese, such as the Philippines, Timor, Angola, Mozambique, Equatorial Guinea, and the former Portuguese settlements in India.

The colonization of the Americas started with the Antilles, the first land discovered, especially the islands of Hispaniola (now the Dominican Republic and Haiti) and Cuba. The native populations of these islands decreased and then died out as a result of the combined effects of their violent ill treatment by the new arrivals, the illnesses that these new arrivals involuntarily introduced, and the destruction of the previous social structure by the invasion. The cities founded included Santo Domingo (1496), Santiago de Cuba (1514), Havana (1514), Panama City (1519), Veracruz (1519), Cartagena de Indias (1533), and they all reproduced simplified models of the Mediterranean city. Although from 1513 onwards the importation of slaves introduced a new component into the population of the new American colonies, many urban centers retained the ways of life of the metropolis for a long time, until intermarriage and the very different environmental and climatic conditions led to the adoption of better-adapted customs. In the countryside, the colonizers tried to introduce the crops they had cultivated in their country of origin, but only a few plants that had previously been cultivated in the hottest areas of the Mediterranean, such as cotton, sugar cane, and the citrus fruits, adapted well to the tropical climate of the first settlements in the Antilles and on the Caribbean coastline. Until the colonists reached and settled the more temperate areas of the Central American highlands, the Andes and the Chilean mediterranean, they had little success with wheat. The vine and the olive were first successfully cultivated in irrigated coastal areas in Peru, which produced the first American harvest of grapes in 1551. The livestock reared in the Mediterranean was transported to the other side of the Atlantic with the same unequal success. As well as their animals, the colonists took with them their techniques of large-scale livestock rearing and transhumance, as well as the use of fire to eliminate shrubby plants and favor the growth of the grasses preferred by their flocks of sheep and goats. In the Americas, no kind of large domesticated livestock existed except for the Andean camelids such as the llama, and so the colonizers introduced horses, cows, pigs, sheep, goats, asses and even rabbits, most of them varieties that had long been reared in the Mediterranean. In the Chilean mediterranean, sheep and goats had the same success (and similar consequences) as in the Old World, and they were also successfully introduced into some sectors of the Andes and the Mexican plateau, from where they later spread to California. Yet in the pampas areas of the eastern slopes of the Andes, the first success was the horse, followed by cattle, which were directly introduced into the savanna areas known as the Llanos in Venezuela and Colombia. Even in Australia, which was colonized by the British, the sheep introduced was a breed that had originated in the Iberian peninsula, the Merino. Probably only in the Cape in South Africa, where indigenous livestock breeds already existed before the arrival of European colonists, did the introduction of breeds from other mediterraneans take place later and on a smaller scale.

1.4 Mediterraneans beyond the Mediterranean

The Mediterranean biome and the Mediterranean culture or cultures coincide relatively closely with the area of the *Mare nostrum* (Roman Mediterranean). Yet this fact cannot be exported to the other regions of the biome, where human history has followed very different paths, and it is thus necessary to consider the existence of the indigenous human populations in the overseas Mediterraneans. The relationships of these peoples with their environment is impregnated with environmental conditioning factors of a Mediterranean nature. This is the ecological dialectic of the mediterraneans beyond the Mediterranean.

The native population of California

The native population of the area that now forms the state of California has been estimated at 310,000 inhabitants in the period before the establishment of the first mission in San Diego in 1769. As this area covers 158,645 square miles (411,000 km²), this represents a density of about two people per square mile (0.75 persons per km²), one of the highest in the world for populations with a pre-agricultural economy. As the areas with a Mediterranean climate were by far the most populated parts of what was then known as Nueva or Alta (New or Upper) California, one may conclude that the Californian mediterranean was already densely populated before European settlement, and was perhaps one of the most densely populated areas in North America apart from the Valley of Mexico.

Pre-Columbian populations

The peoples that lived in the Californian mediterranean mostly belonged to cultures and languages of the Penutian group, although Algonquin, Athapascan, Hoka, Hohokam, and Uto-Aztecan were also present. This diversity of languages and cultures was typical of the entire western coastline of North America, an apparent reflection of the fact that emigrations by several populations of very different origins and sizes terminated there. The only linguistic group that had been considered exclusively Californian, the Yuki, to which only two languages belong (Yuki and Wappo), has since been discovered to be connected to a group of Penutian languages spoken around the Gulf of Mexico and the lower Mississippi more than 1,860 mi (3,000 km) away (i.e. the language of the extinct Natchez tribe and that of the Creek). The Willot and the Yurok, Algonquin-speaking peoples

128 Flowers and fruits of the California bay (*Umbellularia californica*), as drawn by C.E. Faron in the book *The Silva of North America*, by Charles Sorague Sargent. The pre-Columbian Indians of the Californian mediterranean used the branches of this tree to separate layers of fish when smoking them. This species is a Californian vicariant of the Mediterranean laurel.
[Photo: The Huntington Library Collection]

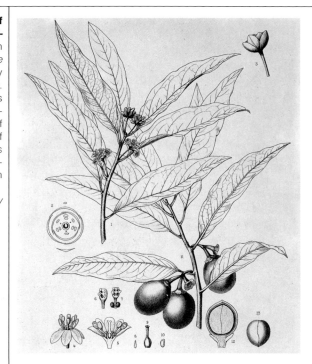

that settled on the Californian coast between Cape Mendocino and the lower basin of the Klamath River, were likewise more than 620 mi (1,000 km) from the nearest inhabitants of another language of the Algonquin group, the Kutenay of southwest British Columbia.

The Californians main food supply other than hunting was the collection of the acorns produced by oaks and of grass seeds. Agriculture was practised by some Hoka-speaking peoples (Halt-Xidhoma, Mohave, Yuma), but only outside the Mediterranean climatic area in the lower valley of the Colorado River in southeastern California. All the coastal tribes north of present-day Monterey, such as those of the valleys of the Sacramento and the San Joaquin, caught salmon, although the most productive rivers were those flowing into San Francisco Bay. This activity was especially relevant in the economy of the tribes of northwest California (also outside the Mediterranean area), whose lifestyles were very similar to those of other tribes along the Oregon coast and as far north as Alaska. It has been estimated that the Californians exploited at least 500 species of animal and plant as food or for other needs. To pass the winter months they stored acorns and grass seeds in baskets and preserved fish by curing it with smoke (salmon in the rivers and blue fish on the coast). Fish too was stored in baskets, each layer of smoked fish separated by a layer of small leafy branches of Californian laurel (*Umbellularia californica*). The Californians collected the acorns of many different members of the Fagaceae, mainly

tanbark oak (*Lithocarpus densiflora*), Californian live oak (*Quercus agrifolia*), interior live oak (*Q. wislizenii*), Canyon oak (*Q. chrysolepis*), blue oak (*Q. douglasii*), and California black oak (*Q. kelloggii*). They also highly valued the chestnuts of the California horse-chestnut (*Aesculus californica*) and the other forest nuts.

The case of the Maidu

The Maidu can be taken as an example. They were one of the five Penutian nations that lived in the central valley of California, specifically in the eastern part of the Sacramento valley, in an area that approximately occupied the basins of the Feather, Yuba and American rivers, from the low ground at the base of the Sacramento valley to the peaks of the Sierra Nevada. They spoke three different dialects; the northwestern (Concow), the northeastern and the southern (Nisenam or Nishinam).

The lowland Maidu formed numerous tribes that built villages on small natural rises so their huts stayed dry when the river overflowed. Their diet was as varied as that of the other California peoples, and was based on acorns and grass seeds. They hunted deer and other quadrupeds, they caught trout (*Salmo gairdneri*), salmon (chinook, *Oncorhynchus tshawytscha*, in spring, and coho salmon, *O. kisutch*, in the autumn) and other fish. They also caught insects (mainly grasshoppers and grubs). They sporadically provoked more-or-less controlled fires in the chaparral to make movement easier, to attract game by improving the pastures, and to increase the quantity of seed collected. In the summer, groups of hunters went in search of better hunting, reaching the highest levels of the mountains where they came into contact with their neighbors from the other side of the range, the Washo from the region of Lake Tahoe. When the first snow fell, the Maidu hunters, together with deer and other game, descended to lower areas below the snow line.

The Maidu, like other peoples of central California based their society on tribal organization. The tribes consisted of groups of villages, and their chief lived in the biggest one. These villages, however, were politically and territorially independent. The houses of the Maidu were in general circular and built with wood, half-burie and covered with earth. Larger structures built using the same technique were used for religious worship; these ceremonial buildings were the largest and most complex architectural structures that the Californian Native Americans built.

129 **Maidu Indians in the Sacramento valley (1905)**. The Maidu lived in the basins of the Feather and American rivers and used the resources their surroundings provided. These included grass seeds, the acorns of the deciduous oaks that grew in all the valleys, deer, elk, and antelopes, as well as molluscs and fish. They set fires in many places to make movement, and thus hunting and gathering, easier. This measure favored pastures and led to increases in the populations of herbivores.
[Photo: J.W. Hudson Photos / Field Museum of Natural History, Chicago]

The religion of the Maidu is only partly understood, but is known to have included the worship of Kuksu (the god of the south); a practice which had not started with them but was shared with other Penutian and Hoka tribes. The worship of Kuksu had started among the Patwin, eastern neighbours of the Maidu in the Sacramento valley, and it was later spread by the Pomo, a Hoka-speaking tribe living in the coastal regions north of modern-day San Francisco. The ceremonies, performed by spectacularly dressed dancers, were intended to ask Kuksu to renew the world and ensure their supply of food and other resources and to protect them from natural catastrophes. The Maidu also celebrated rites to honour the spirits of the dead. They held an annual ceremony where they burnt figures made with vegetables fibers representing the dead.

It seems that when the Europeans arrived life in central California was not difficult, and the resources available allowed the establishment of villages with populations of as many as 1,000 inhabitants, at least on the terraces of the main rivers. Population density 3.9 inhabitants per square kilometer in the Sacramento valley (the highest estimated pre-agricultural population density). It is estimated that in the early 19th century there were about 60,000 Maidu, but the population of native Californians had been declining sharply since Spanish coloniza-tion started, due mainly to the diseases the colonists brought with them to which the indigenous peoples lacked resistance. In the 65 years between the creation of the San Diego mission (1769) and the Mexican government's decree secularising the missions (1834), a total of 81,000 Native Americans were baptised and 60,000 deaths were recorded. But mortality outside the missions was even higher, as it is estimated that California's total population halved in this period. The last great epidemic of the period, an outbreak of malaria (1828-1833) introduced by trappers from Fort Vancouver in Oregon (where the illness had arrived on boats from Hawaii) is calculated to have caused the deaths of 20,000 Californian Native Americans.

The gold rush started in 1848, two years before California became part of the United States, and it completed the disintegration of the Californian Native American populations, especially those of the Maidu and the other Native American populations of the region's central valley. At the beginning of the 20th century, the Native American population dropped to its lowest point (15,000 people in 1900). Eighty years later (1980), the state of California's population of Native American origin grew to between 90,000 and 100,000, about half of whom were members of ethnic groups that lived in California before 1769 and the other half members of ethnic groups from other areas.

The population of central Chile

Until Spanish colonization, the Mapuche peoples of the Chilean Mediterranean were agricultural peoples with an economy based mainly on maize in the lowlands and potatoes in the highlands. Evidence for the domestication of some native plants adapted to winter rains and summer drought (the only cases in America before the arrival of Europeans), especially *teca* (a now unknown plant that disappeared in the 18th century), mango (*Bromus mango*), and *madia* (*Madia sativa*), suggest that agriculture had reached Chile long before the conquistadors. Archeological evidence suggests agriculture had been consolidated 1,500 years earlier, contemporary with the beginning of the Christian era. In the case of of the mango, a grass, the grain was used, while in the case of the esteraceae madia, the oil of the fruits was used as fuel and as food. Other species of the same genus grow in California and their fruits are also used as foodstuffs, but they never appear to have been cultivated. The Incas had conquered all the territory to the north of the river Maule a century before the arrival of the Spanish, and had introduced llama and alpaca rearing into the Araucania region.

The Mapuche territory was much larger than the Mediterranean climate region of Chile: it included all central Chile south of Coquimbo, the Los Lagos region, as well as the Atlantic slopes of the Andes, and much of the modern-day Argentinean provinces of San Juan, Mendoza and Neuquén. North of the 35°S parallel the Picunches, and to the south the Peuenches, occupied the Chilean mediterranean and they put up stiff resistance to both the Incas (who did not cross the river Maule) and to the Spanish (who did not secure their control of the the area to the south of the river Negro until the 18th century, and who had very limited presence between the Bío-Bío and the River Negro). After 20 years of war with the Spanish—until the death of Caupolicán (1558)—the Mapuche were devastated by an epidemic of smallpox that was probably spread through what is now north Argentina and Paraguay by Araucanians fleeing after their defeat. The number of Mapuches is now estimated at 250,000, concentrated basically in the Araucanian region, to the south of the limits of the Chilean mediterranean, but the mestizo population is numerous throughout the country, and has been so since immediately after the conquest.

The population of South Africa

When the Portuguese sailors rounded the Cape of Good Hope for the first time at the end of the 15th century (1487), the inhabitants they found were shepherds who called themselves the Khoikhoi (the Hottentots). According to their traditions they had come to the southern tip of Africa a century earlier from the central great lakes region, after their expulsion by more powerful, warlike peoples. They reared large red oxen with long horns that they used as draft animals and for food, as well as sheep with long tails full of fat and with short hair instead of wool. In fact, archeological remains show they came to the continent's southern tip at the beginning of the Christian era from the northeastern areas of the Kalahari, from where they had spread sheep and cattle rearing to the west (north of modern-day Botswana and Namibia) and later to the south, along the Atlantic coast. They eventually entered the valley of the river Orange and marginalized the previous San populations to the land least suitable for grazing. Some of these populations still lived in these marginal areas when the first European colonists arrived and were called *sonqua*. The first Khoikhoi were not very different from the San and spoke languages related to those of the tribes of the central Kalahari, but probably due to their contact with the shepherds of the Zambezi Basin, they had learnt how to domesticate and rear animals and to make ceramic, knowledge which they spread during their expansion.

When Jan van Riebeeck founded the Dutch colony on the Cape in 1652, the colonists said there were as many cattle as there were blades of grass. In 1661, Akambie, the chief of the Namaqua Hottentots, possessed 4,000 head of cattle and 3,000 head of sheep. It is estimated that the Khoikhoi population in the southwestern region of today's Cape Province at the time of the establishment of the colony was about 50,000, but the Dutch displaced the Hottentots from there in 1673. The spread of colonization, the epidemics it caused (especially the smallpox epidemic of 1713), and the arrival from the northeast of the Bantu tribes gradually displaced the Khoikhoi entirely from the climatically Mediterranean areas of South Africa. At the beginning of the 19th century their population had halved and there were approximately 25,000 whites and 25,000 slaves, some of whom may have been Khoikhoi, but most of whom were Africans from other ethnic groups. Today the mediterranean lands of South Africa are inhabited exclusively by whites and Bantus.

The population of the Australian mediterranean areas

Small groups of Aboriginal hunter-gatherers formed the human populations of the Mediterranean areas

3. HUMANS IN SCLEROPHYLLOUS FORMATIONS

in today's states of Western Australia and South Australia before the first British settlements (1828 and 1830, respectively). The population density in these areas of Australia was very low, possibly due to limitations of food supply, which forced the clans of different tribes living there to exploit large areas. Yet their impact on the environment was not as insignificant as might appear, given their use of fire to eliminate matorral in some areas.

Tribal frontiers often coincided with natural limits. Thus, in the Adelaide region the "frontier" between the Kaurna and the Peramangk followed the crest line of the Mount Lofty Range. The Kaurna were nomads, but lived during the winter on the woody slopes at the base of the mountain in shelters excavated underneath fallen tree trunks or made of branches, and in the summer they moved along the coast or the plain now occupied by Adelaide, always within the territory of their tribe or clan. The upper part of the range was virtually uninhabited all year round and the Kaurna only went there to hunt opossums (*Trichosurus*), bandicoots (*Chaeropus*, *Perameles*, *Macrotis*) and other small animals, and especially to collect the much appreciated wichity grubs of the moth *Xyleutes affinis* from the galleries they bore in the branches of the golden wattle (*Acacia pycnantha*). They also collected the shrub's sweet gum resin, which was an important foodstuff in the Kaurna's diet; furthermore, when treated with water, mixed with lime and heated, the gum was used to glue the handles of their stone instruments into place.

The Peramangk lived on the eastern slopes of the mountain range, and were feared by the Kaurna, who thought they had magical powers. Unlike the Kaurna, the Peramangk lived most of the year in the higher parts of the range. In winter, when the high moist sclerophyllous forests became too cold and moist, they went down to areas below 1,148 ft (350 m). Lacking permanent watercourses, the Peramangk exploited the water resources of the poorly-drained areas that were indicated by the presence of swamp gum (*Eucalyptus ovata*). Their diet was based on foodstuffs similar to those of the Kaurna, and included small mammals, grubs, acacia resin and the tubercles of sedges (*Cyperus*). They almost never went down to the plain of the river Murray or its estuary, Lake Alexander, which was occupied by small almost sedentary tribes (Nganguruku, Ngaralata, Warki, Portaulun, Jarildekald) with whom they traded.

The tribes of the lower stretches of the Murray and Lake Alexandrina lived mainly off the resources provided by the water (fishing and capturing ducks with nets) but they sought to stay on good terms with the Peramangk because they needed the bark of the river red gum (*Eucalyptus camaldulensis*), an uncommon and generally small tree, found on low riverside sites, to construct their boats. Although they feared the evil powers of the "mountain people" they obtained the bark from the Peramangk in exchange for the light, flexible lances they made from the branches of the mallee eucalypts.

It is not known exactly how many Aborigines lived in the areas with a Mediterranean climate before the arrival of the Europeans, but there must have been several thousand. It is known that there were 650 Kaurna in 1842, and they were probably one of the most numerous tribes occupying the coastal plain on which Adelaide was founded. In any case, it is a fact that the Australian Aboriginal population diminished rapidly after coming into contact with the colonists, and most now live in the north and center of the continent, although there are still some groups near to former missions, such as New Norcia, northeast of Perth, in the area known as the wheatbelt, founded in 1847 by the Benedictines Josep Maria Serra, a Catalan, and Rosendo Salvado, a Galician, and also in the periphery of the large cities, especially Adelaide.

1.5 The present day Mediterranean population

The human population of the Mediterranean Basin is a single population, distinct from its neighbors in the north and the south, and plural, in that it has diversified into innumerable variations. It is the result of a long history of migrations, exchanges, and confrontations. The human population reflects the area's early colonization from end to end, but the area has always been a centre of attraction for the neighboring peoples. It reflects its condition of a privileged space for communication and circulation, at the same time as it is fragmented into a multitude of constituent units, with their inflexible, often clannish, solidarity.

The overseas mediterraneans were colonized relatively recently by a variety of Europeans, and their human populations have undergone an almost complete transformation in the last two centuries. In every case, except perhaps for the Australian mediterranean, where Europeans of Nordic types totally dominate, the model of mixture of very diverse populations, even more diverse than those of the old Mediterranean Basin, seems to be repeated.

The actual population of the Mediterranean Basin

The morphological types described in the past as human races may be of use here as a reference to describe more easily the variation of human types in Mediterranean populations. All the Mediterranean peoples are leucoderms (light-skinned), although with relatively dark skins. Most individuals have oval or elongated faces, with elongated, narrow noses, with a straight or convex profile. Most individuals also have narrow skulls that are elongated from front to rear (dolychomesocephalic), although in the eastern half of the basin there are many individuals with shorter heads (brachycephalic), although taller. Most Mediterranean individuals have dark, wavy hair, though there are also people with fair or red hair, especially in the west of the basin.

The main type, coinciding with the first Neolithic settlers of the Mediterranean Basin who colonized it entirely, is the *Ibero-insular* type, with a short build (average 61-64 in [155-165 cm]), mesocephalic or dolichocephalic (with more pronounced dolichocephaly in the populations with shorter individuals), with a thin, straight nose and an elongated face. This type dominates the Iberian peninsula, the islands of the western Mediterranean, Tunisia, and southern Italy, and is also present in Thrace, on the Balkan coastline of the Black Sea and in Egypt. In the southeast of the Iberian peninsula and in the western

131 Typically Mediterranean aesthetic values: Hermes with the infant Dionysus in his arms. This 4th century B.C. sculpture by Praxiteles was found in the temple of Hera on Mount Olympus. Praxiteles was one of the most representative figures of a revolutionary period in classical Greek art, when the population lost interest in the religious and political function of the artists' works (as had happened two centuries before in the Archaic period). Art lost its rigidity and styles were more fluid. In this new conception of art, sculptors sought to reproduce muscles and bones that were anatomically correct, that appeared to move under a soft skin and that gave the effect of a living body. In fact, Greek artists established which images were most beautiful on the basis of knowledge and by excluding the aspects that did not please them. They standardized aesthetic values based on the symmetry and proportions which resulted in human bodies that were practically perfect. In antiquity, the Mediterranean Basin was the kingdom of beauty.
[Photo: The Ancient Art and Architecture Collection]

132 Mediterranean types from the Maghreb. Since antiquity, humans have carried out commercial exchanges: hunters and gatherers already sought articles of value, such as flint, salt, obsidian, ochres, canes, and honey, from other unrelated groups. The market was institutionalized in simple agricultural societies, where consumer articles were exchanged, sometimes with hunter-gatherer peoples. Later, the use of money made it possible to measure the social value of material objects, animals, people, and work, and markets became more diverse. Yet the market was more than the simple place to exchange merchandise, as it also played an important socializing role. The importance of the market is so great that when the geographer Pau Vila divided Catalonia into regional authorities in 1932, the only question he asked the local people was "Where do you do your shopping?"
[Photo: J. Enric Molina]

133 **The towns of the Mediterranean, in the Catalan atlas (1375) of the Majorcan Jew, Cresques Abraham**. The cities of the Mediterranean Basin were a special case, compared with other European cities of the time. After the fall of the Roman empire, many cities in northern Europe declined in population and the nobility became rural, maintaining the feudal structure. In the Mediterranean the number of inhabitants declined in the cities, but the nobility turned to commerce and did not leave the cities. This led to the construction of city walls and towers to defend the city in the bitter wars between nobles. The fact that the Mediterranean nobility turned to commerce opened up communication routes between cities, especially in the late Middle Ages.
[Photo: Bibliothèque National (Paris)]

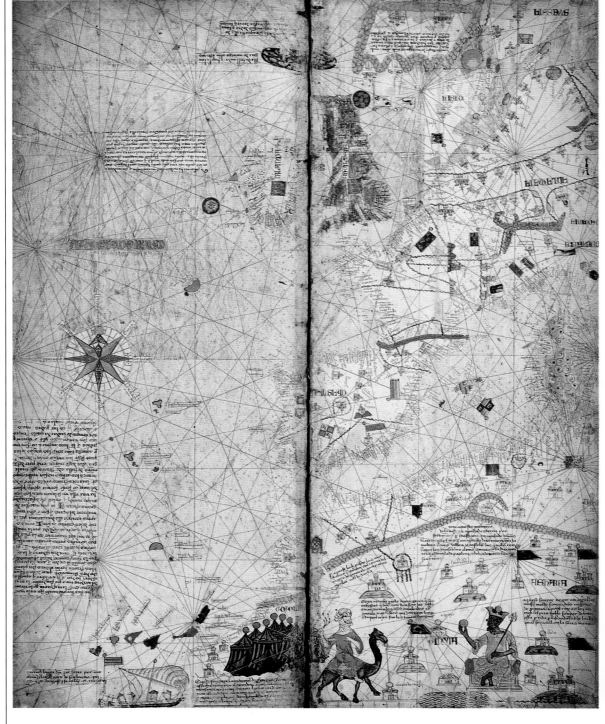

Maghreb there is the type called *Atlanto-Mediterranean*, to which the Berbers belong. The individuals are taller (64-66 in [165–170 cm]), with a more mesocephalic skull and a more prominent nose. The Atlanto-Mediterranean type includes most of the people with fair hair and redheads in the Mediterranean. Only very locally does the Atlanto-Mediterranean type dominate, but it is widespread throughout the Mediterranean. These two types are traditionally included within the Mediterranean race.

In the easternmost end of the Mediterranean Basin, in Palestine, Syria, and eastern Anatolia (as far as the Caucasus and Afghanistan) there is a taller (average height, 65-68 in [165-175 cm]) and more dolichocephalic type, with an elongated face, a large, straight, nose and fleshier lips. This type, known as *Iranoid*, dominates the Near East and is present throughout the Mediterranean Basin, especially its southern areas. In contact with the Iranoids, the type known as Anatoloid, may be found in Anatolia, Lebanon, north-

ern Syria, Armenia, and many locations in the Balkans; unlike the three previous types, this type consists of brachycephalic individuals with wider faces, a high cranial vault, a raised forehead, and flattened occiput. Their average height is between 65-66 in (165-170 cm). This type is dominant in Anatolia, except in Kurdistan, and in some districts of the Aegean coastline. The *Dinaroids* are also brachycephalic, and form most of the eastern Balkan population and are also present in large numbers in Egypt, Cyprus, and some

points in Anatolia. They are the tallest inhabitants of the Mediterranean, together with the Iranoids (average height greater than 68 in [175 cm]). This type has a broad face and high cranial vault, with a long and narrow nose; it was already present in the Aegean and the Balkans during the Bronze Age, as is shown by the many burials from this period that have been found in Bosnia, Croatia, Serbia, and in Cyprus. In some points of the eastern half of the basin and in Albania and Montenegro there is a last type, called *Alpinoid*, which is also brachycephalic, but shorter, with a rounder face and a more sloping forehead than Anatoloids and Dinaroids. Nowhere are they dominant, but this type is especially frequent in almost all the mountain areas to the northwest of the Mediterranean (Alps, Pyrenees, and northern Apennines). All these general types should be understood as a set of models that dominate to a greater or lesser extent in different geographical areas. In practice, it is normal to find a mixture and intermediate types almost everywhere, as well as some atypical types, such as: Nordic types in Sicily, the Balkans, some areas of Andalusia and the Maghreb; Mongoloids in Anatolia and in the Balkans; and Afromelanoids in Egypt and the Maghreb.

The present day populations of the overseas Mediterraneans

In Chile, the Mediterranean-type population dominates but the over time intermarriage with the Amerindians has given rise to more brachycephalic types, with wider faces and noses and longer extremities. The pure Amerindian type is, however, rare. The Nordic European type is more common, the result of recent immigrations by British and Germans. The population of California, especially its urban areas, is one of the most diverse in the world. The leucoderm (light-skinned) type dominates, mainly Nordic, and to a lesser extent the Ibero-insular type, as well as others. There are also several types of melanoderms (dark-skinned), as well as xanthoderms (yellow-skinned) proceeding from the former Native American populations (a small minority of 100,000 in California's population of 30 million), and from east Asia (Sinoids and Mongoloids). There are countless combinations and as much diversity so that there is almost no human type that is not present in the Californian population.

In the Cape province, the population is mostly leucoderm, totally dominated by the Nordic type. In southern Africa there are also considerable minorities of Khoikhoi and of Zambezoid melanoderms. The Khoikhoi are the descendants of the indigenous populations found by the white colonists and many interbred with them (the "Cape colored"). The Zambezoid melanoderms are the descendants of more recent immigrants from the most northerly regions of South Africa and the neighboring countries.

In the Australian mediterraneans the population is almost entirely white and of the Nordic type, although there are many other white immigrants from other populations and types. The population of Australian aborigines forms a minority.

Urban planning comes from the word urbs

One of the features that characterizes the Mediterranean areas is the high relative importance of the urban population and the complex relations that the system of cities in most of the world's Mediterranean areas brings with it. This was true in classical antiquity as it is today. If Rome was the classic example of a city in antiquity, the following cities or conurbations have just as decisive an influence on their surroundings: Perth (with three quarters of the population of Western Australia), Adelaide (with three quarters of the population of South Australia), Cape Town (whose conurbation houses a quarter of the population of the Cape Province), Santiago de Chile (with almost one third of Chile's total population) and the Los Angeles conurbation (almost 40% of California's entire population).

The urban phenomenon

The urban phenomenon is not of specifically Mediterranean origin, but appears to have been linked to the beginning of the production of agricultural surpluses by Neolithic societies in the Fertile Crescent and Mesopotamia, and later in other areas, where this production of surpluses gradually disrupted the closed system of self-sufficient production by primitive agriculturalists and livestock rearers, and the creation of social functions that were not associated with food production by cultivating plants and rearing domesticated livestock. Yet the Mediterranean, and specifically Greece, made contributions as essential to what we now call urbanism, as the economic and political integration of the city with its environment or "region" and the rational organization of the urban layout, based on more-or-less explicit general principles. Also closely linked to the Mediterranean lifestyle is the creation of the public space (the Agora of Greek cities, the market, the main square present in all cities and in most Mediterranean or Greco-Roman villages), where many of the events of general interest take place, from markets and festivals to acts of faith and executions. Phoenicians and especially Greeks transported and spread goods throughout the Mediterranean, with their boats, not only more-or-less perishable merchandise for trading purposes, but also a model of city destined to be widely imitated and which is basically still valid. This consisted of a walled space dominated by a raised citadel (acropolis)—the center of both civil and religious ceremonies and at the same time a defensive bastion—centered on the agora with as regular a layout as the site allowed, and a port, if the sea or a river was close enough to allow one. Hipodamus of Miletus, in the 5th century B.C., took this idea to the limit and introduced the layout based on a straight-line grid, the reason why he is considered the precursor of modern urbanism.

Urban design, to the extent that it submits the construction of the city to rational rules, although with occasional lapses, has always triumphed in the Mediterranean over the spontaneous development of the city. It has even sought to justify itself *a posteriori*, for example in the myths of the foundation of Rome, according to which the city perimeter (*pomerium*) was marked out first by the furrow of a plough and then in the center a pit (*mundus*) was made into which the future citizens threw some of their first harvest and a clod of earth from their land. The myth of foundation, perhaps the idealization of a ritual that really took place, expresses the two basic requirements of every urban space: the establishment of a central space to carry out collective activities and an external limit to distinguish the space of the citizens from that of the foreigners.

The Greek contribution is not limited to the material aspect of the city, but also includes its social and political organization, adopted by many other peoples and organizations over history. Classical Greek civilization was basically a civilization of cities with subordinate territories of greater or lesser size, but in practice this had little effect when it came to defining the political nature of the city-state. The most important thing about the city was not its city walls nor its territorial limits, but its citizens and their ability to provide the city with laws and the democratic organization of political life by means of egalitarian assemblies

The Mediterranean city

The Mediterranean city, the daughter of commerce and trade, has of necessity also been the site of a cosmopolitan culture that not even periods of ferocious intolerance or xenophobia have managed to destroy. The Roman conquest of the entire Mediterranean area, completed shortly after the beginning of the present era, reinforced and even increased the movement of individuals of all types from one end of the basin to the other, and beyond, since the Roman Empire covered a large area outside the Mediterranean Basin. It also furthered the presence of elements of very different origins in each city—above all in Rome, the quintessential city also known as the *urbs*. In many territories, especially in the eastern Mediterranean, the first cities worthy of the name were founded by the Romans, and where there was an urban tradition the Romans extended and consolidated it, or moved it slightly to find a more favorable location. The price of the establishment of the city and its urban growth

in the days of the Roman Empire was, and still is, the exploitation of resources from a greater or lesser surrounding area, sometimes from far away. Plato mourned the loss of the days when Attica was covered by trees, and it was the lack of wood that limited Athens' growth as a sea power. In Italy in the 2nd century B.C., population increase and the military expansion of Rome were the cause both of the breakdown of the peasant and livestock-based societies living there, and of the later migration by small landowners from the countryside to the city. Land ownership was concentrated into large estates that became progressively specialized in production for trade, often at great distance. Similar processes took place in each of the new provinces incorporated by Rome, and, in some of these, the Romans (or the Romanized local populations) were the first to exert significant pressure on natural resources through agriculture or extractive activities.

The fall of the Roman Empire slowed urban development, but most Roman cities, despite the damage caused by barbarian invasions and the ensuing loss of social organization, survived the Middle Ages and have lasted until the present day. Some cites, like Turin, have completely maintained the street layout of the Roman city. But most have modified this layout greatly, even using stones from destroyed ancient monuments as construction materials. Only a few, such as Ampurias and Italica (in Hispania) and Volubilis, Hippo, and Sufetula (in Africa), did not endure and were abandoned. The insecurity of the times made a city wall necessary, while population growth brought with it the development of suburbs outside the walls, and these were then enclosed when it became necessary because of their size or the renewal of temporarily dormant hostilities.

The fall of Rome coincided with the rise of the new Rome, the capital of the eastern Roman Empire which was reestablished by Constantine on the site of the former city of Byzantium—Constantinople. For a millennium, the splendor of Constantinople equalled that of Rome, the model for its construction. It soon became the largest of the Mediterranean's cities, and always has been except for during the catastrophic period when it fell to the Turks. It ceased to be the capital of the Byzantine Empire and became the capital of the Ottoman Empire. The churches were turned into mosques and the markets into bazaars, but Ottoman Istanbul, like Constantinople and to an even greater extent Rome, continued to be the quintessential city. Its modern name, Istanbul, is the Turkish deformation of the Greek phrase, *eis ten polis*, which means "toward

Juniper in Majorca, cypress in California

Miquel Josep Serra i Ferrer was born in Majorca in the village of Petra in 1713. Hardly anyone recognizes this name. Yet many people know who Fray Junípero Serra was (also known as Fray Ginebró). Fray Junípero Serra and Miquel Josep Serra i Ferrer were one and the same person. Junípero, or Ginebró, was the name that Miquel Josep adopted when he took his vows as a Franciscan; years later the name Fray Junípero would become famous in California. The name of a Mediterranean member of the cypress family was very appropriate for someone who died in Monterey in the heart of the area where the Californian cypress grows.

On the Mediterranean Basin's coastline there are many junipers, as well as savins. These small trees, gnarled and twisted by the wind, are members of the same genus (*Juniperus*), and belong to the cypress family (*Cupressaceae*). Miquel Josep Serra must have been fascinated by them because of his Majorcan roots and because he was of the Franciscan

Branch of the Californian cypress (*Cupressus macrocarpa*) with its cone, a galbulus [Norman Owen Tomalin / Bruce Coleman Limited]

The church of San Carlos Borromeo de Carmel, Monterey, in the mid 19th century [The Huntington Library (California)]

order, which is characterized by its love of nature. The name he adopted, Ginebró, refers to *Juniperus oxycedrus* subspecies *macrocarpa*, a coastal juniper with a conspicuously large fruits as its scientific name (*macrocarpa*), indicates. A fitting name for a man who would later die surrounded by Californian cypresses (*Cupressus macrocarpa*), small trees that are also members of the cypress family, like Majorca's junipers, cypresses, and savins, but a product of the Mediterranean landscape far from home. He died in 1784 in the San Carlos mission, now in the modern city of Monterey. Yet before he died, Fray Junípero had played a leading role in a remarkable adventure.

The European colonization of the Californian mediterranean—in present-day (Alta) California—was unusual. The area was colonized by people with Mediterranean roots—by Catalans, Majorcans, and some Valencians—who must have felt at home there. The Baja (or lower) California peninsula, an extra-Mediterranean territory that extends from the 32°N parallel to the Tropic of Cancer (23°N) and that is now part of Mexico, was explored by the Spanish during the expedition led by Diego Hurtado de Mendoza (1523) and colonized by Jesuits at the end of the 17th century. Yet present-day California was not colonized until the second half of the 18th century (in reality, for a long time it was thought to be an island and figured as such on maps!). João Rodrigues Cabrilho, a Portuguese sailor in the service of Castile, was the first European to step on the shores of California. On June 27, 1542 he set out from from Puerto Navidad (near the modern city of Manzanillo, Colima, in southern Mexico [19°N]) on an expedition and entered San Diego Bay on October 7. From then until the middle of the 18th century there were a few sporadic incursions by Europeans, such as the Englishman Sir Francis Drake in 1579 and the Spaniards Sebastián Rodríguez Cermeño (1595) and Sebastián Vizcaíno (1602). California was then fully in the hands of the 300,000 Native Americans who lived there, basically Athapascan, Algonquin, Yuma, Hoka, Penutian, and Hopi.

Californian Navel orange [Ernest Costa]

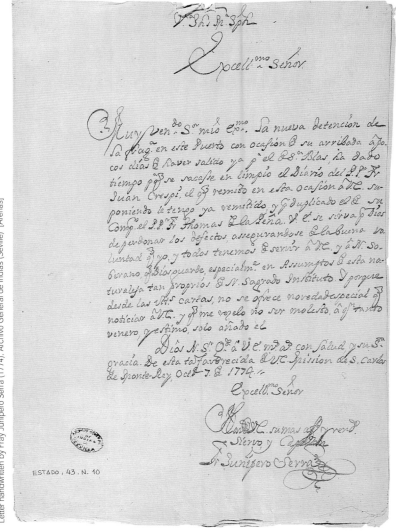

Letter handwritten by Fray Junípero Serra (1774). Archivo General de Indias (Seville) [Arenas]

ESTADO, 43. N. 10

The Russian colonization of Alaska, in the mid 18th century alarmed the Spanish crown which decided to strengthen its role north of its Mexican territories, which were then completely under subjugation. A Catalan expedition, almost the only one in the entire colonization of the Americas, was given this task: Gaspar de Portolá was in charge and Fray Joan Crespí and Fray Junípero Serra were responsible for evangelization. They did more than evangelize, as they founded many missions, starting with San Diego de Alcalá in 1769 and San Carlos Borromeo de Carmelo (in Monterey or Carmel). These missions became the population centers which later grew into the large cities of San Diego, Los Angeles, San Francisco, San José, etc. Thus, Fray Junípero Serra, by assuming missionary and urban development, anthropologically "mediterraneanized" a territory that was already ecologically mediterranean. He is considered the father of modern California, which joined the United States of America in 1850.

However, the indigenous population did not survive unchanged or unscathed. In 1834, when the Mexican government secularized the missions, 81,000 Native Americans had registered as Christianized, that is to say, they had partly or totally assumed the Christian faith and at least 60,000 had died from illnesses brought by the Europeans (mainly smallpox, diphtheria, malaria, and measles). Livestock and a series of introduced animals, such as the wild boar (*Sus scrofa*) and the rat (*Rattus rattus*, *R. norvegicus*) modified the territory's structure and physiology. But not without a response: 200 years after the beginning of the colonization, the same cypress that Fray Junípero found by the sea in his mission at San Carlos has taken over many parks and gardens in the Mediterranean Basin itself, while almost all the orange groves from which the first cuttings and seeds were sent to California have now been replanted with the Navel variety developed in California. So that what can go from Majorca to California can go from California to Majorca.

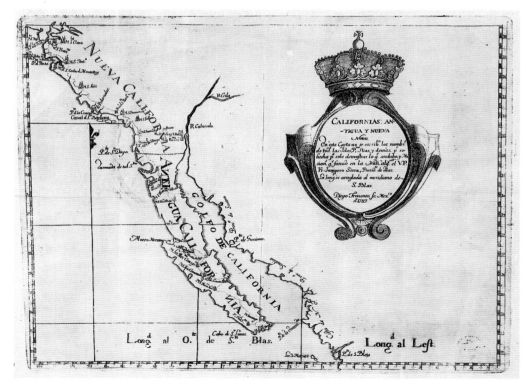

134 The grid plan for the city of Concepción (Chile) in 1752, one image of the layout of the Roman city that European colonists exported to the other mediterraneans. The Mediterranean city planned by the Romans was arranged and laid out in a rectangular grid. Rome, however, was different, since it had formed as the result of the fusion of several pre-existing population centers. The Roman layout was maintained over many years, but not forever, as the lack of public control did not prevent new buildings invading the street, and the ruins of old buildings remained for a long time and hindered circulation. Note that population growth in cities is due more to immigration than to urban birth rates, and this is why their populations are constantly being renewed.
[Photo: Archivo General de Indias (Seville) / Arenas]

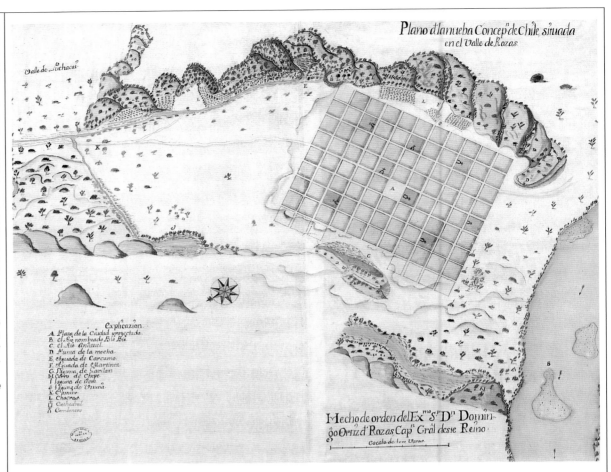

the city." None of the cities of the medieval western Mediterranean, not even the impoverished Rome ruled by the popes, or Venice, which owned three eighths of Constantinople at one stage, or the rising city of Florence could be compared with the capital of the Byzantine Empire. Only Cordoba under the caliphate in the 12th century could compare, for a time.

The Renaissance and modern era saw the rebirth of the town planning ideals of antiquity. Main squares became monuments, and public buildings were surrounded by open spaces. The Mediterranean city represented the ever greater projection of its inhabitants' social relationships and collective projects. Yet the social integration implied by the Mediterranean city is transmitted from larger population centers to smaller ones. A village of few hundred inhabitants living precariously by exploiting barren, unproductive land can, in the Mediterranean, be a "city" with all the rich social relationships, solidarity, and conflicts that urban life implies. To speak of the city in the Mediterranean is to talk of associative life, of social relationships organized on the basis of group solidarity on all scales and in all fields of activity. And also as an ordered and intelligible public space that encour-

ages communication and exchange. In the southern and eastern Mediterranean, the Muslim cities, with a fortress playing the role of the acropolis, show their own peculiarities which cannot hide their points in common with the cities of the rest of the basin. The dense housing leaves no space for public squares, but its social roles are played by the courtyard of the mosque (where men meet for prayer, and where the decisions taken by those in power are announced), the bazaars (where commercial activity takes place), and the city gates (the scene of links and exchanges with the outside world).

The tradition, if it is possible to speak of a tradition, found in the world's other Mediterraneans has been very different. They are much less densely populated and in geographical locations less favorable for exchange, and all their cities were founded relatively recently. Santiago de Chile was founded as a frontier fortress town in 1540. Cape Town was founded in 1652 as a staging post on the long route from the Netherlands to the Dutch East Indies. The cities of the Californian mediterranean (San Diego, Los Angeles, San Francisco, etc.) were founded as Franciscan missions between 1769 and 1780. Perth was founded in 1829, and Adelaide in 1835 (by the colonists of Edward G. Wakefield's South Australia Association).

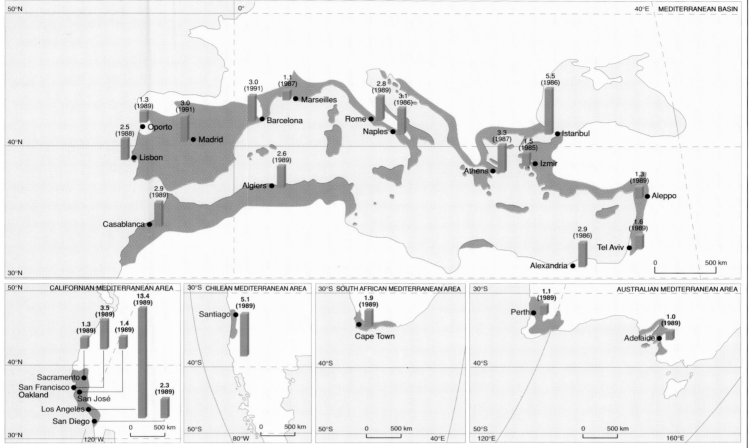

These cities are very different from those of the Mediterranean Basin, although some preserve some features in common, especially their sociability.

As a result of the long Spanish colonization, Santiago de Chile is probably now the city that has the most features in common with the cities of the Mediterranean Basin. It was created as a walled city, a reminder of its militarized past, which lasted until well into the 17th century, because of the instability of the Araucanian frontier. The old town is on a hill (Santa Lucía) and has a large square in its center, the Plaza de Armas. Its straight-line grid pattern is found in many colonial cities in the Americas, and it inherits something from the Greek and Renaissance tradition. Yet what most identify it as a Mediterranean city are the large number of plazas in the different districts. These plazas apparently repeat the classical feature of the public square, sometimes monumental and sometimes intimate, that is so characteristic of the Mediterranean city.

The cities of the world's other mediterraneans are closer to the model of cities created *ex novo* during the last century and a half in areas of Anglo-Saxon colonization. The straight-line structure of their road network is probably the only feature remotely connecting them to the Mediterranean tradition. Although Los Angeles, for example, has a historic center with a Spanish colonial period appearance, it is now probably one of the models for a city that is furthest from the Mediterranean tradition of a public space for socializing. On the contrary, public space appears to be intended mainly for the automobile and is hostile to human beings not surrounded by a car's bodywork. Public space is dedicated to the transportation of people as they go from the suburban areas where they live to where they work in the city center (dedicated to the service sector) or the industrial outskirts. An explanation, or perhaps one of its effects, is the enormous area covered by the city, about 463 square miles (1,200 km²)—almost as big as the island of Rhodes and one third of the size of the island of Majorca—with an average width of 22 mi (35 km). There are clear and large social imbalances between different districts, ranging from the luxury villas in Beverly Hills to the most marginal and degraded districts inhabited by Latin American and Asian immigrants. It is the center of the American film industry and thus probably the city that has appeared most often in films and on television. Yet it has no typical space that visually represents the city, except the bends leading up to the heights of Hollywood and Beverly Hills with illuminated letters spelling out the name Hollywood, the Mecca of the film industry. Sunset Boulevard, Los

135 Urban areas in the mediterraneans with more than one million inhabitants (showing the number of inhabitants in millions and the year of the census).
[Drawing: Editrònica, from several sources]

Angeles' best-known street, does not appear any different from any street in any American city.

1.6 Mediterranean illnesses

The Mediterranean Basin is famous as a healthy and invigorating area, especially among its northern neighbors. Although the Mediterranean area lacks the large number of parasites that cause diseases in humans living in more tropical latitudes, it is no illness-free paradise.

Leaving to one side the epidemics and plagues mentioned in the Bible (the descriptions of which almost never allow even speculation about the type of illness), the Greek doctors, especially the father of medicine, Hippocrates, left unmistakable descriptions of an epidemic of mumps on the island of Thasos in the northern Aegean, and of cases of a kind of fever recurring every third or fourth day that may well have been the tertian or quartan fevers caused by malaria or Mediterranean fever. Since he never described anything resembling smallpox, measles, or bubonic plague, these illnesses probably had not reached the Mediterranean when Hippocrates was alive (*c.* 460-*c.* 377 B.C.). Malaria has long been one of the main causes of death in many low-lying regions with wetlands throughout the Mediterranean, especially in coastal waters occupied by marshes.

Yet since the days of Hippocrates illnesses have been known that, while they are not specifically Mediterranean, are most common in this area. The most important ones—thalassaemia, Mediterraanean fever, and favism—are genetic and so it may be considered that they are related to mutations that occurred within some Mediterranean populations in the more-or-less distant past. There is also an infectious disease, Malta fever, whose vectors are domesticated animals that are not exclusively Mediterranean, but which are widespread there.

Thalassemia

Of these illnesses, the different forms of thalassemia affect the largest number of persons, especially in Greece, North Africa, Cyprus, Sardinia, and other Mediterranean islands. An estimated 7-8% of the Greek population are carriers of the genetic defect responsible for the illness; and in the Ionic islands

and in Rhodes this percentage may reach 14-16%. Thalassemia also occurs outside the Mediterranean Basin, especially in India, China, and southeast Asia, but at much lower frequencies, and in some cases associated with individuals or populations of partially Mediterranean origin, such as the descendants of Greeks and Italians in the United States.

The cause of the illness is a genetic defect that prevents the correct synthesis of some of the polypeptide chains of hemoglobin in the blood. Homozygotic individuals show the most severe form of the disease (thalassemia major) which sometimes directly causes the death of the fetus. If this does not occur, the child develops symptoms between the ages of three months and two years that include severe anemia (with a very low of production of hemoglobin), the presence of abnormal erythrocytes (called target red blood cells due to their coloration in concentric bands—the center and a band at the edge are darker than the intermediate band—like a target), major bone deformations, and pathological growth of the liver and spleen. Untreated, death normally occurs before six years. Survival is only possible with frequent blood transfusions, but this therapy may lead to an excess of iron that is also toxic and causes life expectancy to decline sharply after the age of 30. It seems impossible that this disease should cause such a high mortality and yet have been able to maintain itself in Mediterranean populations. As in the case of other hereditary forms of anemia, this is because the individuals that are heterozygotes for the defective gene that causes the disease (thalassemia minor) obtain some protection from the malaria parasite.

Favism

Favism is another typically Mediterranean genetic illness. It takes the form of an allergic reaction after eating broad or fava beans (*Vicia faba*) or coming to contact with bean pollen, and it leads to a hemolyt-

136 Photograph of a blood smear of a patient suffering ß-thalassemia. The geographical distribution of thalassemia in the Mediterranean coincides with the areas where malaria used to be endemic. It has been shown that heterozygotes are resistant to malaria, as the presence of some tetramers of hemoglobin in the red blood cells creates an environment unfavorable for the the agent that causes malaria, the protozoan *Plasmodium*. Thus, one part of the population is favored by the gene, but the victims of thalassemia pay the high price.
[Photo: Josso / CNRI, Paris (x 1,000)]

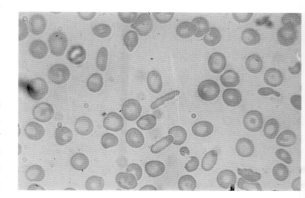

ic anemia of variable severity. In some cases, favism may lead to death. The cause of favism is an anomaly in a locus on the X-chromosome (and thus much more common among men than women), that leads to a lack or insufficiency of the enzyme glucose-6-phosphate-dehydrogenase (G6PD) in red blood cells. This enzyme plays a role in the breakdown of glucose in the red blood cells and prevents the oxidation of glutathione, and this protects the hemoglobin in its gas transport function. The lack of G6PD leads to the destruction of the red blood cell in when it comes into contact with the oxidizing agents present in the beans.

G6PD deficiency is common throughout the Mediterranean Basin, especially in Greece, Sicily, Calabria, and Sardinia. It is also present in Egypt and in the rest of North Africa, the rest of Italy, and the Near East. Islands appear to be particularly affected, especially Rhodes, where more than 30% of males show a G6PD deficiency. Very high frequencies have also been found in Cyprus, Sardinia, Sicily, and Corsica, perhaps the result of greater selection pressure. Despite being an illness leading to severe hemolytic anemia, the general geographical distribution of the affected populations and its high frequency show, as in thalassaemia, that there is a link between favism and protection against malaria.

The pre-Socratic philosopher Pythagoras recommended that his disciples not eat fava beans or even touch them. Legend states that when he was pursued by his enemies he was caught when he had to resolve the dilemma of having to cross a bean field in order to save his life. Although the illness as such does appear in the clinical descriptions of the classical Greek doctors, and was not "discovered" until the 19th century, Pythagoras's precept shows that the ancient Greeks knew the distress that fava beans may cause. However, since they lacked a rational framework to explain why eating a common foodstuff that was innocuous to most people was noxious to some, they could only consider it in the framework of a magical-religious doctrine such as Pythagoreanism or the traditional medical practices of the Greek peasants.

Mediterranean fever is a genetically transmitted illness with an unknown etiology (cause), found mainly in some populations in the eastern half of the Mediterranean Basin. It causes bouts of fever lasting two or three days, accompanied by muscle, joint, lung, and chest pains. The fevers are separated by periods of between two and four weeks without fever. The illness may lead to peritonitis.

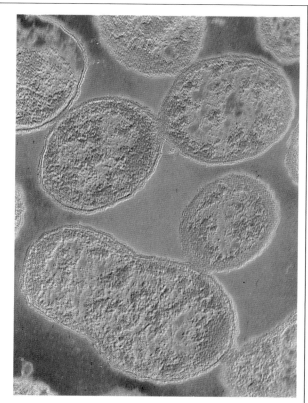

137 **Micrograph of *Brucella abortus*,** a gram-negative bacteria. This bacteria along with other bacteria of the same genus (*B. melitensis* and *B. suis*) can cause Malta fever or brucellosis, a disease endemic to the Mediterranean Basin. The disease's first symptoms are not very specific, such as tiredness and loss of weight. This is followed by the beginning of the most characteristic feature, undulant fever with temperatures of 102°F (39°C) followed by brief periods of fever. Patients respond well to treatment with antibiotics and more than 90% of cases recover without after effects. The natural habitat of *Brucella* is farm mammals. Human infection is due to contact with sick animals, and especially to the ingestion of contaminated milk or milk products that have not been treated by pasteurization. [Photo: CNRI Science Photo Library / AGE Fotostock (x 25,000)]

Malta fever and other infectious diseases

Malta fever, better known as melitococcic brucellosis, is an infectious disease caused by the gram-negative bacteria *Brucella melitensis* found in sheep and goats. In humans these bacteria produce a fever that follows an intermittent (undulant) course, accompanied by constipation, abundant generalized sweating, and swelling of the spleen and liver. It is spread from livestock to humans through milk and milk products and it appears that it can also be transmitted through the blood by fleas and ticks. It has long been endemic in the Mediterranean Basin, and the spread of goat-rearing has led to its transport throughout the world. Other diseases endemic to the Mediterranean Basin include "pappataci" fever, which is restricted to the coasts of the Adriatic and is caused by a virus transmitted by the sand fly *Phlebotomus pappataci*, and recurrent fever (also known as Mediterranean eruptive fever or rickettsial pox), which is caused by a rickettsia (*Rickettsia conorii*) transmitted by the tick (*Rhipicephalus sanguineus*). Pappataci fever takes the form of general malaise, with muscular pains and headaches followed by a violent fever lasting three days (it is also known as "three day fever"). Other symptoms include reddening of the skin, digestive disorders, and bradycardia (abnormally slow heartbeat). Mediterranean recurrent fever is characterized by a necrotic patch at the site of the tick's bite, accompanied by shivering, fever, and skin rashes that appear on the second to fourth day.

2. The use of plant resources

2.1 Harvesting without planting

Like all biomes, the mediterraneans provide a variety of native plant resources that can be used by human populations directly, as they are found in nature, or after using simple techniques. These plants may provide food, fibers or other raw materials, such as essencial oils or chemical products of different types, or they may simply be ornamental. Some may have symbolic significance or magic properties may be attributed to them. The use of these plants and the resources they provide was much greater when humans were hunter-gatherers, but diminished as they gradaually turned to agriculture and stockraising. Today many of these resources come from cultivated plants or they have lost their symbolic or economic importance. In other cases they are still collected, but only on a very small scale, and are of little economic importance. They are still harvested as leisure activities or marginal inputs to the rural economy in poorer areas, although some of them are quite profitable.

Fruits, shoots and other foodstuffs

The fruits of many wild shrubs and trees have been gathered since time immemorial, although today mainly on a local, small scale. In the past, however, they must have been an extremely important food resource. This is the case of the acorns produced by many Californian oaks, collected and treated to make them edible by the population of the territory before European colonization (see section 3.1.4). The acorns collected were a basic food resource for the human population and their exploitation was so intense that it is thought that it may have affected their regeneration, as in bad years, with low acorn production, the harvest could have taken the entire crop.

In the Mediterranean Basin, the acorns of different species of *Quercus* were also collected, especially those of *Quercus ballota*, and the cork oak (*Q. suber*), which are less bitter than those of the holm oak (*Q. ilex*). They are still used as animal fodder in many rural Mediterranean areas. Other fruits are still collected, such as the fruits of the strawberry tree (*Arbutus unedo*), berries produced by different species of bramble (*Rubus*), the fruits of the myrtle (*Myrtus communis*), and sloes (*Prunus spinosa*). They are collected all over the Mediterranean Basin and sometimes reach markets as a minor seasonal product. Some products are of great economic importance, such as the pine nuts of the maritime pine (*Pinus pinaster*) and sweet chestnuts (*Castanea sativa*), and some woodlands are managed to obtain these products.

In Chile, some fruits are also collected such as the Chile nut (*Gevuina avellana*), locally called "avellanas"; the fruit of the "queule" (*Gomortega keule*), which make delicious jams; the fruit of the "peumo" (*Cryptocarya alba*); the fruit of the "algarrobo" (*Prosopis chilensis*), not to be confused with the Mediterranean carob (*Ceratonia siliqua*) which has the same name in Spanish, and the fruit of the "boldo" (*Peumus boldus*). "Coquitos," the fruits of the palm *Jubaea chilensis*, are of economic importance, as are the fruits of the "maqui" (*Aristotelia chilensis*), now used to give color to wine. It is known that the Araucan tribe used "maqui," mixed with the fruit of the pepper tree (*Schinus latifolius*), to make a fermented drink, which was totally replaced by the wine introduced by the colonists. Also in Chile, a syrup-like liquid known as palm honey is obtained from the liquid that oozes from the apex of the stem of the Chilean palm, *Jubaea chilensis*. The palm is felled, and once it is on the ground successive discs are cut from it to make the liquid flow out. A good specimen may produce 318-424 quarts (300-400 liters) of sap, which is converted into 110-132 lbs (50-60 kg) of palm honey. This rather drastic method of exploitation is now prohibited, as the populations of this magnificent endemic palm have been severely decimated.

In some cases, it is not the fruit of the plant that is consumed but another part. Asparagus, for example, is the tender shoot of the asparagus (*Asparagus acutifolius*), and the heart of some

138 Chestnuts, the fruit of the sweet chestnut (*Castanea sativa*), are collected in the autumn, when the leaves are beginning to wither. They are used to make purées and sweets and to fatten up livestock. The Greeks and Romans used to cultivate them to make flour for use in bread when cereal harvests were bad, and they were responsible for increasing the chestnut's range into land formerly occupied by beech and oak. The wood of the chestnut also has many uses. (See also figures 143 and 234.) [Photo: Age Fotostock]

species of palm is eaten, such as that of the dwarf fan palm (*Chamaerops humilis*) found in the warmer parts of the Mediterranean Basin. The flower shoots of some Chilean plants of the genus *Puya* are used to make a sweet, and some herbaceous plants are simply consumed in salads or boiled, such as borage (*Borago officinalis*), which grows on waste ground, and water cress (*Rorippa nasturtium-aquaticum* [=*Nasturtium officinale*]), which grows on humid riverbanks, and others.

Mushrooms

Mushrooms are a special case. Not only because the Fungi form a kingdom distinct from the plants and animals (and from the monerans and bacteria), but also because they are still of great economic importance in some parts of the Mediterranean Basin, and they have been eaten by hunter-gatherers in all the mediterraneans. Mushrooms are harvested and sold, and in some cases they are even cultivated. Some Mediterranean woodlands that produce very little wood generate considerable economic profits through the harvesting of mushrooms.

The case of mushrooms is especially interesting, as it shows how humans, on the basis of a combination of factors such as hunger and curiosity, have learned to use many food resources offered by the biosphere, even in such perilous fields as that of the mushrooms. As is well known, mushrooms are not limited to good ones and bad ones, but there are also delicious ones and poisonous ones, some of which cause temporary sickness and others that can cause death. It is easy to imagine the large number of unfortunate events necessary to acquire our detailed knowledge of degree of edibility of mushrooms.

This is even more true if we consider that some species poisonous to humans can be eaten safely by other animals: trials of toxicity have been taken on by the human species. The risk associated with eating mushrooms makes it easy to understand why some cultures avoid these disconcerting organisms. Wherever mushrooms are abundant some species are consumed, but mycophobic cultures only consume a small number of species that are easy to identify, and give the others general names that are not very precise or even derogatory, with references to wolves, devils, or witches.

Mycophagia, the desire to eat a variety and quantity of mushrooms, is harder to understand. Mycophagous cultures have a good knowledge of the different species and their qualities, which has led to the creation of a large vocabulary dealing with the subject. In a mycophagous culture, as many as a hundred different species may be distinguished by having their own common name, sometimes requiring an expert eye to distinguish between similar species. A good example of this are the mushrooms of the genus *Lactarius*. In some Mediterranean areas, not only are clearly distinct species of *Lactarius* identified, such as "rovelló" (*L. sangifluus*) and the "pinetell" (saffron milk cap, *L. deliciosus*) in Catalunya, but there are even common names for specimens parasitized by the ascomycete *Hypomyces lateritius*, which are called "rovellola" and "pinetella" and are considered to taste better.

Surprisingly, these two attitudes do not reflect a different cultural heritage, but can coexist within a single cultural setting. In the European Mediterranean mycophagia is present in the Catalans, the habitants of the Languedoc, and the Italians who

139 Grilling "pinetells" (*Lactarius deliciosus*) in Catalonia (the northeast of the Iberian Peninsula), a social event and ceremony that occurs every autumn in the Mediterranean. All over the world, there are cultures that are mycophilous, who recognize and enjoy eating different kinds of mushroom, and mycophobic ones, who reject them. The mycophilic areas include Catalonia, eastern Andalusia, and coastal northwest Africa, and northeast Italy. Mycophilic peoples value them more for their flavor than for their nutritional value. As there are different ways of preparing mushrooms, a given species may be more appreciated in one area than in another. In Catalonia, for example, the "rovelló" is greatly appreciated because it is cooked over hot coals. In other places, where food is stewed, this species is rejected and other mushrooms, which could not be cooked over coals, are preferred.

[Photo: Oriol Alamany]

coexist with mycophobic peoples like the Galicians, the Greeks, and to a lesser extent the Castilians in very similar geographical, climatic, and cultural conditions.

Mycophagia is even more surprising if we consider the poor nutritional value of most mushrooms. A typical mushroom contains 90-95% water, 3-5.5% protein, and 0.1-5.2% glucides as well as lipids, vitamins and mineral salts. Their food content is thus acceptable, but only a small part is absorbed, due to the low digestibility of their materials. Thus, in spite of their danger, mushrooms are consumed more for their smell and taste than for their food value, and might be considered a condiment rather than a food. The risks and misfortunes that have allowed the accumulation of modern knowledge were not related to the search for a food resource as such, but for a pleasure related to the consumption of other foodstuffs.

Mycophobes are all in basic agreement that mushrooms are dangerous and lacking in interest. Mycophages think exactly the opposite, but they do not agree at all on which mushrooms are good and which are mediocre. It is clear that in something so closely related to taste, a consensus should not be expected. Species considered excellent in one area, where they are collected and sold for high prices, may be considered mediocre in other areas. People in the Languedoc and in Italy consider that the edible boletus (*Boletus aereus, B. aestivalis*) is one of the best mushrooms, suitable for consumption raw or cooked, or dried for use, as strips or crumbled, sprinkled over pasta or into the accompanying sauce. In Catalonia, it is considered a mediocre mushroom, as the "rovelló" (*Lactarius sangifluus*) is preferred. Regions of the same country may also show differences, as may people in the same region. Only some mushrooms, such as Caesar's mushroom (*Amanita caesarea*), are generally accepted, possibly because they combine an exceptional taste with great scarcity.

Mycophagous cultures have seen their daring rewarded by a wide variety of textures and tastes, multiplied by the different culinary possibilities discovered for each edible variety. In any case, mushrooms are delightful even though they are not very nutritious. Very few command high prices. The clearest example of ones that do are the truffles (mainly *Tuber melanosporum* and *T. brumale*), subterranean fungi present in Mediterranean forests of pine and holm oak, and among the most expensive of all foodstuffs.

Aromatic plants

Plants rich in essential oils are very common in the Mediterranean biome, and this has been interpreted as an adaptation to the water regime and to fire. Essential oils accumulate in the leaves and stems, so the smell does not come from the flowers, as is commonly held to be the case, but from the plant as a whole. These aromatic plants may be used for a variety of purposes, such as in the manufacture of perfume, or as aromatic herbs or

140 A stall at the fair of Saint Ponç, in Barcelona. The plants on sale include thyme (*Thymus vulgaris*), peppermint (*Mentha x piperita*), and chamomile (*Matricaria* [=*Chamomilla*] *recutita*). Culinary herbs are used to "dress" foods. Even ancient cookery books described how to season a salad with herbs to make it more agreeable to the palate, or how to flavor stews with herbs to make them more tasty. These herbs are added in small quantities to simple or elaborate dishes. They can be collected from the countryside, cultivated in market gardens, or grown in a pot on a windowsill. Many of these species are not native but have been introduced. In some cases, once they have been dried, cut and/or ground, they are important commodities. [Photo: Ernest Costa]

infusions. They are of economic importance, and even though many of them are picked in the wild, some species are cultivated in order to assure sufficient production near to the sites of processing.

Kitchen herbs

Aromatic herbs used for cooking come in many varieties and fill Mediterranean cuisine with the smell of the matorral. Many culinary herbs belong to the labiate family, like thyme (*Thymus vulgaris*), oregano (*Origanum vulgare*), savory (*Satureja hortensis*), mint (different species of *Mentha* and basil [*Ocimum basilicum*]), to name but a few. They are used to season cooked foods and salads and some have other properties. Basil is said to repel mosquitos, and is grown in pots on windows to repel these pests that transmit malaria, formerly common in marshy areas of the Mediterranean Basin. Some herbaceous members of the family Umbelliferae, like fennel (*Foeniculum vulgare*) and parsley (*Petroselinum crispum)*, are also used in cooking.

Some of these herbs, combined with other plant products like flowers, leaves or the peel of citrus fruit, the fruits of the myrtle (*Myrtus communis*) or juniper "berries" (*Juniperus communis*) are used to flavor a wide variety of spirits, called stomachics, because they are thought to favor digestion. Special mention should be made of the Greek retsina wines that are aromatized with pine resin, giving them an intense Mediterranean flavor.

Aromatic herbs may be used to make infusions, some for medicinal purposes and others simply because they taste good. Examples are thyme (*Thymus vulgaris*), peppermint (*Mentha x piperita*), the Chilean "boldo" (*Peumus boldus*) or the South African "rooibos tea." This is obtained from the leaves of a bush (*Aspalathus linearis*) that occurs in the fynbos; annual production reaches about 4,405,000 lbs (two million kg).

Perfumes

A third use of aromatic plants is as a source of essential oils, which are extracted and used in the manufacture of perfumes, cleaning products, and pharmaceuticals. Examples include lavender (*Lavandula angustifolia*), rosemary (*Rosmarinus officinalis*), and sage (*Salvia officinalis*) which provide essences for perfumery, or the South African bush *Agathosma betulina*, a source of volatile oils used in the pharmaceutical industry. In addition, there are many Mediterranean plants that provide essential oils that are used to some degree by humans, ranging from trees like the carob (*Ceratonia siliqua*) to lichens.

Ornamental plants

The proteaceous flowers of southern Africa, are an economically important plant resource, though their use is ornamental rather than as a food or for their essences. They are greatly appreciated for their striking appearance and their long life as cut flowers; many are sold as dried flowers and last even longer.

Other ornamental flowers from the Cape region are also commercialized, like some varieties of

From the matorral to Grasse

Tanning leather is not the most sweet-smelling of trades, and it is even worse if performed using traditional old-fashioned techniques. This has had the apparently paradoxical result that the small Provencal town of Grasse is surrounded by fields of lavender and is the headquarters of many companies devoted to the extraction of plant essential oils and to the manufacture of perfume. It all began centuries ago, when the city was a major center for the treatment of skins and hides.

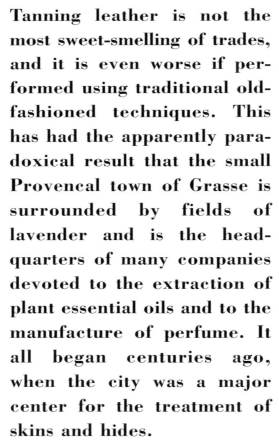

Field of lavender (*Lavandulus angustifolia*), Provence, France [Herbert Kranawetter / Bruce Coleman Limited]

Enamel and gold perfume case (18th century), France [Museu del Perfum (Barcelona) / Albert Masó]

The hides had many uses, including making items of clothing, such as gloves. When these gloves become fashionable among the court in Paris and the upper classes, the tanneries in Grasse started working to eliminate the strong smell of the tanned hides, which made them useless for items with pretensions of elegance. This is how this town started to work on what is now its main activity, the extraction of essences from the plants of the Mediterranean matorral. The stink of the former tanneries has given way to delicate perfumes.

The important essence extraction industry in Grasse has led to the prosperity of an unusual form of agriculture in the surrounding area. Many of the plants that are used in the extraction processes are collected directly from woodland and scrub, but other come from crops that are cultivated to ensure supply and to increase quality. The fields of lavender (*Lavandula angustifolia*), and those of some other aromatic plants, are a major component of the landscape around this city of perfumes and their smell impregnates the air.

Detail of half-title page of the book *Nouveau recueil de secrets et curiosités*, Amsterdam (18th century) [Museu del Perfum (Barcelona) / Jordi Vidal].

Another traditional method is "dry solution," the transference of the essences on to a fatty substrate. The dry, shredded plant is left on a bed of animal fat for several days. When the extraction is considered to have finished, the aromatic essence is chemically separated from the fat.

The essential plant oils that form the basis of perfumes are especially abundant and aromatic in many plants that grow in Mediterranean matorrals, and they are extracted and purified in a variety of different ways. The oldest procedure, still in use, is distillation in an alembic. The process consists of placing the plant in water and boiling it, so that when the water evaporates it carries with it the essential oils which are then condensed and separated by decanting. The almost pure product obtained this way is highly concentrated, and is called absolute essence.

A hydria jar for essences from the Roman Empire (3rd century A.D.) [Museu del Perfum (Barcelona) / Albert Masó]

A 1st century B.C. fresco, National Roman Museum (Rome) [Photothèque Stone International]

The most modern method in use is extraction using volatile solvents. The dried, ground vegetable material is not treated with water but with an organic solvent, like hexane, ethyl alcohol, or benzene, to dissolve the plant's aromatic components. The second stage consists of evaporating the volatile solvent to obtain a concentrate consisting of essential oils and vegetable waxes. The third stage is treating this product, known as concretion, to separate the waxes and leave the absolute essential oil in a pure state.

These processes have very low yields, due to the small quantity of essential oils contained by even the most aromatic plants. For example, 2,200 lbs (1 metric ton) of a highly aromatic plant like lavender only produces about 22 lbs (10 kg) of essence. Other less aromatic plants produce even less, and 8,800 lbs (4 metric tons) dry weight of rose petals are necessary to obtain 2 lbs (1 kg) of essence.

Flask of perfume by Christian Dior (20th century) [Museu del Perfum (Barcelona) / Albert Masó]

Obtaining the essence is only the first step, as far as the perfume industry is concerned. Essences are not the same thing as perfumes at all. Essences are pure, concentrated aromas, while perfumes are carefully studied mixtures of essences in an alcoholic (or other solvent) base. The intense aroma of pure essences is the basis of a perfume's subtlety, but inadequate blending or incorrect treatment may ruin the entire batch. Even the most perfect perfume cannot be compared to the pleasure of the smell of rosemary when walking across a Mediterranean matorral. Grasse, after all, is the same aroma subtly decanted into a bottle.

heather. They constitute an important resource, most of which is for export. Approximately two thirds of South Africa's flower exports are fresh, and one third are dried flowers. More than three quarters of the total harvest is obtained by gathering wild plants, and only a quarter is from cultivated plants.

2.2 Using forests

The forests and woodlands of the Mediterranean biome are not very productive because of climatic limitations. In addition, the intensive humanization of the Mediterranean biome has relegated forests to the worst land, unsuited for agriculture or pasture, reducing their already low productivity. The different regions of the biome vary greatly. Tree cover, for example, has always been scarce in southern Africa, but it is still an important part of the landscape in the wetter parts of the Australian areas.

There has, however, been a large reduction of tree cover in Chile, California and, above all, the Mediterranean Basin where, one has to remember, intense human activity dates back for millennia, especially the eastern part where some of the first civilizations emerged. This has led to serious deforestation, with woodland now occupying only 5% of its original area. In fact, of all the forest systems in the world, those in the Mediterranean Basin and China are the most degraded by human activities.

Complicated management, low yields

The long period of human colonization implies an equally long use of forests, or their simple destruction and then replacement by other ways of using the land and its resources. The extraction of firewood, grazing, and repeated fires over many centuries have led to heavy erosion and soil deterioration, further reducing the potential production of the remaining forests. The final result is woodlands that produce on average 35 cubic feet (1 cubic meter) of wood, about 2,200 lbs (1 metric ton) dry weight, per hectare (1 hectare=2.47 acres) per year. In favorable sites yields may reach 4-5 metric tons per hectare per year, reflecting the potential of this biome, but values like this are so uncommon as to be merely anecdotal.

The management of Mediterranean woodlands presents a series of related problems that are very specific. The first is derived from their mixed use. They are not only used to produce wood but there is also stock-raising activity that may be continuous, or restricted to the dry season. This dry season activity tries to take advantage of the herbaceous plants, shrubs and fruits, at the time when the meadows are dry and unproductive.

The problems, however, do not end here. When forests are relatively large and uniform, such as in the temperate biome, forest exploitation is technically easy. This is uncommon in the Mediterranean biome, where woodlands generally form small patches and are very heterogeneous spatially.

Probably the only exception to this is in Australia, especially in the southwest of the continent. In the Mediterranean Basin there are 40 main forest trees species and 50 secondary species, a very different situation to temperate central Europe where there are 12 forest tree species and 20 secondary ones, forming a forest that is simpler to manage and more productive.

Wood and wood-producing trees

Mediterranean woodlands do not produce large quantities of wood, and even less wood of high-quality. The Mediterranean biome, unlike tropical or temperate biomes, shows a deficit in wood production, although it produces more firewood than wood for construction, and this deficit is especially acute as regards the production of high-quality wood. This has not always been so. Until recently, excellent wood for building boats, so important in Mediterranean civilization, came from the mountains around the Mediterranean.

Whatever the present limitations, there are many different Mediterranean species that provide wood, some with special uses that make them highly appreciated.

Resinous woods
The most important Mediterranean trees that produce wood are the conifers, especially the pines and cedars. The wood of Mediterranean firs is of litle importance, as their range is highly restricted and they are strictly protected because they are in danger of extinction. To these have to be added

the cypresses and other members of the cypress family, *Sequoia* and other members of the Taxodiaceae in California, and, in the southern hemisphere, some trees belonging to the Araucariaceae and Podocarpaceae. The most important Mediterranean pines, especially for wood production, are black pine (*Pinus nigra*) from the Mediterranean Basin, and Monterey pine (*P. radiata*) from California. The maritime pine (*P. pinaster*) also produces reasonable wood, but it is more valued for its resin production.

The different varieties of black pine (*Pinus nigra*) grow wild in relatively isolated populations from one side of the Mediterranean Basin to the other, from the Iberian Peninsula and northern Morocco to Anatolia and the Crimea, and including the Cévennes, Corsica, Sicily, the Italian Peninsula, the western Alps and the Balkans. It is widely used in plantations and has extended its range to include many mountainous regions of the Mediterranean biome. It was important in naval construction, and it is still used for beams and planks in construction and in making furniture. It has also been exploited as a source of resin.

The Monterey pine (*P. radiata*) was originally an endemic species of the California coast near Monterey, the reason for its name, but it is widely used in plantations throughout the mediterraneans, mainly in the western Mediterranean Basin and South Africa and also outside these areas. It has a soft, light, whitish wood suitable for making paper pulp, and is occasionally used in furniture and packaging.

The cedars are also typical of Mediterranean mountains, but in many places the natural cedar forests have been heavily exploited for so long that few trees are left. This is what has happened in Lebanon, where the cedar of Lebanon (*Cedrus libani*) is the national symbol. Today, its range is mainly restricted to Anatolia, although it used to be found in plantations throughout the eastern Mediterranean and in some countries in southern Europe. Cedars are large trees that can reach 131 ft (40 m) in height (except *C. brevifolia* in Cyprus, often considered a smaller variety of the cedar of Lebanon).

They have been greatly prized since antiquity for shipbuilding. Cedar wood is light and soft with an intense resinous smell and it is long-lasting, because its aromatic oils make it resistant to insects and fungi. It has many applications and is used in construction and all types of carpentry. The Atlas cedar (*C. atlantica*) is found throughout the Middle Atlas, the High Atlas, the Rif mountains and in the north of Algeria, and it is cultivated in Mediterranean Europe. Its reddish wood is used in construction and carpentry, and also for cigar boxes, furniture, and planks for building boats used for sports or for recreation. Although it is outside the Mediterranean area, the Himalayan cedar (*C. deodara*), native to the western Himalayas in Afghanistan, is widely cultivated in southern Europe, and produces a wood that is highly appreciated in cabinet-making. All the cedars have been planted as ornamentals in parks and gardens throughout Europe and other regions with temperate climates.

141 **A forest of Atlas cedars (*Cedrus atlantica*) in Morocco.** The Mediterranean climate does not permit excessive plant growth, and in the case of woodlands, this is further slowed by the cost of producing wood. The most suitable trees for timber production are the resinous woods of cedars and pines. Cedars were very abundant in the Tertiary and have a long-lasting, very aromatic wood. This has led to such intense human exploitation that their range has diminished; the Cedar of Lebanon (*Cedrus libani*), for example, is almost extinct in the wild. [Photo; Jordi Bartolomé]

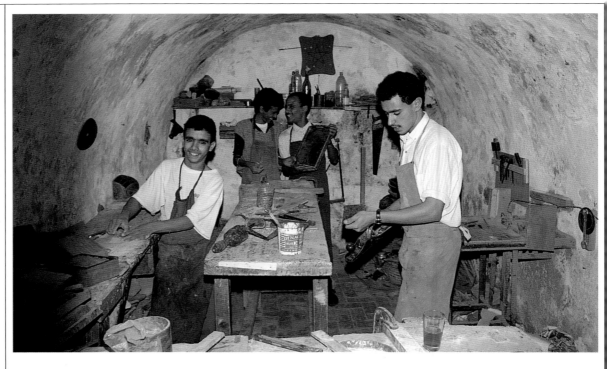

142 Moroccan cabinet-makers in their workshop in Essaouira, Morocco. Wood was the first material that people used to make dwellings and implements. Wood in its many forms can furnish an entire house. Uses include chairs, tables, beds and wardrobes, as well as bowls, plates and eating implements. In medieval Europe carpenters had to complete a tour of their country, staying in the houses of other carpenters, to learn the skills used in woodworking in other places. This gave them an extensive, practical education. [Photo: Josep Germain]

The Mediterranean firs include the Spanish fir (*Abies pinsapo*), found in Andalusia and the Rif; *A. numidica*, found in the mountains of Algeria; *A. nebrodensis* from north Sicily; the Greek fir (*A. cephalonica*), found in southern Greece; and *A. cilicica*, found in the Taurus mountains in southern Anatolia. These and the other firs have very restricted distributions, and although they have been used for wood, today their importance is purely local, and many are protected. Of more importance, although somewhat marginal to the Mediterranean, is the white fir (*A. alba*), found throughout the mountains of south and central Europe (as far north as the Vosges, the Black Forest, the forests of Bohemia and the Tatra mountains), and from the Pyrenees in the west to the Rhodope mountains in the east. The second is the Caucasian fir (*A. nordmanniana*), found in the Caucasus, the mountains of Armenia, and the Pontic mountains in northeast Turkey. Both are large trees, with a light, white wood that is not very resinous. The wood is suitable for carpentry, facings, and for producing veneers for furniture and musical instruments. The Caucasian fir has been used in aeronautical construction and to make paper pulp.

Although the cypress (*Cupressus sempervirens*) is now used mainly as an ornamental tree, or in hedging and windbreaks, its light, knotty, yellowish wood which smells like cedar, is still greatly appreciated in cabinet-making, turnery, and sculpture. It is rot-proof, resistant to attack by fungi and insects, and long-lasting. The original distribution of the cypress is not clear, as it has been planted widely throughout the Mediterranean Basin since antiquity. It may have originated in the eastern Mediterranean, but it appears to have covered large areas of North Africa before the Roman period and before the area's progressive desiccation displaced it. Some Californian cypresses, such as the Monterey cypress (*C. macrocarpa*), have similar uses to those of the cypress, although they are mainly planted as ornamentals.

The redwoods, or sequoias (*Sequoia sempervirens*, *Sequoiadendron giganteum*), grow in mountains of California and are the tallest trees in the world. Some of them are more than 328 ft (100 m) tall, and they are among the most long-lived of all trees. Some felled specimens are estimated to have been over 4,000 years old. Their red wood, of mediocre quality, was heavily exploited in California in the last century. Today, in addition to the respect these giants deserve as natural monuments, they are widely cultivated for ornamental purposes in Europe and North America.

The wood of the members of the beech family
Holm oaks (*Quercus ilex*, *Q. ballota*) are the most important wood-producing trees of the beech family in the Mediterranean Basin. Their wood is dense and compact, hard and resistant to immersion and friction, and has been used to make all sorts of objects (carpenters' planes, handles, shuttles), in construction (especially in hydraulic works), and for other purposes, such as agricultural implements and cartwheels, which are now almost obsolete.

143 A cooper at work in Vilafranca de Penedès (Catalonia), making a barrel from the wood of the chestnut (*Castanea sativa*). This is a medium-quality, fast-growing wood. In addition to its uses in cabinet-making to make drawers and to line furniture, this wood also makes good barrels for wines and spirits. They contain the liquid and also retain fermentation gases. Barrels are difficult to build, as it is a skill that relies little on theoretical models and fixed rules. Coopers rely on intuition and the careful study of models and examples, the only useful guides. [Photo: Ernest Costa]

Although they are not so highly-valued as the oaks of central Europe, the wood of some Mediterranean or sub-Mediterranean deciduous oaks, especially *Q. humilis*, is of comparable quality. Their hard, compact wood is used in construction or shipbuilding, as well as quality cabinet-making. Several Californian deciduous and evergreen "oaks" have been used for similar purposes.

The sweet chestnut (*Castanea sativa*) provides wood, used in hoops and casks, as well as its fruits (chestnuts) and tannin products. It originated in the eastern sub-Mediterranean area, and was planted in the Iberian Peninsula before Roman colonization. Its wood is firm, flexible, light, medium-hard, fine-grained, and not very porous. It is greatly appreciated in both in cabinet-making and in the construction of frames and hedge-posts, as well as in making barrels and hoops, one of the most highly valued traditional uses. The spread of chestnut blight (*Phytophthora cambivora*) has led to attempts to introduce other species into the Iberian Peninsula, such as the far-eastern species *C. mollisia* and *C. crenata* and the North American species *C. dentata*.

The wood of eucalypts

The many tree and shrub species of the genus *Eucalyptus* cover large areas of Australia. The most important species in the Australian mediterraneans are jarrah (*E. marginata*), karri (*E. diversicolor*), and marri (*E. calophylla*). In their natural habitat they grow relatively slowly, especially marri and jarrah, and thus their wood is denser than that of faster-growing eucalyptus, such as the riverside species *E. camaldulensis*, the mountain species *E. dalrympleana*, and the Tasmanian blue gum (*E. globulus*), which grow in Australia in regions with wetter climates. In the Mediterranean Basin (and other regions) they are widely cultivated for paper pulp.

Some of the eucalyptus from the Australian mediterraneans are also cultivated to a greater or lesser extent in areas with similar climates, such as karri which is cultivated in Cape Province, and around Viña del Mar in Chile, and *E. occidentalis* which is cultivated in the more arid areas of the Mediterranean Basin (northwest Africa, Israel, Cyprus). The hybrid form *E.* x *trabuti*, is unusual as it originated in Sardinia through the fortuitous crossing of *E. camaldulensis* and *E. botryoides* and is unknown in Australia, even though it is now widespread in Italy and northwest Africa.

Less important woods

"Litre" (*Lithrea caustica*), an endemic tree from Chile and member of the laurel family, has a hard, heavy wood like that of the deciduous and evergreen oaks of the Mediterranean Basin and California. It is used for similar purposes.

The wood of the olive tree (*Olea europaea*) is yellow with dark brown streaks and is dense and very hard. It is used in cabinet-making, turnery, sculpture, handicrafts, woodwork and, like the wood of the holm oak and other Mediterranean trees, to make charcoal. The hard, compact wood of the

144 **Harvesting the wood of the river red gum** (*Eucalyptus camaldulensis*). Eucalyptus are fast-growing, disease-resistant trees and most species can sprout from their rootstock. These characteristics mean that they are often used in reforestation and plantations, although, especially in the case of reforestation, their effectiveness is dubious. The wood of eucalyptus is used in construction and especially in industrial uses: coat hangers, telephone poles, pit props and, once milled, chipboard and paper pulp. Eucalyptus also provide essential oils used in perfumery.
[Photo: Wayne Lawler / Auscape International]

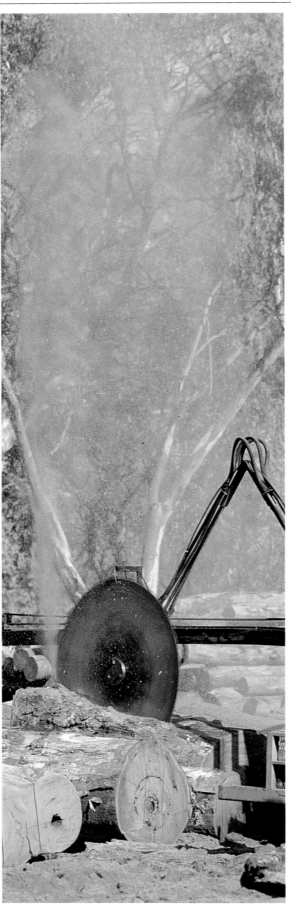

carob tree (*Ceratonia siliqua*) is used in cabinet-making, turnery, and woodworking (carts, wooden benches and shoe blocks), in construction (parquetry), and for making charcoal.

Some Mediterranean shrubs, as well as supplying low-quality firewood, also provide wood with limited but important uses. The wood of some Mediterranean heathers, especially that of the rootstock of the tree heath (*Erica arborea*) is used in the craft production of pipes; it is reddish, hard to work and almost incombustible, and is also used for other tobacco-related items. Other hard, high-density woods are boxwood, from the box (*Buxus sempervirens*), the wood of the strawberry tree (*Arbutus unedo*) and the wood of the terebinth (*Pistacia terebinthus*), and the very hard wood of the mastic tree (*Pistacia lentiscus*); they are all used in cabinet-making, turnery, sculpture, and woodwork.

White poplars (*Populus alba*) and black poplars (*P. nigra*) grow well by the sides of rivers, but now are mostly found in plantations on alluvial terraces with a high water table. They produce wood of mediocre quality that is widely used for packaging and paper pulp.

Wood production
Although many Mediterranean trees produce wood, there is a deficit of wood and forestry products in general, except in Australia. Normally, what the forests produce is used for firewood, sometimes in the form of charcoal, rather than for construction. The difference between producing timber and firewood is usually nothing more than the diameter of the tree's trunk, and thus depends on the frequency of felling and the management regime. The more frequent the felling, the smaller the diameter of the trunks produced. To obtain high-quality timber it is necessary to control the tree's shape carefully. Trees with many branches and full of knots are the result of excessive thinning that opens up large clearings, and they produce wood of lower quality than those that have grown in adequate conditions of density, the result of gentle thinning to stimulate forest regeneration without encouraging excess branching. Whether clearing is performed rarely or frequently, and whether the density is greatly or only slightly reduced, determine not only the quantity of the timber produced but also its quality and price.

Most Mediterranean woods are used for a specific purpose, related to their high density. Mediterranean and Californian oaks and Chilean "litres" were traditionally used for making the handles for implements and parts, such as wheels, for carriages, because of their hardness and resistance to wear. Such uses have lost their importance in the 20th century, leading to the abandonment of many woodlands or their conversion to firewood production. The conifers generally produce wood of poorer quality that is suitable for some purposes in the construction industry or for making paper pulp.

The case of the jarrah in Australia is quite different. Unlike most eucalyptus, generally trees with wood consisting of long fibres suitable for paper pulp, its slow growth, due to the harsh Mediterranean climate and poor soils, means its wood is dense enough for many uses. The magnificent forests found by the first colonists consisted of trees with trunks with large diameters and up to 131 ft (40 m) tall, and were intensely exploited for the construction of buildings, telegraph poles and railway carriages and sleepers. The progressive reduction in trunk diameter caused by the gradual increase in intensity of exploitation, together with the decrease in demand for many former uses, has led to attempts to increase the value of its wood by favouring high-quality uses, such as furniture construction.

Most of the wood now produced in the Mediterranean biome comes from plantations dominated by conifers, eucalyptus, and some deciduous broadleaves, such as chestnuts and poplars. The species planted may be native or introduced. The Monterey pine (*Pinus radiata*)

145 **The stages in the manufacture of a briar pipe**. The wood of the rootstock of the tree heath (*Erica arborea*), because of its resistance to fire and its round shape, is an excellent material for making pipe bowls. All the world's best pipes are either made of meerschaum (an incombustible mineral) or briar (a fire-resistant wood from the Mediterranean world).
[Photo: Ernest Costa]

and some eucalyptus such as the river red gum (*E. camaldulensis*) are good examples of Mediterranean species, or species from partially Mediterranean areas that have spread from a restricted range within one part of the biome and now grow in an area larger than the entire biome, as they have been widely planted. Native species are also planted in the Mediterranean Basin, such as the chestnut (*Castanea sativa*), pines (*Pinus pinaster* and *P. insignis*), and the Atlas cedar (*Cedrus atlantica*). Other trees planted are extra-Mediterranean, such as the Douglas fir (*Pseudotsuga menziesii*), originally from the Pacific coast of North America with a range stretching from southern California to British Columbia. The Tasmanian blue gum (*Eucalyptus globulus*), from Tasmania and southeast Australia, has been planted in large plantations in California and the Mediterranean Basin. Poplars (*Populus* spp.) are different, as most of the speci-

146 **Harvesting osiers from a plantation of willow** (*Salix fragilis*, *S. viminalis*) in Priego, in Castile near the center of the Iberian Peninsula. Osiers are the straight new branches that sprout up vigorously after the trunk has been cut back. They may also be obtained from *S. alba*, *S. triandra*, and *S. purpurea*. In the picture the trunks are only a few centimeters tall, so the stems make a tight thicket of shoots. Once the osiers have been peeled, or in their raw state, they are the raw material for Mediterranean wickerwork.
[Photo Lluís Ferres]

147 Harvesting a plantation of Monterey pine (*Pinus radiata*) in Concepción, Chile. This pine is originally from the coastal area of southern California, the reason why it is called the Monterey pine, although it is not very abundant there. Its rapid growth has led to it being planted, preferably on acid soils, in the other mediterraneans (Chile, Australia, and the Mediterranean Basin). It is often used for wholesale reforestation of areas for wood production. This is why a species that was not very common in its place of origin is now so widespread. In some places it now grows more or less spontaneously. [Photo: Claudio del Río / Fotobanco]

mens planted are hybrids of European and American species.

Plantations of wood-producing trees have been established on abandoned pastures or croplands, common enough in the Mediterranean Basin, or as substitutes for unproductive woodlands, also seen in the Mediterranean Basin and to an ever greater extent in the Chilean and South African mediterraneans. These agriculturally marginal soils are much better than those occupied by most remaining forests. When plantation occurs in a wood, soil ploughing and terracing improve the soil, although this is often transitory because of the increased risk of erosion.

The good quality land used for plantations, the correct selection of trees and their correct treatment, such as pruning and thinning, mean that plantations show much higher production than Mediterranean woods. The modest wood production, 1-2 m³ per hectare per year in most woods in the biome, can reach 5-10 m³ per hectare per year in some plantations, and may 15-20 m³ per hectare per year in riverside poplar plantations and in plantations of Monterey pine (*P. radiata*) on deep soils. If one takes into consderation the fact that their wood is less dense than that of Mediterranean sclerophyllous broadleaves, the difference is not so great when measured in tonnes per hectare. Even so, if we take the extreme case of a difference in density between these woods of 1.0 and 0.5 g/cm³, the production in plantations is clearly much greater than that of woodlands.

Wood from plantations can be used for many purposes, such as obtaining paper pulp, and to make packaging and planks. Overall, they are poor quality woods, unlike those of some trees of temperate and tropical biomes and even some trees from the same Mediterranean biome.

Organic fuels: firewood and charcoal

In practice, most Mediterranean woodlands produce firewood rather than timber—firewood as such or charcoal, both biologically-produced fuels.

Firewood

Faggots—bundles of the branches of pines and possibly broadleaf trees—were the main fuel in the Mediterranean Basin for centuries. Often trunks have been used as firewood because their slow growth produced twisted trees with dense, knotty wood that is excellent firewood but is of little use in construction. This is true for most species of *Quercus*, the deciduous and evergreen Mediterranean and Californian oaks, and many of the trees in the Chilean mediterranean, such as "boldo" (*Peumus boldus*), "litre" (*Lithrea caustica*), soapbark tree (*Quillaja saponaria*) and espino-cavan (*Acacia caven*). Most conifers grow rapidly, but their light-weight wood is not a good fuel, although it has been used in times of scarcity.

Firewood used to be an important fuel, covering domestic and industrial needs until fossil fuels and electricity became widespread. This has meant that many woodlands have ceased to be exploited or their management has changed, at least in the more developed areas of the biome, since obtaining firewood by taking advantage of the woodland's ability to regenerate from stumps allowed intense felling over short cycles, clearly inconsistent with management for producing quality wood. Over the last few years, changes in fossil fuel prices have meant that firewood is again seen as an important source of energy, increasing interest in the exploitation of Mediterranean forests, and even encouraging the exploitation of degraded woodlands or scrub, as current shredding technology allows combustible materials to be obtained from branches and scrub.

Charcoal

Dense woods with a high calorific value can be made into charcoal, which is a lighter and more easily transported fuel than firewood. Charcoal is better because it maintains the wood's energy content but loses most of the volatile substances, meaning it burns with much less smoke.

Charcoal-making is the conversion of firewood into charcoal. The slow, controlled combustion of the firewood causes it to lose almost all the volatile substances, while conserving almost all the original carbon, i.e. its energy content. The loss of water and volatile substances means that charcoal weighs only one fifth of its original weight and is brittle, making it easier to package and transport. These properties have made charcoal a widely used fuel, especially before coal and other fossil fuels became widely available.

Traditionally, charcoal was made directly in the wood itself. The process began in the autumn with the felling of the trees, normally holm oaks, but also deciduous oaks, carobs and olives, and some others. Wood is accumulated over two or three months, then trunks and branches are cut into suitable sizes and in mid-winter they are placed in piles.

The correct combination of length and thickness and careful arrangement are necessary to ensure the stability of the pile, maintained by the friction between the rough barks of the trunks. Once built, the pile is covered with branches, then a thick layer of earth is added to isolate the wood from the exterior and control its combustion. Red hot coals are introduced through the hole at the top, which acts as a kind of central chimney. Once combustion has started, the hole is covered and a long, delicate process begins

148 **The wood of the holm oak (_Quercus ilex_)** has been the most important firewood in the Mediterranean basin for centuries. Wood becomes firewood once it has been cut into smaller pieces and has dried out, so it will burn more easily. As it says in Ecclesiastes "as is the wood thrown on the fire, so it burns." In fact, light woods burn quickly with a lot of flame, while heavier woods burn more slowly and produce more concentrated heat. Light woods are used for strongly heating objects at a distance from the fire (in kilns for porcelain or clay, for example). Heavier woods are used when heat is to be concentrated in a small space, such as in stoves, boilers, and some heating devices. [Photo: Josep Maria Barres]

149 **The different stages of making charcoal**, in an engraving by H.L. Duhamel de Monceau (from the book *Treatise on the care and use of thickets and forests* published in Madrid in 1773). Since the first use of metal humans have known how to make charcoal by burning wood with too little air for complete combustion. The residue, called charcoal, burns without a flame but reaches a higher temperature than wood. Charcoal is made from wood, normally near where it has been felled, and ideally, close to a stream. The most common method of making charcoal is to make a pile of wood around a post, that is then covered by a layer of sand and ash. The stake is then removed, leaving a central chimney, and small air intakes are made at the bottom. The fire started at the top of the pile is controlled by opening and closing these windows, and eventually reaches the base. The charcoal-maker monitors the process by watching the smoke.
[Photo: Biblioteca de Catalunya (Barcelona)]

that must be carefully controlled. From time to time it is necessary to introduce firewood through the hole to revive the combustion process.

The color and smell of the smoke given off allows the charcoal maker to tell how the process is going, and if the inflow of air needs to be reduced or increased, which is done by making or closing holes in the earth layer. This process, called "cooking" by charcoal makers, may last eight to ten days, depending on the pile and the water content of the wood. When the process is considered to have finished, the holes are sealed and the fire put out. It is necessary to wait a couple of days for the pile to cool off, then it can be taken to pieces and the charcoal extracted. This process is more complex than it appears, as it attempts to achieve an even carbonization of several metric tons of wood, using only the smoke given off as an indicator.

Until this century charcoal-making was common throughout the Mediterranean Basin, and records show that wood charcoal was used in the 13th century. After the First World War coal replaced charcoal in many industries, and since the 1940s it has been replaced for domestic purposes by electricity and fossil fuel derivatives, such as diesel and butane gas. Until then, charcoal manufacture was a seasonal activity involving many people, seasonal workers who lived in the woods for several months. The remains of their rudimentary cabins and of the stone-edged sites prepared for building the piles in the middle of the woods can still be seen in many deciduous and evergreen oak woodlands, remnants of an almost lost activity. The high demand for charcoal meant that they used even low quality wood, such as the branches left after felling pines. In this case the process was simpler, consisting of partial combustion in trenches or holes. The abandonment of charcoal-making allowed many holm oak woodlands to recover after long periods of being cut every 15 or 20 years, and regenerating from stumps. Now demand for charcoal is low and the traditional method coexists with the modern techniques for cooking in kilns, meaning some of the volatile compounds emitted can be used.

Intensive charcoal-making has also been practiced in the other mediterraneans. In California, the oak woodlands supplied charcoal for household and industrial use from 1850 to 1960, when the general use of fossil fuels and the low price of charcoal imported from Mexico caused this activity to cease. The Chilean Mediterranean area is also rich in hardwood trees suitable for burning, and charcoal is still manufactured following the old Mediterranean technique. Wood from "litre" (*Lithrea caustica*), soap-bark tree (*Quillaja saponaria*) and "boldo" (*Peumus boldus*) trees is used, although the most valued because of its density is from the shrub, espino-cavan (*Acacia caven*). Unlike in the Mediterranean, the wood-piles, are covered by a permanent structure made of clay that hardens and sets as a result of the heat. The holes made in this cover can be covered or uncovered to regulate air flow and control combustion.

Charcoal continues to be used as fuel, especially in the rural areas of the Mediterranean Basin and in Chile, but it is trivial in comparison with its use in the past. It is used in some rural kitchens and in urban barbecues, but a century ago it was the most important fuel for domestic and industrial purposes. The current state of many Mediterranean woodlands is closely related to the many tonnes of charcoal extracted from them. As an example, in 1950 Italy produced 3 million cubic meters of charcoal, about 60-70% of the total production of wood by Italy's Mediterranean woodlands.

Cork

Cork oak (*Quercus suber*) stands are typically Mediterranean woodlands that are exploited by removing the bark of the trees, rather than felling them. The main product of cork oak woodland is cork, although wood and acorns, formerly highly valued, are also produced. Cork is the bark that covers the trees, insulating them and protecting them from fire. Cork oaks regenerate this layer after its removal, meaning that these woodlands can continue to produce cork after being harvested. These woodlands occupy wet zones with acid soils in the western Mediterranean Basin. There are about 4 million hectares, spread between the Iberian Peninsula, France, Italy, and North Africa, as well as on some of the larger islands like Sicily, Corsica, and Sardinia.

The removal of the bark takes place between the end of spring and the middle of summer, when the tree is actively growing and the soft layer of the current year's cork means the entire bark can be easily peeled. The process begins with the trees at lowest altitude, as they start growth earlier in the season. The peeling operation is delicate, because mistakes may wound the cambium, with the consequent risk of infection leading to poor quality new cork. The first part is making an incision across the upper part of the trunk, the "neck," and then opening the bark longitudinally using a special tool. The tool's bevelled handle is used to separate the mass of bark from the trunk, preferably in a single sheet. Once the cork has been removed the base and "neck" are polished and some individual trees are marked with the year of harvest. Removing the bark must be performed by a skilled operator who knows when to stop: a sudden change in the weather, such as a thunder storm, may cause the bark to separate very badly from the trunk, with the risk of breaking the sheets of bark and wounding the tree. It is also necessary to exercise judgement about the height to which the tree should be peeled, because, if this is excessive, the tree suffers and the cork produced subsequently is of lower quality. Recently peeled cork oaks are bright yellow, turning red in a few weeks, and dark red over a few months. After a year there is a

150 **Stripping a cork oak** (*Quercus suber*) in Andalusia, in the south of the Iberian Peninsula, to obtain its bark. Note the bright orange color of the trunk when its corky bark is removed. In Andalusia and Portugal, cork oaks are peeled up to the topmost branches. In Catalonia and Sardinia only the cork from the trunk is taken, meaning the stripping can be more frequent, as the damage to the tree is less.
[Photo: Romano Cagnoni / Zardoya]

new layer of cork, which increases in thickness every year until it is ready to be peeled again.

The yield of a cork oak may be may be changed by modifying the time between successive peelings, as the thickness, and thus quality, depend on the cork's rate of growth. Production of 100 kg/ha per year is considered very good. In sites with rapid growth, the trees are peeled every eight to 10 years, while in others they can only be peeled about every 12-14 years. Between peelings it is necessary to carry out a series of operations to maintain production and quality. These operations are selective felling to improve and ensure regeneration by eliminating diseased, old, or damaged individuals, and clearing the canopy before peeling to ease the work of the cork strippers, and the treatment of the productive trees to ensure the quality of the cork. This treatment basically consists of cutting the cork, making a vertical incision in the growing cork so as to make the next removal of the cork easier and to ensure that the cork does not split irregularly, but opens along the incision as it grows from the inside outwards. This cut is made in mid-winter, four to five years after the last peeling, and on a day after abundant rainfall which wets the cork and makes the operation easier.

Cork oaks are peeled for the first time when they are 30-40 years old, and this cork is rough and of low quality. The best quality cork is obtained from the seven or eight peelings after this one, when the tree is between 50 and 120-130 years, if the process is performed every 10-12 years. After this the quality declines and the trees are felled to improve and regenerate the woodland, as it is necessary to remove old individuals and stimulate the appearance of new trees. The quality of the cork depends on the rate of growth, which determines its density and porosity. The habit of peeling the cork off high up the tree, beyond the fork, also reduces the quality of the product, since the tree is badly affected. Sites that favor rapid growth produce lower quality cork than sites where growth is slow. The best cork comes from sheets 1.2-1.6 in (3-4 cm) thick that have taken 12-14 years to grow, such as the cork from the northeast of the Iberian Peninsula.

Other forest products

Some types of bark, including those of conifers (especially those of pines), holm oaks (*Quercus*

151 **Tanners preparing skins in the traditional way**, using natural tannins, in Fez, Morocco. Properly treated leather is a long-lasting material that is rotproof and almost impermeable, yet smooth and flexible. Tanning skins is basically a chemical process that changes them into leather, and is traditionally performed using plant products. Tannins are most widely used and may come from many plants, such as oak bark, wood and galls, and the roots and fruits of several other plants. The discovery of chestnut tannin (in 1845) meant that tanning, which had changed little since the days of the Romans, became quicker. Today, it is mainly performed using industrial chemicals, such as chromium salts.
[Photo: Jordi Bartolomé]

ilex), carobs (*Ceratonia siliqua*) or alder (*Alnus glutinosa*), are not waste but products used in the treatment of hides, although this use is disappearing as the chemical industry can supply better alternatives. The bark of the holm oak, like that of other oaks, has been used as a tanning agent for a long time, and gives the hides a yellowish brown color that is greatly appreciated. The case of the soap-bark tree (*Quillaja saponaria*) in Chile is similar: its saponin-rich bark is used in making soap and toothpaste. However, the whole tree or shrub must be felled to obtain the product, and its exploitation is little different from felling for firewood or timber.

In coniferous woodlands, especially those of the maritime pine (*Pinus pinaster*) a resin is obtained by tapping the trees. It is rich in oleo-resins, used in the preparation of the solvent turpentine. This resin is largely responsible for the high inflammability of pines and flows down the bark in the form of transparent drops that are collected in cups like those used to collect rubber.

152 **Resin production** on a plantation of *Pinus pinaster* in Alentejo, Portugal. Conifer resin is obtained when it exudes from wounds caused in the bark for this purpose. The resin used to be extracted from trees with insect-damaged bark, as the damage meant the trees exuded resin. The resin is soluble in alcohol, and on distillation it gives oil of turpentine, which is used as a solvent, and rosin, a raw material for paints, varnishes, and soaps.
[Photo: Ernest Costa]

Light, insulating, impermeable

Opening a bottle of champagne [J. Myrdal / Index]

The Benedictine monk Dom Pierre Pérignon's name entered history in 1681, in the French abbey of Hautvillers. After observing that the wines of the Champagne district tended to form foam spontaneously, he discovered that adding the right quantity of sugar to the wine ensured this effervescence. This was the origin of the *méthode champenoise* that would later lead to the production of the first *champagne*, followed by *blanquette* in the Languedoc, *spumante* in Italy, and *cava* in Catalonia. Yet this discovery would have been useless if he had not thought to cork the bottle with something that was effectively hermetic, yet permeable enough for some gas exchange. If the bottles were not sealed with true cork, there would be no champagne.

A cork float for use in fishing [Jordi Vidal]

Corks for bottles of still wines [Firo Foto]

Corks for bottles of sparkling wines [Firo Foto]

Dom Pérignon found a new use for the cork made of true cork, but he had not invented it. In the accounts of a 17th century English wine merchant, there is an entry dated October 28, 1662 for the purchase of a shipment of corks. In fact, the material's properties have been known since antiquity, and is mentioned in Theophrastus, Cato, Varro, Pliny the Elder, Columella, and Plutarch, classical authors who lived and wrote between the 4th century B.C. and the 2nd century A.D. They all praised its insulating properties and lightness, although they assigned it an unimportant role among the humble materials useful for making floats for fishing nets, beehives, shoe soles, in addition to corks for amphoras. Corks started to spread after Dom Pérignon's discovery, replacing the wood and oiled hemp formerly used to seal bottles and barrels, and became well-known.

Sheets of cork, São Bras (Portugal) [C.C. Lockwood / Bruce Coleman Limited]

Cork is an inert material formed by the accumulation (a thickness of 0.08 inches or 2 mm a year) of the suberized cellulose membranes of the protective cortical tissue formed by the phellogen (or cork cambium), which is responsible for secondary growth, not the epidermis. The phellogen is like a second bark consisting of empty dead cells, containing small bubbles of air within the former cell membranes. This unusual structure is responsible for cork's characteristic properties. It is light and floats on water, as its density is only 0.24 g/cm^3 in comparison to water's density of 1. It is an impermeable insulator (it is a very bad conductor of heat and vibration, and thus sound). It is long-lasting and almost rot-proof, as it is resistant to most environmental agents that could degrade it, whether biological or chemical). It is compressible and elastic (after 10 years in the bottle, a cork may expand to 75% of its original volume). These properties are due to the reaction of cellulose with suberin (the polymeric ester of phellonic acid and suberinic acids, among others), and are not restricted to the bark of the cork oak (*Quercus suber*), although no other plant structures shows them so clearly. Cork is a truly unique material.

The bark of the cork oak appears to have been designed to make corks for bottles. This is why, at the beginning of the 17th century demand for corks began to rise in France. This was stimulated by the fortunate circumstance that the range of the cork oak (now grown over about 4 million hectares) coincided more or less with that of the vine, as both species are Mediterranean. During the 18th and 19th century, Catalonia and specifically the Empordà region, was the center of the world cork industry (with a strong presence of British capital and technology). Catalonia has shared this leadership with Portugal since 1870, and with Sardinia since the early 20th century. For many years most of the cork processed has came from cork oaks in Andalusia, Portugal, and Morocco, as their cork oak woodlands cover a much larger area than those of Catalonia. The following data show the growth of Empordà factories' sales and the increase in world demand: in 1845 they produced 700 million corks, in 1881 they made 1,800 million corks, and in 1900 the sold 3,300 million corks on the world market.

Decorative item using cork [Josep Loaso]

At the end of the 19th century these corks were still made by hand. A great deal of skill is necessary to cut a perfect cylinder with just a knife! Before the cork can be formed, a series of preliminary operations must be completed. First, the sheets of cork must be boiled to eliminate any residual sap and tannic acid, and to increase the sheets' flexibility so they can be flattened. The sheets are then sliced to remove the cracked and broken edges and they are cut into narrow strips (or into dice-like cubes when corks were made by hand). Once this operation is finished, the actual making of the cork starts, although today this is done by machines. After this the corks must be washed with a mixture of salt water and oxalic acid (or calcium chloride, or even potassium permanganate) and the brandname is stamped with hot metal (pyrography).

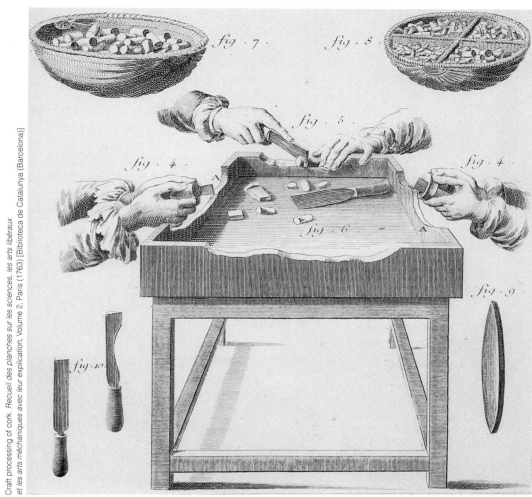

Craft processing of cork. *Recueil des planches sur les sciences, les arts libéraux et les arts méchaniques avec leur explication*, Volume 2, Paris (1763) [Biblioteca de Catalunya (Barcelona)]

Corks for sparkling wines usually have a more complex structure, consisting of several sheets of cork glued together to prevent gas escaping through any pores. Corks are now cut perpendicularly to the direction of the lenticels (pores) that run through the bark, so that these natural perforations are blocked by the glass of the bottle's neck. One need only look at a cork to see this. The pressure generated by sparkling wines, however, requires additional precautions. These corks often have an axis of cork agglomerate, an industrial product obtained by gluing many small pieces of cork together. Cork agglomerate has so many other uses that, except for corks for non-sparkling wines, most modern applications of cork use agglomerate, such as roof and wall insulation, flooring tiles, etc. Cork has even been used in the fashion industry, and this use requires unblemished sheets of the best cork, suitably treated for use in decorative goods, coverings and even garments. All this from the remarkable bark of the cork oak.

153 Mediterranean agro-silvo-pastoral strategies, according to the main vectors determining them: the investment of capital and labor, the diversity and size of the space available, intermediate consumption, and the price of the final products. In the traditional integrated strategy, extensive stock raising was an additional activity that restored the fertility of the agro-silvo-pastoral system (the Roman *ager, saltus et silva*) by manuring crops with the excrement of animals grazed in forests and woodland. The extensive strategy arises where agriculture has been abandoned, when pastures increase in size but are less diverse (as in marginal agricultural land in the European Mediterranean). Market strategies appear when the value of milk or meat products makes extensive livestock raising profitable, despite the low productivity of Mediterranean pastures (as is now happening in some Mediterranean islands or the mountains in northwest Africa). The opportunist strategy is a combination of the two above strategies where the annual cycle permits it, or where it is hoped that grazing will provide the additional service of controlling the biomass, as happens in Mediterranean forests in Europe.
[Drawing: Editrònica, based on Joffre, 1991]

2.3. Pastures with hardly any grass

The predominance of woody over herbaceous vegetation in the landscapes of the Mediterranean biome is a reflection of its climatic conditions, because the fact that the hot season and the dry season coincide imposes severe restrictions on herbaceous vegetation, which thrives in climates with wet summers. This characteristic climate greatly conditions livestock farming as, generally speaking, grasses are much more suitable than trees and shrubs for feeding livestock, in terms of both digestibility and nutritional value.

Herbaceous grazing formations

In the Mediterranean biome, herbaceous formations are restricted to dry places with poor soils

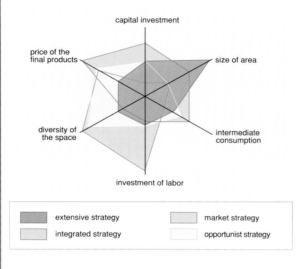

that can not support woody plants. They are dominated by annual plants that get round the problem of aridity by spending the dry season in the form of seeds. In the drier areas of the biome, the dominant communities are low, open shrub communities, such as thyme scrub (*tomillares*), with an almost negligible herbaceous presence. Nevertheless, human degradation of woods and thickets can lead to formations dominated by herbaceous plants, and many Mediterranean pastures are the result of a long history of human exploitation and modification.

This degradation from thickets to meadows is the result of a whole series of modifications ranging from excessive felling, to an increase in the frequency of fires and the ploughing and subsequent abandonment of land. However, the main factor

has been uninterrupted livestock farming: in effect, using wasteland for stock raising was the beginning of a long history of favoring herbaceous vegetation over shrubs.

The meager grazing offered by the pastures in the Mediterranean biome has meant that maximum use has had to be made of the food potential of woods and thickets, either by reserving them almost exclusively for livestock farming or by combining them with forestry activities. Farmers have often had to use their agricultural production to feed their livestock. All of this means that in the Mediterranean stock raising activity is closely linked to the other local forms of land use.

The origins of the pastures

Those regions of the biome situated in the Mediterranean Basin were the birthplace of ancient pastoral civilizations which have led, on the one hand, to a notable and extensive degradation of the landscape, and, on the other hand, to a very rich stockraising culture. This is based on the existence of a genetic heritage of livestock breeds that are well adapted to the biome's distinctive conditions, but selected for outdated features, such as wool production or strength. Paradoxically, the same exploitation model that has caused the problems contains possible solutions, as long as the good and bad points of centuries of stockraising experience are correctly separated.

In the other mediterraneans outside the Mediterranean basin, modification of the land due to human exploitation is a relatively recent phenomenon. Even so, their intensive exploitation has led to major or even drastic changes, by acting on landscapes with few pre-adaptations to tolerate such use. Although grazing is very recent in these areas, it has played a major role in changes to the landscape. To illustrate this, the six merino sheep introduced into South Africa in 1789 had 50 million descendants by the year 1930.

In fact, the landscapes in the Mediterranean Basin are the result of long co-evolution with human exploitation, and this has made them more resistant, bearing in mind their intrinsic fragility. This explains the success of species from this area as invaders of the other regions of the biome, whether in Australia, Chile, California, or South Africa. The herbaceous plants that dominate in the Mediterranean Basin, after centuries of livestock farming, have become resistant to grazing and this

154 A herd of goats (*Capra hircus*) grazing in the Sierra Nevada mountains (Andalusia), the highest range in the Iberian Peninsula. Pastures are a complex system of relationships among herbaceous plants, animals, and human management. Mediterranean pastures are mostly poor, apart from the *dehesa* but there is a wide range of herbaceous communities that may be explained by the exceptionally difficult environment for plant growth. The plants have to grow in infertile soil in a climate in which, unfortunately, the hot season and the dry season occur simultaneously.
[Photo: Jordi Bartolomé]

has helped them to dominate a large part of the Californian, Chilean, South African and Australian mediterraneans, as well as spreading to temperate Europe.

In California, for example, the European colonizers who arrived at the end of the 18th century found landscapes that were almost untouched by human activity, as the original inhabitants knew nothing of agriculture and stock raising. Furthermore, there were only four species of large herbivores, none of which fed exclusively on grass, as they had a very varied diet of plant resources derived from trees and shrubs. All in all, the largely unspoilt landscapes were not adapted to heavy grazing pressure. The modifications introduced by the colonizers, including felling, ploughing and intensive grazing, opened the door to an invasion of species from the Mediterranean Basin. European and North African annual plants rapidly displaced the native herbaceous plants, and meadows originally dominated by herbaceous perennials were replaced by formations of annuals. Over-grazing and climatic fluctuations from year to year led to the abandonment of marginal cultivated land in exceptionally dry years, and these factors favored the more resistant and prolific introduced annual species, which were able to produce larger seed banks. Between 1769, the year of the first settlement, and 1869, 134 European and North African species were introduced, many of which were herbaceous.

Some authors consider that the original Californian herbaceous formations that the colonists found may have been relicts from a climate with wetter summers. Their persistence might have been favored by the absence of exclusively grass-eating herbivores over the last 10,000 years and by the presence of browsing herbivores which kept the growth of woody vegetation under control. In California, and as a consequence of colonization, an area of meadows relatively similar to the original area (about 8 million hectares) has been maintained, but its distribution is clearly different. Most of the land which was dominated by herbaceous vegetation is cultivated or built-up, while today's meadows occupy severely degraded former tree or shrub formations. These changes have had a variety of consequences, one of the most disadvantageous of which is that pasture production, dominated by introduced annuals and not native perennials, is much more dependent on rainfall; it is, therefore, much more variable than production by perennial species with a summer pause regulated more by the photoperiod than by rainfall. Furthermore, serious problems have arisen with respect to the regeneration of oaks in the savanna oak regions, as the summer pause imposed by the photoperiod in the case of the native perennial grasses allowed good germination and growth of oak saplings in years when plentiful spring rains maintained soil humidity during the summer. Today, however, herbaceous annuals remain active as long as there is moisture, preventing the germination and growth of the oak seedlings.

In Chile, the replacement of native herbaceous plants by introductions from the Mediterranean Basin was equally fast and intensive. Mediterranean Chile's long isolation and the lack

155 Goats (*Capra hircus*) grazing on grass covered in hoarfrost in the Monegros (in the middle of th Ebro Basin, in the northeast of the Iberian Peninsula). The variable Mediterranean winter, like the dry summer, limits the productivity of pastures. This means that cold winters lead to overgrazing, as the sparse winter growth forces the herds to share resources. This over-exploitation of pasture and unproductive land is basically due to the pressure exerted by goats in areas where regeneration is difficult. [Photo: Josep Germain]

of pre-adaptations to intensive grazing favored the rapid invasion of introduced herbaceous plants, since they had been co-existing with the pressure of grazing livestock for centuries. This led to the total disappearance of the native evergreen grasses.

The South African and Australian mediterraneans have also seen invasions by herbaceous plants from the Mediterranean Basin. The absence of grazing pressure in the case of Australia, the long isolation of both floras, and their lack of influences from colder climates, being limited to only those of warmer climates, whether drier or wetter, meant they were at a disadvantage with respect to the herbaceous flora introduced from the Mediterranean Basin. Intensive grazing by livestock introduced by the Europeans finished off the process, even though in both regions the fact the colonists came from outside the Mediterranean Basin introduced a new factor of diversity into the range of invader species.

Low productivity

Mediterranean herbaceous formations generally have a low level of productivity, regardless of whether they consist of native or introduced species, and this is independent of whether they are the result of degradation of woody formations or not. Their production ranges from a few hundred kilograms to 10 tons per hectare per year, extremes which correspond to small meadows of annual plants on very poor soils, and meadows in valley bottoms on deep soils with good reserves of soil water. The most normal values, though, are around 2 t/ha per year.

This low productivity of itself limits livestock production and, furthermore, it is irregularly distributed over the year, concentrated in the spring and autumn months, with a little in the winter, if temperatures are not so low as to limit plant growth. The cold in winter can become an even more serious factor if the autumn rains arrive late, as then the prevailing humidity may make the grass rot. This brings us to another adverse factor of Mediterranean climatic conditions, which is that yearly variations may be very large, and this makes it even more difficult to plan appropriate stocking levels, leading to overgrazing and degradation during particularly dry or unfavorable years.

High floristic diversity

The considerable floral diversity of Mediterranean herbaceous formations was originally concentrated in the regions of the Mediterranean Basin itself, but it has now spread to other regions of the biome. This floral diversity leads to great diversity over time because of the great variety of seasonal phases which leads to the prolongation, to some extent, of the peaks of production, which never occur in the summer period. Moreover, this also leads to considerable spatial diversity, resulting from the combination of floristic diversity with the highly heterogeneous Mediterranean landscape.

All in all, this variety is really a series of advantages for stock raising as long as it is based on a thorough knowledge of the environment, because this allows very precise management geared to making successive use of different seasonal phases in different micro-environments. This type of management is feasible as long as the knowledge accumulated during centuries by pastoral cultures is preserved and improved. This does not always happen as changes in management models often go hand-in-hand with a gradual loss of more traditional techniques.

Fallow land and stubble

Cultivated farmland supplies a whole series of by-products that are used as resources in stock raising, whether for their intrinsic nutritive value, or because they represent an input of food when herbaceous production is low or has stopped altogether.

Stubble on cereal fields, for example, is used as summer pasture when dry conditions halt the growth of herbaceous communities, forcing the livestock to consume surplus dry production, if there is any. In regions where cereals are widely cultivated, this can represent an important food source during the dry period, with an input of around 500-1,000 kg/ha.

156 **Sheep (*Ovis aries*) and goats (*Capra hircus*)** grazing on stubble in Serrans, in the Autonomous Community of Valencia, in the east of the Iberian Peninsula. Cereal field stubble is used by the animals as food at the height of summer, when only limited pasture is available. In the worst years, the entire cereal crop may be used to feed the animals. An alternative to using cropland exclusively for livestock production is to move to areas where more resources are available. This practice of seasonal migration, called transhumance, has long been practiced in the Mediterranean Basin.
[Photo: Jordi Vidal]

Fallow land and abandoned fields colonized by agricultural "weeds" are also a source of food for livestock. This agricultural practice, employed especially on low quality land incapable of supporting permanent cultivation, often seeks to create and maintain pastures rather than agricultural production as such. Regular ploughing for cultivation in fact prevents shrubs from developing, thus keeping the land suitable for pasture, as well as regularly providing a modest quantity of agricultural products. In certain cases, the crop is not even harvested, but is directly grazed, clearly showing that this practice is basically livestock-related.

Cultivated land supplies other types of resources for use as fodder. Leaving to one side products such as the fruit of the carob tree which is grown for this express purpose, and concentrating only on by-products, the leaves of the almond tree, in summer, and those of the olive tree, in winter, are fed to livestock when the trees are pruned after the harvest. Although they are of mediocre quality, they may play an essential role in maintaining livestock levels in fluctuating environments like the Mediterranean. Animals can also graze on the

herbs growing under woody crops in unirrigated fields, such as olive groves, carob groves, almond groves, vineyards, and also on those growing in irrigated land such as orange and lemon groves.

Woods and thickets that can be used for grazing

In the Mediterranean biome, woods and thickets have been relegated to the least productive areas, with poor soils and often on steep slopes. They supply wood, fuel and other products, and are often also used as a source of food for livestock, either exclusively or in combination with forestry uses.

The herbaceous layer of Mediterranean woods and thickets is poor and sparse. It is for this reason that in these areas the livestock basically consume the foliage and tender shoots. Many woody Mediterranean plants have sclerophyllous leaves, in other words leaves which are small and hard, with thick cuticles, a waxy coating and abundant hairs and trichomes. This type of leaf contains much structural material of limited nutritional value, such as lignin, and is a low-quality food, but it is a reliable source of fodder in the dry or cold seasons, as well as a form of insurance against yearly climatic fluctuations. This is because sclerophyllous trees and shrubs, with their hard leaves and deep root systems, are less sensitive than herbaceous plants to very severe conditions of cold or dryness.

Cleared woods and thickets are often found, in which the elimination of some of the woody structure favors the development of a herbaceous layer which can supply good quality fodder, while the remaining trees and shrubs provide a structure that stabilizes the supply of resources. This stabilization reduces potential herbaceous production, as the woody structures use up part of the water resources and significantly shade the herbaceous layer. The total elimination of trees and shrubs often leads to a spectacular, but often transitory, increase in herbaceous production.

This explains why shepherds have shown a marked tendency to eliminate woody structures, thus increasing the presence of herbaceous communities in a biome that was originally poor in them.

Maintenance of all or part of the tree or shrub mass is usually because of the need to obtain other products such as firewood, resin, and cork, or to protect the soil from erosion. Unfortunately, the desire to protect the soil has not been very high in the Mediterranean biome, and this is shown by the severe degradation of many of the region's woods and thickets caused by over-grazing.

2.4 Dehesas: woodlands with pastures and crops

In Mediterranean climates, the maintenance of a certain density of woody plants in pastures has several advantages that, in the long run, make up for the disadvantages derived from reduced herbaceous production. Mediterranean forests are also used for grazing, and so it is not uncommon to find pastures with trees, although their density may vary greatly. This is the case of the Californian savannah oak formations, with occasional large oaks in dry meadows, the "espinales" in Chile, with a woody layer made up of the small tree *Acacia caven*, or certain Australian herbaceous formations dotted with eucalyptus, as well as many other wooded pastures or open pasture woods which can be found throughout the Mediterranean Basin. In many cases it has been shown that the presence of trees has many advantages, ranging from soil enrichment to extending the vegetative cycle of herbaceous vegetation.

These systems are basically pastures with trees, where the trees play a secondary and minor role. In the Mediterranean environment, though, a combined crop-growing, forestry and livestock farming model has appeared that totally integrates the trees into the pastures and arable fields, giving them a very important role. These integrated systems are the result of a cultural process stretching back centuries that has transformed the Mediterranean forests into open woodland formations, including a range of domesticated resident species that make them more productive and controllable. They are

157 Chilean dehesa with a soap-bark tree (*Quillaja saponaria*) and "espinos" (*Acacia caven*) in the background. The dehesa is a good example of human natural resource management, and has changed the landscape. In pasture lands, forestry is less important, and the pasture is dedicated to livestock, which also take advantage of the fruit and shoots of the trees. The shade cast by the crowns of the trees also reduces summer water loss and helps livestock withstand the hot summer.
[Photo: Jaume Altadill]

only found in the southwest of the Iberian Peninsula and are called, "dehesas," when they occur on the plains of Estremadura and "montados" in Portugal.

The types of dehesa

Dehesas are characterized by a combination of climatic and soil conditions that is very uncommon anywhere on Earth: the superimposition of the semi-aridity of the Mediterranean climate on acidic soils. Dehesas occur in environments where limits to agricultural intensification have led to a management model with many agricultural, livestock, and forestry products. These limitations, if not the specific environmental combination mentioned above, are present in many areas of the Mediterranean biome, and this is why the dehesa management model is a useful starting point for creating other, similar models. To sum up, this model of land-use is multiple and extensive, adapts to the low productivity imposed by the environment, but functions with only minimal external inputs. It is thus an extremely useful guideline for managing a limiting environment, such as the Mediterranean biome.

Within the dehesa agro-silvo-pastoral system, different production and management units can be distinguished: permanently cultivated land, land alternately used for crops and pastures, permanent pasture and areas occupied by shrub or tree communities. Trees play a leading role in the dehesa system and occur, in varying densities, in all the above units.

Dehesas with permanent crops

The land set aside for permanent crop cultivation corresponds to areas with better soils, in other words, well-drained, deep, fertile soils in areas with little, if any, slope. In this unit, although there are some trees, their density is very low, and is always less than in the other units. The traditional form of cultivation alternates three crops (a fallow period, wheat, barley, oats or rye, and then fallow again) or four crops (fallow, wheat, barley oats, or rye, vetch, and then fallow again). These crops give a maximum grain production of 2 t/ha per year, complemented by a straw production of 0.5-1 t/ha per year, used as fodder. When cultivation is for fodder, production may reach 6 t/ha per year. The agricultural weeds that grow during the fallow period can also be used as pasture.

Dehesas with alternation of crops and pasture

Land where crop-growing and pasture are alternated, is regularly worked, at intervals of between two and 10 years. The most common form of alternation consists in using the land for cereal crops for 3-5 years and then for pasture for the same number of years. This periodical working is not performed to increase agricultural production, but to maintain the pasture land in good condition, as it prevents shrubs growing and thicket formation. On the poorest land, the crops are often not harvested but are directly grazed.

Dehesas with permanent pastures

Permanent pastures can be subdivided into different categories according to their productivity. The best quality ones, called "majadales," have been improved

158 A dehesa of holm oaks (*Quercus ilex*) with an annual crop of oats (*Avena sativa*) in the Montes de Toledo, La Mancha, in the Iberian Peninsula. The use of dehesas to cultivate cereals, as in the photo, can be carried out in areas with rich, deep soils, with good drainage and a slight slope. Although production may be lower than in open spaces it is a method of landscape management that integrates livestock and agricultural management. In fallow periods they provide additional livestock fodder.
[Photo: Ernest Costa]

159 **Female fighting bull** (***Bos taurus taurus***) in a pasture in Estremadura (southwest Spain). The use of pastureland for breeding fighting bulls is special to the Iberian Peninsula. It is known that the inhabitants of the Mediterranean have paid tribute to bulls since time immemorial, probably as a fertility symbol. This is shown by the representations of sacred bulls in most of the area's polytheistic religions, and in contests confronting animals and humans (which appear in the Minoan frescoes at Knossos). [Photo: Oriol Alamany]

by good pasture management and penning which keeps the grass in good condition and improves the soil thanks to the organic manure from excrement.

In some years, if the rainfall leads to a very high production of grass, a part of this can be harvested and stored as fodder for times when productivity is lower. This surplus production usually only occurs in pasture located in very special conditions, in valley bottoms receiving additional water, forming what are called *vallicares húmedos* and *prados semiagostantes* (meadows that only partially dry up in high summer). In the majadales, maximum production is around 4,000 kg/ha dry material per year, while in the unusual sites mentioned it may reach as much as 7,000 kg/ha per year.

The most common category of permanent pastures, i.e. the ones occupying the largest area and therefore playing the most important role in the system as a whole, have a maximum production of 2,000-2,500 kg/ha per year. The least productive pastures correspond to wasteland on very poor, dry, stony soils; they are dominated by annual plants and their maximum production is around 2,000 kg/ha per year, although in most cases they do not reach even half this amount. Finally, the marginal areas of poorer quality, such as rocky ridges, are occupied by shrub or tree formations that also provide fodder, although this is not very important. The dominant species in the shrub layer are brooms (*Retama sphaerocarpa, Sarothamnus scoparius*), different species of cistus (*Cistus ladaniferus, C. hirsutus, C. salviifolius*), lavender (*Lavandula pedunculata*), lavender coton (*Santolina rosmarinifolia*), different species of thyme (*Thymus*), and various heaths and heathers (*Erica*). These marginal areas play a very important role when herbaceous production is low (mid-

summer and midwinter), and livestock is thereore forced to subsist on these resources.

In general terms, production is greatest in pastures at the bottom of a valley than on the slopes, and lowest on rocky slopes. On the other hand, open land has a greater production than areas under trees or shrub formations. These pastures show great diversity in space and time. This makes their management more complicated, but at the same time it allows their resources to be used thoroughly by the wide range of livestock, as long as those responsible for putting it into practice have a thorough knowledge of their environment. The alternative is to apply an extensive pasture model with a large surface area per head of livestock. But, in any case, management models for homogenous pastures are not valid in this system. On average, dehesa pastures produce 900-2,300 kg/ha of dry material per year, which can support a livestock load 1.9-4.6 breeding sheep per hectare. (1 hectare =2.47 acres). It should be borne in mind that inadequate pasture management or changes in management model, such as the current tendency to reduce livestock diversity, working with lower stocking levels and not exploiting the tree and shrub formations as a source of fuel, lead to the increasing spread of woody vegetation to the detriment of other types of pasture, and this in turn causes the dehesa system to degenerate.

Areas occupied by shrub or tree communities

As mentioned before, trees are present in all the previous units, although with varying densities. In marginal areas, they may reach high densities and form a closed forest. In modified areas occupied by permanent or semipermanent crops and pastures, tree density tends to be greater at higher levels than in valley bottoms, and is also higher in permanent pastures than in permanently or semipermanently cultivated ground. As a guide, average density might be 30-50 trees per hectare, and this gives the dehesa its special, savannah-like appearance.

The treatment of dehesas

The trees of the dehesa belong to a group of species of the genus *Quercus*, such as holm oak (*Q. ilex*), *Q. ballota*, cork oak (*Q. suber*), *Q. faginea*, and Pyrenean oak (*Q. pyrenaica*). The role of these trees is to protect, enrich and structure the soil, to shade the herbaceous layer, and to maintain air and soil humidity and thus the soil fauna. Furthermore, they supply forestry products such as firewood, charcoal, cork, and products that can be used as fodder, such as leaves and acorns. Dry leaves are useful as bedding for animals and as a fertilizer for crops.

The trees are treated as if they were fruit trees; they get regular pruning to renew the crown, to allow light to reach the grass, and to stimulate acorn production. More drastic prunings, known as *desmoches* (pollarding) are carried out every 12-14 years and provide a good quantity of wood, either used directly as fuel or transformed into charcoal or drawing charcoal. On average, each tree yields some 550 lbs (250 kg) fresh weight of wood when it is pruned roughly every 10 years. The green leaves from pruning are an important food source for livestock, especially in years when the winter is particularly cold, or dry or when the combination of late autumn rains and cold make the dry grass rot. About four or five years after this severe pruning, a milder pruning known as *enfaldado*, is carried out. This thins out the young branches that have sprouted after pruning. The aim of this lighter pruning is to open the crown and remove the inner branches as they produce little or nothing. The branches from this pruning are made into small pieces of charcoal and used as fuel.

When the dehesa tree layer consists of cork oaks, the bark is removed every 10 years and gives about 44-66 lbs (20-30 kg) of cork per tree, representing about 1,000-2,000 kg/ha for the normal densities mentioned above. Dehesa cork oaks are peeled high up the tree and little care is taken to renew the tree mass, giving rise to an exploitation model closer to mining than to forestry, that usually leads to a lower-quality tree mass and final product.

The fruits from the trees, acorns, are used as food for pigs of the *ibérica* breed. The process of using acorns (*montanera*), starts in autumn and continues throughout winter. The acorns are made to fall from the trees by hitting the branches with long poles, because their gradual, natural fall would prevent their thorough usage.

In order of decreasing importance, the livestock farming resources provided by the dehesa are pastures, acorns, stubble, foliage from tree pruning, the plant resources from shrub formations, and unproductive cultivated ground that is usually used directly as pasture without being harvested. These resources are not usually sufficient to maintain livestock, and so they have to be complemented by other resources, either from crops in the same system or acquired from outside the system. This additional input accounts for 18% for goats, lambs and cows, and 75% in the case of pigs. For livestock as a whole, it represents about 25%, which shows just how self-sufficient these systems are, particularly if we bear in mind that part of these additional resources comes from cultivated ground in the dehesa itself.

In addition to agricultural and livestock resources, there are other less important ones that contribute decisively to diversifying production and help make the system almost self-sufficient. These include garden produce, the meat and eggs of chickens, turkeys and pigeons, cheeses, butter, and honey.

In traditionally managed dehesas, renewable energy resources account for 98% of the total used, while a mere 2% comes from fossil fuels. Furthermore, 90% of the resources obtained are returned to the system while the remaining 10% are sold. External resources only account for 20% of the total used, representing a overall global efficiency of transformation of around 10%. If we measure the efficiency of production with respect to external inputs, this figure reaches 45%, and if we only refer to the quantities of fossil fuels used, we reach an efficiency of 530%. These figures show the highly self-sufficient nature of the traditional dehesa systems and, at the same time, their great efficiency in transforming the few external resources that they require.

Today, the dehesa management model has changed towards a less self-sufficient model, and fossil fuel inputs now account for 14% of the total energy consumed, while 65% of production is reused within the system and the remaining 35% is exported. Overall efficiency of production may reach higher values than those of traditional systems, up to 35%,

160 Unirrigated plot, with almond trees (*Prunus dulcis*) in flower on the red oxidized soil of Les Garrigues (Catalonia). Arboriculture is one of the distinguishing characteristics of the Mediterranean landscape, as the range of crops that can be grown without irrigation is very restricted. The irregular rainfall and poor soils mean agriculture on unirrigated land yields little, and this has always led to attempts to compensate for this by the cultivation of fruit trees.
[Photo: Ernest Costa]

while efficiency with respect to external inputs can be as high as 100%. On the other hand, if we refer to fossil fuels, values are only half of those of the traditional model, at around 260%.

2.5 Dry land farming

One of the most characteristic features of the agrarian structure in the Mediterranean is the lack of continuity, caused by the mountainous and varied relief. Tertiary folds and faults gave rise to today's fragmented landform: steep, but not very high mountains encircle occasionally marshy plateaus or plains. In the Mediterranean Basin there are hardly any large, flat areas of arable land, and those that exist are difficult to cultivate.

Another environmental feature that has been of prime importance to the development of traditional Mediterranean agriculture is the scarcity of water resources. This situation is further aggravated by the highly seasonal climate: periods of summer drought alternate with rainy periods mainly limited to the colder half of the year. The hot, dry summer climate

only allows the cultivation of crop species that can be harvested before the summer (cereals) and species of xerophilous trees (the olive, almond, carob, fig, and vine) whose deep roots allow them to obtain underground water during the summer.

However, when irrigation is possible in the summer (thanks to the rains and the winter snowfalls in the mountains nearby), the Mediterranean plains boast as wide a range of crops as can be found anywhere in the world. Native species are complemented by perennial plants from northern Europe, some annual tropical plants, and a large number of subtropical trees (mulberry and orange trees) which make the most of the long hot summers. Both fertile and barren tracts share the Mediterranean plains. Cultivated areas, oases among vast expanses of woods or scrub, are bordered by mountainous, uncultivated areas, where livestock graze. This is the contrast between the *ager et saltus* of the Roman agronomists. In other parts, however, the plains are marshy and unhealthy, with their margins densely settled by peasants who have terraced the slopes to create areas of intensive multicrop cultivation. In these areas, trees usually dominate and irrigation is common.

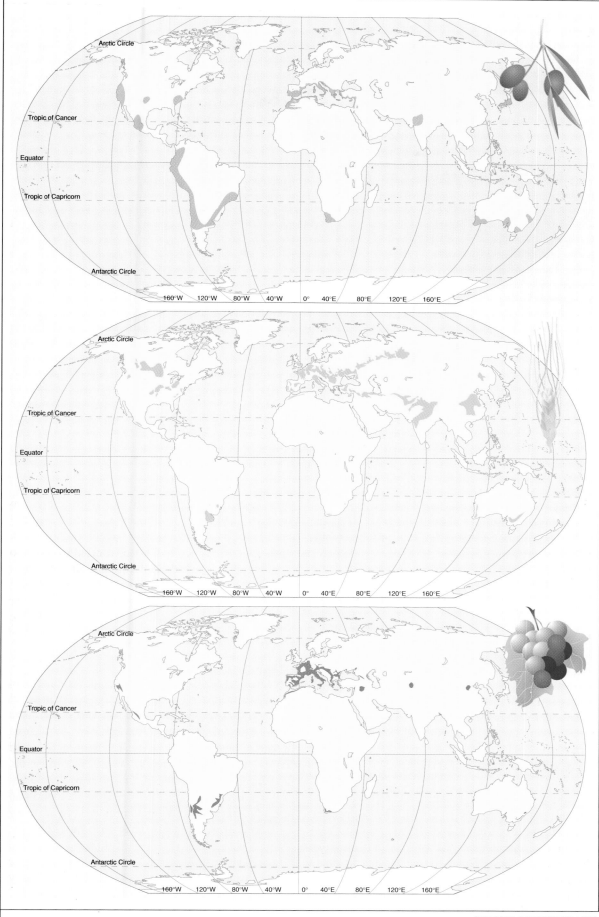

161 Distribution of cultivation of three Mediterranean crops, olives (*Olea europaea*), wheat (*Triticum*), and grapes (*Vitis vinifera*). Plants originally cultivated around the Mediterranean Basin have been dispersed to the other mediterraneans. Species adapted to the hot summer climate of the old world, where many colonial empires originated, were carried by explorers on their voyages and by colonists to their settlements.

[Maps: Editrònica, from several sources]

3. HUMANS IN SCLEROPHYLLOUS FORMATIONS

The Mediterranean biome is thus mostly a world of trees and shrubs, a woody rather than herbaceous landscape. This characteristic is derived from the specific climatic condition of hot and dry seasons that coincide, which places serious limitations on the growth of herbaceous plants, while favoring woody plants with deep roots. Woody life-forms also appear in the agricultural landscape, which is often dominated by dry farming of tree crops: these are fields with occasional or frequent trees, i.e., woody unirrigated areas that are well adapted to the Mediterranean environment. Apart from wheat, other dry land crops include grapes and olives, and other secondary, but very relevant, crops, such as almonds and carob beans. There is also a whole range of minor crops, covering limited areas and with a relatively modest production, although they may be of great local or regional importance. The pomegranate, the fig and the hazel are all examples.

The Mediterranean trilogy

The most important crop in the Mediterranean is wheat (*Triticum*), followed by the olive (*Olea europaea*) and the grape vine (*Vitis vinifera*). These three have been the most important crops in the region since before 6000 B.P. Cultivation began on low-lying, flat areas with wheat and barley (*Hordeum*), while olive trees (the wild form of the olive trees known in Italy and the south of France since the Paleolithic) and grape vines (which are native to the Mediterranean area and whose fruits have been eaten also since the Paleolithic) were grown in hilly areas. The combination of these three crops in the Mediterranean area did not result in an even distribution; on the contrary, there were all kinds of combinations.

The island of Sicily has always been an important supplier of wheat, and in the first centuries A.D., it was called was the granary of Rome. By the 14th century it also supplied Barcelona and Florence, since the rich commercial cities of Europe were large consumers of grain. In Catalonia the fragmented landform complicated overland transport of the wheat produced in the Ebro valley. In rural Tuscany the main crops were grape vines and olives, products with a high added value, much greater than that of wheat.

Many centuries earlier, in Attica during the classical period, forests and pasture land had been exhausted, and because the farming land available was not very fertile, the region specialized in the cultivation of olives. The olive oil was exchanged for Scythian grain in the Black Sea ports, and as a result earthenware jars were produced (as oil containers), tech-

niques of navigation were improved, and payment systems were developed that used coins minted with silver from the mines of Attica.

Wheat, olives, and grapes are the three basic products of Mediterranean culture. That is why wheat, olive trees, and the grape vine are the agricultural trilogy that symbolize the biome.

Wheat and other cereals

The first two plants cultivated in the Mediterranean Basin were wheat and barley. These plants grew beside the early Neolithic communities of the Near East and they have played a very important role in the history of Mediterranean culture and civilization.

Wheat in the Mediterranean's history

Since Neolithic times, wheat has been a staple food in the Mediterranean; carbon-14 dating indicates that wheat was already cultivated in Israel, Syria, Anatolia, and in the north of Mesopotamia between 9000 and 8000 B.P. From this initial center, the cultivation of wheat spread west to the whole of the Mediterranean area and into central and eastern Asia.

Even when it was first grown, wheat was highly productive and could sustain a sizeable urban population. It rapidly become a staple food. The early river civilizations of the Near East depended on the success of their wheat harvest. It was so important that control of wheat soon passed into the hands of priests. Since then, trading in wheat has had a decisive influence upon price trends in general, and the control of the wheat trade has conferred enormous social and economic power. The state has always attempted to control wheat distribution to minimize the effects of oscillations caused by the cyclic nature of its production.

Wheat, in the form of bread, was the staple food of the Greeks and Romans, and many of the economic problems of the time were related to wheat. Greece suffered a permanent wheat deficit, therefore Greeks imported it from the Black Sea region. Rome's main wheat-producing provinces were Egypt (the most important), Sicily, the Mahgreb, and Baetica (Roman Andalusia); it was transported by sea, since this was the cheapest method. The need for wheat was an important factor in the entry of Germanic peoples into the Roman Empire.

Throughout the Middle Ages, bread's importance as a staple food throughout the Mediterranean Basin caused the area of land under wheat to increase: agri-

CHROMOSOME COMPLEMENT (n=7)	COMMON NAME	SCIENTIFIC NAME	MAIN CHARACTERISTICS	WHERE IT IS GROWN
2n	wild spelt	T. aegilopoides		spontaneous in the eastern Mediterranean region
2n	einkorn	T. monococcum	small spike, fruit that adhere to the husks, a single grain per spikelet, a fragile articulated axis; not very productive	Middle east and Mediterranean countries, in steep areas with thin soils
4n	wild emmer	T. dicoccoides		spontaneous in the eastern Mediterranean
4n	emmer	T. dicoccum	flattened spike with awns, fruits that adhere to the husks, fragile axis; not very productive	Middle East and the Mediterranean countries, on poor dry soils
4n	Persian wheat	T. persicum		Middle east
4n	Polish wheat	T. polonicum	Flattened spike, spikelets surround by dry membraneous bracts; not very productive	Middle east, Mediterranean countries and northeast Africa
4n	—	T. timopheevi	highly resistant to all diseases	Georgia
4n	durum wheat	T. durum	long spike 4-6 in (10-15 cm) with long awns, white, amber or reddish grain that is hard and rich in gluten; very productive; best for sites with a dry climate	Mediterranean and eastern European countries, and India
4n	turgidum wheat	T. turgidum	thickened pendulous spike, hard grain; not very productive; undemanding	Britain, southern Europe and northeast Africa
6n	bread wheat	T. aestivum	spike with awns or without awns, soft grain, white or red, rich in starch; very productive; very diverse requirements	all over the world
6n	club wheat	T. compactum	short awnless spike, small grain; not very productive	southeast Asia, southern Europe and southern United States, on poor soils
6n	spelt wheat	T. espelta	flat spike, distichous and loose, hard white grain, retaining the coverings, articulated axis; not very productive	Mediterranean countries
6n	—	T. sphaerococcum		India
6n	—	T. macha		Georgia

162 The main types of wheat (*Triticum*) and where they are cultivated. (Einkorn, emmer, and *T. polonicum* are now rarely cultivated.) The existence of different names for wheat in the world's oldest languages shows how long it has been cultivated. There are names for it in Chinese (*mai*), Sanskrit (*sumana* and *godhuma*), Hebrew (*khittah*), Egyptian (*br*), and guanche (*yrichen*). There are also many other names in the languages derived from early Sanskrit. The Basque word *gari* may go back to the Iberians.
[Source: data collected by the author]

cultural investments were directed towards clearing woods and ploughing wasteland, rather than to selecting improved seed strains. In the time of Charlemagne, wheat-growing was regulated and controlled by public authorities. The wheat grown in the fields was stored and traded by feudal lords, boroughs, and monasteries. Despite these efforts, in the 9th and 10th centuries, wheat-growing declined considerably, apparently due to harsh climatic conditions. Peasants, bonded to the land they worked, supplied the labor in a social system where work was considered something dishonorable by the dominant classes, the nobility and clergy. It has been estimated that around the year 1500 in Mediterranean Europe there were about 30 non-productive people, such as nobility, clergy, patricians, urban traders and beggars, for each peasant.

The arrival of the modern era led to changes in the control of agricultural production, which passed from the hands of the nobility and the monasteries to the commercial and financial bourgeoisie of the cities. In the last quarter of the eighteenth century many societies for agricultural improvement were established to introduce into the Mediterranean the innovations in agriculture that had been developed in the north, especially in Britain. The price of wheat never lost its social influence, and unrest connected to subsistence crises occurred from the Middle Ages to the 19th century.

Spelt, emmer, and wheat

Ten thousand years ago, at the end of the Würm glaciation, conditions in Anatolia, Persia, and Syria, today a highly degraded and partly desert region, favored the development of grassy meadows. This led to the birth of cereal farming. It began with the selection of plants with seeds rich in food reserves. These included two species of grasses with large spikes full of grains: wild spelt (*Triticum aegilopoides*) and einkorn (*Triticum monococcum*).

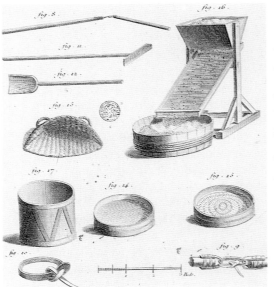

Instruments for threshing and sifting. *Recueil des planches sur les sciences, les arts libéraux et les arts méchaniques avec leur explication*, Volume 1, Paris (1763) [Biblioteca de Catalunya (Barcelona)]

Durum wheat (*Triticum durum*) showing awns [Firo Foto]

Before 10000 B.P., the primitive wheats were insignificant grasses not the fine plants we know today, but they crossed with another wild grass and the resulting hybrid (presumably due to genetic chance) was also fertile. It is thought that at the end of the last ice age similar events must have occurred often among the emerging vegetation. The first ancestors of modern wheat are very probably two very old diploid species. They are spelt (*Triticum aegilopoides*), which still grows wild in Armenia, Georgia, Turkey, and nearby areas, and einkorn (*T. monococcum*), a wheat with a single poorer grain and little grown. Charred remains of einkorn 6,700 years old have been found in Jarmo, a village in eastern Iraq. These diploid wheats (only 14 chromosomes) were probably cultivated in Turkey, and then cultivation spread to western Europe, but it never reached further east.

Turgidum wheat (*Triticum turgidum*) from the book *Le Règne Végetal* [Sánchez-Durán Aisa]

Archeological and botanical data show the cultivation of both wheat and barley began in the Near East, and that the process of transformation of wild wheat into modern-day wheat has included genetic mutation, hybridization, and polyploidy. The cultivated wheats (*Triticum*) are classified into three groups. There is the diploid group of spelts, with 14 chromosomes (2 sets of 7), the tetraploid group of emmer wheats and durum wheats, with 28 chromosomes (4 sets of 7), and the hexaploid group of the true wheats (6 sets of 7). The last two groups, the tetraploids and the hexaploids, have originated from the first group by doubling the number of chromosomes.

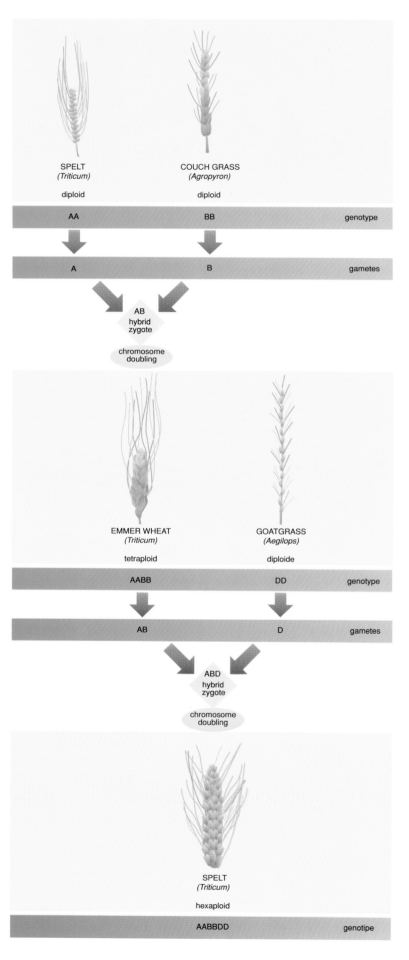

SPELT
(Triticum)

diploid

| AA | BB | genotype |

| A | B | gametes |

AB
hybrid
zygote

chromosome
doubling

EMMER WHEAT
(Triticum)

tetraploid

GOATGRASS
(Aegilops)

diploide

| AABB | DD | genotype |

| AB | D | gametes |

ABD
hybrid
zygote

chromosome
doubling

SPELT
(Triticum)

hexaploid

| AABBDD | | genotipe |

Later, it appears that einkorn (*T. monococcum*) hybridized with a related grass with 14 chromosomes, perhaps a species of goat grass (*Agropyron*). This gave rise to emmer wheat (*Triticum dicoccum*). The new, more robust species, became stabilized by doubling its chromosomes, thus creating the first tetraploid wheats with 28 chromosomes. Emmer was for centuries the most cultivated of all wheats, and spread through North Africa, Europe, Arabia, and Egypt. Alexander the Great even took it with him, and it was known as "soft wheat." It is still possible to find native varieties of wild emmer (*T. dicoccoides*), a species that is very similar to *Triticum dicoccum*, in the south of Armenia, northeast Turkey, west Iran, Syria, and northern Palestine. Other tetraploid species include: durum wheat (*T. durum*) which was already cultivated in the Greco-Roman period (around the year 2100 B.P.) and its high gluten content means it was—and still is—used for the production of semolina and pasta; Persian wheat (*T. persicum*) found from the Caucasus to the north of Turkey; Polish wheat (*T. polonicum*) which is cultivated from the Iberian Peninsula to Turkestan and Ethiopia; turgidum wheat (*T. turgidum*), rounded, with a thick stem and a pith full of a spongy paste, which is cultivated in southern Europe, Ethiopia, and Great Britain; and timopheevi wheat (*T. timopheevi*), whose origin is not yet certain, which is cultivated in a very restricted area of Georgia. Timopheevi wheat is of great interest for its disease resistance and because it possesses a genome unlike that of any other wheats.

The double hybridization in the origin of wheat
[Editrònica, using data supplied by the author]

Once it had entered cultivation, emmer (*Triticum dicoccum*) crossed with another wild goat grass and by a stroke of genetic luck it produced fertile offspring. this gave rise to a hexaploid hybrid (with 42 chromosomes) that was larger. Hybrids are not often fertile, even in plants, but for the hybrid of a hybrid to be fertile is even less common. Today's wheat could not have been fertile if emmer (*Triticum dicoccum*) had not previously suffered a mutation on a particular chromosome. The fact is that a further chromosome doubling

Bread wheat (*Triticumn aestivum*) [R. Wanscheidt / Bruce Coleman Limited]

trymaking. There are two more hexaploid wheats, spelt (*T. spelta*) and *T. macha*, both easily threshed. After such a complex history, the wild ancestors of modern-day wheat have now disappeared. The brittle, shattering spikes of spelts and emmers favor seed dispersal, and so modern wheats—which may or may not have awns but always have non-shattering spikes—have to rely on humans to sow their grains. Wheat needs the human race to perpetuate itself, just as humans need wheat for food.

occurred with viable offspring and this led to the many modern hexaploid wheats. The hexaploid species of wheat had, for the first time, a non-shattering stalk and grains that were easily threshed (separated from the husk). This is the case with the many varieties of bread wheat (*Triticum aestivum*) that provide flour suitable for making bread; with *T. sphaerococcum*, which has been cultivated in central and north India since 4500 B.P.; and with *T. compactum*, which produces a protein-poor flour that is widely used in pas-

163 Frequency of corn imports to Valencia between 1401 and 1500. The thickness of the line represents how many times (shown in brackets) grain was imported. Wheat has always played two roles in human societies. It was a subsistence food for rural peasants and landholders, while the surplus, and that of other cereals, was of great political importance. Proximity to the sea conditions the development of the regions that produce wheat surpluses. Apulia, Sicily and Sardinia experienced economic development as a result of exporting grain to Rome. This did not happen in the fertile Po valley, because transport overland was much more difficult.
[Drawing: Editrònica, from several sources]

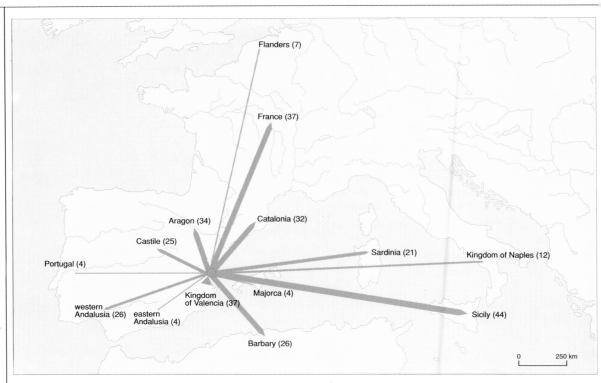

164 A revolt by reapers, in the painting *Corpus de Sang* painted by Antoni Estruch in 1907, now in the Museu d'Art de Sabadell (Catalonia). If this led to the Corpus de Sang events in Catalonia, followed by a long war, in Elche civil strife was caused by the auctioning of the flour supply, and in Alicante there were revolts demanding the grain stored to supply Madrid. In Madrid (1766) the people revolted against the Marquis de Esquilache. The leading role of the grain harvesters in the events in Barcelona in 1640 and the successive separation of Catalonia from the spanish crown (1640-1659) were the origin of the popular song that gave rise to the national anthem of Catalonia "Els Segadors" (the reapers), represented in the song as "defenders of the soil."
[Photo: Jordi Gumí]

The French Revolution, although it was a very complex phenomenon that cannot be reduced to a subsistence crisis, was preceded by several years of shortage and it broke out in the year when wheat reached its highest price in 60 years.

Industrialization changed the geography of wheat-growing due to technical progress in agricultural machinery, but subsistence farming was so engrained in Mediterranean society that it has never been completely abandoned: one still finds fields divided into small plots with a rotation of two to three years that were so typical of medieval rural life. But there are sectors that present sharp contrasts: some farmers use traditional techniques and procedures that have changed little over the course of time (especially clear in the countries of the Mahgreb), while others employ modern techniques. Part of the traditional economy has remained largely untouched by the development of the modern economy This means that the area as a whole has undergone a dual economic process. In general, this economic, historic, institutional, and physical differences divide the Mediterranean countries into two sub-groups: southern Europe and Israel in one group, and North Africa and the rest of the Near East in the other. Although, from a historical point of view, these differences may be partly attributable to climate and to the influence of Islamic culture in the south and east of the basin, they are now due to different levels of economic development, industrialization, urbanization, and income. To summarize, agriculture has progressed further in the areas to the north and west of the basin, where it is now more specialized and intensive, than in the south and east. To a great extent, this division is reflected in the percentage of the population dependant upon agriculture, which is much lower in the more developed countries.

Even today, nearly 50% of the arable land in the Mediterranean is used for growing cereals, while more than 40% lies fallow, leaving little more than 10% for fruit, vegetables, and other crops. Together with rice, wheat is the cereal that feeds the greatest number of human beings. In many countries it is still the staple food of the population, and as a result, wheat cultivation and trade are still of great importance in the international agricultural economy.

Cereals in the other mediterraneans

Wheat has spread across the world along with European colonization since the 15th century. By 1440, it had already arrived in the Azores and Madeira, the first steps in Europe's expansion, and reached the Americas in 1495. Wheat was already known to the Guanches of the Canary Islands, who were heirs to the Neolithic revolution of the Fertile Crescent. The cultivation of wheat in the New World led to a change in the eating habits of the natives and improved the situation of the colonies allowing the development of local industries such as mining. Wheat thus ensured the success of colonization. It became so well established in the overseas colonies that it was the leading export of the Venezuelan Andes from the sixteenth century to the eighteenth century, and it is still grown there by peasants who inherited the Mediterranean tradition.

In the other areas of the Mediterranean biome, wheat prospered, but the attempt to introduce it to other climatic zones was not always successful: in tropical areas, where the seasons are not as well defined as in the Mediterranean, it was a total failure; generally, the

165 Harvesting, by hand in the case of a Berber peasant in the Moroccan Atlas, and by machine near the Sacramento Valley in California. Reaping the cornfield, harvesting the grain has long been performed with sickles. The first sickles were made of stone, then of metal. With the passing of time this chore is now performed by harvesting machinery. In early agricultural societies cults worshipped the countryside as a provider of harvests. Also, there were different rites and offerings that sought to ensure the harvest. Harvesters have only recently begun to be used for cutting grasses, especially cereal crops. There are many types of mechanical harvesters: scythes and scything machines, small grass-cutting machines and larger ones with accessories for other harvesting activities.
[Photos: Ernest Costa and Age Fotostock]

166 Corn poppy (*Papaver rhoeas*) in a cereal field in the northeast of the Iberian Peninsula. The corn poppy is one of the typical annual plants of cultivated ground. Cornfield weeds are other plants that can be found growing among crops. These consist of species that form seeds that are very similar to the cultivated species, making them difficult to separate. Examples include bearded darnel (*Lolium temulentum*) which is inseparable from wheat and wild oats (*Avena sterilis, A. barbata,* and *A. fatua*) that mix in with oats (*A. sativa*). The presence of weeds that compete with the crops may make agricultural production unviable in some circumstances.
[Photo: Francesc Muntada]

plants did not grow synchronously due to the absence of seasonal differences in the photoperiod.

Wheat is grown in all parts of the Mediterranean biome as well as in the Mediterranean Basin. The role it plays is most important in Chile, the only South American country where wheat production exceeds that of maize. In Chile, cereals occupy three quarters of the cultivated land, and wheat occupies four fifths of the area under cereals. It is cultivated by dry farming, with fallow periods. Harvests are low and variable, and mechanization has hardly been introduced, even on the biggest estates. In Western and South Australia and in South Africa, wheat and barley occupy much of the cultivated land. The fields have fallow periods and the harvests are not very big except in the wheat belt in Western Australia, even though farming is on a large scale using a mechanised agriculture that requires little human labor input. In both the Cape Province and the two Australian mediterraneans wheat production has diminished due to shifts in land use towards sheep grazing. Until recently wheat was a large-scale and highly mechanized crop in California, but these large areas have now been replaced by intensive crops in which cereals (barley is the most important) occupy only a third of the arable land.

Wheat cultivation

Most grain crops are adapted to those latitudes of between 25° and 45°N where the winters are cold, the summers are hot and dry, and the intervening seasons are rainy; the production of seeds suitable for human consumption is typical of climates showing sasonal changes. There are fourteen species of wheat (*Triticum*), all of them cultivated species, and most have arisen as the result of an explosive kind of evolution.

Wheat is not very demanding in its cultivation requirements, as it is viable in different climates and in a very wide range of soils. It can be grown at a longitude of between 30° and 60°N and between 25° and 40°S; in tropical areas it can survive at an altitude of up to 10,500 ft (3,200 m) and in central Europe up to 3,280 ft (1,000 m). The temperatures at which it can survive depend on the variety concerned. They range from 41°F (5°C)—below which it cannot germinate or grow—to 86°F (30°C) — above which transpiration is too intense; no wheat will form ears below 61°F (16°C) or ripen below 68°F (20°C). Wheat will root deeply if the land has been sufficiently ploughed. This cereal does not grow well in marshy areas, in soils that are not very compact, or those that contain a high concentration of salt. It also requires considerable quantities of nitrogen in the soil.

Wheat is usually a dry farming crop, allowing the farmer to make the most of the area's scant water resources, but in suitable conditions, it may also be grown in irrigated areas. In dry farming wheat is sown when the first autumn rains begin to fall, and is harvested at the beginning of summer (it is usually reaped between the end of May and the end of July). To germinate, the grain needs humidity, a high temperature, and air around it. The growing embryo lives off the reserves provided by the seed until it sprouts its first leaf and starts to photosynthesize.

When the seeedlings have grown to a certain size, the lower part of the stem sprouts secondary stems (called *tillers*) and roots; wheat forms a tussock that may consist of up to 400 tillers. If the seeds are sown further apart, or rainfall is heavy, wheat and other cereals will produce a larger number of tillers. This is much more obvious in dry farming varieties than in varieties grown on irrigated land.

If the wheat was sown in autumn, when spring arrives and the days become warmer, the plant continues to grow and the stalk begins to form. While the plant grows, the stalk is wrapped in a sheath, from which it will not emerge until the plant has produced a spike. The wheat spike consists of a central stalk (called a *rachis*) with short internodes, each of which bears a spikelet protected by a rather coriaceous bract (or glume) on each side; each spikelet can bear nine flowers. In bearded wheats the flower is protected by glumes, the outermost of which is prolonged to form an awn.

The period during which the ear is growing is the period of most intense physiological activity, showing maximum values for transpiration, and water and nutrient uptake. Until the ear forms, transpiration and photosynthesis only take place in the leaves, but once the ear is formed, the glumella, the beards, and the ear itself all take part in carbohydrate production. Fertilization occurs about a six or seven days after the ear has emerged from the sheath. Pollen falls from the anthers onto the stigma and fertilizes the ovules in the ovaries (this means that wheat is self-fertilizing). As the plant matures, the leaves gradually cease photosynthesis and die while the sugars produced by the lower leaves of the plant are translocated to the developing wheat grains. With the exception of a few varieties, wheat is much less susceptible to shedding its grains before reaping than other cereals, and for this reason its harvest is less pressing than that of other cereals.

Wheat is liable to suffer various afflictions: flattening (when the stalks bend due to an imbalance in nutrition, or because of excessively dense sowing, or as a result of persistent rain or wind), scorching (due to an excess of heat and dry, warm winds), and stunting (failure to develop properly). Wheat does not like cold, wet springs or frost (especially at times when the spikes are forming). Wheat is attacked by various fungi including *Ustilago tritici* (which causes smut); the rust fungi *Puccinia graminis*, *P. rubigovera*, and *P. glumarum* which attack the leaves and stalk; *Ophiobolus graminis* which attacks the stalk and roots causing them to become brittle and die; and some species of *Erysiphe* which produce powdery mildew in wheat. Many insects also attack wheat: the larvae of the dipteran Hessian-fly (*Cecidomya destructor*) attack the first two internodes of the stalk, which breaks (the result is referred to as "dashed wheat"); the larvae of the thrip *Haplothrips aculeatus* tunnel into the stalks; the lepidopteran *Anacampsis cereanella* lays its eggs on top of the ear and the caterpillars bore into the grain; and the larvae of the

granary weevil (*Calandra granariae*) tunnel in the grain once it has been stored. The granary weevil also attacks the seeds of maize, rye, and other cereals.

It is possible to introduce a high level of mechanization into every stage of cereal growing. The tendency is always to attempt to reduce the labor required by the exhausting tasks involved. Soil preparation was improved with the introduction of the wooden plough. Much later, when cheap cast iron became accessible to most of the population, the local blacksmiths lost their monopoly of making agricultural implements which were then produced industrially. However, progress was not uniform: in 1850 there were still many wooden ploughs in the Mediterranean Basin. Although seed drills have been used for a very long time (the Chinese have used them since the distant past), they appeared much later in the Mediterranean: the first was presented to Philip IV of Spain by the Italian Guiseppe Lucatello, but although it was well received it was forgotten until the eighteenth century, when Jethro Tull rediscovered and perfected it. Since then many improvements have been made to seed drilling machinery. Reaping the harvest used to be the cereal farming operation requiring the most labor. However, harvesters were invented in non-Mediterranean regions, North America and northern Europe, where fields are larger and cultivation is on a larger scale. Surprisingly, there was a 50-year delay between their invention and their general use.

Barley cultivation

The cultivation of barley (*Hordeum vulgare*) arose in the same areas as wheat. Barley has a dense, bearded spike and single-flowered spikelets, arranged in threes; it is lighter in color than wheat and grows earlier. Many varieties are cultivated, some of which are able to grow in very harsh climatic conditions, such as those of Siberia and the Andes. Barley requires less nitrogen than wheat and is more resistant to droughts. In many parts of the Mediterranean, especially North Africa, a larger area is now under barley than wheat. It is grown for fodder, and may be turned to malt for making alcohol, whisky, and beer.

Flour, bread and pasta

In prehistoric times, people everywhere prepared wheat grain by cooking it whole, but its consumption in the form of flour gradually spread. Forms of porridge and gruel made their appearance in prehistoric times and are still enjoyed today. The oldest archaeological evidence of the grinding of wheat into flour dates from 8,000 years ago. This process was at first performed by pounding wheat between two stones, a task usually performed by women. Later, milling took place in specially constructed flour mills. The invention of the vertical water wheel, used for many centuries in the Mediterranean, occurred in the 22nd century B.P. Wheat has mainly been consumed in the form of flour and its derivatives since classical times.

Flour was soon used to make bread. Bread is made by baking a dough, containing a mixture of wheat flour (sometimes mixed with rye or other grain flours), water, salt, and yeast. This mixture may be left to rise in order to give it a more or less spongy texture. The dough was originally baked unleavened and took the form of a biscuit. The discovery of leavening is attributed to the Egyptians, and there are references indicating that bread was made in 4600 B.P. in very much the same way as it is now. The oldest known piece of bread was found in the excavation of a 5,000-year-old lakeside city in Switzerland and there are references to bread in several chapters of the Bible. Other sources on the history of bread indicate that 72 different types were made in Athens in the 22nd century B.P., although wheat bread was only eaten on feast days. In the excavations of Pompeii, a batch of bread was discovered that had been abandoned because of the vocanic eruption. Bread became a standard item on the table: during the Renaissance a basket full of bread was laid for every two diners which they shared (the Latin root of the word companion means someone with whom we share bread).

Bread is normally made with wheat flour since this is the only flour containing an insoluble protein, gluten, that absorbs water, tripling its weight, and forming a sticky, elastic material. Gluten binds the dough in the process of bread making, helping to retain the carbon dioxide gas that forms the bubbles that make the dough rise before it enters the oven. The quality of the bread depends greatly on the physical properties of the gluten, and these depend in turn on the variety of wheat from which the flour is made. There are many varieties of bread: wholemeal, wheatmeal, white, unleavened, etc. The whiter a wheat bread, the less its nutritional value, since most of the vitamins and mineral salts remain in the bran, which is often fed to animals. Modern bakers usually include additives that will improve the properties of the flour, making the bread keep longer, accentuating its smell and texture, or increasing its nutritional value.

Bread was not the only flour product people have consumed: pastry-making (the combination of flour with butter, sugar, jams, dried fruits, eggs and a multitude of other ingredients depending on the imagination of the pastry cook) existed in ancient Egypt, classical Greece, and imperial Rome. During the reign of Augustus there were more than 300 pastry shops in Rome, and their products were eaten at the circus. In North Africa a form of puff pastry was made and cooked with cheese and honey. In other parts of the Mediterranean Basin flat oval masses of dough were (and are) baked and sweet or savory ingredients added to them. Pizza was developed in Italy. It is a flat base of wheat-flour dough to which olive oil and various farmhouse products are added before baking in a very hot oven.

Another product of wheat flour is Italian pasta, or *asciutta*. It is made by blending flour with semolina made from durum wheat and water. Salt is added to the resulting dough which is not left to rise and finally the dough is formed into different shapes. The pasta is then boiled and served with one of many different sauces. It forms the basis of meals in households throughout the Mediterranean, especially in Italy.

The tremendous importance of bread in the diet of the early Mediterranean peoples has led to it being given great symbolic importance since ancient times. In Greece, different types of bread were made as offerings to different gods: a bread made with flour, honey, and sesame seeds in the shape of the female pubis was offered to Demeter and Persephone, and crescent-shaped pastries were offered to Artemis. At weddings in both Greece and Rome the bride and groom offered wedding cakes to the gods, that were then consumed by the more earthly mouths of the guests at the wedding banquet. The Jews normally eat bread made from a wheat-flour dough fermented with yeast (*zymi*), but to symbolize purity, on holy days they consume an unleavened bread (*azymi*). This tradition was inherited by Muslims and Christians. Christians still celebrate the Eucharist with an unleavened bread that represents the body of Christ. In the Mediterranean Basin there are breads and pastries to celebrate every religious and secular festival: there is a Christmas bread (which is eaten after the midnight mass), an Epiphany cake, Lenten bread, Easter bread, All Saints bread (for All Saints' Day), breads to celebrate the harvest, and other types of breads for the initiation ceremonies of different professions.

The olive tree

The olive tree (*Olea europaea* var. *europaea*) is perhaps the most typical tree of the Mediterranean biome; some authors even define the boundaries of the Mediterranean climate on the basis of the distribution of the olive. The cultivated olive is an agricultural variety of the wild olive (*Olea europaea* var. *sylvestris*), a shrub of maquis vegetation and a

167 **Olive grove in Málaga** (Andalusia), in the south of Spain, where olive plantations reach their largest size. The olive trees are planted in rows about 33 ft (10 m) apart to make harvesting easier, creating a distinctive landscape. The olive (*Olea europea*) is basically a tree of dry, unirrigated land. In fact, it is no more than a wild olive without spines and with bigger fruit.
[Photo: Josep Maria Barres]

close relative of other very similar species, such as *O. laperrinii* found in the mountains of the Sahara and *O. cuspidata* from the interior of Asia.

From the wild olive to olive groves overseas

The wild olive tree produces a small fruit, a drupe with a fleshy mesocarp rich in lipids, which ripens at the end of autumn and beginning of winter. These fruits have been gathered since ancient times, as shown by the remains of olive stones found in archaeological sites. The cultivation of the olive tree began some 6,000 years ago, in the eastern Mediterranean Basin. This agricultural practice implied the selection of varieties that produced larger, more edible fruits. This continuous improvement in olive quality ended up giving rise to the present day cultivated olive. The new varieties and their cultivation techniques spread over the basin and arrived on the shores of the Aegean Sea some 5,000 years ago, reaching the western shores of the basin 4,000 years ago. All the cultures that developed around this sea contributed to the cultivated olive's expansion, as the vegetable oil obtained from its ripe fruit became one of the mainstays of the Mediterranean diet.

The great importance of the olive tree in the Mediterranean Basin is clearly shown by the large surface area it occupies and also by its presence in the mythologies of ancient Mediterranean cultures. The Egyptians attributed the invention of the olive oil extraction process to the goddess Isis, the wife of Osiris, the supreme god of Egyptian mythology. In the case of the Hebrews, the olive tree features in many episodes in their sacred books, from Genesis to the New Testament: an olive branch borne by a dove was the sign that revealed to Noah that the Flood had abated. As for the ancient Greeks, the olive tree appears in a great number of mythological episodes and it was considered to be a present to humans from Pallas Athene, the goddess of wisdom. Even today, an olive branch is considered a symbol of peace. The olive tree now occupies an area which stretches beyond the Mediterranean biome. Its presence and importance is hardly surprising, since the tree supplies an almost extraordinary liquid that was used for food, curing wounds, and lighting. The fact that it is a source of food and light, and also has healing qualities, has ensured the olive a position of honor in the culture and landscapes of the Mediterranean Basin.

Today, olive farming has spread to all areas of the biome and has even reached some places outside the biome. This is the result of colonization by people of Mediterranean origin used olive oil and eager to continue cultivating it in their adopted countries. Nevertheless, most olive trees continue to remain within the limits of the biome where the climatic conditions suit them best. There are roughly 800 million cultivated olive trees in the world, of which only 20 million grow in extra-Mediterranean regions.

Olive farming is concentrated mainly in the northern hemisphere, between latitudes 30° and 40°N. Further north than this, the winter cold limits this evergreen tree's growth, while to the south the excessive length of the dry period hinders its growth. The olive tree is very sensitive to autumn and spring frosts and infrequent cold spells may cause widespread damage to the crops, even in locations where the climate is otherwise ideal. It is easy to find written references to the mass death of olive trees because of cold weather. There are historical records of the three great cold periods in the 16th century, two in the 18th century and one, the last, in 1956, all of them on the western side of the Mediterranean Basin. As for droughts, there are also references to severe dry periods which have led to the death of some olive trees, as happened in the Iberian Peninsula in some periods in the 19th century. However, in almost all these cases and in the case of fire, only the aerial part of the tree dies; the tree then regenerates by sprouting from the stump and roots. This allows a rapid recovery of the damaged olive plantations, even though grafting may have to be repeated.

Olive cultivation

As a fruit tree, the olive tree has some very specific characteristics. It is very long-lived, easily reaching several centuries, and it is not unusual to find specimens over a thousand years old. It is very robust, and can tolerate dry, poor soils, although production may be very low. This has meant that it can be used to cultivate areas where the soil is extremely poor and stony, and in the worst conditions it may even be treated as semi-cultivated. Its root system is very adaptable and can grow in very different soil conditions. Vegetative reproduction is very simple, making it easy to propagate the varieties obtained by selection. If specimens of the cultivated varieties are not pruned from time to time they may attain great size, reaching 50-65 ft (15-20 m) in height and 6.5 ft (2 m) in trunk diameter.

There are a great many cultivated varieties of olive tree, the result of over 6,000 years of selection; and as the wild and cultivated forms are interfertile, there are also spontaneous hybrids. There is a certain degree of confusion about the distinctions between the different varieties, but it is thought that there are more than 100 distinct varieties. Reproduction is by cuttings, although some varieties do not root very well from cuttings and must be grafted onto wild olive trees or other cultivated varieties.

In general terms, after the olive trees have been planted, there is an unproductive period lasting 1-10 years, even though flowering and some fruit production begin in the fourth or fifth year of life. Their production increases between 10 and 35 years of age, and from 35-150 years it reaches its peak. From 150-200 years, production gradually diminishes as the tree approaches old age.

Olive tree plantations have an average density of around 32 trees per acre (80 trees per hectare), although this may vary greatly depending on the climate and the characteristics of the site. If rainfall is 8-16 in (200-400 mm) per year, density may range from 7-28 trees per acre (17-70 trees per hectare); where rainfall is between 16-27 in (400-700 mm) per year, density is 28-53 trees per acre (70-130 trees per hectare). When rainfall exceeds 27 in (700 mm) per year, density may reach around 80 trees per acre (200 trees per hectare). Olive trees need a reasonably deep soil to act as a water reserve for the dry season, if they are to maintain a good level of production. The best conditions are found where the soil is more than 3 ft (1 m) deep and where the water table does not reach the roots, which have very little resistance to flooding. In areas with a high rainfall that is well spread out over the year, or that are irrigated in summer, densities of 80-160 trees per acre (200-400 trees per hectare) may be reached.

In general, olive groves may be found in areas where rainfall varies from 8-39 in (200-1,000 mm) per year. In the first case, the groves will be situated at the bottom of the valley, while in the second they will be on the slopes on well drained soils. Furthermore, because of the tree's susceptibility to mists and high atmospheric humidity, which lower its resistance to fungal attack, the best cultivation conditions are relatively dry places not too close to the coastline. Nevertheless, its robustness means it can survive in places where the conditions are very unfavorable, so olive groves can be found in sites with far from ideal conditions, even though under these conditions fruit production is limited and irregular. The optimum conditions include 17.5-31 in (450-800 mm) of rainfall per year, a site that does not freeze, and average relative humidity.

Like all fruit trees, the olive tree is regularly pruned to rejuvenate the crown and stimulate fruit production. Normally, the crown is thinned every 1-3 years. For fruit production, a shorter period between prunings is better, but in the past this period was often extended to produce more wood, a high-quality fuel, but little used nowadays. The arable weeds growing between the trees in the large free surface area of an olive grove, must be eliminated either mechanically or with herbicides. These plants, as they absorb part of the water from the soil, compete with the trees and so reduce production.

168 Harvesting olives in **Jaén** (Andalusia), in the south of the Iberian Peninsula. This task is traditionally performed by groups of seasonal farm laborers. The harvest begins in mid-winter, when the branches are weighed down by the weight of the ripe olives. It is simply performed by hitting the trees with sticks. Green olives, on the other hand, are picked and pass from tree to hand and from hand to basket. The fact that the trees are separated and there is no crop between them makes harvesting easier.
[Photo: Lluís Ferres]

169 An olive branch (*Olea europea*) laden with olives. Olives can be collected when ripe or while still green, but when they are picked from the tree they are very astringent and bitter. They need to be treated, by marinading and lactic fermentation, before they are edible. After this treatment, the unripe olives will become "green" olives, while the ripe ones will become "black" olives. Black olives do not come from different varieties from the green ones; each variety can give green or black olives depending on whether they are eaten ripe, unripe, or artificially blackened by oxidation.
[Photo: Josep Maria Barres]

The fruits of the olive ripen at the end of autumn and beginning of winter and reach an individual weight of 1-15 grams, depending on the variety. Harvesting is carried out manually, either by running the hand along the branches to make the unripe green olives fall, or by hitting them with rods if the fruit are harvested when ripe. One worker can harvest about 176 lbs (80 kg) of olives a day, roughly equivalent to the production of eight trees. The 800 million trees currently in existence thus involve a harvest labor requirement equivalent to 100 million working days. Attempts are being made to mechanize harvesting, although there are technical difficulties that stem from the fact that the structure of the plantations is planned for manual harvesting, as well as the social problems relating to the disappearance of this important seasonal activity.

Every year about 7.8 million metric tons of olives are produced of which 7.2 million metric tons (93% of the total) are used for oil production, while the remaining 0.6 million metric tons are for table use. From 1940 to 1980, production increased 60%, although olive oil, the main product derived the trees, is having serious problems competing with other vegetable oils obtained from plants that are herbaceous, and thus more productive and faster-growing crops, that are easier to treat mechanically.

Olive oil

Olive oil is the sixth most important vegetable oil in terms of the yearly volume of production, after those obtained from soya, peanut, cottonseed, sunflower, and rapeseed. About 92% of production is consumed in the producer countries, clearly showing how typically Mediterranean olive oil is.

The oil is extracted from ripe olives which, once cleaned and milled, are pressed in traditional oil mills or submitted to a continuous centrifugal extraction process, the technical term for which is pyralisis. The solid residue is called olive oil cake (*orujo*, *grignon*), and a small quantity of oil can still be extracted from it using chemical procedures, although its main use is derived from its considerable caloric energy which makes it a very good fuel.

170 The olive oil production process, as seen in a Renaissance engraving. Once harvested, the olives are taken to the vat. After removing the leaves and branches, the olives are milled in warm water and made into a thick paste (left). Afterwards, the paste is placed in layers, separated by fiber mats and then pressed. The milling and pressing of olives has hardly changed since time immemorial. In many areas it is still performed mechanically or by animal or human strength. Modern processing uses machines that perform the entire operation.
[Photo: Archiv für Kunst und Gesichte (Berlin)]

2.
OLEVM OLIVARVM.
Decuſſæ oliuæ adhuc acerbæ, ex arbore, Preſſæꝗ, pinguis dant oliui copiam.

The liquid product obtained is a mixture of vegetable fats, water, and other organic products. It is left to settle or is centrifuged to separate the fatty fraction from the watery one. The watery fraction contains a great deal of organic material and is of no use whatsoever, and the large quantities produced make it a serious environmental problem. The fatty fraction is known as *virgin olive oil*, in other words, oil which has not been physically or chemically treated. Oil with an acidity of less than 3° is suitable for direct human consumption, even though an acidity of less than 1°, called *extra virgin olive oil*, is considered better. If the acidity is higher than 3° it is not suitable for direct human consumption, and has to be physically and chemically treated to eliminate the acidity, color, and taste. This refined oil, which is very insipid, is used to lower the acidity of low-quality virgin oils, a blend called *pure olive oil*.

Olive oil's nutritional qualities are due to the high quantity of unsaturated fatty acids (88%) and low quantity of saturated fatty acids (12%). There are more monounsaturates (oleic acid, 80%) than polyunsaturates (linoleic acid, 8%), thus reducing somewhat its nutritional value in comparison with other vegetable oils which are richer in polyunsaturates, such as sunflower or maize oil.

This minor deficiency is more than compensated for by the fact that it is easily digestible, and it does not promote the the formation of plaque in blood vessels as oils derived from saturated animal fats do.

Table olives

A small proportion of the olives harvested are consumed as table olives. Table varieties tend to have larger fruits, with a lot of pulp and a small stone, high sugar levels, and low oil content. This is because oil negatively effects storage, since the inevitable oxidization of the fats turns olives rancid. This is why many olives eaten as fruits are green—they are harvested before the ripening process has finished.

The preparation of green olives involves, in the first instance, a treatment with caustic soda to remove the bitter taste of the glucoside, oleuropeine. Next they are thoroughly washed to remove the product from the surface of the fruit, and then they are stored in brine. In some cases, they are then sterilized, but others undergo a lactic fermentation to improve the quality of the product.

The process is often completed by macerating the olives with aromatic herbs, or by removing the stone and then stuffing the olive with pimento, almonds, anchovies, etc.

Black olives may come from two different sources. They may be ripe olives, which have the disadvantage of having looser flesh that is richer in oil, thus making storage difficult. Treatment to eliminate bitterness is almost unnecessary in this case. Sometimes, however, black olives are just green olives that have been blackened by artificial oxidation. This makes them look like ripe olives but they maintain the texture and storage properties of green ones.

While olive oil is suffering from competition from other vegetable oils that produce greater yields or are less costly, table olives have no such problems as there is no alternative product. Whatever the case, the low percentage of olive production destined for consumption as fruit is not enough to resolve the problems facing olive cultivation. An effort is needed to improve and rationalize production by increasing planting densities, and using fertilization and adequate irrigation. Improvements in parasite control methods and pruning techniques are also needed. The mechanization of harvesting, though, raises other problems related to the socioeconomic structure of olive-farming areas, since it would involve the loss of many jobs in areas with a poor, rural economy. This would be unfortunate even if the jobs are only seasonal.

The grape vine

The grape vine (*Vitis vinifera*) is a typical plant of the Mediterranean biome, but it has some unusual characteristics, the most surprising of which is that it is a winter deciduous plant. In a biome dominated by evergreen trees and shrubs with small leathery leaves, the grape vine is a woody plant with a climbing growth-form and large tender leaves that fall at the beginning of autumn.

From the wild vine to all vineyards

The grape vine is one of the 60 species of the genus *Vitis*. Almost all members of this genus originated in and are restricted to the northern hemisphere and they are especially abundant in North America. Many of these species produce edible berries in racemes, or bunches. Wherever they occur naturally their berries have been harvested since prehistoric times. The cultivation of grapes, though, began about 6,000-8,000 years ago in the eastern half of the Mediterranean Basin with wild

171 Refined (0,4° acidity), virgin (1°) and semi-virgin (0,7°) olive oil (from left to right). The color and smell of virgin olive oil are not to the liking of some consumers, and so the oil is usually refined to reduce free fatty acid levels.
[Photo: Josep Loaso]

172 An olive stall in a Provence market, in the southeast of France. Table olives may be green, ochre, or black because the olives are collected at different stages of maturity. When harvested they are not edible, but must be treated to make them edible, by methods such as soaking in brine, caustic soda, or a salt solution. After this initial processing, they are stored in a weaker salt solution for a longer period. There are many ways of treating olives, such as splitting them, or pitting them to stuff them with garlic, onions, pimentos, or mixtures of the aromatic herbs that are so widespread in the Mediterranean biome. The resulting enormous diversity is visible on olive stalls in the different mediterraneans, although the Mediterranean Basin still has the widest range.
[Photo: Joe Cornish / Photothèque Stone International]

vines from the species *Vitis vinifera*. This agricultural practice rapidly spread through the eastern coastal areas and the Nile delta, until it reached Greece, Crete, Italy, Corsica, and Sardinia some 3,000 years ago and the Iberian Peninsula 2,000 years ago. Grape vines are now cultivated in all the regions of the Mediterranean biome and in many other places outside it. This expansion was due, not so much to a need for the fruits as a food source, but to the discovery of the fermentation of grape juice to produce wine, an alcoholic drink which has always been very popular in the world influenced by western culture.

It could be said that today grape vines are cultivated everywhere in the world where the environment is favorable. Viticulture is most successful in the area originally occupied by the different wild species of the genus, between 34° and 49°N although it does grow outside these limits, making the best of unusual local conditions. In tropical climates, for example, the grape vine behaves like an evergreen, but gives a sparse and poor quality crop, although in Bolivia, at altitudes of almost

173 Bunches of black and white grapes. The grape vine is one of the few Mediterranean species to lose its leaves in the winter. In autumn, when the leaves take on reddish shades—an unmistakeable sign of autumn—harvest time has arrived. The different colors of the grapes are due to the different varieties, and not to how ripe the grapes are. They depend on the color of the grape skin: for example, white wine can be made from gently pressed black grapes. In winter, the fields are weeded. In spring the leaves appear and by summer they will be dark green. In the autumn, the grape pickers return.
[Photos: Josep Maria Barres and Jordi Vidal]

9,800 ft (3,000 m), grape vines can be grown at less than 20ºS. At high latitudes, cold starts to act as a limiting factor and forces farmers to use the lowlands and hillsides facing south or westwards, as happens in the Rhine valley, which is around 51ºN.

The cultivation of the vine

The best climatic conditions for cultivating grape vines occur when the summer is long with a temperature between warm and hot, and the winter is cool. The vine does not thrive where there are humid summers—it is prone to fungal diseases—or exceptionally cold winters. The first autumn frosts, and especially late frosts in spring, are very damaging to this plant. Temperatures only just below 30ºF (-1ºC) can kill the tender growing shoots.

As regards the soil, the vine tolerates all sorts of conditions, even though in the poorest soils it becomes difficult to cultivate and is not very productive. Except for very compact clay soils, saline soils, very poor soils, or badly drained soils, the vine can grow and bear fruit in a very wide range of conditions. However, it may display certain particular preferences in each of the regions where it is cultivated. The depth of the soil is an important factor, as the vine needs dry summers for the fruit to ripen

well, and there requires good reserves of soil water or additional irrigation. In average conditions, it is accepted that for dry farming the vine needs soil about 59 in (150 cm) deep, while if it is going to be irrigated, this depth might be reduced to 23 in (60 cm).

The morphology of the grape vine, a climbing plant with a woody stem, means that special cultivation techniques have to be applied, as it needs some form of support. In the ancient Roman Empire, trees used to be planted to support the vines, but the most widespread technique is to train the vine over a fixed structure, a trellis, designed specifically for this purpose. It is not uncommon in some regions of the Mediterranean Basin to see cultivated vines without any kind of support. They are pruned very severely every year until they produce a very thick, knotty stump that supports the mass of leaves that grow every year. The training technique and the shape of the trellis vary greatly according to the climate and the use the grape is put to. Vineyards planted for wine production have very different ripening requirements to those planted to produce grapes for eating. The shape of the plant is also very important, since it greatly influences the plantation's microclimate and the fruits' ripening process.

174 Vineyard with grape vines trained on trellises in the Napa Valley, California. Vines can be left to grow freely, or trained on a support. When the leaves sprout in the spring, so do the tendrils that allow the vine to climb. Then the vineyard turns into a plantation in mid-air. Vines are not demanding in their soil requirements. They can grow on siliceous, calcareous, or even volcanic soils that are stony or porous. *[Photo: AGE Fotostock]*

175 **Harvesting black grapes in the Autonomous Community of Valencia, Spain**. The work related to the grape harvest lasts throughout the year. In the winter, the plant's dormant period, the vine is pruned to ensure correct growth and fruiting; the pruned shoots are normally used as fuel. In the spring, hoeing gets rid of weeds and allows the rain to enter the soil, and a second pruning ensures good flowering and fruiting. In the summer, the vine is threatened by the dry weather and so sometimes drip-irrigation is installed. In the autumn, when the leaves begin to fall and the fruit is ripe, the seasonal work of the grape harvest takes place. The grapes may be prepared for sale as fruit or for producing wine.
[Photo: Jordi Vidal]

The grape vine is easy to propagate vegetatively by cuttings; this greatly facilitates the propagation of varieties obtained by selection and by hybridization with spontaneous forms or other cultivated varieties. The first cultivated vines were forms of the wild Euroasiatic vine *Vitis vinifera*. Although more than 90% of world grape production comes from varieties derived from this species, other varieties also exist which have been obtained by hybridization with North American species, most commonly *V. lambrusca*, *V. aestivalis*, *V. lincecumii*, or *V. riparia*. Hybrid varieties have been investigated, especially in search of plants resistant to various pests, such as phylloxera and nematodes, that attack the roots. The resistant variety serves as a rootstock on which to graft a variety that produces better fruit.

The 8,000 cultivated varieties of grape vine in the world produce grapes with white or black skins and are distributed unevenly. Certain varieties are very common, and cover about 25 million acres (10 million hectares) of cultivated land. The grapes produced are put to four different uses: they may be eaten as fruit, dried, or made into juice or wine. The consumption of grapes as fresh fruit or in the form of raisins is important, but grape juice consumption is less important.

Most of the grapes produced are used in making wine—7,950 million gal (30,000 million liters) are produced every year. Wine is the main product obtained from grape vines and the one that explains the crop's wide distribution.

176 **Grapes spread out to dry to make raisins**, in la Marina (Valencia). In la Marina, the grapes are left to dry after scalding or bleaching. This is called bleaching because the water used to scald the grapes contains bleach or caustic soda. This speeds up the process of drying, but is unnecessary in hotter or less rainy places. Raisin-making is common where grapes are grown. Famous examples include raisins from Málaga (Spain) and currants from Corinth (Greece), but raisins from Italy and California also exist.
[Photo: Ernest Costa]

Wine

Wine is obtained by fermenting the sugars in the grape juice, or must. The yeasts on the surface of the grape convert the sugar into ethyl alcohol. Alcohol concentration increases, slowing the yeasts' growth and finally it brings the fermentation process to a halt. But the transformation process can continue if aerobic bacteria take up where the yeast leaves off, transforming the ethyl alcohol into acetic acid, and ultimately into wine vinegar. This transformation may be deliberate or it may accidentally occur in a bottle of table wine.

Wine-making has really changed very little over the centuries. The process has become more technically sophisticated and a great deal has been learnt about the process, but the yeast continues to work just as it did thousands of years ago. The wine's fermentation and aging are now controlled much more precisely than they used to be. It all begins when the grapes, harvested at the peak of their ripeness, are pressed. If the pressing is gentle, it produces what is called *flor* must, a better quality than the normal must obtained by a more vigorous pressing. If the fermentation of the grape juice is started immediately, it will produce a *white wine* (even if the grapes are black), but if the juice is left to macerate for some time with the solid remains of the grape it becomes enriched with colored substances and turns a color that may range from slightly pink to almost black depending on the variety of grape and, above all, the length of maceration: they are know as *rosé* or *red* wines respectively.

Fermentation is a process that generates heat and this raises the temperature of the must, leading to the loss of several volatile compounds. To avoid these losses and obtain a more fruity wine, wine makers now prefer to cool the fermentation tanks. If fermentation is halted by adding sulfide or by filtering the wine before it stops spontaneously, a sweet wine with a low alcohol content is obtained. If the wine is bottled before fermentation has finished, fermentation continues in the bottle and sparkling wine is produced, since the carbon dioxide released by fermentation accumulates inside the hermetically sealed bottle (as happens in *champagne* and *cava*). If the fermentation is allowed to continue until it comes to a halt naturally, a normal wine is produced with an alcohol content and characteristics dependent on the grapes used. Some wines, such as sherry (*jerez*), are fortified with alcohol and have a higher alcohol content, while others are distilled to obtain spirits.

The wine obtained can be consumed immediately as a young wine or it may be aged for a shorter or

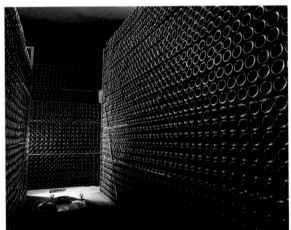

longer period, which profoundly changes its characteristics. The aging process begins in a wooden cask where the wine absorbs a range of products from the wood, and undergoes complex oxidation processes. Once this phase of the aging process is over, the wine can either be consumed or bottled to continue the process. Inside the bottle, the wine is only in contact with the residual air, isolated from the world by the glass and the cork. It now undergoes a series of changes dominated by reduction processes.

The optimum aging time depends on many variables and is not even fixed for wines from a given region made with a specific type of grape. Furthermore, wines are often the product of mixtures of must or wines from different varieties (*coupage*) in an attempt to reinforce and accumulate the positive

177 The forced fermentation of must in cooled steel tanks, in Valencia. The wine fermentation starts with the must, obtained by pressing grapes using traditional methods or hydraulic presses. This separates the solid part of the grape from the liquid part, a frothy, cloudy liquid. Fermentation is the conversion of the grape's fructose into alcohols by the action of yeast that grow spontaneously on the skin of the grape. Today, this fermentation, which may take a few weeks, is carried out in stainless steel tanks at low temperatures to avoid overheating. Overheating leads to the loss of the volatile compounds that give wines their fruity taste and character.
[Phot: Jordi Vidal]

178 A cellar full of bottles of new wine in Catalunya. New wine obtained by fermenting must is transferred to wooden barrels, preferably of oak, to age. There, the wine undergoes physical and chemical changes, mainly the oxidation of some compounds and other changes caused by contact with the wooden barrel. During this process, all wines do not age in the same way, but give quite different results. Once aged, the wine is appropriately blended and bottled. Inside the bottle, isolated from contact with air, the wine continues to change, mainly by reduction processes. Tests by wine experts make it possible to study the wine's development over the entire aging process. A complete study includes the wine's color, aroma, taste, and body.
[Photo: Josep Pedrol]

The shipwreck of the vineyards

Cultivated grape vines are susceptible to many diseases, especially fungi and the nematodes (parasitic worms) that attack their roots. Many of these diseases cause a reduction of the harvest, or even its total loss due to the destruction of the fruit, but they rarely cause the death of the plant. Yet at the end of the 19th century European vineyards were almost totally devastated by a small insect that ceased being harmless and became extremely virulent—phylloxera.

Vine leaves parasitized by phylloxera (*Phylloxera vastatrix*) [Lluis Giralt & Joan Reyes / Department of Agriculture, Livestock and Fisheries, Generalitat of Catalonia]

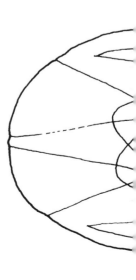

Leaf galls with phylloxera larvae (*Phylloxera vastatrix*) [Lluis Giralt & Joan Reyes / Department of Agriculture, Livestock and Fisheries. Generalitat of Catalonia]

Phylloxera (*Phylloxera vastatrix*) is a small insect belonging to the aphid family in the order Homoptera that sucks the sap from the roots. The species comes from the Mississippi valley, where it attacked all the different species of wild vine, although behaving as a relatively mild parasite. In fact, almost all American wild vines easily withstood the insect's attack, but when it was accidentally introduced into Europe it became highly aggressive, attacking the cultivated European variety, which had never before come into contact with the insect.

Like almost all the members of the aphid family, phylloxera has a very complex life cycle, with phases in which the insect attacks the roots alternating with other phases where it lives in leaf galls—which the plant develops in response to the insect's attack. The insect's life-cycle also has generations of parthenogenetic wingless females that alternate with generations of sexually reproducing males and females that may be winged or wingless. The insect's life cycle shows some differences when in its original area, where it attacks wild vines, and when its new areas where it attacks cultivated vines (*Vitis vinifera*). In the original area, the more complicated cycle includes aerial phases that develop on the vine's leaves, while in its new areas it spends almost the entire life-cycle underground. The major difference between the two cases lies in the degree of sensitivity of the species under attack. The American species of wild vine withstand the insect well, but the forms of the European vine suffer a process of rapid root degeneration— due to the galls that form on the root hairs—and the plant quickly dies.

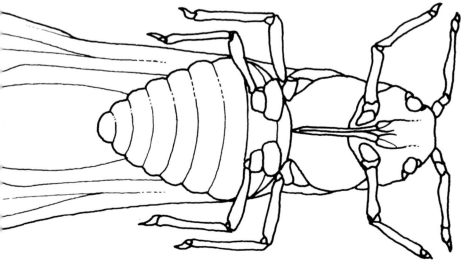

Phylloxera (Phylloxera vastatrix) [S.F. / Roman Montull /ECSA]

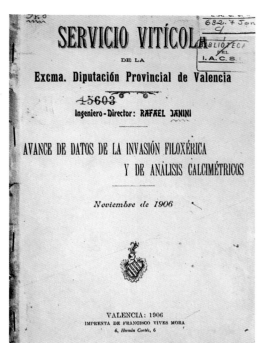

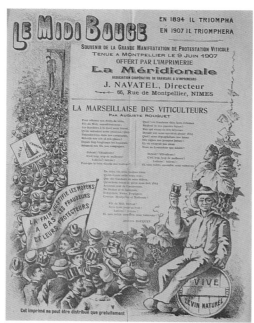

Leaflet commemorating a winegrowers' demonstration in Montpelier
[Archive of the Centre de Documentació i Animació de la Cultura Catalana (Perpignan) / J.L. Valls]

Leaflet on phylloxera
[Institut Agrícola Català de Sant Isidre / ECSA]

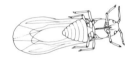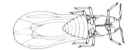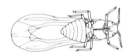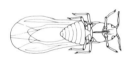

Phylloxera reached the European continent around 1860, as a consequence of the importation of American stocks as museum specimens and for trials to find hybrids resistant to a range of fungal diseases. Its spread was devastatingly rapid, and between 1860 and 1900 it annihilated almost all Europe's vineyards, except for a large part of the Iberian Peninsula. However, this area succumbed a mere 10 years later, although it did have a brief period of splendor while it was the last area of Europe still to produce wine. This advantage caused an intense but short-lived boom in vine cultivation until it occupied even marginal areas. Once phylloxera had arrived many of these marginal areas were not reoccupied, so these abandoned non-irrigated cropland were rapidly colonized by forests and scrub that covered the dry-stone walls and the remains of the former terraces.

One of the consequences of phylloxera's arrival in Europe was the almost total disappearance of wild examples of the species *Vitis vinifera*, the origin of all the cultivated varieties of grape vine. Only a few small populations now remain in spots in the Balkans and Pyrenees that the insect has not yet reached, but this formerly common species is now in danger of extinction in the wild.

The great economic importance of vine cultivation ensured that the reaction was rapid. As controlling the expansion of the disease was almost impossible with the methods then available, the solution was to use the rootstocks of resistant American species as stocks on which to graft European varieties. The most widely used American species were *Vitis riparia*, *V. rupestris*, and *V. berlandieri*, all of which are highly resistant to the parasite. These and other species were then hybridized in search of a resistant stock that was also adapted to the environmental conditions best for the crop. The appearance of strains with names like *AxR1*, *E.M.41*, *Rupestris St. George*, *C.3309* and *5BB* allowed European viticulture to recover.

 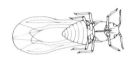 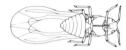 

The Chilean mediterranean is isolated and so it was untouched by the last century's phylloxera outbreak. In California, however, the parasite was first detected in 1873 and it caused serious crop losses. It is still not known if the insect reached California from France, as a result of the importation of European varieties, or if it arrived from its original range in eastern America as a result of trials carried out with some wild species from the area.

A hundred years later, at the end of the 20th century, phylloxera is once again in the news, this time in California. California viticulture expanded greatly between 1960 and 1980, and this formerly unimportant region became one of the world's largest wine producers. Many vineyards were planted with rootstock of the resistant *AxR1* strain, but unfortunately a new strain of phylloxera has appeared that attacks this variety and can multiply extremely fast. The first symptoms were detected in 1983. After almost 15 years of declining harvests it seems that there has been a recovery, and it is hoped recovery will be complete by 2010.

characters of each one. All wines varies from year to year, giving a clearly distinct product that continues to change over time as it matures.

The final quality of a wine depends on many factors, the first of which, and the least important is the soil the vine grows in. Soil is an incidental factor in comparison with the two main factors that determine the characteristics of a wine—the climate and the variety of grape. A good variety growing in unfavorable climatic conditions will produce a poor or mediocre wine, as it will not reach its full potential. As for the climate, it is accepted that moderately cold summers that slow the ripening processes tend to produce better quality wines than those obtained in hotter places where the ripening process is rapid. This is why many oenologists, while recognising that almost all wine-producing regions have some wines of exceptional quality, consider that the finest and most complex wines are produced in areas where the climate is not Mediterranean.

In the case of Europe, for example, the wines produced in the Mediterranean region are mild and full-bodied, with low acidity, a high alcohol content and a not very fruity flavor, while wines from temperate regions are fruitier, have a lower alcohol content, higher acidity, and greater aromatic complexity. A vine variety adapted to a temperate climate planted in the Mediterranean region would produce a wine with clearly Mediterranean characteristics. This shows how important climate is in determining a wine's final characteristics.

The almond

The almond tree (*Prunus dulcis*) flowers in winter before its new leaves have appeared. It is well adapted to the Mediterranean's mild winters and dry summers. It needs to be planted in sites with good air flow but protected from strong winds and from heavy frosts, as both may destroy the flowers. Its origin is unclear, as the examples found in the wild in the Mediterranean biome may be derived from cultivated plants. This is true of the wild almonds found in Sicily, Greece, Italy, France, and Spain, while the ones found in Mesopotamia, Turkestan, and some parts of Algeria appear to form part of natural wild populations.

In any case, almond cultivation is very ancient in the Mediterranean Basin, as is shown by the wide range of uses made of its fruit, both raw and roast-

ed. There are two basic varieties, depending on the characteristics of the fruit: some trees produce sweet almonds suitable for consumption, and the others produce bitter almonds, which are not edible but are used in pharmacy. There are more than 300 varieties of sweet almonds and a single field may often contain several varieties since the self-sterility of many varieties makes this necessary.

The carob tree

The carob (*Ceratonia siliqua*) is a typically Mediterranean tree that, in the wild, forms part of the coastal maquis, as does the wild olive. This evergreen tree produces carob beans, which have long been eaten by humans, and extensively used as fodder. Carob fruits are legumes, with a fleshy pod, the part that is consumed. Harvesting carob pods is performed in the autumn, manually or mechanically, but care has to be taken not to damage the inflorescences that will give rise to the next year's fruits, since they are present on the plant at the same time as the mature fruits.

The carob has long been cultivated in most countries in the Mediterranean Basin, and more recently, it has

179 The almond tree (*Prunus dulcis*) has been cultivated since time immemorial in the Mediterranean Basin for the almonds it bears. It is thought to have originated in the Middle East and North Africa. The almond flowers earlier than any other species in its genus, and its fruit takes the longest to mature, about eight months. As it ripens, the fleshy part of the fruit thins, loses its fleshy appearance, hardens, turns leathery and falls away from the kernel. Unlike other drupes, such as peaches and plums, what we eat is the seed. In all the other members of the same genus, what we eat is the fleshy mesocarp.
[Photo: Ernest Costa]

spread to areas outside the biome and to other non-Mediterranean temperate areas. It is not very demanding in its soil requirements, but is very susceptible to frosts and thus needs a climate where winters are extremely mild, such as those found on the coastal strip of the Mediterranean Basin. Its cultivation has suffered ups and downs, but it expanded greatly at the end of the last century when it was used to replace many of the grape vines destroyed by phylloxera. Later, as a result of the progressive disappearance of draft animals, its cultivation ceased to be profitable and many carob groves were abandoned. Since the beginning of the 1980s its cultivation has again become profitable thanks to the sale of the gums derived from the seed of the carob. It has several uses in the food and pharmaceutical industries. Its robust, resistant nature makes it a useful crop in the dry, warm areas of the Mediterranean biome.

The fig tree

The fig tree (*Ficus carica*) is another typically Mediterranean tree, although it is deciduous. It is an excellent crop for small-scale dry farming. Yet, it may play an important role in the economy and in

180 **A branch of a carob tree** (*Ceratonia siliqua*) with female flowers and fruit. The carob tree can be monoecious or dioecious. This means that there may be stems bearing only female flowers, or stems bearing only male flowers, or stems bearing hermaphrodite flowers. Carob beans only grow on female or hermaphrodite trees. The carob beans remain on the tree for a year, making it possible to see the female flowers side by side with the previous year's pods.
[Photo: Josep Maria Barres]

181 **A fig tree** (*Ficus carica*) about a century old, on the island of Formentera (Balearic Islands). The fig tree is a deciduous broadleaf that lives in the warmer areas of the Mediterranean. The fig tree's fruits, pruned branches, and leaves are used as forage. Figs usually produce two types of sweet, fleshy, composite fruits. About the time of the summer solstice the brebas figs appear, while the main fruits appear once a year at the end of the summer. What we call a fig is a multiple fruit termed a syconium. The syconium is a hollow, pear-shaped receptacle with fleshy walls that contains hundreds of flowers. They are pollinated by a tiny wasp (*Blastophaga psenes*) that enters through a little pore at the bottom. The relationship between the fig tree and the insect is an example of coevolution.
[Photo: Ernest Costa]

182 The pomegranate (*Punica granatum*) is a common fruit tree throughout the Mediterranean Basin (the photo is of a specimen in the east of the Iberian Peninsula). It was cultivated by the Egyptians and later spread by the Phoenicians and the Romans. Its fruits ripen over the summer, and in the autumn they have formed a pomegranate. The fruit of the pomegranate, known as a balausta, is unlike that of any other flowering plant. It is a large indehiscent sphere, consisting of lots of seeds, each in a translucent, juicy, red pulp. When the fruit is ripe, it opens unevenly, spontaneously exploding. For this reason they have given their name to small explosive shells called grenades.
[Photo: Ernest Costa]

183 Caper (*Capparis spinosa*) buds and flower. This plant grows wild in the Mediterranean Basin in cracks between rocks, on cliffs, and on the walls of old buildings. The part that is used is the bud; after pickling in vinegar they become capers. As the flowering period is very long, the buds can be picked once a week, so that each plant will be harvested about a dozen times each season. This frequent harvesting prevents the buds from developing too much, and stimulates flower formation so that each plant may provide 22-26 lbs (10-12 kg) of buds per year. Once collected, they are sorted by size. The best capers are the smallest, the most tender ones. They are pickled in brine and stored in vinegar for consumption.
[Photo: Josep Maria Barres]

the landscape in some areas of the Mediterranean Basin. The many varieties of cultivated fig trees are all derived from the wild fig, but they are clearly distinct from it. The wild fig has male trees, and female trees that produce inedible fruit; strictly speaking, the fig is really a compound fruit (syconium), with a fleshy receptacle containing the true fruits (achenes) inside. Cultivated figs only bear female flowers and thus need wild male trees nearby for the female flowers to be fertilized and the fruit to develop. If wild *caprifigs* are lacking, branches of the wild tree are collected and are hung over the female plants to encourage pollination. It is necessary to ensure both the presence of the pollen and the small hymenopteran pollinating wasp—a method called caprification. Some cultivated varieties develop fruit without needing to be fertilized, which has made them independent of wild figs. There are varieties that fruit at the end of summer, while others produce two crops a year, a *breba* crop at the beginning of summer and the true figs at the end of summer. There is a third type that produces two crops of breba figs—in May and July— and then true figs at the end of the summer. Figs may be consumed fresh or dried and are widely eaten in the countries of the Mediterranean Basin.

The pomegranate, the caper, and the hazel

There is a range of dry farming fruit trees, not of great importance in the biome as a whole, that are cultivated to a greater or lesser extent, such as the pomegranate (*Punica granatum*) and the hazel (*Corylus avellana*). The pomegranate comes from Asia but it is widespread throughout the Mediterranean Basin, because it has long been cultivated and can grow almost wild. The hazel is a Euro-siberian deciduous shrub, that prefers to live on the shady side of low, humid mountains, and is successfully cultivated throughout the Mediterranean Basin, especially in Catalonia. Its cultivation increased notably after the massive destruction of the grapevines caused by *Phylloxera* at the end of the 19th century. In the eastern part of the Mediterranean Basin, especially in Turkey, the Turkish hazel (*Corylus colurna*) is cultivated. The caper (*Capparis spinosa*) is a small, deciduous Mediter-ranean shrub with straggling branches that grows in cracks in rocks or ledges on cliffs, in warm, sheltered places. The product of these shrubs, often planted on poor, stony soils, is not their fruit but the flower bud. To make best use of the shrub's long flowering season, the buds are collected every 6-7 days, and over the summer the total crop from one plant may reach 22-26 lbs (10-12 kg). The few fruits that manage to ripen form caper berries about 2 in (5 cm) long; these are also used, although the buds or *capers* are preferred. Capers are made edible by treating them with brine and then they are stored in vinegar, which makes them lose their very bitter original taste. They are used in salads and pizzas. They are a far from unimportant crop, since production is centered in the drier areas of the western Mediterranean, and reaches almost 20 million lbs (9 million kg) per year, half of which is consumed in Italy.

2.6 Irrigated vegetable gardens and orchards

The areas known as hortas in Catalan and huertas in Spanish are areas of regularly irrigated, arable land, usually smaller than fields dedicated to cereal cultivation, where vegetables and fruit trees are intensively cultivated. This intensive cultivation is possible both because of irrigation (with water being diverted from rivers or nearby mountains) and because of fertilizing the fields. This is one of the reasons why, in those areas where vegetables are grown, there is often a mixed economy based on vegetable gardening, or "horticulture" and raising livestock, whose manure is used to fertilize the fields.

The intensity of cultivation in a huerta can be quite extraordinary. It is often a two-tier form of agriculture, that is to say, fruit trees and herbaceous plants are grown on the same land, and it can support population densities of some 600 inhabitants/km². This type of agriculture, (known as "truck farming" in the United States and "market gardening" in Britain) uses techniques to work the land that are different from those used to cultivate larger areas. For example, ploughing by animals, or now by tractors, typically seen in larger fields, is replaced in these vegetable gardens by a whole series of tools that are used directly by humans. There are tools for levelling the field, devices to ensure even water distribution, as well as digging and weeding implements. The painstaking cultivation of a vegetable garden requires a radically different approach to that required in large-scale agriculture. The different vegetable crops were usually introduced from areas with warmer, wetter climates. Therefore, in the dry Mediterranean summer this cultivated land required special treatment to supply the water that is needed by the roots, especially those of the the trees. Even though flat or easily levelled fields require less water and areas with many vegetable gardens were crisscrossed

184 An irrigated vegetable plot in the Mediterranean, of onions (*Allium cepa*), a plant that originated in central Asia. A vegetable plot often requires more water than is provided by the Mediterranean climate. Historically, this required strict control of the available water to sustain and promote the crops. The Romans spread improved forms of agriculture all over the Mediterranean Basin after acquiring them on the eastern shores of the Mediterranean Basin. Mesopotamia, and to an even greater extent Egypt, were civilizations dependent upon irrigation, and were accustomed to rationing water use. The expansion of Islam, ranging from India to the western Mediterranean, led to the acquisition of improved systems for irrigating vegetable plots, which was often done by drip systems, as in oases.
[Photo: Josep Pedrol]

with irrigation channels, influential groups were formed in some areas whose job it was to regulate the fair distribution of water. A good example of the distribution of irrigation water in an area with vegetable gardening is the plain of Valencia where in the thirteenth century a tribunal called the Tribunal de les Aigües de València (Water Tribunal of Valencia) was established to resolve any conflicts that might arise between different farmers in the plain—known as the *Horta* or *Huerta* for its many market gardens. The Water Tribunal, which still meets today every Thursday morning from eleven to twelve outside the Apostles' Door of Valencia Cathedral, regulates the fair distribution of available water, especially in times of drought, and is empowered to impose large fines on those who disobey the orders of this democratically elected court. The products of vegetable gardens and orchards provide the basis of a range of industries. Fruits and greens may be specially treated to preserve them, like conserves for example; or they can be distilled to make liqueurs and other spirits. Fruit pulp, when cooked with sugar (except in the case of peaches and plums which are already sweet enough), produces long-lasting jams of varying consistency. Liqueurs are sweet, aromatized alcoholic drinks, prepared by the maceration, infusion, or distillation of various fruits and other plant material with alcohol. Cider, which is made from fermented apple juice, was known to the Greeks and Romans.

Vegetables and legumes

Many species from the Mediterranean Basin are cultivated in these vegetable gardens, but there are many more species introduced from other biomes. Among species cultivated in the Mediterranean basin are chard (*Beta vulgaris* var. *cicla*) and beetroot (*B. vulgaris* var. *rapa*), both chenopods and cultivated in the Mediterranean area more than 2,000 years ago; carrot (*Daucus carota*), the root of an umbellifer; lettuce

(*Lactuca sativa*), the tender leaves of an Asteraceae; cabbage (*Brassica oleracea*), a crucifer native to the coasts of England and Brittany that was introduced into the classical world 3,000 years ago, and its variations such as Brussels sprouts, broccoli, and cauliflower. The cultivation of the cabbage probably followed the tin routes in the Iron Age. Some vegetables require special growing techniques, as is the case with asparagus (*Asparagus officinalis*) and young onions—the individual shoots of white onions (*Allium cepa*)—specially grown to be cooked over hot ashes. In order to obtain blanched white onions or blanched asparagus, which are preferred for their flavor and appearance, the plants are covered with mounds of earth to prevent light reaching them. The cardoon (*Cynara cardunculus*) is cultivated for chymosin, which is used to curdle milk when making cheese. It is a relative of another Mediterranean composite, artichoke (*C. scolymus*), which is appreciated for the fleshy receptacle of its inflorescence. Among the first crops to be cultivated in vegetable gardens in the Mediterranean were the legumes that originated in the Mediterranean basin or in the Near East. Indications of the cultivation of broad beans (*Vicia faba*), possibly originating in the southeastern part of the Fertile Crescent, were found in the ruins of Troy. They were also grown by the Egyptians. On the banks of the Nile, as well as broad beans, Egyptians also cultivated lentils (*Lens culinaris*), a native of the Near East and the Mediterranean region. They were clearly cultivated with some success, as the Egyptians were the most important lentil exporters of the ancient world. Beans and lentils were grown during the Neolithic in northern Greece and the practice was continued by the ancient Greeks. The pea (*Pisum sativum*), a Middle Eastern plant, was cultivated by the ancient Greeks and the Romans, although they both considered them suitable food only for the poor. The Romans also cultivated some varieties of the chickpea (*Cicer arietinum*), which are natives of the Mediterranean Basin. The

185 **Field of artichokes** (*Cynara scolymus*). This plant is grown for the fleshy flower head, which is eaten before it opens. The artichoke is, in fact, an edible inflorescence. It originated in the Mediterranean Basin and in the Canary Islands, and has spread all over the world. It is widely cultivated, especially on the southern shores of the Mediterranean, as it is often used in couscous stews. [Photo: Aisa]

186 Tunisian half-Dinar note, showing some of the most widely cultivated Mediterranean crops. Some are introductions while others originated in the area. The crops shown include olives, dates, oranges, lemons, watermelons, and grapes, as well as sheep and sardines. This shows how exclusively cultural products, such as banknotes, may reflect a country's natural resources. [Photo: Jordi Vidal]

generic term, legume, comes from the Romans, from the Latin *legumen* derived from the verb *legere* meaning "to gather."

The expansion of vegetable gardens and citrus and fruit cultivation

It is difficult to know how important market gardening cultivation was in the Roman world However, it is known that the largest cities were surrounded by a great deal of land given over to horticulture, and that this land was irrigated, especially in the summer. The Romans learned how to cultivate many species from the Babylonians and the Egyptians—empires that based part of their agriculture on horticulture. Many of the crops grown in ancient Mesopotamia were imported from the wetter, more easterly areas of Asia (some from monsoonal regions) and therefore required the development of effective water supply systems. Some vegetables and fruits spread from Mesopotamia to Egypt and the eastern Mediterranean Basin, such as onion (*Allium cepa*), garlic (*A. sativum*), and cucumber (*Cucumis sativus*) as well as fruit trees from cold temperate zones like apple (*Pyrus malus*) and pear (*P. communis*) or warm temperate zones like peach (*Prunus persica*), apricot (*P. armeniaca*), and cherry (*P. avium*). The effects of the barbarian invasions and the fall of the Roman Empire on agriculture, and especially on horticulture, are difficult to measure, although there are indications that agriculture declined in northern Africa before the Roman Empire did. Agricultural dominance then passed to the Byzantine Empire. The mulberry tree (*Morus alba*), originally from central Asia and whose leaves are the food of the silk worm, was introduced into the Mediterranean Basin via Greece in the first half of the sixth century

A.D. during the reign of Justinian. It was cultivated initially for its fruits, and later on to supply the silk industry. Despite the collapse of the Roman Empire, there was no serious decline in agricultural techniques in the western part of the Mediterranean basin. The Visigoths did not dismantle the agricultural infrastructure left behind by the Romans in the Iberian Peninsula, even though pastoral activities increased in inland areas during the centuries of their rule. In Italy, however, the setback seems to have been greater. Another important factor in explaining agricultural decline in the Mediterranean is demography: the reduction in the number of people in the early Middle Ages led to less intensive cultivation of arable ground.

187 Preparing apricots for drying, in South Africa. Throughout the winter, women in South Africa prepare apricots for drying and preservation. The Khoisanids, or more specifically, Khoikhois, are natives of the South African mediterranean. The apricots are split, pitted, and placed on trays to dry in the sun. Apricots are the fruit of the apricot tree (*Prunus armeniaca*) and are a summer crop in the mediterraneans, in this case the South African. It is a small tree that tolerates the dry, stony soils of the Mediterranean Basin and the other mediterraneans. [Photo: Ramon Folch]

188 Branch of a lemon tree (*Citrus limon*) bearing fruits, in the east of the Iberian Peninsula The lemon tree originated in southern Asia and is widely cultivated throughout the Mediterranean Basin. The fruit, called a hesperidium, is divided into juicy segment while the epicarp (or peel) is full of glands that secrete aromatic oils. It has many uses, including giving flavor to jams and making fruit juice, as well as in cookery. Its yellowish wood with pale stripes is hard and of medium density; it can be used in cabinet-making and woodworking.
[Photo: Ernest Costa]

The growth of Islam to cover a territory stretching from India to the western Mediterranean led to the introduction and spread of a wide range of irrigation techniques from the east to the west. On the Valencian plain, on the eastern coast of the Iberian Peninsula, the Muslims established an elaborate irrigation system and spread the use of the waterwheel. This led to improvements in both intensive veg-etable gardening and orchards. The Arabs did not just bring new techniques, they also brought the new species that were widely cultivated in the huge territories they ruled. The lemon is the fruit of a tree (*Citrus limon*) native to Southeast Asia. It was already known in the Mediterranean Basin in the time of the Romans, but its cultivation spread with the arrival of the Arabs. The bitter orange (*C. aurantium*) did not arrive in the Near East until the end of the 10th century, and the sweet orange (*C. sinensis*) was introduced from India by Portuguese explorers in the late 15th century. The Arabs also introduced the cultivation of rice (*Oryza sativa*), a native of monsoonal Asia, sugar cane (*Saccharum officinarum*) from Southeast Asia, and cotton (*Gossypium*), whose origin is still uncertain.

Other crops were spread throughout the Mediterranean by Crusaders returning from the Near East. The date palm (*Phoenix dactylifera*) is cultivated in the southern and western part of the Mediterranean basin. It originated in North Africa and southwestern Asia and is an important crop in the southern part of the Iberian Peninsula, principally in Elche. Palm cultivation is an ancient practice, and almost all its parts can be put to some use. This is why it was a symbol of fertility for the ancient Egyptians, the tree of life for the Chaldeans and Arabs, and a sacred tree for the Jews. Its fruits ripen in autumn or winter and are eaten dry by the Arabs, while in Europe they are preferred when picked green and sprinkled with a little vinegar. Date palm fields actually started because palm trees were used to enclose vegetable gardens in the southern Iberian Peninsula, and therefore they were irrigated with the vegetables.

Overseas exchanges

The fifteenth century saw a major event, which occurred in the European western countries of the Mediterranean: the beginning of long sea voyages and the colonization of the New World, until then unknown. The sailors spread new farming techniques and transported new species that would be cultivated by both conquerors and conquered. The year 1492 saw the beginning of an important exchange of plants between two very different worlds. The tomato (*Lycopersicon esculentum*) originally from South America; the potato (*Solanum tuberosum*) from the high plateaus of South (and Central) America; *Zea mays*, known as maize in the United Kingdom and corn in the United States, another native of the American continent; tobacco (*Nicotiana tabacum*) a native of Central America; and the bean (*Phaseolus vulgaris*) from Central and South America were all brought to Europe by the Spanish colonists. The Mediterranean exported its own native crops and acted as a bridge for the exchange of crops between Asia and the American continent. The cultivation of cotton and sugar cane was exported to the New World, and was immediately successful. However, the demanding nature of their cultivation had a secondary effect that was to play a decisive role in transforming the demography and the economy of the entire planet: the phenomenon of slavery. Slavery had existed since antiquity, but agricultural innovations in the New World led to a new and more intensive traffic in human beings, unrelated to war. Over a period of four centuries many thousands of slaves were taken, mainly from the western coasts of Africa. The increase in population in the fifteenth century was accompanied by new improvements in the irrigation of vegetable gardens in the Mediterranean region, especially in the Po valley in northern Italy and on the southeastern coast of the Iberian Peninsula. Later, starting about 1850, there was another important change in Mediterranean agriculture—the increase in the area of land under wheat. In 1758, the Mesta lost its royal authority over seasonal migration routes in Spain. From the middle of the 19th century, other crops spread, including olive trees, grape vines, fruit trees (especially citrus fruits and apples), and vegetables. The cultivation of rice, cotton, sugar beets, and tobacco also increased. The north of Italy and the Lebanon were the two most important apple-growing areas. Cultivation of citrus fruit, especially oranges, has been very important in Lebanon since the 1930s. The famous Valencia oranges have displaced the olive trees from the *horta* of Valencia since their introduction in 1873, and soon after they were being exported to the United States and Germany. Most of these vegetable garden crops required a greater water supply, and this led to agriculture becoming more intensive and the use of a greater input of human labor. At the same time, the spread of the railways played an important role in aiding the commercialization of fruit and vegetables.

The market garden model of farming has also spread to the other mediterraneans and is still widely practiced. Fruit and vegetable farming in California is of great importance, using 30% of the state's agricultural land. Fruits like the pear and apple, and vegetables such as tomatoes, lettuces and asparagus are grown on the Los Angeles plain, while the cultivation of citrus fruits has moved to the San Joaquin Valley as a result of urban expansion. Horticulture is less important in Chile, South Africa, and Australia. In Chile, there has traditionally been almost no demand for vegetables from the urban markets, and their cultivation did not

become important until exports increased a few years ago. Some fruit trees are grown, like the pear, the apple, and more recently the peach and the apricot. In the South African mediterranean, peaches, apples, and apricots are cultivated. Few fruit trees and vegetables are grown in Western Australia; and arable land is largely dedicated to the cultivation of wheat and sheep farming.

3. HUMANS IN SCLEROPHYLLOUS FORMATIONS

190 Harvesting palm leaves (*Phoenix dactylifera*) in Elche, in the southeast of the Iberian Peninsula. The date palm originated in North Africa, Arabia, and the Persian Gulf. It is not known when it arrived in the Iberian Peninsula, but one of the oldest references is the Nerva coin (dated between 96 and 98 A.D.). It shows a female palm laden with dates. The introduction of the palm tree might have occurred when the Phoenicians or Carthaginians brought dates with them, as the fruit can be stored for some time. Almost the entire palm can be used. In Catholic countries the leaves, or fronds, are commonly used to make "palms" that feature in the Christian liturgy on Palm Sunday. These leaves have never been exposed to light—they are tied together to make them pale.
[Photo: Ramon Folch]

Edible yellow

It is said that saffron is the child of the Sun and of poetry. This spice, highly valued since antiquity, consists of the stigmas of the flower of *Crocus sativus*, a small bulbous plant, 4-12 in (10-30 cm) tall, that belongs to the iris family. Its attractive flower has a deep violet corolla and the three orange-yellow stigmas. The generic name *Crocus* is a latinized version of the Greek word *krokos* which means thread. The English word saffron comes from the Arabic word *zahafaran*, a derivation from *assafar* which means yellow. Saffron is thus the yellow that can be eaten.

Field of saffron (*Crocus sativus*), Castile [Sebastián Bellón]

A Greek legend attributes the creation of saffron to the god Hermes. After mortally wounding his friend Krokos unintentionally, Hermes transformed the blood trickling from his head into a beautiful flower with yellow stigmas. The oldest references to the cultivation of saffron go back to about 4300 B.P., locating it in Accad in Mesopotamia. The oldest representation of the gathering of saffron flowers is a polychrome fresco found in the palace of Knossos (Crete), dating from between 3700 and 3600 B.P. In the fourth century B.C., the main zone of saffron cultivation was in Corycus in Cilicia on the coast of what is now Turkey. Its cultivation spread to the east and to the west. It spread west throughout the entire Mediterranean, and was introduced into Spain by the Arabs between the 8th and 10th centuries, while in the rest of southern Europe it was introduced as a result of the Crusades. It reached as far east as India, where Buddhist monks use it to dye their robes a distinctive yellow color.

Flowers of saffron (*Crocus sativus*). Castile [Sebastián Bellón]

Saffron is now cultivated in Europe and Asia. Saffron fields look surprisingly different from other crops and it is cultivated from Spain, France, and England to Kashmir and western China. In the New World it has also been introduced into Pennsylvania in the United States. *Crocus sativus* is probably derived from the wild species *C. cartwrightianus*; it is known in cultivation only as it is a sterile triploid plant that only reproduces vegetatively. Saffron plants grow in very diverse climates. In the southeast of the Iberian peninsula they grow in almost tropical conditions, while in Kashmir it is cultivated more than 6,560 ft (2,000 m) above sea level. The corms are planted at the end of June and the flowers appear from September to October; the corms last for three years, after which they are replaced. The flowers are collected every day, as they are very short-lived and their stigmas lose their color and aroma with exposure to air and light. The stigmas are extracted from the flowers the same day and then dried; and the skill with which this is done influences their quality.

Separating the stamens from the flowers [Sebastián Bellón]

Sorting saffron, Castile [AGE Fotostock]

Throughout history saffron has been used for therapeutic and magic purposes, as a perfume, as an aphrodisiac, and above all as a spice. The oldest reference to the use of saffron is as a medicinal plant. Although the plant is not native to Egypt the Ebers papyruses contain more than 30 ways of using it. In the year 3000 B.P. it was used to dye the cloths used to wrap embalmed corpses. Later, Cleopatra used it in her make-up. The Greeks, who represented Zeus as sleeping on a bed of saffron, considered its aroma to be sensual. The Romans used the flowers to prepare an alcoholic tincture used to perfume the entrances to public places. Both in Greece and in Rome it was used to dye the clothes of newlyweds. In India married women still paint their arms and breast with saffron to show they are "real women." The Romans also appreciated another of its virtues, and sought to combat drunkenness by taking an infusion of saffron before drinking alcohol.

Flowers of saffron (*Crocus sativus*), Castile [Sebastián Bellón]

Saffron is the world's most expensive spice. Its very high price is no doubt due to the very low yields obtained from saffron cultivation. A single kilogram of saffron requires 5 kg of stigmas, obtained 300 kg of flowers, or about 150,000 flowers. This explains why until recently saffron cost more than gold. This high value has meant that since antiquity people have fraudulently tried to pass off a wide range of things as saffron. Pliny the Elder (1st century A.D.) said "nothing is more often faked than saffron."

Saffron's color and aroma really come into their own in the kitchen, where it is widely used in the cookery of Europe and the rest of the world. Saffron is widely used in Mediterranean cookery and in many Asian recipes. It can be used to flavor soups, eggs, fish, mollusks, game, birds, rice, as well as sweets. Among the best-known dishes using saffron are: in India, the sauce accompanying some curries; Iranian and Hindi yoghurt sauces; the Middle Eastern shish kebab and the Lebanese plaited bread or the Sabbath challah which goes back to Roman times; the special cake with which Jews celebrate Yom Kippur; the Moroccan "tagine," the Mediterranean rice dishes, including Italy's risotto, Africa's rice dishes, the paella from Valencia and others from Provence; in Marseilles and Provence people also ate young salmon (smolt) with saffron cooked in a copper vessel; saffron is used to make jams (the preparation of which must begin on Saint John's night); saffron is also used to prepare an aphrodisiac wine which requires more than a gram of the precious stigmas for three quarters of a liter.

Saffron [Ernest Costa]

In the Middle Ages, saffron was held to be a genuine heal-all. In small doses, it was used as an antispasmodic, tranquilizer, and digestive stimulant, while it was used in high doses to induce abortion. According to Roger Bacon it delayed aging. The papal court was also concerned about aging and not only replaced the gold of the priest's robes with saffron but also included saffron in many of the dishes they ate. In the 19th century it was still used as a cosmetic and hair dye, and in cabinet-making varnishes. Nineteenth century microscopists used saffron to stain preparations, and it is still used for this purpose. It is still used in the pharmaceutical industry, but only as a coloring agent, although it contains some interesting active principles.

3. The use of animal resources

3.1 Bloodless collection

A large part of the animal protein consumed by the inhabitants of the world's Mediterraneans comes directly from the wild, from non-domesticated species. This is true of the marine fauna—almost all the fish consumed comes from the sea—and of some terrestrial animal species or products, and also of the aquatic fauna of the land. A first step in this direction is the bloodless collection of animals or their products.

191 A female honeybee (*Apis mellifera*) flying above the inflorescence of a *Callistemon* in eastern Australia. On its rear legs are the yellow pollen masses she has gathered, and after regurgitation, the nectar sucked from the flowers will form the honey. That honey has long been used as a foodstuff is shown by the many cave paintings showing collection from primitive beehives. During antiquity, the Middle Ages and much of the modern era, honey was the only substance with which to sweeten food or drink, until the appearance of products refined from sugar beet and cane sugar. When refined sugar reached Europe, honey largely disappeared from kitchens. Honey is a highly concentrated solution of glucose and fructose, with small quantities of sucrose, maltose and other minor components, such as mineral salts, organic acids, proteins, amino acids, vitamins, and enzymes.
[Photo: Reg Morrison / Auscape International]

Beekeeping

Beekeeping was initially a predatory activity and has been practiced since prehistoric times. The oldest image of this activity is probably the cave painting in the La Araña Cave in the Autonomous Community of Valencia, in the east of the Iberian Peninsula, which shows that honey and wax were collected about 9,000 years ago. There are historical records, for example, in hieroglyphics from lower Egypt that are 7,500 years old and it is known that Hittite laws punished those who stole beehives. The bee had already been domesticated by around 4400 B.P. in Greece, as is shown by archeological finds of beekeeping installations dating from this period. In Thebes (Egypt) and in Greece clay objects have been found that served as beehives. There are also references to honey and beehives made of straw, or of plaited canes (often covered with cow dung mixed with clay and dried in the sun), or the cork of the cork oak in the western Mediterranean, before Roman domination. In 27 B.C. the Latin scholar and writer Marcus Terentius Varro, in his work *De re rustica*, described a beehive that was the ancestor of the movable-frame hive.

Bees and beehives

Various hymenopteran insects of the bee family, the Apidae, produce honey, but only the honey bee (*Apis mellifera*) produces enough for it be collected. The bee is an insect from the tropics that cannot live in the nordic countries, where the winters are too cold and long, but it has adapted very well to the much milder climate of the Mediterranean regions. The races of bees in the Mediterranean area are the result of natural selection without human intervention, unlike truly domesticated animals. In each Mediterranean region natural selection has resulted in a different race of bee, adapted to the types of flowering plants that occur there and their flowering pattern. There are four main subspecies or races of the European bee, the black, the Italian, the Carnolian, and the Caucasian. The black races are smaller, dark in color and good honey producers although very aggressive, and they have been introduced into various parts of America. Italian bees have a high reproductive capacity, and they are less aggressive, and good producers of honey and wax, and

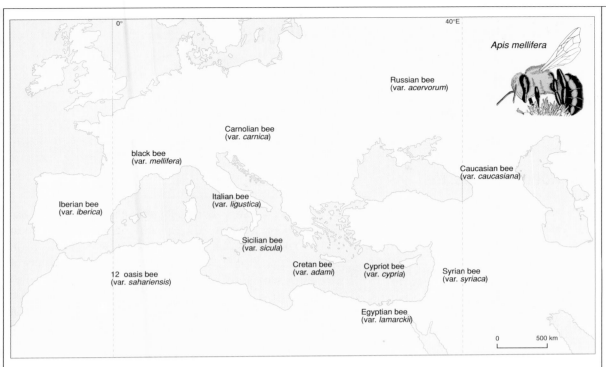

192 The distribution of the varieties or breeds of bee (*Apis mellifera*) in the Mediterranean Basin. The bee is from the tropics, but it has adapted very well to the climate and vegetation of the Mediterranean areas. The bee has since evolved into different races adapted to each area's flowering plants and their flowering season. Each race has different uses: the black bee is kept for its honey, the Italian for its honey and wax, and the Caucasian bee for its good production of propolis.
[Drawing: Editrònica, on the basis of data collected by the author]

have been exported to North America, China, and many other places. The Carnolian ees are good honey producers, but not such good wax producers. The Caucasian bees are not good honey producers, but they produce propolis, a resinous substance that they collect from the buds of some trees and use as cement to repair the hive.

Until the mid 19th century the only beehives used were the type known as fixed, which were inconvenient to handle and which had to be broken when the honey was collected. The beehives now in use are known as movable-frame hives and consist of a series of movable compartments arranged on top of each other, and they are easier to handle. They were originally made of wood, but may now be made of aluminum, cement, glass, glass fiber and other materials, and they make it possible to watch the bees, their development, the sanitary conditions within the hive, and the process of honey-making. The hive consists of different overlapping series of parallel frames (normally in sets of 10, 12, or 14, but there may be as many as 18 or 24 or more) where the wax combs are placed, which can be removed at will. To remove the honey the frames are removed from the hive, fixed to the central axis of an extractor and then spun. The centrifugal effect forces the liquid honey from the wax cells of the honey comb without breaking them, and they can be replaced in the beehive.

Beekeeping production
The traditional main products of the bee were its wax, an important source of energy for lighting, and to a lesser extent, the honey which was collected a sweetener. Honey production is now more important than wax, and as a result of studies of the bee's biology, other products are now collected, such as pollen, royal jelly, propolis, and venom.

The main European honey-producing countries are the Mediterranean countries, except for Portugal, while the other European countries import about 80% of the honey they consume. The most important product collected from Mediterranean bees is their honey. The region's varied flora includes many members of the thyme family, and, together with the favorable climate (dry periods and little rain), this gives the honey an exceptional flavor that is greatly appreciated, as is the pollen obtained which is also of very high quality. Yet the southern shores of the Mediterranean Basin do not produce as much honey, pollen and royal jelly as the northern shores or the tropical and sub-tropical regions of America and Asia.

Honey
The quality of single-flower honey depends mainly on the climate, the rainfall regime, cultivation methods and, above all, the installations and procedures followed (the type of beehive used, the preparation of the apiary, the classification of the honey during extraction, etc.). The honey collected immediately after a period of abundant flower production maintains the flavor and smell of the flower whose nectar the bees have taken. Not all plants flower at the same time, and so with adequate control it is possible to obtain as many as five different types of honey. The honeys that

193 A beekeeper extracting the honeycomb. The beekeeper manages the bees using protective equipment, such as a veil or mask, bee repellants, a fumigator to make them drowsy and other accessories to keep them off the beekeeper when collecting the honey. Movable hives, the basis of modern beekeeping, were invented in the 18th and 19th centuries and greatly improved beekeeping. It became possible to make transparent beehives to observe the bees and manipulate the colony without disturbing it.
[Photo: Sandra Prato / Bruce Coleman Limited]

are produced in the Mediterranean Basin vary greatly. The honey produced from rosemary (*Rosmarinus officinalis*) and thyme (*Thymus vulgaris*) are typical of lowland areas, and taste and smell excellent. The honey of fruit trees, such as the orange and the apple have a very delicate flavor. Areas where brassicas are extensively cultivated yield a large quantity of an insipid honey useful as a sweetener. The honey of the French lavender (*Lavandula stoechas*) is collected in sites where there are acid-loving matorrals, and is of high quality, very sweet and a magnificent golden color. Heaths on acid soils produce heather (*Erica*) honey, sold either as a single-flower honey or mixed with French lavender honey, which is dark red, slightly bitter and has an intense aroma. The robinia, or false acacia (*Robinia pseudoacacia*) produces a large amount of a very pale, translucent, white honey that is improved by blending with the honey of the labiates that flower at the same time. The honey of the lime (*Tilia platyphyllos*) is a slightly greenish grey, not very dense and has a very characteristic aroma and taste. The honey of the alfalfa (*Medicago sativa*) is highly valued, but is only obtained from the flowering of the last harvest. The honeys known as wildflower, meadow flower or mountain flower, depending on where it is collected, are also of high quality. Honey is also harvested in the mountainous Mediterranean areas where firs and other conifers grow, and although is of high quality, it is not obtained every year and is not of commercial importance. The honey of the sunflower (*Helianthus annuus*) is an intense yellow color and pleasant tasting, and is exploited in large quantities in

the areas where the plant is cultivated, although it is not considered very good unless blended with the mountain wildflower honey or that of labiates. Another type of honey with an exceptional flavor and aroma is lavender (*Lavandula angustifolia*) honey, either pure or mixed with other types of honey. Other types of honey include savory (*Satureja montana*), high quality and very smooth, that of ling (*Calluna vulgaris*), which has a very pleasant aroma and flavor, that of the holm oak (*Quercus ilex*), and other oak species, which are dark in color and highly astringent.

At the end of autumn it is possible to obtain honey from the strawberry tree (*Arbutus unedo*), which is bitter. Eucalyptus plantations produce a lot of honey, although different species produce honeys of different qualities. This is true in Australia (not just in the Mediterranean regions but in a large part of the continent), the country of origin of the eucalypts and now one of the world's major honey producers.

Other bee-keeping products

Pollen is mainly collected in the months of June and July, the flowering period of the cistus (*Cistus*) and other species, which the bees visit almost exclusively for their pollen. Spain is the largest pollen producer in the Mediterranean Basin.

Royal jelly is another substance produced by the bees who use it to feed the queen and the larvae, but commercial production is only worthwhile if there is a constant abundance of flowers. Except for the

Balkan countries, royal jelly production in the Mediterranean Basin is insignificant.

Propolis is a substance produced by the bees from resins and balsams of certain trees and shrubs. It has been used since antiquity, but it has only begun to be studied and used in medicine in the last 30 years. Its best-studied effect is the benefit it confers on the the hive since it is an excellent bactericidal, bacteriostatic, and antiseptic agent. Current studies may reveal further beneficial qualities. The Mediterranean countries are suitable for production of propolis, both in terms of quantity and quality.

Beeswax, the wax produced by bees, is extracted from the old combs but is now almost entirely returned to the hive, saving the bees the work of manufacturing it. Only a small part of the production goes to cosmetic and pharmaceutical laboratories. It was formerly used industrially to make candles.

Bee venom is used to treat some illnesses, such as rheumatism, arthritis, and arthrosis, but it is used most in the prevention of allergy to bee stings in hypersensitive persons. Its properties and potential uses have still not been thoroughly studied, so there is no demand to justify large production.

Gathering activities

Gathering activities include collecting animals and their products. Some livestock operations, such as wool collection, milking, and egg production are bloodless forms of gathering. Several types of bloodless collection may however lead to a relatively cruel end (snails are collected nondestructively, but are killed by cooking and then eaten). The horns of deer are used as ornaments or as tools. Bird feathers (sometimes after killing the animal, but often after the plumage is shed during moulting), are also used for decoration, while their clutches of eggs may be used as food.

Sea-bird colonies have long been exploited as a food resource, especially those of the gulls (*Larus cachinnans*, *L. argentatus*, *L. audouinii*, *L. genei*) of the Mediterranean coastline and many islands. This exploitation has traditionally contributed to controlling the birds' population, and its cessation is one of the causes for the increase in some gull populations over the last few decades. The eggs of some species of fish are also collected, and in some places they are eaten as aperitifs or as food-

stuffs. This type of exploitation also includes the capture of animals (mainly passerine birds and birds of prey) as pets or for falconry.

The most spectacular and economically important cases of bloodless collection are of invertebrates, such as snails (the genera *Helix*, *Iberus*, *Otala* and *Eobania* among others), which are highly appreciated in some parts of the Mediterranean Basin, but especially bees, for their production of honey, wax, and other products.

3.2 Hunting activity

It is clear that the first humans that lived in the Mediterranean Basin were basically hunters or hunter-gatherers. Even so their impact on the fauna was generally slight and they played an eminently regulatory role as a predator. During the Paleolithic, in Mediterranean ecosystems and elsewhere, the most important irreversible effect of humans on the fauna was the extinction of the animals in competitions with humans (for shelter, food resources, mutual predation, etc.).

Until recently, humans have continued to get rid of the species that caused them serious problems, mainly the larger predators. None of these animals had been, however, especially abundant in the Mediterraneans, and this, together with the relatively low population density of humans, their inefficient hunting techniques and the limited economic use of the carcasses, merely meant they were progressively marginalized to the less populated areas, generally the most mountainous areas, at least in the Mediterranean Basin, California, and Chile.

Eating snails

Pseudotachea splendida [Marti Dominguez]

Terrestrial snails, like mushrooms, are controversial foodstuffs. Whether people eat snails or not is a basically cultural question. There are some peoples that greatly appreciate snails, such as the Chinese, many peoples in western Africa, Australian Aborigines, and the Bushmen. Other peoples rarely eat them, for example some Native American tribes, the Arabs of North Africa, and most peoples of central and northern Europe. Gastronomic opinions on the different species of snail available vary greatly from culture to culture.

Snails cooking a la losa [Jordi Gumi / Firo Foto]

Among the most dedicated snail-eaters in the world are the peoples of the Mediterranean Basin, at least those living on its northern shores (and the Berbers of the Mahgreb), who eat them with great enjoyment, cooking them in a variety of ways and often make a party of eating a large quantity of these gastropods. These open air collective festivities are held, at the slightest excuse, by more-or-less numerous groups (a family, a group of friends or an entire village), and are a deeply-rooted custom throughout the Mediterranean. What is actually eaten varies greatly, but towards the end of the summer, especially in the second half of August, many snail-parties are held in countries on the Mediterranean coastline, especially Catalonia and Languedoc.

Snails on plant stem [Jordi Bartolomé]

Unlike other collective meals, such as those for carnival, the eating of sardines at the beginning of Lent or chestnuts on All Saints day, snail-feasts lack a fixed date and depend more on a local summer donwpours falling from convective clouds, and these are more frequent in particular from mid-August onwards. A true snail-feast begins just a few hours after the storm, when the snails that had been dormant since the beginning of summer wake up and start moving around the fields in search of food, the best moment to collect them. They are starved in dry conditions for a few days inside a snail basket (an elongated basket with a relatively narrow mouth, covered by a rag or metallic cloth instead) and the snails lose all that they had eaten as feces, which are got rid of, and return to their lethargy; this is the moment to organize the snail-feast. Any piece of wasteland, near a spring, or a hermitage, with a roof or the hut on a vineyard will do. All that is needed is to invite a few friends and to decide how to cook the snails.

Eobania vermiculata [Martí Domínguez]

Normally, the way the snails are cooked is a rudimentary affair. The snails are soaked (to make them come out of their shell, rather than to clean them). They are then salted abundantly to get rid of the slime and they are cooked *a la brutesca*— burying them in a pile of straw that is then lit, and the snails then have to be fished out of the cooled ashes. Or they may be cooked *a la lata* or *a la losa*—placing them with the shell opening facing upwards on a metal surface or a flat brick, heated from below by a fire of vine prunings. (See picture.) In both cases, it is essential to accompany the snails with an *aïoli* or *alioli* (garlic mayonnaise), and abundant wine. This is not to say all snail dishes are so simple, as some involve cooking the snails in sauces with many spices, especially paprika, and apart from *alioli* can have all sorts of accompaniments.

Helix aspera [Marti Dominguez]

195 The hunting of birds in wetlands, such as these black-tailed godwits (*Limosa limosa*) shot in the Ebro delta (Catalonia), usually takes place in the winter period, when these groups of migrants from the cold countries are concentrated in Mediterranean wetlands. Firearms are normally used, together with dogs to recover the dead birds. The impossibility of identifying a bird in flight, or the bad faith of some hunters, means that many legally protected species fall victims to the hunters' guns. For example, since 1981 it has been illegal to hunt black-tailed godwits anywhere on Spanish territory.
[Photo: Xavier Ferrer]

Small game: rabbits and hares, vermin predators and partridges

After the Neolithic, the early agriculture and stock-raising based on sheep and goats created a socioeconomic model in the eastern Mediterranean that later spread to the rest of the Basin. The deforestation and the increasingly artificial ecosystems favored the proliferation of what is known as "small game:" rabbits (*Oryctolagus cuniculus*), hares (*Lepus capensis*, *L. europaeus*), squirrels (*Sciurus vulgaris*, *Atlantoxerus getulus*), red-legged partridges (*Alectoris rufa*), rock partridges (*A. graeca*), Barbary partridges (*A. barbara*), chukars (*A. chukar*) and grey partridges (*Perdix perdix*), quails (*Coturnix coturnix*), pheasants (*Phasianus colchicus*), wood pigeons (*Columba palumbus*), rock doves (*C. livia*), stock doves (*C. oenas*), turtle doves (*Streptopelia turtur*), collared doves (*S. decaocto*), Barbary doves (*S. risoria*) and laughing doves (*S. senegalensis*), pintailed sandgrouses (*Pterocles alchata*), black-bellied sandgrouses (*P. orientalis*), stone curlews (*Burhinus oedicnemus*), little bustards (*Tetrax* [=*Otis*] *tetrax*), and Houbara bustards (*Chlamydotis undulata*). These species, whether flesh or fowl, are all relatively abundant medium-sized animals. They thus represent large stores of material (biomass), protein, and energy that are relatively easily available to a top predator like the human being. Even today, fewer than 30 species receive the attentions of most hunters in the Mediterranean.

The rabbit and the hare

The rabbit (*Oryctalagus cuniculus*) is the most typical vertebrate of the Iberian Peninsula's Mediterranean ecosystems and those of southern France, the areas where it originated. It is now widespread, not only in the world's other mediterraneans (such as the rest of the Mediterranean Basin, Chile and Australia), but also in much of the rest of Europe and areas of other continents (especially non-Mediterranean Australia). This lagomorph has since antiquity formed part of the diet of the inhabitants of the western Mediterranean, as shown by the remains in archeological finds. This is not surprising as it was the most abundant animal and easy to catch. It can attain considerable population densities, greater than 5,000 individuals/km^2, representing about 5 kg/ha (1 hectare=2,47 acres) of rabbit at any given moment (annual production may be considerably higher as a result of the effect of reproduction). Apparently the presence of large rabbit warrens, as well as water, was a factor influencing the siting of human settlements.

When the basic needs of human populations had been covered by agricultural and livestock activities, rabbit hunting often changed from hunting for food to hunting for recreation, although this did not exclude enjoying a tasty additional foodstuff. This recreational aspect has undoubtedly helped to make hunting rabbits so popular, although rabbit populations in the Mediterranean Basin have declined due to the introduction of the myxomatosis virus.

The story of the rabbit can be interpreted in other ways, however. The development of agriculture and its intrusion into the rabbit's territory, often respecting the natural divisions of the vegetation, led to the appearance of a conflict between rabbits and agriculturalists, as the rabbits ate the herbaceous harvests (cereals, etc.) of the former. The result of this was that rabbits got a bad name and were relentlessly pursued by all means available (ferrets, fire, smoking out the warrens, etc.). As a result, rabbit-hunting has now become so deeply rooted culturally, and it is often unclear if it is for food, for fun, or to eliminate a competitor. The outlook is even more complex if it is borne in mind that this species can now be reared as livestock.

196 A desert cottontail (*Sylvilagus audubonii*) in the shade of a xerophyllous plant in California. The different species of rabbits and hares are frequent in all the mediterraneans. They are found in thickets with low scrub and clearings.
[Photo: Jeff Foott / Bruce Coleman Limited]

In Australia, and thus its Mediterranean ecosystems, the rabbit caused one of the most important ecological disruptions ever caused by human intervention. Starting from a few individuals, they spread throughout the entire territory, thanks to the lack of natural predators. It is worth pointing out that the rabbits that were introduced were not wild-type individuals, but domesticated ones, and this is why their markings vary so much. Yet they show some ecological differences, for example they climb trees (in places where there are no herbs or shrubs) and they show no fear of humans. In a short time this animal became the main problem of agriculturalists and administrators who were unable to control it effectively.

Unfortunately in the search for methods to control the rabbit, pathogenic agents have been tried out, particularly viruses. Myxomatosis, for example, was introduced to Australia (where rabbits now tend to be resistant) and it also caused a dramatic reduction of the number of rabbits in Europe during the 1950s. The rabbit has still not recovered its former population levels over much of the European continent, in spite of local recoveries. The situation was worsened when a new viral infection (viral haemorrhagic fever) which has been superimposed itself on myxomatosis since 1989.

The hares (*Lepus capensis*, *L. europaeus*), in the Mediterranean regions of the Old World, are species that are complementary to the rabbit, with similar but not identical habitats. In California there are also hares and cottontails of the genus *Lepus* and *Sylvilagus*. The flesh of hares is also valued, although hunting hares is different from hunting rabbits, as it generally requires the use of

dogs. The dogs are pointers, although the Iberian hare typical of wasteground is usually hunted with greyhounds.

Predators

Predator species are very abundant in Mediterranean ecosystems although, as has been seen, they often rely on a small number of prey species. Due to lack of information about the biology of these species, they were almost always considered enemies of human beings, because of the damage they cause to agriculture or cattle-raising; in many cases totally imaginary supernatural powers were attributed to them. They are also often hunted for sport (and often represented in coats of arms: lions, bears, wolves, eagles, etc.) or to use their fur for coats. Yet the few species with high-quality furs—the otter, (*Lutra lutra*), pine marten (*Martes martes*), and stoat or ermine (*Mustela erminea*)—have been gradually replaced by more northerly species and subspecies as trade has increased. Cold wet conditions favor high-quality furs, and this is why Mediterranean ecosystems are not very suitable for the commercial use of the furs of their wild animals.

Thus, the persecution of Mediterranean predators has been based on their presumed status as competitors of humans, and because they are despised for deeply rooted cultural reasons. It has not only affected the larger species but also the smaller ones, both in the open countryside and in some species' incursions into farms. Trapping methods have developed from stones and the use of branches in pit traps to snares, traps, dogs, and firearms. Unfortunately the spread of poison and the gassing of rabbit warrens was common in the 19th and 20th centuries. Intended to eliminate rabies

197 In the Californian mediterranean, the coyote (*Canis latrans*) is the great trophic competitor of the fox, to such an extent that is the king of the scavengers. Its area of distribution, however, spreads through all the herbaceous and sub-desert formations in North America. [Photo: Erwin & Peggy Bauer / Bruce Coleman Limited]

and birds of prey, it was at its worst from the 1940s to the 1960s. It was counterproductive, because this onslaught against predator populations has indirectly favored precisely the anthropophilous and generalist species that do most damage to hunting and are most active in the transmission of certain illnesses and parasites.

The fox (*Vulpes vulpes*) is an important predator in the Mediterranean Basin and the rest of the Holarctic region, as well as Australia (where it was introduced). It is very abundant in Mediterranean ecosystems, especially as a result of modifications of the environment and the reduction of the predators that control its populations (wolves, lynxes, eagles, owls). In California, the coyote (*Canis latrans*) occupies the place of the fox, which is relegated to less favorable environments, where it does not reach the high population densities that it might elsewhere. Both the fox and the coyote have a large impact on small game when favorable circumstances lead to high population densities (possibly exceeding 10 adults/km^2). This occurs mainly in highly humanized areas, where foxes can easily find a lot of food (rubbish bins, waste tips, farm waste, etc.), and this disproportionate population growth is due more to human causes than to the foxes or coyotes.

Fortunately, since the end of the 1970s, and especially in the 1980s, this situation has changed significantly, although not entirely. Thus in western Europe (but not in the Balkans, Asia Minor or North Africa) relatively effective protection has been introduced for birds of prey (diurnal and nocturnal) and for most carnivores (brown bear, lynx, wild cat, mongoose, otter, European mink). Furthermore, the fall in fur prices, mainly due to market saturation by farm-produced furs, has eliminated many of the economic reasons for persecuting species, such as the pine marten (*Martes*), weasels and stoats or ermine (*Mustela*), and genets (*Genetta genetta*). Now the only specimens caught, especially of the pine marten, fox, and badger (*Meles meles*), are the ones that damage hunting or agricultural interests or, in the case of pine martins, beehives.

Partridges

There are five species of partridges in the Mediterranean region between Europe, Asia Minor, and North Africa. Four species belong to the genus *Alectoris* and a single species to the genus *Perdix* (only found in peninsular Italy) and they form the most characteristic group of small game birds. They are typical of open spaces, although their populations depend to a large extent on the abundance of ecotones, such as the edges of fields, bramble patches, various crops, fallow land, etc. This is precisely why modern farming practices do not favor partridges, mainly because individual fields and plots are too big, a highly prejudicial factor. There are, however, other factors, such as the use in the countryside of agricultural chemicals that have led to the proliferation of dis-

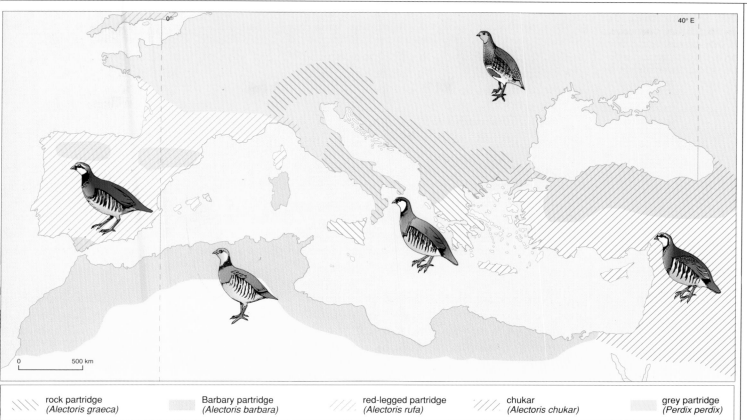

| rock partridge (Alectoris graeca) | Barbary partridge (Alectoris barbara) | red-legged partridge (Alectoris rufa) | chukar (Alectoris chukar) | grey partridge (Perdix perdix) |

ease/illnesses, parasitic infections, and reduced reproductive efficiency. Furthermore, breeding in farms for hunting purposes and the constant hybridization between the four species, or with domesticated forms, is seriously altering their genetic intergrity and weakening their resistance to illness and life in the wild. Even so, these species continue to be a favorite quarry and are hunted both individually and in groups (the celebrated beats). The density of partridge populations varies greatly, but it is not unusual to reach springtime densities of more than 8 brace per acre (20 brace [pairs] per hectare), which represents 40-120 individuals per acre (100-300 individuals/ha) at the beginning of the shooting season. In spite of everything, densities are usually 1-8 pairs per acre (2 and 20 pairs/ha).

Waterfowl

In 1986 a census of the Mediterranean Basin (including North Africa and the Black Sea), more than 11 million ducks of the genus *Anas*, 2 million ducks of the genus *Aythya*, more than 250,000 geese (genera *Anser*), and 2.5 million coots

198 Distribution of Mediterranean partridges, and the large number of red-legged partridges (Alectoris rufa) killed in a single shoot in Toledo (Castilia, in the middle of Spain). In some estates in the Iberian Peninsula the Greek partridge (*A. graeca*) has also been introduced. *[Drawing and cartography: Editrònica, from several sources; photo: AGE Fotostock]*

(*Fulica atra*), among many other waterfowl, were counted. These data refers to winter censuses in wetlands. Since they do not include every wetland and watercourse, the real figures are surely higher, especially the numbers of mallards (*Anas platyrhynchos*). This duck is the main species shot in the Mediterranean, as the hunting season coincides with the arrival of winter visitors. Until recently all species of waterfowl were hunted, including herons, waders, and others. Today, fortunately, in numerous countries the most endangered species are protected.

On the other hand, ducks (in the broad sense) and coots are the most common prey caught in wetlands. Their capture is made easier by their gregarious winter behavior—they may form groups of tens or even hundreds of thousands of individuals. Ducks and coots are usually hunted at night or at dusk when they move between their feeding and resting areas. Shoots (when hunters hide in the natural vegetation or in specially built blinds) are social and recreational events, like boar hunts. Outside wetlands, hunting waterfowl is mainly restricted to mallards and a few other species, such as the moorhen (*Gallinula chloropus*), which is generally hunted with the help of dogs or by lying in wait.

Big game

However, this model of transformation of the environment left some fragments of natural systems more or less intact. These included woodlands that were difficult to penetrate or excessively marginal, inaccessible mountainous areas, inhospitable areas (those with endemic diseases, such as many swampy areas, and places with unpleasant climates, such as islands, excessively windy hilltops or very dark, shady places), areas that presented difficulties for agriculture and stock-raising, such as steep slopes or stony soils, and especially areas selected for the enjoyment of kings and other powerful people.

Undoubtedly the feudal structure, in both the Christian and Muslim civilizations of the Mediterranean in the Middle Ages, favored the existence of these areas reserved for the powerful and their friends and family, where the fauna was partially protected by the mere prohibition of hunting. Apart from the powerful, only a few shrewd gamekeepers and poachers, sometimes at the risk of their life, presented a threat to the ani-

mals living there. These areas were known as big game reserves, i.e. reserved for the species especially appreciated for their large size, relative rarity, and difficulty to capture. Big game included the brown bear (*Ursus arctos*), wolf (*Canis lupus*), lion (*Panthera leo*), leopard (*P. pardus*), cheetah (*Acinonyx jubatus*), lynx (*Felis pardina* and *Felis caracal*), deer (such as the red deer [*Cervus elaphus*], fallow deer [*Dama dama*], and roe deer [*Capreolus capreolus*]), wild boar (*Sus scrofa*), mountain goats (*Capra*), aoudads or Barbary sheep (*Ammotragus lervia*), some bovids (like the gazelles [*Gazella*] and different antelopes [*Taurotragus, Tragelaphus*]), cranes (*Grus grus* and *G. anthropoides*), and the bustard (*Otis tarda*).

There are many surviving written, painted, or engraved references to hunts by the powerful, often accompanied by a large retinue. The hunting activities practiced by royal and nobles families have been responsible for the conservation to the present day of some of the best fragments of Mediterranean vegetation. In Spain, this is true of the Doñana Reserve in the southwest, the Coto del Rey (King's Reserve) in the southeast, and the Monte del Pardo in the center. Hunting in these areas was for big game, the larger animals that hide and are difficult to catch, and other more complex activities such as falconry (hunting using birds of prey). Falconry is deeply rooted and widespread in North Africa and Asia Minor. All of this has led to an elitist conception of big game hunting. It was considered a noble and elevated art practiced by the powerful and almost always forbidden to other social classes, while small game hunting was considered inferior, common, and plebeian. Many rulers, such as Alexander the Great, were famous for their love of hunting. In a single hunt Sancho VI of Navarre killed 22 red deer, 15 roe deer, 12 chamois, 16 wild boar, and 14 bears (almost as many as the number of wild boar!) and in 1345 Alfonso VI of Castile wrote a book, the *Libro de Montería* on hunting big game, especially bears.

The wild boar

In many texts of the 18th and 19th centuries and even the 20th, the wild boar was treated as a rare species, and in many cases comparable to those in danger of extinction, such as the bear or the wolf, but in the last four decades its population has expanded, recolonizing areas where it had not been known for many years. In addition to the undeniable role of socioeconomic changes in Mediter-

199 Boar hunts (*Sus scrofa*) normally take place in winter in the entire northern Mediterranean Basin, where the relative abandonment of forested areas has led to an increase in its numbers since the 1960s and 1970s. Boars are shot after being flushed and surrounded by packs of hounds. It was formerly common to hunt wild boar with a lance from a horse, as shown by this 15th century miniature belonging to the Grimany Breviary kept at the Biblioteca Marciana in Venice. *[Photo: Index]*

ranean forest usage in this expansion, it is still not known exactly what role hybridization with domesticated pigs may have played in many areas, considered possible even though the latter have a different number of chromosomes from the wild boar. In any case, the fact is that the litters of piglets are larger, 6-8 offspring, or even more, is common. As a result of this change, the density of wild boar in Mediterranean forests usually exceeds 10 boar/km². In the last few years, catches in several areas have reached 7 or 8 boar/km², implying the existence of much denser populations. This is even happening in Mediterranean islands like Corsica, where there are at least 20,000 to 30,000 wild boar (although their genetic purity is questionable).

The increase in wild boar populations has arrived at just the right time to channel the energy of the hunters frustrated by the sharp decline in rabbit numbers. The wild boar is considered big game and is mainly hunted by groups; this has become very popular in rural areas, largely due to the social activ-ities developed around the hunt (parties, meals, etc.). The wild boar is now usually hunted by beating, preferably with 10-30 hunters aided by a pack of hounds. Dogs pursue the wild boar, flushing it towards places the boar must pass, where the hunters lie in wait. About 75,000 wild boar are hunted every year in Spain, between 75,000 and 100,000 on mainland France and another 10,000 in Corsica. Despite the high large number of individuals which ensure high population densities, they are in good health. However, the menace of African swine fever hangs over the wild boar like the sword of Damocles. It could lead to a sharp decline in wild boar populations in Mediterranean Europe, as happened to the rabbit after the release of myxomatosis and other viral diseases.

Falconry and ferreting

Falconry is the art of hunting with specially trained birds of prey. This activity became established in the Mediterranean countries in the Middle Ages, especially among the nobility. Many people of high rank in the north and south of the Mediterranean Basin possessed large numbers of falcons, and it is still practiced today. This activity developed in the Mediterranean and the surrounding regions for two reasons: the large diversity of birds of prey, especially those that hunt in open spaces, and the abundance of deforested open spaces suitable for falconry.

The species of birds of prey that give their name to falconry are the peregrine falcon (*Falco peregrinus*), lanner falcon (*F. biarmicus*), and the saker falcon (*F. cherrug*), which are the fastest birds alive (recorded in dives at speeds exceeding 300 km/h). Eagles were also once used in Europe and the western—the species used include eagles of the genera *Aquila* (golden eagle [*A. chrysaetos*], imperial eagle [*A. heliaca*], tawny eagle [*A. rapax*], spotted eagle [*A. clanga*], lesser spotted eagle [*A. pomarina*]) and *Hieraaetus* (Bonelli's eagle [*H. fasciatus*] and booted eagle [*H. pennatus*]). Especially in the northern part of the Mediterranean, forest species such as the goshawk (*Accipiter gentilis*) are also used. Nocturnal birds of prey such as the eagle-owl (*Bubo bubo*) were once used.

Another activity using an animal to hunt involves the hunting of rabbits with ferrets (*Mustela furo*), known as ferreting. The ferret is a domesticated animal, frequently albino, that has still not been definitively classified. It is generally considered to be a domesticated and selected form of the polecat (*Mustela putorius*). The polecat is a predator that in natural conditions mainly preys on amphibians and rabbits, and the ferret is thus suitable for hunting rabbits.

The ferret is tame and trusts humans. It is placed in rabbit warrens and it either directly kills some rabbits or causes them to flee the warren to be caught outside with sacks, dogs, or shot. Ferrets are trained with rabbits to ensure that they do in fact force the rabbits out of the warrens and do not remain in the warren eating their catch. Those rabbits that die are usually young rabbits, and so ferreting only makes sense if the rabbits are a pest (and damage crops), or in situations of overpopulation. This is why in most countries of Mediterranean Europe ferreting is now controlled and regulated, and only authorized in special cases.

The capture of passerine birds

As already mentioned, Mediterranean countries are located in middle latitudes between northern and tropical or sub-tropical systems and are, therefore, in a privileged position in terms of populations of small passerine birds. This situation encourages a great diversity of migratory species to stop over and, to a lesser extent, winter in these countries. In many species, overwintering individuals join groups of reproductive birds (which are sedentary). In spring, autumn, and winter, many species form large groups, and some, such as starlings, form huge flocks (murmurations) of hundreds of thousands, or even millions, of individuals. As a result, they may be hunted on a large scale, especially in Italy, where their flesh is appreciated. Other passerine birds form very large groups, for example house sparrows (*Passer domesticus*), Spanish sparrows (*P. hispaniolensis*) and tree sparrows (*P. montanus*); in city centers, they may become troublesome in houses and gardens. Smaller groups, consisting of at most a few hundred or thousand birds, are more normal; these groups may consist of fringillids (finches, etc.) and emberizids (buntings, etc.), which form mixed groups containing several different species. The most important members of these two families are the chaffinch (*Fringilla coelebs*), brambling (*Fringilla montifringilla*), goldfinch (*Carduelis carduelis*), siskin (*Carduelis spinus*), greenfinch (*Carduelis chloris*), corn bunting (*Miliaria calandra*), rock bunting (*Emberiza cia*), yellowhammer (*Emberiza citrinella*), cirl bunting (*Emberiza cirlus*), and black-headed bunting (*Emberiza melanocephala*). These birds are the most often captured species, being caught for a variety of purposes (pets, songbirds, food, etc.). However, although they are not the only ones, since other groups, such as the muscicapids (flycatchers) and sylviids (old world warblers), are also much sought after.

Due to birds' migration patterns, most hunting takes place along the eastern coast of the Iberian Peninsula, the Italian peninsula, and the migratory route running through Anatolia, Israel, and the Nile corridor. Diverse methods are used, and include traps (often spring-loaded traps), nets (both mist nets and collapsible nets), and liming. Liming, the spreading of a sticky substance on branches to trap birds, has long been traditional along the eastern coastline of the Iberian Peninsula (from southern Catalonia to northeast Andalusia). It is practiced from *barracas*, stone constructions built around a tree (normally an olive or a carob tree), where hunters wait with bird whistles (generally imitating thrushes and finches) to attract birds of these species to settle on branches prepared as traps. These naked branches are covered in bird lime and jut out from the *barracas* so that birds stick to them. Many *barracas* are centuries old and

200 A passerine bird caught in a spring trap, in this case a robin (*Erithacus rubecula*), a small member of the turdids that is of no hunting interest. These devices capture and kill small birds indiscriminately, and are thus a shameful method of hunting. [Photo: Javier Andrada]

3. HUMANS IN SCLEROPHYLLOUS FORMATIONS

are handed down from father to son. Cassette recordings of bird calls are now used to attract birds and the number of birds killed may be in the millions. In Spain alone, it has been estimated that more than 25 million birds (not just passerine birds) were killed every year between 1970 and 1980 by these techniques. This activity has a despicable effect on the bird population and is now often banned, although it is still regularly permitted in some regions where it is a tradition. In the north of Africa, taking advantage of migration corridors, the capture of passerine birds is still widespread (especially near the Strait of Gibraltar and in the eastern Mediterranean).

Apart from the species already mentioned and members of the crow family, the other passerine birds that are really considered as small game in the Mediterranean Basin are thrushes—blackbirds (*Turdus merula*), ring ouzels (*T. torquatus*), mistle thrushes (*T. viscivorus*), redwings (*T. iliacus*), song thrushes (*T. philomelos*), and fieldfares (*T. pilaris*). In many cases, especially on the coast, thrushes are so common that they are blamed for damaging agriculture (especially the olive and almond crops), and this is why they are hunted. According to folklore, these birds can transport olives by impaling them on their beaks, by carrying them under their wings or between their feet, and when they are numerous, if a gun is fired, they drop the fruit and take wing. But the truth is that these alleged abilities to transport olives and release a hail of olives are false, although these birds do damage crops when they are present in large numbers.

Hunting in the overseas mediterraneans

In the other mediterraneans the situation is very different, as already mentioned in the sections on small and big game. It should be pointed out that the type of hunting has been directly determined by the type of legal system in operation in the colonizing country. This was Roman Law in Chile and in California (until 1848) and Anglo-Saxon common law in South Africa, Australia, and California (after it joined the United States).

In the Chilean mediterranean, the low population density and the limited number of species worth hunting has prevented problems similar to those of the European Mediterranean appearing. Rather to the contrary, the growing population of wild rabbits (descended from domesticated rabbits introduced in the mid 19th century) created problems for livestock beginning in the 1940s until myxomatosis virus was introduced in 1953 to control their numbers. Yet the carnivores have declined, including the puma (*Felis concolor*), kodkod (*F. guigna*), and Geoffroy's cat (*F. geoffroyi*), which is highly valued for its fur. Herbivores such as the guanaco (*Lama guanicoe*) and the Andean deer or huemul (*Hippocamelus bisculus*) are now restricted to some more southerly locations.

The Cape region was originally exploited by Bushmen, some of whose descendants still survive in the deserts and sub-deserts of southern Africa. The large population increase over the last few centuries has greatly affected the species hunted, especially big game, now almost totally restricted to the national parks. The most important species that are hunted are the Cape hunting dog (*Lycaon pictus*), spotted hyena (*Crocuta crocuta*), serval (*Felis serval*), lion, cheetah, caracal (*F. caracal*), African black rhinoceros (*Diceros bicornis*), bontebok (*Damaliscus dorcas*), a typically Mediterranean species, white-tailed gnu or black wildebeest (*Connochaetes gnou*), and the sable antelope (*Hippotragus leucophaeus*), also a strictly mediterranean species. Within the African context, it is the large fauna communities of the Cape mediterranean ecosystems that have been most altered.

201 Male Geoffrey's cat (*Felis geoffreyi*), a wild south American cat from the Chilean mediterranean. It is highly prized for its skin, and is thus subject to excessive hunting pressure, like other members of the cat family. Every year between 10,000 and 20,000 cats of this species are killed in Bolivia and Patagonia, its area of distribution.
[Photo: Rod Williams / Bruce Coleman Limited]

In California, rapid population growth is even more recent, as are the changes with respect to the hunting activities of the original Amerindian inhabitants. Yet increasing population density and farming intensity have marginalized the large fauna and some animals have disappeared, such as the wolf (*Canis lupus*), brown bear (*Ursus arctos*), swift fox (*Vulpes velox*), pronghorn (*Antilocapra americana*), wapiti or red deer (*Cervus elaphus*), the bald eagle (*Haliaeetus leucocephalus*), and the Californian condor (*Gymnogyps californianus*). But some important animals are still relatively abundant, such as the puma (*Felis concolor*), bobcat (*Felis rufus*), and deer (*Odocoileus virginianus* and *O. hemionus*).

In southern Australia the white settlers have greatly modified the landscape. Except for the bird species similar to those in Europe, there is hardly a single species of interest for hunting. The native mammals are almost all marsupials, and are not greatly appreciated by hunters. Perhaps it is worth pointing out the hunting of the dingo (*Canis familiaris dingo*) and kangaroos, the most exploited of which is the red kangaroo (*Macropus rufus*). In

202 Hunting the red kangaroo (*Macropus rufus*) is habitual in the Australian mediterranean, especially in the western area. It is practiced on a massive scale, both to reduce their pressure on crops and for recreational purposes and even for profit. The red kangaroo's skin is sold and high-quality leather goods use the male's scrotum, one of the more unusual wildlife products.
[Photo: Jean-Paul Ferrero / Auscape International]

this region it is often difficult to distinguish between the hunting of animals for recreational purposes and the hunting of animals considered agricultural and livestock pests. This lack of species worth hunting has led to the introduction of many foreign species for hunting. They include the rabbit (*Oryctalagus cuniculus*), fallow deer (*Dama dama*), red deer (*Cervus elephas*), sika or Japanese deer (*C. nippon*), sambar deer (*C. unicolor*) from southeast Asia, hog deer (*Axis porcinus*) which is also from Asia, European hare (*Lepus europaeus*), and fox (*Vulpes vulpes*).

3.3 River fishing

A single family of fish dominates the Mediterranean rivers of the Palaeoarctic, the cyprinids. It includes many species adapted to all conditions imaginable and forms the most important part of the river biomass. Mediterranean river ecosystems are not at all homogeneous and show great differences. Bearing in mind that droughts periodically affect the Mediterranean regions, the rivers are a factor of the greatest biotic interest and at the same time a type of habitat that is generally not very common.

The river regime and characteristics

On the Atlantic slopes of the Iberian peninsula, where there are many central-European-type areas, the average flow is 0.25 hm³/year/km². In the Mediterranean Basin, it is 0.16 hm³/ year/km², little more than half. If just strictly central-European-type habitats are considered, the difference is even greater, with a ratio of 4 to 1. In other words, Mediterranean water resources are scarcer and more concentrated than those of the neighboring temperate regions, with their higher rainfall and lack of summer drought.

Due to the irregularity of the rainfall regime there are also large and marked variations within the year and between years. This is why freshwater species, especially fish, have to be adapted to fluctuating conditions, with dry years and wet years, dry summers and wet springs, and autumns with storms and overflowing rivers. They also need to be adapted to the lack of water during certain periods, when they must take refuge in isolated pools where increasing temperatures and decreasing oxygen are a problem.

This is without taking into consideration human factors. Before the intensification of industrial activities, sewage, fertilizers, and livestock rearing had already intorduced large amounts of organic matter and some chemicals in the watercourses affected. This, together with the higher temperature, generated large fish production, normally associated with a tendency to water eutrophication. Some human activities now gravely affect watercourses, whether it be pollution (chemical contamination is especially negative), or the alteration of riverbanks or the river bed. This is extremely serious in Mediterranean rivers, where the lower stretches (and often the middle and upper stretches) have been dredged and channelled and their routes altered in a misguided attempt to protect people and their property from surges and floods. There is also the additional problem of the over-use of water resources, as the watersheds of the Mediterranean areas are often densely populated, and intensively farmed.

As an example, consider two Mediterranean watersheds corresponding to two rivers in the Iberian peninsula, the Segura and the Júcar. Their respective annual flow are 1,100 hm³ and 3,966 hm³. At the end of the 1980s, human demand for these resources amounted to 1,354 and 2,916 hm³, i.e. 23.1% more than the resources available in the case of the Segura and 73.5% of the available resources in the case of the Júcar. Official estimates of future demand for these watersheds is 1,977 and 4,806 hm³ i.e. 79.7% and 21.2% more than the resources available. Situations like this have led to the exhaustion of the watersheds, especially in the lower stretches, and most notably in rivers with low volumes of flow as they are now

204 River fish traps, not unlike the fykes used in the marine fisheries, are used to catch eels and other fish. They consist of hoops of iron or wood holding open a net with a mouth shaped like a funnel and they are normally set with the mouth facing the current.
[Photo: Miquel Monge]

dry almost the entire year, and therefore without fish. This situation has led to decrease in the rivers' self-cleaning capacity (as there is a lower quantity of solvent), which has been accompanied by an increase in the quantity of contaminants.

Fishing in marshes, pools, and lower watercourses

The lower stretches of rivers house the most productive Mediterranean environments, with the greatest abundance of organic material and chemical elements and compounds, together with higher temperatures; yet these zones are also the ones that have been most altered. Thus, the lower stretches of many rivers, streams, and torrents in Mediterranean Spain, France, and Italy are completely lifeless, especially in densely populated or industrial areas. They are sewers rather than rivers and do not contain a single living fish. This does not even include the cases already mentioned in which the lower stretches are completey dry.

However, where alterations have not occurred or where they have been moderate, productive environments may be found between salt water and freshwater that benefit from the ecotone effect favoring greater production. Furthermore, estuaries, marshes, deltas, and the very ends of the rivers contain the richest fish communities, since they have the fauna elements of sea water, freshwater, and estuaries. Here, in addition to the cyprinids represented especially by the genera *Barbus*, *Cyprinus*, *Gobio*, *Rutilus*, *Leuciscus*, and *Aulopyge*, there are other highly characteristic fish, such as the eel (*Anguilla anguilla*), lamprey

(*Petromyzon*), shad (*Alosa*), sand-smelt (*Atherina boyeri*), sturgeon (*Acipenser*), flatfish (such as the plaice [*Platichthys flesus*]), and members of the Mugilidae (mullets such as *Mugil*, *Liza*, and *Chelon*). These are the most typical fish of these environments, and they often replace the Cyprinidae on many Mediterranean islands, such as the Balearic Islands, Corsica, and Crete, where the groups of land fish are absent. However, members of the Salmonidae, such as the Adriatic salmon (*Salmothymus obtusirostris*), are found on Mediterranean islands.

In these ecosystems, fishing has always been a very important economic activity. It is worth noting, as an example, the abusive exploitation that has led to the virtual extinction of the common sturgeon (*Acipenser sturio*) and the Adriatic sturgeon (*A. naccarii*). Eels have also been adversely affected as a result of the proliferation of reservoirs, which interfere with their migration. This eliminates them from the middle and upper stretches of most rivers. In the lower stretches they have been eliminated as the result of excessive catches of the juvenile phases (elvers) for human consumption.

All these areas have developed a craft fishery that was first based on the use of trammel nets, river traps, fykes, and nets, and then changed to an approach based on taking advantage of the movements of fish, by means of sluices (such as that in the Laguna de la Encañizada in the Ebro delta) and other mechanisms that channel the fish during their movements towards very narrow straits. Since the 1970s, captures in these areas have decreased considerably. It is not uncommon now for catches to be

205 Angling is now a very common recreational activity in the mid-mountain Mediterranean areas of the Iberian Peninsula. Traditionally caught species are gradually being replaced by introduced cosmopolitan species, greatly reducing the original fish populations. The pike (*Esox lucius*) and the American perch (*Micropterus salmoides*) are the most troublesome, together with other members of the salmon family such as the rainbow trout (*Salmo gairdneri*) and the river trout (*Salvelinus fontinalis*). [Photo: Ernest Costa]

only 10-20% of those of 15 or 20 years ago, forcing many countries in western Europe to take administrative measures to slow this reduction down, and in some areas this has had positive results.

Fisheries in middle and upper stretches

The fisheries in lower stretches of rivers are mainly for economic purposes, but this is not true of the middle and upper stretches, where fishing has generally included a substantial recreational component. In the middle stretches of the Paleoarctic Mediterranean rivers, cyprinids are predominant (*Barbus, Chondrostoma, Leuciscus, Alburnus, Rutilus, Phoxinus, Gobio*, etc.). Other typical species include cobitids (loaches, such as *Noemacheilus, Cobitis, Sabanajewia*), the blenny (*Blennius fluvatilis*), and the eel which has practically disappeared. In general, these fish are taken in a relatively intensive way with lines, hooks, and nets. This important food source has often been ignored by people in the Mediterranean, who have always shown a certain tendency to turn their backs on rivers. (Ironically, there are now plans for precisely this type of use, when fish populations are at their lowest and the high levels of contaminants in the fish make them not very suitable for human consumption.) This neglect is surprising, considering that normal densities may reach 4,050 fish/acre (10,000 fish/ha) and it is very common to attain a biomass of 10,000 kg/ha (and often much greater). In the middle stretches there are also very productive tributaries, although they are subject to major variations in flow. In many years, the summer drop

in water level leads to massive fish death, normally due to oxygen deficiency. However, the great ability of the fish to adapt to these conditions, already mentioned, is shown by their enormous reproductive capacity; for example, barbels (*Barbus bocagei*) can produce 20,000 eggs per female, *Chondrostoma toxostoma* can lay more than 15,000, chub (*Leuciscus cephalus*) can exceed 60,000, and *Carassius* release up to 380,000. As can be seen, drastic declines in population can be compensated for relatively easy, if adequate living conditions are restored in the rivers.

In the upper stretches fishing is recovering its importance as a leisure and sports activity, and increasingly as a source of income. In these stretches, the most abundant and most typical family is the Salmonidae, the salmons and trouts. The trout (*Salmo trutta*) is the most typical species of upper stretches, although there is also the salmon (*Salmothymus ohridanus*) found in Lake Ohrid and its affluents on the frontier between Albania and Macedonia. In salmon-bearing rivers, other families gradually decline as the altitude increases until the trout is the only native fish that is found in the coldest waters. In these normally oligotrophic waters, fish production is usually 5-20% of that in the middles stretches of river and this makes it rather surprising that fishing for sport should be carried out here. The reasons are the transparency of the water, the closeness to nature, the greater difficulty of catching trout (which are considered trophies), and their better organoleptic (taste, aroma, etc.) qualities. It is difficult to establish exactly how many people catch trout in the countries around the

Mediterranean, but it is increasing. As an example, the trout has been introduced to many new sites, where it would not occur naturally. Thus, in addition to mountain lakes, affluents, and streams outside the Mediterranean area it has been introduced into many rivers, often in isolated areas in Mediterranean mountains, for fishing purposes. In many cases, this is a serious ecological problem, since the trout is a predatory fish. It most commonly preys upon amphibians (especially newts *Triturus* and *Euproctus*), certain endangered species of fish, and some endemic invertebrates.

Introduced species

The fact that native Mediterranean species were often not very suitable for fishing purposes (generally as a result of the absence of large predatory fish), led to the introduction of other species chosen by fishing enthusiasts. These species were often quite different from Mediterranean species. The Mediterranean Basin has seen the introduction into both southern Europe and North Africa of species mainly from central Europe and North America. These include the pike (*Esox lucius*), tench (*Tinca tinca*), some salmonids (*Hucho hucho*), and wels (*Silurus glanis*). The more important American introductions include the members of the centrarchids (sunfish), such as largemouth bass (*Micropterus salmoides*), sunfish (*Lepomis gibbosus*), brown catfish (*Ictalurus nebulosus*), and members of the salmonids such as *Salvelinus fontinalis* and *Oncorhynchus mykiss*. These large fish can reach more than 6.5 ft (2 m) in length in the case of the wels (*Silurus glanis*) and they need a lot of food. This is why they have destabilized aquatic ecosystems in many places.

Fish have been introduced for other reasons in other cases, but the effects have been just as negative. The topminnow (*Gambusia*) was introduced to combat mosquitoes, but in many places it has caused a decline of more valuable species, and some are now endangered such as *Aphanius iberus*. In some cases, fish such as the minnow (*Phoxinus phoxinus*) have been introduced for use as bait for the capture of other fish.

Some species native to Mediterranean rivers have been introduced into sites where they did not occur naturally. These include the common trout, introduced into many islands in the Mediterranean Basin (e.g. Corsica) and into the other mediterraneans. (The trout is very common in the rivers of the Chilean mediterranean and some cyprinids, such as the carp, are also widely spread throughout the rivers of all the mediterraneans.)

3.4 Stock-raising

Most areas with a Mediterranean climate have never been typical stock-raising areas, due to the scarcity of areas suitable for grazing by large herbivores, and thus the rearing of cattle. The Mediterranean climate's hot, dry summers do not favor the growth of high-quality meadows for grazing. Accordingly, a livestock production model based on the sheep has evolved, and to an even greater extent a model based on livestock rearing for subsistence, mainly goats. When there is only low quality grazing of low food value, no animal can cope better than the goat.

Extensive sedentary herding: the goat

Tough, resistant, and sober, the goat (*Capra hircus*) digests the coarse, cellulose-rich Mediterranean vegetation better than any other ruminant. As early as the middle and late Paleolithic it was eaten by the hunter-gatherers in the Mediterranean Basin, as shown by many archeological remains. These include sites at Grimaldi (Liguria), near the French-Italian border; Terra Amata (Provence) near Nice; Hortus (Languedoc), not far from Montpelier; the Aragó cave, near Perpignan in France; Arbreda in Catalonia, near Girona; the Romaní grotto, near Capellades (Catalonia). In the Natufian settlements of the Fertile Crescent the remains of numerous wild goats have been found. When Neolithic humans learned to domesticate animals, goats, along with sheep and pigs, were among the first to be domesticated.

Goat herding

It appears that the present-day domesticated goat (*Capra hircus*) is the result of hybridization between two species of wild goat at the end of the Tertiary and the beginning of the Quaternary. There were then three different species of wild goat: the pasang (*C. aegagrus*), the markhor (*C. falconeri*), and, according to some authors, a third species, *C. prisca* (now extinct), that many authors consider was just a Balkan and southeastern European variety of the domesticated goat. Studies of the cranium and arrangement of the horns suggest the domesticated goat is derived from the pasang (*C. aegagrus*), still found in the wild in Crete, Asia

206 Billygoat, nannygoat, and kids (male, female, and young of *Capra hircus*). Goats have been reared by humans for as long as sheep and pigs, although the number of heads of goat has never been as high as that of sheep. The goat's advantage over the sheep is that, while it needs to eat the same amount, what it grazes can be of lower quality. As a result of its agility and ability to climb rocky slopes, it needs more attention from the shepherds.
[Photo: Jordi Camardons]

Minor, and the Caucasus. This species almost certainly hybridized with *C. falconeri* or *C. prisca* and gave rise to some contemporary breeds of domesticated goat. The most reliable hypotheses date this domestication to around 9,000 years ago somewhere in Anatolia. From Anatolia domestic goats rapidly spread to Palestine (and later Egypt and North Africa), and towards the Zagros Mountains (and then to Iran and India). Later, the goat spread throughout the Mediterranean Basin as an integral component of the Neolithic agricultural and livestock complex in the Fertile Crescent, starting with the coastlines of the Aegean and the Balkans.

Later human selection of the variations most suitable for stock-raising and those giving the best yields, gave rise to a large number of breeds that can be broadly spilt into three groups, European, African, and Asiatic. European goats have very diverse coats, uniform or with patches and there are many breeds, such as the granadino, serrano, greek, and murcian, etc. African goats are less robust, with a uniform coat, long ears, and no horns. Asiatic goats have long coats and have no horns or narrow, spiral horns that point backwards. Today, a wide range of goat breeds are reared in the areas around the Mediterranean, but the most typical is the Malaga breed, also known as the coastal goat, which dominates the whole of the coastline of the Iberian Peninsula and Morocco. The Malaga type has a uniform light red coat, tending towards beige or brown, depending on the particular strain. It is mainly kept for its milk.

Herding practices

Goat-raising is almost always extensive, though they can be kept in stables, too. Extensive livestock rearing is a system in which the livestock lives permanently in the fields making direct use of the natural resources by grazing. A lot of land is needed for this system. This is why when flocks of domestic goats begin to increase, they displace the wild goats, whose numbers decline and are restricted to the highest areas of the mountains where goatherds do not take their flocks. Traditionally the herds graze or browse freely in one place until the site is exhausted and then they move on. They are almost never enclosed in stables, and almost all their food comes from grazing. Sometimes they are given fodder, but for most of the year they live on what they eat in dehesas, fallow or non-cultivated land, or what they can find on mountains, since they are very agile and can reach apparently inaccessible places to eat even the shoots that are hidden among the rocks. At the beginning of spring, when grass is scarce, they eat the tender new branches and leaves of trees and shrubs.

Extensive herding implies irregular and often deficient feeding regimes, since it only makes use of the available natural resources whose abundance depends on the season. In the Mediterranean climate, the abundant spring grazing means there is more food than the livestock needs, but for the rest of the years the grass and

207 The goat (*Capra hircus*) shows synergy in its absorption of nutrients. If the goat only eats leaves of the holm oak, for example, it fulfils nutritional requirements for maintenance but not for milk production. If molasses, urea, and minerals are added to the natural fodder, the absorption of the holm oak foliage increases and the food value of the diet is high enough for milk production. [Drawing: Editrònica, from de Joffre, 1991]

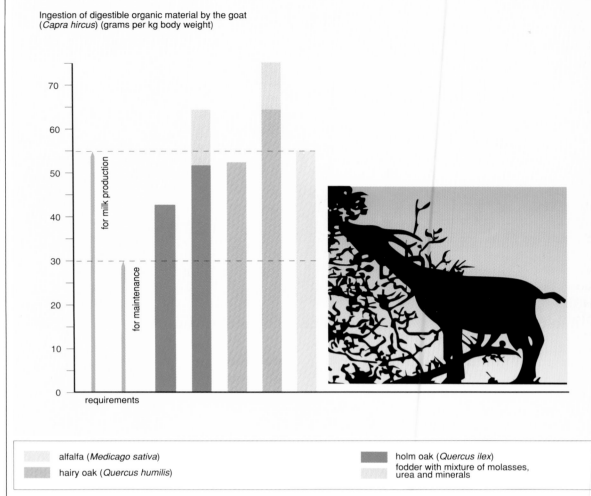

Ingestion of digestible organic material by the goat (*Capra hircus*) (grams per kg body weight)

for milk production

for maintenance

requirements

alfalfa (*Medicago sativa*)
hairy oak (*Quercus humilis*)

holm oak (*Quercus ilex*)
fodder with mixture of molasses, urea and minerals

shoots do not cover the animals' requirements. But, in addition to the seasonal differences in grazing, the quantity of food also varies depending on the autumn rains, the winter temperatures, and the summer reserves. A further problem is that the Mediterranean climate often suffers periods of intense drought that reduce the food value of the plants, and also limit the quantity of water available for the animals, which may be insufficient.

As well as these problems related to the irregular climate, the soils of the Mediterranean are also poor in nutrients, especially phosphorus, so that the vegetation is not as nutritive as it might be in other climates. Soil, in much of the world, is a limited and irreplaceable resource, and the Mediterranean regions are no exception. The number of animals in the herd has to be established in accordance with the availability of food in the ecosystem, and this in turn depends closely on soil conditions, which are often variable. The scarcity of food during much of the year also has

the disadvantage of forcing the stock-raiser to ensure the animals' peak food requirements (late pregnancy and lactation) coincide with the greatest availability of food, although there is always the possibility of giving additional food to the animals in the form of previously cut grass or animal feed.

Goat products

The goat's main product is its milk, which is rich in fats, casein, and lactose. Of all female mammals, the goat produces most milk for its weight: a 66 lb (30 kg) goat may give 80-106 gallons (300-400 liters) of milk a year, i.e., 12 or 13 times its own weight, while a cow only produces five or six times its own weight. Goats milk can be drunk or turned into cheese. The flesh of the goat is also eaten, mainly as the young kid—which is very tender but not very nutritious—because the flesh of the adult is hard and stringy, and thus considered of low quality and is only eaten when sheep are scarce.

In California and South Africa the goats reared are Asiatic (Angora and Kashmir) from the mountains of Kashmir (north India) and Afghanistan. Their long, fine, silky coat is of great value as a textile fiber used for high quality fabrics. The Angora goat, which is also reared in Greece and Turkey (in fact, it is named for the former name of the Turkish capital, Ankara), is periodically sheared for its hair. Kashmir goats are not sheared, the hair used is the hair naturally shed in the spring, when the goats are combed every second day with a fine-toothed comb and the hair is collected. Less importantly, the goat's skin is used for leather and the horns for turnery.

From the productive point of view, the goat is not a very profitable animal, and thus farmers prefer where possible to rear other more profitable livestock. Developed countries that can afford to supplement the deficient food supply of the natural Mediterranean vegetation with animal feed and hay have, for this reason, opted to rear other more profitable animals, and so the number of goats has declined greatly.

Yet goats still dominate in many Mediterranean areas, especially on the north African coast, where the number of goats has slightly increased in the last few years.

Extensive transhumance herding: sheep

Another typically Mediterranean animal is the sheep (*Ovis aries*), a ruminant related to the goat. Although it is not as tough as the goat, the sheep is an undemanding animal that requires little attention and uses marginal land, where it eats grass that other ruminants such as the cow would not touch. The sheep also adapts well to hot, dry climates, where the lack of pasture limits the cattle-raising. Sheep farming is now spread throughout almost the entire world, and the most suitable areas are the regions on either side of the 40th parallel. Around this parallel, approximately between 30° and 45° of latitude, are precisely the Mediterranean climates from which the sheep appears to have originated.

Sheep-raising

The origin of the sheep is uncertain, and its living relatives in the Mediterranean Basin are moufflons (*Ovis musimon* and *O. orientalis*), which are similar to sheep but have more highly developed horns. The moufflon lives in herds and *O. musimon* is abundant in Cyprus, Sardinia, Corsica, while *O. orientalis* occurs in some areas of Anatolia and Iran. They are so closely related to sheep that they can cross and produce fertile offspring. Available archaeological and chromosomal data seem to show that the sheep is derived from the moufflon and that its domestication

208 A flock of Merino sheep (*Ovis aries*) grouped in an Iberian dehesa of cork oaks (*Quercus suber*) and holm oaks (*Q. ilex*) in Catalão, Portugal. The Merino breed is from the interior of the Iberian Peninsula, and is derived from primitive breeds brought from North Africa or the East. It is prized for its wool and its robustness, the reasons why it has spread throughout the world. It has diversified greatly and Australia's large flocks of sheep are of Merino origin. Over the centuries, the exploitation of the Merino's wool was an Iberian monopoly protected by a strict ban on the export of sheep of this breed.
[Photo: Ernest Costa]

209 **The main Mediterranean breeds of sheep and goat and their original distributions**. The Mediterranean Basin has been the breeding center for many types of domesticated animal. These have been selected for their suitability for different types of production, mainly either for milk or for meat. Livestock selection has included adaptation to a highly seasonal climate, the ability to consume woody or poor quality plants, and resistance to some illnesses.
[Drawing and cartography: Editrònica, based on de Joffre, 1991]

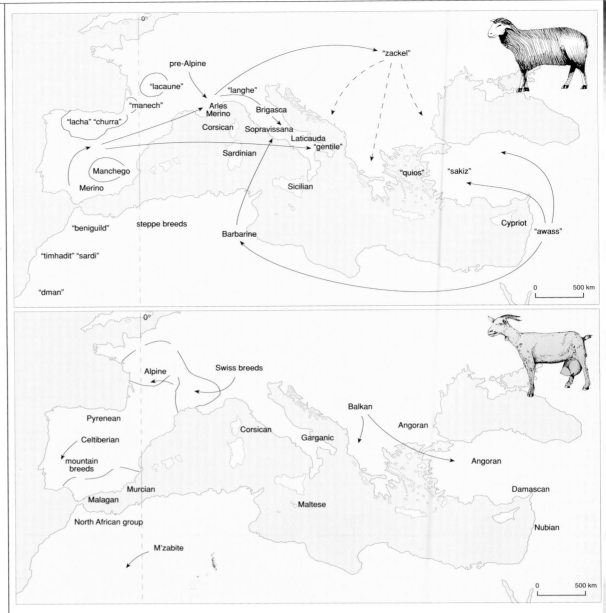

took place somewhere in the eastern Mediterranean between 10,000 and 11,000 years ago. Afterwards, as part of the Fertile Crescent Neolithic agricultural and livestock complex it spread throughout the Mediterranean Basin and beyond. One curious feature of the genus (*Ovis*) is that, with a few exceptions, all the wild species have a short tail, while domesticated breeds have very long ones. This appears to be a degeneration produced by domestication.

Herding practices

The main characteristic of sheep rearing is that it is traditionally based on seasonal migration (transhumance). Transhumance is a method of livestock rearing that implies the regular migration of livestock and their owners (or just their guards, the shepherds) to sheltered valleys in winter and mountain pastures in the summer, in order to take advantage of the sequence of different grazing land that is available. It is, therefore, an efficient way of exploiting the productivity of regions that are uninhabitable and unproductive for most of the year. Seasonal migration has been practiced since time immemorial and is the oldest system of farming sheep in Mediterranean regions, where it is often the only means of making use of land that is unsuitable for other purposes. But seasonal migration is not a technique that is exclusive to the Mediterranean regions. It is practiced in many other parts of the world, wherever there are high mountain pastures with a climate making it too cold to graze there at any time of the year other than the summer.

There are several types of transhumance. The form practiced in Mediterranean environments consists of keeping the animals in the lowlands in the winter and taking them up the mountain when the summer heat arrives. The winter grazing in lower areas is poorer but there is more of it and there are many farmers, so there are usually fallow fields and stubble to graze. The dung from the grazing sheep also helps to improve soil fertility. The summer grazing is usually in nearby mountains, and is richer but there is less of it. Although seasonal migration has many advantages, it also has some disadvantages; one of the most serious problems is the high risk of transporting diseases from one place to another.

Seasonal migration of livestock involves bringing together the livestock of several different owners when the time comes to migrate and they are all herded together by one or more shepherds. To recognize each owner's sheep afterwards, they are branded before they are combined. Different marking systems are used, e.g. a mark may be made on the sheep's back with pitch or with a branding iron or a mark may be cut in the sheep's ear.

Shepherds are responsible for leading the flock during the transhumance. The shepherd is an important person—respectable and learned—who has a large stock of knowledge not only about the animals, but also about medicinal plants and other remedies to treat his and the livestock's most common accidents and illnesses on the annual migration. Although it is not essential, on the seasonal migration it is better to have a sheep pen or enclosure to keep the sheep overnight. A pen has the advantage of concentrating the animals in an enclosed space, where they can be vaccinated, given additional fodder, inspected, and treated for wounds and illnesses. Furthermore, bringing the animals together means a certain mutual protection against the weather, be it inclement weather or excessive sunshine. It also has the advantage of allowing the use of the dung produced during the night. Another useful function of the pen is that the sheep can be given salt, something that is very necessary because not all plants contain a sufficient quantity. The salt required depends on the food, but on average a flock of 100 sheep requires 66-88 lb (30-40 kg) of salt every 8 days.

Most Mediterranean stock-raisers lead a sedentary life and only the shepherd moves with the livestock, but there are also much more mobile human groups, such as some nomadic shepherds of the Near East that raise large mixed flocks of goats and sheep. In the winter they live in the lower areas to the south where there is little grazing for the sheep, but where they can graze over large areas. In the meantime, the

210 Sheep are shorn and branded, as seen here in the Pallars district of Catalonia, before seasonal migrations, although removing the wool is not directly related to the migration, and there are many sedentary flocks. Branding avoids losses when the flocks of several proprietors get mixed up, and is usually painlessly applied with paint or soot. Permanent marks can be made on short-haired livestock, such as cattle, but the growth of the sheep's wool means marking must be repeated every year.
[Photo: Antoni Agelet]

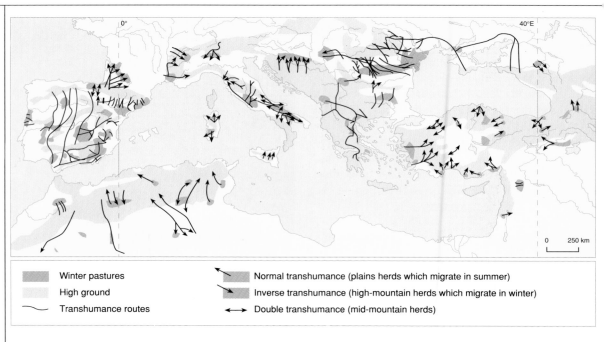

211 The major transhumance (seasonal migration) routes in the Mediterranean Basin. The regular migration of domesticated livestock, mainly sheep, in search of seasonal pasture in the high mountains in the summer, and in the winter to the sheltered valleys. In the interior of the Iberian peninsula, the seasonal migration routes consisted of main routes, known as Cañadas Reales and by secondary routes, called *cordeles* and *veredas* that led into them or interconnected them. The Cañadas Reales of the Mesta were 90 Castilian *varas* wide (246 ft [75 m]) and the large flocks could follow them along routes up to 500 mi (800 km) long. The *cordeles* were 45 *varas* wide (125 ft or [38 m]) and the *veredas* were 25 *varas* (69 ft [21 m]) wide and connected the main routes. [Drawing: Editrònica, based on Grigg, 1970]

Winter pastures	Normal transhumance (plains herds which migrate in summer)	
High ground	Inverse transhumance (high-mountain herds which migrate in winter)	
Transhumance routes	Double transhumance (mid-mountain herds)	

mountain pastures are covered in snow. In the spring and summer the lower pastures dry up and the shepherds start to migrate north. When the grazing in the high mountains has dried up completely at the end of August, they return to the south and the livestock grazes on the stubble of the last crop.

Transhumance is disappearing from the zones where large-scale irrigation has been introduced, as ground formerly unproductive due to the lack of water in the summer can now be cultivated using artificial inputs of water and fertilizer. This limits the area of uncultivated ground for the sheep to graze, as well as eliminating the fallow, which means that there are no uncultivated areas left for the livestock to graze. In the northern part of the Mediterranean Basin transhumance began to disappear in the second half of the 19th century and the sheep were replaced by cattle or pigs that did not undertake seasonal migrations. But intensive livestock raising is a recent system and until the 1970s transhumance was of great importance throughout the world, and it still is in some areas.

Sheep products

The sheep provides flesh, milk and wool. There are breeds of sheep raised specifically for milk production and others for wool production, although in fact production is always mixed (wool and flesh or milk and flesh). There are also breeds that are typically reared for their flesh, whose milk and wool are considered byproducts. The exploitation of sheep for milk mainly occurs in the eastern Mediterranean, in Greece, and to a lesser extent in the Balkans. The races raised for their flesh are mainly those from the

south of the Mediterranean Basin, especially from North Africa. In the Mediterranean regions of western Europe, South Africa and Australia, the main product is wool.

The best sheep for wool production is the Merino breed, which provide the best wool. The sheep of this breed have been crossed to improve the wool of many breeds of sheep throughout the world. The breed's origin is uncertain, but there is now a consensus that in its present day form it is the product of a mutation introduced into Al-Andalus by the Arabs of northern Morocco in the 14th century, although going further back its origins probably lie among the breeds of sheep with fine wool of the eastern Mediterranean or Iberian Peninsula that were introduced into North Africa during Roman times or before. The excellent characteristics of the Merino breed—resistance to heat and cold, effective heat regulation, high-quality wool, willingness to walk long distances, etc.—mean it was widely raised and politically protected, and this led to its spread throughout the Mediterranean. From the 18th century onwards, it spread to countries with more continental climates (it reached Sweden in 1723, Germany in 1765, Hungary in 1769, and France in 1776). The Merino produces excellent wool everywhere and conscious selection has led to great improvements in some countries, although in others this aspect has been neglected. Today, the best specimens are those raised in Mediterranean climates, especially those of Australia or South Africa; in both regions, Merino sheep were introduced by Europeans in the 18th century and they immediately began to thrive. Merino sheep were

212 A cellar for the fermentation and maturing of sheep cheese. Livestock that performs transhumance provides flesh, wool, and also milk. Mediterranean livestock has never been a great producer of milk. Milk keeps badly, and so traditionally it was made into cheese. To make cheese, it is first necessary to curdle the milk, i.e. to separate the liquid fraction (whey) with an adequate curdling agent (rennet). The curd that is obtained is fresh cheese. A process of salting, fermentation, and aging give the cheese its proper flavor and long life.
[Photo: AGE Fotostock]

spread from the Mediterranean regions of southern Australia throughout the continent, without taking into consideration the effects of overgrazing and the damage caused to the natural vegetation when the flocks reached the drier, arid areas in the center. The wool of Australia's Merino sheep was intended, above all, for the British textile market. Wool quality does not only depend on the breed, but also on the individual's age and sex, and the part of the animal it comes from. While wool used to be essential for the production of fabrics, today artificial fibers mean there is less demand for wool from the textile industry and production has greatly declined, because stock-raisers prefer to raise breeds of sheep for their flesh as they are more profitable.

The milk of the sheep is also used, especially in Turkey and in the Balkans. It is used to manufacture cheese, such as the well-known Manchego cheese from the La Mancha region of Spain, or Roquefort, which has been produced in the Languedoc region of France since Roman times. Sheep flesh (mutton) is appreciated, but does not fetch a high price. In fact, it is the most profitable product derived from rearing sheep due to its low cost of production, no higher than that of rabbit. Sheep flesh is succulent, juicy, and easy to cut up and cook. Sheepskin is used in the leather industry, and is known as suede if it is matt, and napa leather if it has a shiny finish; it is also a source of parchment once it has been skinned and left to dry. In many places the intestines of sheep are still used to make the strings for musical instruments. A further important byproduct of sheep-rearing is their dung. Sheep dung is greatly appreciated as a natural fertilizer, since it is not washed away by water and plants absorb it easily. One sheep produces about 5,500 lbs (2,500 kg) of dung per year.

Livestock in dehesas: pigs

Dehesas are a typically Mediterranean landscape characterized by large areas of ground where natural vegetation, including trees, is allowed to grow subject to livestock exploitation. They are mainly intended for grazing, although timber and firewood are also extracted, and they are occasionally used partly for cultivating crops and partly for hunting activities.

Pigs

Stock-raising in dehesas is mainly based on free-ranging pigs. In the mediterraneans and in the rest of the world, there has been a recent tendency to fatten pigs in farms, because this gives a better yield of meat, but pig-raising continues in the dehesas because the pigs reared there are used in craft industries which use the flesh to manufacture highly appreciated products that fetch high prices for their excellent and distinctive aroma, flavor, texture, etc. The pig is thus another typical example of Mediterranean stock-raising, and was one of the first livestock animals to be domesticated after the sheep and goat.

The pig (*Sus scrofa*) was first reared in order to convert otherwise unusable resources into flesh. It is a very common animal in the Mediterranean area, although it can adapt to very different climates. Pigs

The Mesta and the Dogana

Herd of goats (*Capra hircus*) and sheep (*Ovis aries*) in the Pyrenees [Ernest Costa]

Pedro Rodríguez de Campomanes, Count of Campomanes, detested the Mesta. Yet this enlightened politician, economist, and servant of the Spanish crown, ran the Mesta's management body from 1779 to 1796. His advanced ideas were totally opposed to the principles of a medieval organization. (The principles of this medieval organization were incompatible with the concepts of the new industrial society that was beginning to form in Europe.) And so, paradoxically, he became the institution's president and used his authority in order to weaken the organization, which finally disappeared in 1836. What was the Mesta? And why did it seem so bad to the advanced minds of the late 18th century?

The Mesta and transhumance have formed part of history ever since large flocks of sheep started roaming the Iberian Peninsula on annual migrations in search of seasonal pastures. The organization of these movements, their rights of way and grazing rights, the maintenance of their routes, and regulation of the wool trade were all placed under the authority of the Honrado Concejo de la Mesta (the Honorable Council of the Mesta) by Alfonso X of Castile and Leon in 1273. Thus, the Mesta was an organization of stock-raisers whose origins lay in medieval Castile, and whose name, the Mesta, was the Castilian word then used for pasture. It was an appropriate response to the needs of its time, but as time went by it became a structure that served to maintain the privileges of the large stock-raisers and was a major impediment to agricultural and economic development. In fact, the rural world of the 13th century bore little relation to the needs of the industrialized society that would arise in the 19th century.

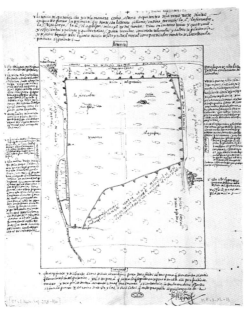

Drawing of the "Dehesa del Espadañar" (Extremadura, 1576)
The routes of the Mesta [Archivo General de Simancas (Valladolid)]

Sheep (*Ovis aries*) in transhumance in the Pyrenees [Ernest Costa]

Sheep bells for migratory herds [Ernest Costa]

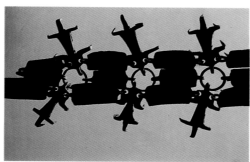

Collar with branching design [Ernest Costa]

Collar with spikes [Ernest Costa]

Collar with branching design [Ernest Costa]

Transhumance was, and still is, a common stock-raising practice throughout the Mediterranean Basin, from the Maghreb to the Balkans and the Near East, but nowhere has it been as important as in the Iberian and Italian peninsulas, perhaps because their geography and relief inherently favor seasonal livestock migration. This is why the other major kingdom in the Iberian Peninsula, the kingdom of Aragon, also had a body like the Mesta, although less important, which lasted until the 19th century. It was called the Casa de Ganaderos de Zaragoza (the House of Stock-raisers of Zaragoza), which was given legal powers in 1129 by Alfonso I of Aragon. Later, in the 15th century, the Dogana della Mena delle Pecore was created by the kingdom of Naples. The Dogana was protected by the crown, which wanted to increase wool production. Thus, even though stock-raisers had to pay the taxes set by the Dogana's magistrates, they enjoyed large advantages over farmers. The need for more croplands and the constant expansion of stock-raising in the kingdom of Naples, especially after the second half of the 15th century, caused the deforestation of the center and south of the Italian Peninsula. The sheep and the Dogana, in fact, prospered under the kingdom of Aragon's administration of the kingdom of Naples, but in the third decade of the 16th century it began to decline as a result of the pressure from smallholders. The Dogana was officially dissolved in 1806.

The Mediterranean regions of northern Italy and the Languedoc also had medieval migratory herding models that were as large as those in Castile, Aragon and Naples, but their organizations never became as powerful as the Mesta or the Dogana. The small local livestock associations, such as the Esplèche in Arles, had some powers over animals and grazing, but never became major institutions. This may well have been because they lacked a strong central government like those in the kingdoms of Castile and Naples.

Pyrenean shepherd [Ernest Costa]

In addition to its importance in land use and delimitation, the Mesta created a structure articulating the wide open spaces of Castile around a complex network of livestock routes, the widest called *cañadas*, followed by *cordeles*, *veredas*, and *coladas* in order of decreasing width, and many stretches of these former routes can still be seen in the landscape of the Iberian peninsula. The main routes, the *Cañadas Reales* (the Royal Cañadas), through the provinces of Soria, Segovia, Leon, and Cuenca were busy routes where the flocks could travel 12.5-18.5 mi per day (20-30 km/day) on journeys of more than 500 mi (800 km). Seasonal migrations in search of pasture, undertaken by flocks guided by half a dozen shepherds with their dogs and beasts of burden, could thus have taken up to 40 or 50 days for the longest journeys. As there were hundreds of these flocks moving around, these migratory routes were clearly major communication pathways, with corrals to provide shelter and a variety of basic services: land use acquired a grid-like structure based on these network of routes and was subordinated to grazing requirements.

The Mesta, the Casa de Ganaderos, and the Dogana have disappeared. Only small fragments of the cañadas remain in use. Most have been invaded by forests or crops, and the rest are now main roads. Today livestock are taken to seasonal pastures by train or truck. Yet the countries that practiced this livestock strategy until the 19th century have still not totally recovered from the ill effects that Campomanes saw clearly.

Tratturi, from the Dogana [Courtesy of the Archivio de Stato de Foggia / Foto d'Autore /Ariston de Mitococcio]

213 Pigs (*Sus scrofa*) of Iberian strain in Salamanca, Autonomous Community of Castilia-Leon (western of the Iberian Peninsula), above; near the town of Sonoma (California), below. This black color is typical of the different breeds of Iberian pigs, such as these semi-penned pigs in a dehesa. Today pure-bred specimens are very scarce, because most individuals are the result of hybridization with breeds of white European pig, which is less robust and whose flesh is less tasty. The pig's comparatively recent domestication means that the primitive races revert to the wild if they are set free, as has happened to these pigs that have have become wild. They are probably the descendants of pigs taken to California by Spanish colonists in the 18th century.
[Photo: Xavier Ferrer and Larry Minden / Minden Pictures]

are not raised in the areas of the Mediterranean that follow Judaism or Islam, although at least one of its centers of domestication was in the Fertile Crescent. More than 10,000 years ago people began to domesticate different animals, and since then they have selected the species that do not compete with them for food. Most domesticated animals are herbivores that consume natural vegetation that human beings cannot eat. The Arabs and Jews who now occupy Mediterranean areas came from regions that were much hotter and more arid—probably from the sandy deserts of Arabia—where water is scarce and where the rearing of pigs (which need a lot of water to survive) would have competed directly with human beings.

The oldest traces of the domesticated pig date from 9,000 years ago in southern Anatolia (Çayönü), it is generally accepted that this domestication took place independently in different areas where the

wild boar lived. There are now many races of pig, all of which are descended from crosses between the European wild boar, (*Sus scrofa*), and the wild boar (*Sus vittatus*) that European explorers brought from Asia. In the south of the Iberian Peninsula (especially in Estremadura and eastern Andalusia) and in the Balearic Islands (especially in Majorca) there is still a special breed of pig that some authors consider to be an independently domesticated form of a hypothetical ancestor *Sus mediterraneus*, the Iberian pig which has a typical dark skin.

The different breeds of pigs of the Mediterranean regions are closely linked to traditional methods of exploitation. For example, the Majorcan pig, is black and belongs to the Iberian stock and is thus related to the dark-skinned races of the south and west of the peninsula. It is raised on a small number of farms in a mixed regime based on penning and grazing. Its diet consisted of figs and the fruit of the prickly pear, both now abundant on the Balearic Islands (the figs since antiquity, while prickly pears have only been abundant since their introduction from the Americas in the 16th century). This breed of pig was used to manufacture *sobrassada* (a spiced preparation of minced lean pork with *pimentón*), although this is now produced from other types of pig because this breed is now almost extinct. In the last 20 or 25 years, the dramatic increase in tourism in the Balearic Islands has led to a reduction of the number of fig trees and prickly pears, and the number of pigs has also decreased. Another Mediterranean pig is found in Corsica, and is a very small, robust, pig similar to the wild boar. It ranges freely and eats beechnuts, acorns, and chestnuts. Like the Majorcan breed there are now only a few specimens of the pure race.

Pork production

The pig is very productive because it is highly fertile, and this makes it the most profitable animal to rear, and it is now the leading meat-producing species in the world. It is an omnivore—it eats all sorts of foodstuff of animal or plant origin—and can thus be fed easily on waste scraps although this is not the best way to obtain a high yield of meat. It can be taken to graze, and it also eats potatoes, sugar beet, turnips, and malt from beer production. Of all the domesticated animals, it is the fastest and most efficient converter of food into meat. Over its life, the pig can transform 35% of the energy content of its food into meat, compared to 13% for sheep and 6.5% for cattle. For every 4 lbs (2 kg) of food eaten, a piglet can increase approximately half a kilo in weight, while a calf would have to eat 11 lbs (5 kg) for the same increase in weight. In addition, the pregnancy of a cow lasts nine months and she only bears a single calf, while a sow's pregnancy only lasts four months and she bears eight piglets or more.

The transformation of the pig's flesh into meat products and cured hams has always been a craft production. Traditionally more importance is attached to product quality than quantity. Even so, during the last few years an industry has grown up around the rearing of pigs in styes, which is discussed in the following pages. This increase has been greater in temperate biomes than in Mediterranean ones, and is partly explained by the fact that the pig is very suitable for production on an industrial scale, due to its short reproductive cycle and its high fertility. Yet industrial exploitation raises the problem of dependence on external food inputs and the adequate disposal of the feces, both solved by producing some of the pigs' food on the same farm, and fertilizing it with their excrement.

This meal does not contain pork

Ce plat ne contient pas de viande de porc

Esta comida no contiene cerdo

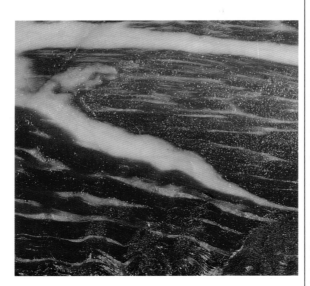

The dehesa and the Iberian pig

The Iberian pig, probably the most interesting Mediterranean breed, is black, with visor-like ears, that are not very big, with the snout elongated subnasally for rooting and a short neck with a double chin. There are two varieties, one black with or without bristles, and the other is pink with black patches and is almost extinct. But the Iberian pig has been crossed with other races, mainly the Duroc-Jersey, and there are now almost no purebred Iberian pigs.

The natural habitat of the Iberian pig is the dehesas of holm oaks and cork oaks. For a long time, this pig was the main source of meat, consumed directly or

214 Information sign saying there is no pork in the prepared dish, in this case the meal served on board an airplane. The ban on eating pork is observed throughout the Semitic area by the followers of both Judaism and Islam, both widespread throughout the southern and eastern Mediterranean Basin. [Photo: Jordi Vidal]

215 The flesh of the Iberian pig is mottled red with veins of fat and little white dots, which are concretions of casein. It has a distinctive appearance and flavor. Known as *jamón de pata negra* (black foot ham; because of the Iberian breed's black coat and hoofs) or as the *jamón serrano* (mountain ham; because the diet of mountain pasture of the Iberian pigs). They are prepared using the shoulder blade or preferably the "ham," which is not cooked but cured by salting and drying (for 1-2 years) in a cool, dry environment. The spontaneous growth of a layer of mold is part of the curing process. The flavor largely depends on the food the animal ate in the last two or three months of its life —the time taken for complete turnover of muscle protein. This is the time when it is most important to feed acorns to the pigs. Since the pigs put on about 175 lbs (80 kg) in the last fattening phase and the yield of flesh and fat after transforming the foodstuff does not exceed 10%, about 1,750 lbs (800 kg) of acorns are needed per pig—the effective production of 2.5-5 acres (1-2 hectares) of dehesa, in addition to all the feed previously consumed to obtain the delicious hams (and the rest of the pig). [Photo: Genin Andrada / Cover / Zardoya]

cured in the regions where it was reared. It grazes and makes use of agricultural subproducts, such as acorns and chestnuts, and is not exploited intensively. During the spring it grazes on the herbaceous plants that cows and sheep ignore. It also eats the acorns of the holm oak (*Quercus ilex*) and to a lesser extent, those of the cork oak (*Q. suber*) and *Q. faginea*. The flesh of the Iberian pig has a much better flavor than that of the white pig. The smell of the natural foodstuffs the pig eats in the dehesa permeates the muscle, giving the flesh a distinctive flavor and exceptionally high quality to the ham. (After a dry autumn when grass is scarce, the cured hams are of lower quality.) There are no other products like these in the entire world. The preparation and curing of an Iberian ham lasts a long time, mainly because the Iberian pig's greater quantity of intramuscular fat slows down water loss and salt uptake. To obtain a good ham it is recommended that the pig should have been rested before slaughter, and care should be taken to drain as much blood as possible, which is ensured by slitting the animal's throat. After dismemberment it has to be washed and salted by hand, and then it is left to dry naturally in a cool, dry place for about two years.

The Iberian pig has always been closely associated with the dehesa and the acorns it eats. To protect this form of stock-raising, laws were passed in the 16th century that harshly punished the theft of acorns, and forbade the entry of persons and seasonally migrating livestock into dehesas without the owner's permission. The Iberian pig now only lives in the southwestern coastal regions of the Iberian Peninsula, that is to say in western Andalusia, Estremadura, Salamanca, and in Portugal's Algarve and Alentejo. Within this region we can distinguish two distinct areas: the areas that produce acorns, where the Iberian pig is reared, and those that do not produce acorns but cure the products. There are some areas where both overlap.

In the last few years, many different types of factors have contributed to the gradual reduction of pork production in the Iberian Peninsula. After 1960, a severe epidemic of African swine fever (against which the Iberian pig is not protected by its management system) struck the Iberian Peninsula, the last foci of which were not eliminated until 1986. During this quarter of a century the association Iberian pig-dehesa has become less important as a result of the need to control this disease, which is easier to control in stabled pigs. On the other hand, Spanish and Portuguese pig products were not allowed to enter the other EEC states until mid 1989, and this meant that raising Iberian pigs became less profitable, though internal demand remained strong. Now both sheep and cattle are raised extensively in dehesas, even though raising pigs is more profitable because pigs are able to peel the acorns (thereby decreasing the amount of fiber eaten), and thus increase their digestibility. Finally, the traditional nature of this form of exploitation should also be borne in mind, in contrast to the intensive nature of pig-raising in all the developed countries. Intensive pig-raising aims at producing large quantities of meat by selecting breeds that are easy to fatten and raise in the pens of highly technical farms, while raising Iberian pigs is still more or less a craft activity. Unfortunately, despite the quality of it Iberian products, they are difficult to sell, though the situation improved greatly after the ban on Spanish and Portuguese exports to the rest of the European Union was lifted.

Livestock in enclosures

For a long time, humans have been more interested in the yield provided by an animal than in the source of this yield. At first they only took an interest in the milk, flesh, wool, and other livestock products but they did not consider exactly what the animal required for it to produce them. Later, new stock-raising techniques took greater care of the animal to ensure peak production and the minimum investment necessary in the animals' requirements. This led to the appearance of intensive stock-raising, i.e. the animals are kept permanently, or part of the day, in stables where they are fed. Now the trend is to adapt the livestock to the demands of the industrial production process.

The historic process

Traditionally, in the Mediterranean region stock-raising was either complementary to farming, activities in which the main interest was on products obtained from the land, or was carried out on an extensive scale in pastures and dehesas. But at the beginning of the 20th century this situation began to change, mainly due to a crisis caused by a farming glut that led to an alarming decrease in the prices of farm products, which renewed interest in stock-raising. Livestock production was increased, partly by selecting the most economically productive breeds, and partly by encouraging stabling, as keeping the animals in enclosures made it possible to obtain much higher yields than rearing them in semi-freedom. Since the early 20th century, when livestock enclosure became the norm, many farms ceased to treat stock-raising as a complementary activity and it became as important as agriculture. Many agricultural smallholders managed to survive—some even prospered and bought the land they worked—thanks to the profits obtained from keeping animals in stables or pens.

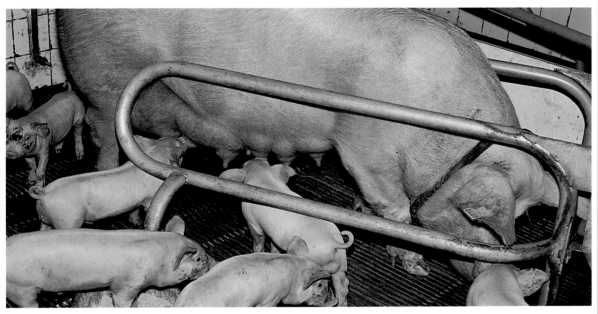

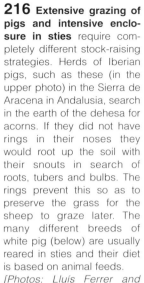

216 Extensive grazing of pigs and intensive enclosure in sties require completely different stock-raising strategies. Herds of Iberian pigs, such as these (in the upper photo) in the Sierra de Aracena in Andalusia, search in the earth of the dehesa for acorns. If they did not have rings in their noses they would root up the soil with their snouts in search of roots, tubers and bulbs. The rings prevent this so as to preserve the grass for the sheep to graze later. The many different breeds of white pig (below) are usually reared in sties and their diet is based on animal feeds. [Photos: Lluís Ferrer and Josep Maria Barres]

3. HUMANS IN SCLEROPHYLLOUS FORMATIONS

Keeping livestock in sheds or pens represented an important change in stock-raising. The animals had formerly depended on their environment they lived in for their food, a natural environment that humans could not control and which had given rise to a characteristic form of stock-raising in each climate. In the Mediterranean area the stock-raising model that had spread was based on sheep and goats which ate the scarce pasture. It also received seasonally migrating livestock from cooler areas where pasture was much more abundant but not available all the year round. The dehesas were used by pigs, which were perfectly adapted to this environment. As penning meant the creation of artificial stock-raising conditions, its extension went hand in hand with the reduction of traditional stock-raising activities in the different mediterraneans.

Conditions under enclosure in sheds or pens

Unlike free range stock-raising, where the farmer depends on the local environment to feed and tend his livestock, enclosure in sheds or pens allows the artificial creation of the environmental conditions most suitable for the animals. It is possible to control everything from the temperature to the food, as well as to monitor their hygiene. Initially, enclosure allowed the monitoring of the livestock to increase production, and technological advances have allowed the creation of environmental conditions that are separating the livestock so far from the environment they live in, or more accurately the area the farm is located, that it has been possible to introduce animals that are not typical of the area. Thus, dairy farms have now been introduced in many sites in the Mediterranean, a type of stock-raising quite unsuited to the region. In any event, although the immediate surroundings are no longer so critical a factor in farm location, they should be taken into account as there may still be some advantages for species that have lived in these environments.

The application of science has led to improvements in breeds of enclosed animals by means of selection and applied genetics. As enclosure in sheds or pens increased, it became necessary to decide which breeds were best for rearing on farms, and there were two distinct points of view, one arguing it was better to improve local breeds in each area, while the other considered it better to find the most productive breeds, even those from other areas, and then to try and acclimatize them. The second tendency won out and now it is considered more profitable to spread more productive breeds than to seek the breeds native to each location.

The production of livestock in enclosures

Enclosure of livestock in sheds or pens has changed Mediterranean stock-raising, which has gradually lost its traditional wide-ranging model. Shepherding, seasonal migration, and stock-raising in dehesas are mere shadows of the past, surviving by producing distinctive products that are far removed from those of modern factory farming.

The changes that followed the arrival of modern industrial society have also changed Mediterranean stock-raising, which has had to change with the times. There has been a loss of interest in traditional stock-raising methods as industrial stock-raising is more profitable. Sheep have lost much of their former importance, as less and less wool is used to make textiles: wool met stiff competition after the introduction of first cotton and then synthetic fibers. Despite attempts to modernize production, production of the Iberian pig is still a declining craft activity, while other breeds that are more profitable meat-producers have spread. Only in depressed areas of the Mediterranean are herds of goats still important, and areas where they are still present are considered backwards. Increasingly rational methods of stock-raising, combined with the application of technology, have given rise to farms with high levels of production. Innovative trends now seek to turn these centers into self-sufficient farms, with the intention of producing in them all the stock-raising inputs needed by making use of the resources available and without damaging the surrounding environment.

Yields and investments from enclosed livestock

When considering the development of stock-raising in the Mediterranean from the point of view of its relation to the environment, it is clear that traditional stock-raising was in balance with the habitat. The animals obtained all the food products they needed from their environment, without damaging it seriously. Enclosure in sheds and pens has increased the distance between the livestock and the environment by creating favorable environmental conditions and by giving animal feeds, partly breaking the relation between the livestock and the places where livestock has always been reared. In extensive stock-raising the flocks must not completely exhaust the limited resources of the environment, or they will be left without forage. Increasingly intensive stock-raising has changed this, since increased production requires large inputs of food and energy, and this has an impact on the environment from which they are extracted. Furthermore, intensive livestock raising produces enormous quantities of organic wastes that are difficult for the environment to absorb.

Self-sufficient production systems have been developed with a view to seeking a batter balance between the environment and intensive-type farms. Apart from the economic benefits an activity of this type may generate, these farms aim to reduce the harm to the surrounding environment. So at the peak of technological progress a new form of exploitation has arisen that seeks a more balanced relation with the environment without diminishing the levels of productivity. These self-sufficient production systems have focused on the products needed for feeding the livestock, the energy used in the farm and, lastly, dealing with the dung the animals produce. To attain high productivity for the animals intensive exploitation systems require large quantities of animal feed inputs from the exterior. Self-sufficient farms seek to produce on-farm as much as possible of the feed the livestock needs, mainly through crops cultivated on-farm. Some farms also produce their own animal feed. Agricultural research continues into new methods to feed the livestock.

With reference to energy requirements, alternative energies have been introduced to replace oil and electricity. One form of energy production that has been tried, using residual agricultural waste products, is the burning of prunings from trees or straw. Solar power has also been successfully developed in the hot Mediterranean climate. So far, energy from the sun has mainly been used to produce hot air and water. Animal dung has also been used to produce biogas. These methods have not yet become widespread, but they show the right track to follow to optimize intensive stock-raising practices of high quality by trying to minimize the negative impacts on the local environment.

Integrated exploitations

Starting from enclosures, it is possible to distinguish between two different types of intensive stock-raising: agricultural and industrial. Agricultural stock-raising uses food for the livestock that is collected on-farm. Industrial livestock stock-raising, however, is independent of agriculture, as the food for the livestock is produced by others, usually in the form of animal feed. These two types of stock-raising, together with seasonal migration, coexist, although stock-raising was essentially based on transhumance until the beginning of the 19th century and industrial stock-raising is very recent. Over the years, there has been a change from stock-raising based on transhumance to one based on agricultural stock-raising. Now the trend is toward industrial stock-raising.

As a consequence of penning or enclosures, there has been an increase in demand for fodder, such as maize, cereal residues and other foodstuffs. Originally, penned livestock consumed products coming directly from the fields, but intensive production has given rise to compound feeds which are now essential in all agricultural stock-raising farms. Some farms use them to feed the animals at some point in their life, others give feeds containing all the nutrients required for their growth, and finally some farms use feeds as an essential concentrated supplement to balance food resources collected on-farm. In all three cases, the feeds are essential for successful intensive stock-raising.

Since the beginning of the 20th century enclosure through sheds or pens has spread throughout the European countries of the Mediterranean Basin, and there has been a trend towards the construction of factory farms with the intention of adapting them as far as possible to industrial production methods. Many farmers became stock-raisers, and others, whether linked to the countryside or not, sought to rationalize livestock exploitations to obtain peak production, since it seemed a good source of income. The arrival of factory farming is an example of the introduction of new technologies. To help to intensify the yield of penned animals, machin-

ery was introduced and interest was taken in the livestock's food regime and their hygienic conditions. After penning became common, in both intensive and agricultural stock-raising, the livestock or agricultural technical expert also became common, helping to modernize the rural economy, in line with the economic spirit of the age. Since then agricultural and livestock technicians have become an essential part of both agriculture and stock-raising, and it is they who have introduced scientific approaches in these activities.

In the growth of factory farming there has been a tendency to consider the animals within the productive process as if they were just another machine, and comparisons have even been made between penned animals and the internal combustion engine. Reducing an animal's life to simple formulas that express variations in their energy balance has led to the complete dehumanization of agriculture. Yet without these formulas it would have been impossible to establish the modern, rational stock-raising model now practiced in the western world.

The main animals exploited intensively in the Mediterranean regions are pigs for their meat, and chickens for their eggs. Attempts have also been made to pen other animals and rear them intensively, but these efforts have always met with less success, either because the animals are delicate and difficult to breed in cages (rabbits) or because they are not a Mediterranean-type species and thus needs more attention (milk cows). Attempts have been made to enclose the milk cows in sheds, following in the footsteps of the northern countries, with a view to introducing industrial milk production. On the other hand, few attempts have been made to pen the goat, originally from the Mediterranean Basin and a good milk producer, although it has been shown that it would adapt to such conditions.

Disposing of the livestock's organic wastes is a serious problem that integrated exploitations have sought to resolve. Wastes were formerly used as fertilizer, but the intensity of production on farms has led to such an increase in the quantity of dung that it is difficult to use it all as fertilizer. To avoid contaminating the environment with this quantity of excrement other applications have been sought. After dehydration and desiccation it can be used as fertilizer or recycled for use in animal feed. The excrement can also be used to produce methane, since it ferments rapidly.

217 **Cave canem** (beware of the dog), the warning to strangers on many Roman villas. This tradition has hardly changed over the course of time, and this mosaic clearly shows how old the dog's role is in daily life in the Mediterranean. The mosaic reproduced is from the National Archeological Museum in Naples.
[Photo: Scala]

3.5 Pets

Over the last few years, there has been an increase in the number and diversity of pet species, especially in developed countries. The most common animals kept as pets are dogs, cats, caged birds and aquarium fish, but others, such as rodents and reptiles, are also kept and bred. It is now very difficult to talk of the pets specific to a given area, as international trade means it is now possible to find any species or breed in any country in the world; some breeds and species have even been changed by selection outside their country of origin, and then returned to it. All that can be said with certainty, and not even this in all cases, is the area that they originally came from. The Mediterranean region, an early center of plant and animal domestication, was also the site where many breeds and varieties of pet originated.

Mediterranean breeds of dog

The most typical of all pets is the dog (*Canis familiaris*). There are many contradictory opinions on the dog's origin but many researchers now consider that modern dogs, except those in the Nordic countries and those from China and Japan, are derived from wolves that lived in the eastern Mediterranean. The first precise evidence for a close relation between humans and dogs come from the Natufian remains of the coastal plains of northern Israel, mainly the skeleton of a young dog together with a human skeleton, dated by archeologists at between 10,000 and 12,000 years old. There are now about 400 recognized breeds of dog in the world, and this diversity is due to two influences —that caused by the environment and and that caused by humans who in the process of domestication and selection have sought to develop breeds adapted to several different purposes. Some breeds are originally from the Mediterranean biome. Probably one of the oldest uses of the dog was to help shepherds to watch and guard livestock, especially sheep. The sheepdogs of Mediterranean areas belong to very different types and they cannot be said to belong to a single, well-defined breed like northern European sheepdogs. It is, however, possible to distinguish breeds, such as the *berger de Brie* and *berger de Beauce* from France, and the *abruzzo* and *bergamasco* from Italy. The two most common races in the Iberian Peninsula are the Spanish mastiff and the *gos d'atura català* (Catalan sheep dog). The Spanish mastiff is strong and robust and was essential to protect livestock from wild predators (wolves, bears, and lynx). The Catalan sheepdog is smaller than the Spanish mastiff and was not used in areas with wolves, since it would have been the first victim. Today, the replacement of seasonally migrating livestock by intensive stock-raising in pens means the sheepdog is disappearing and has almost ceased to exist as a pure breed. In North Africa, especially in the territories of the Berbers, in the Middle Ages there was a breed known as the Arabian harrier, descended from the famous Egyptian harrier that Arabic nobles used to catch gazelles, wild boar, and hares. The Arabian harrier, without a doubt the most intelligent of all harriers, has a long head, a slightly pointed muzzle, large, dark eyes, hanging ears, and a drooping and pointed tail with a sharply curving tip. Its short sandy coat is dense and smooth and its face is black with a dark circle around the eyes, although other colorings are also known. The Balearic hound, from the Balearic Islands and especially Ibiza, where it is known as the *ca eivissenc* (Ibizan hound), is a direct descendent of the Egyptian hunting dog or the Pharaohs' dog, said to have been taken by the Phoenicians to the island 3,000 years ago. However, it is more likely to have been brought by the Arabs during the Islamic period (10th-13th centuries). The Ibizan hound is used to hunt small game, either alone or in groups, as unlike other harriers it has an excellent sense of smell. Its head and muzzle are very long and narrow, the light amber or caramel eyes have a lively expression and it has long, robust limbs and a long tail. The coat may be short or long and may be white and red, white and tawny or a single color, either reddish white or tawny. It is now spread throughout much of the western Mediterranean and the Canary Islands. On the island of Crete there is also a similar breed of dog. Another race from the Balearic archipelago is the Majorcan sheepdog or *gos d'atura mallorquí*, known there as the *ca bestiar* or *ca garriguer*, whose distribution is now restricted, and it is probably the result of crossing of the island's native sheepdogs with Balearic harriers, pointers, and Spanish mastiffs.

218 The *gos d'atura* (Catalan sheepdog, upper photo) and the *ca eivissenc* (Ibizan hound). These two breeds of dog (*Canis familiaris*) are from the western Mediterranean. The Catalan sheepdog is long-haired, and highly skilled at herding livestock. The Balearic harrier—a skilled hunter—is an elegant Ibizan breed with short hair. [Photos: Xavier Ferrer and AGE Fotostock]

Some of the breeds of dog in the Mediterranean Basin were not selected by utilitarian criteria, such as their suitability for specific tasks, but for their small size. Long before the Christian era (there are written records from ancient Greece), dogs had already been selected for purely aesthetic and recreational motives. These dogs were small with long hair and almost always white. At least five breeds with these characteristics were developed—the Maltese, the Bolognese, the Havana, the Bichon and the Lion dog. One of the first breeds selected from these kept dogs was the Maltese, which appears to be derived from the Egyptian guard-dog. Although the Romans called it *canis melitae*, no ancient remains or records of it have been found on Malta. As a result several hypotheses have arisen over its origin—its may be named for a small city in Asia Minor called Melita, the Adriatic island of Melita (now Mljet) on the Dalmatian coast near Dubrovnik, or according to Strabo, a promontory of the island of Sicily. All these hypotheses relate the breed to the Mediterranean and Greek colonization. Maltese dogs have a long, dense, silky, white coat, a flattened cranium, a very short muzzle, dark eyes with a lively expression, triangular hanging ears, short legs, and an upright tail.

Cats among humans

The cat (*Felis catus*) is another very widespread pet. This species was domesticated after the dog, probably about 4,000 years ago, and this is believed to be the reason why it is still relatively untamed. Apparently it was first domesticated in ancient Egypt and sailors and traders in the eastern Mediterranean, mainly the Greeks, were responsible for its initial expansion through their sea and land voyages; as they were aware of the animals skill at hunting rodents, they took it with them. This means that the cat is as cosmopolitan as the dog, although its domestication took place later and is incomplete.

The distant ancestor of the domestic cat was the African bush cat (*Felis silvestris libyca*), although the European wild cat (*F. silvestris silvestris*) may also have played a role in its evolution, especially in the case of the European tiger-like cat, the breed most widespread throughout the world. This breed was known for a long time as the alley cat, because its owners made suitable holes called catholes in the doors and windows, allowing the cats to enter and leave at will. It should be pointed out that, although it was well-known and valued (even as a foodstuff) in antiquity, during the Middle Ages the domestic cat was almost absent from Europe except perhaps for some areas of the Mediterranean. This disappearance is an enigma, although some authors suggest that the cat was not then widespread and that only rich families bought cats as mousers. Wars finished the trade in cats in Europe until the Crusades once more brought them from the Near East.

Yet in the Mediterranean the reduction of cat populations could not have been so great. A study of different cat populations of the Iberian Peninsula's Mediterranean coastline has shown that the cats of Barcelona are genetically very close to those of Athens and other eastern Mediterranean cities. On the other hand, this population is very different from

219 Stray domestic cats (**Felis catus**) socializing in the internal courtyard of a city block in the Eixample district of central Barcelona. Mediterranean cities have large populations of stray cats, eating refuse and hunting rodents and birds.
[Photo: Joaquim Reberté & Montserrat Guillamon]

220 A flock of budgerigars (*Melopsittacus undulatus*) in the wild in Australia. In the open spaces of arid Australian regions it is common to see groups of budgerigars in the air. Budgerigars are very gregarious birds, and some flocks have been estimated to contain more than a million birds.
[Photo: Denis & Therese O'Bryne / ANT / NHPA]

the other populations of the Mediterranean coastline of the Iberian Peninsula, which are closer to the cats of western Europe and North Africa. This is due to Barcelona's commercial links to the eastern Mediterranean Basin during the first half of the 16th century, when the human population of the city reached 30,000 for the first time. This threshold in the human population appears or be related to the founder threshold (about 3,000 animals) above which a cat population become resistant to genetic changes. This explains why cities like Alicante and Murcia, which did not reach 30,000 inhabitants until the end of the 19th century and which then had commercial relations mainly with France's Mediterranean ports and Dutch ones, have cat populations closer to those of Marseille, Amsterdam, and Utrecht than to those of Barcelona. The European cat belongs to the purest of the breeds, since its selection took place without human intervention; its coat is short and dense and shows a great diversity of colors. The different varieties are distinguished by variations in the color of the fur (single color, two-color, smoke, tortoiseshell, and tabby), the length of the hair, and the color of the eyes.

Caged birds

Another widespread pet is the caged bird. In fact, there are few species of bird that have not, at some time, been caged, but not all of them have adapted to this life and only a few breed in captivity.

Pigeons and budgerigars
Pigeons (*Columba*) were probably the first domesticated birds, about 6,000 years ago in the eastern

221 **Songbirds of the Mediterranean Basin**. The linnet (*Carduelis cannabina*), serin (*Serinus serinus*), goldfinch (*C. carduelis*), chaffinch (*Fringilla coelebs*), and greenfinch (*C. chloris*) are often kept in cages as pets. The songs of these birds may have different melodies and different meanings. Songbirds are usually born with a predisposition to sing, but they often need to be in contact with adults of the same species in order to learn how to sing properly. Songs also differ depending on whether the bird once lived in an open or a woody area. In open areas the song is higher. Their songs serve to mark territories, to find mates, or to warn of possible danger. The more precise the function of the song, the more specific the song. For example, alarm calls are registered at the same time by several species which share the same predator, while those calls aimed at a potential mate, or territorial calls, are only directed towards individuals of the same species.

[Drawing: Marisa Bendala]

Carduelis cannabina

Serinus serinus

Carduelis carduelis

Fringilla coelebs

Carduelis chloris

Mediterranean. Classical and biblical mythology show people's interest in these birds. Humans probably started by hunting the rock dove (*C. livia*) for food, and later, domesticated them for food. Subsequently it was discovered that some varieties had the ability to return from long distances and they were used as messenger pigeons.

Although budgerigars (*Melopsittacus undulatus*) are now very well adapted to life in captivity in Mediterranean regions throughout the world, they are, in fact, from the arid and semi-arid regions of Australia, where they live in relatively large groups, depending on the food they can find. They also live in the Mediterraneans of south Australia. The budgerigars are the only members of the parrot family (Psittacidae) that are truly domesticated, and they have spread over virtually the entire world, although they prefer warm climates, since they cannot tolerate the cold. They were introduced to Europe by John Gould who brought a single pair to Britain in 1840, and their introduction into the Mediterranean appears to have started in the last quarter of the 19th century through Gibraltar and other ports frequented by British sailors. These birds' popularity is attributed to their intelligence and the relative ease with which many, but not all, learn to imitate a few words or simple phrases. They are small, about 7 in (18 cm) long, and show great variation in coloring, although the most common is the light green wild type with a yellow head and upper parts with yellow and black transverse markings. Some other Australian caged birds of the same family are from regions with a Mediterranean climate, such as the Adelaide rosella (*Platycercus adelaidae*) and western or Stanley rosella (*P. icterotis*).

Songbirds

In many places, humans hunt and domesticate songbirds merely to enjoy their song and their company. Among the most frequently caged birds are the chaffinch, goldfinch, linnet, serin, and greenfinch. Generally these birds are captured with lime or with nets, an activity that is now controlled by regulations that vary from country to country, and even from region to region.

The chaffinch (*Fringilla coelebs*) is the most common fringillid. In autumn the sedentary population is reinforced by individuals arriving from higher latitudes. In the wild it is a partial migrant and nests in all types of forests in not very high branches in trees and shrubs as well as in gardens and parks. It is distinguished by its double white wing bar and white outer tail feathers. It sings when perched. Its song is short, melodious, and consists of about twelve notes finishing in a "chu-wee-o." It also has many regional variations, since it is a bird with a very broad geographical distribution.

In the wild, the goldfinch (*Carduelis carduelis*) is common in gardens, fruit and vegetable orchards; it nests in trees, generally very close to the tip of a branch and occasionally in hedges. Like other finches it follows a seasonal migratory pattern; in winter it is more common in Mediterranean areas. It is highly appreciated as much for its beautiful coloration as for its song, and shows almost no sexual dimorphism (the size and color is similar in both sexes). Its song, which also serves as its call, is a frequently repeated liquid trill "tswitt-witt-witt," while it also has another more guttural call like that of the canary that incorporates variations on the notes in the song.

The linnet (*Carduelis cannabina*) in the wild lives in open fields with hedges and shrubs, gardens and parks, with a clear preference for Mediterranean-type matorrals. In winter it flies in groups, generally above open fields. Its call is a "tsooeet," while its normal song is agreeable and varied, mixed with pure and nasal notes, generally delivered from the top of a shrub.

The greenfinch (*Carduelis chloris*) in the wild is common in open cultivated fields, gardens, parks. It nests in hedges, shrubs, and small, mainly evergreen trees. Human alterations of its natural habitats have brought it into close contact with human beings. Its call is a strong and rapid with a short, repeated "chup" or "teoo;" in the breeding season it emits a prolong nasal "dzhweee" noise; its song is delivered from the top of a tree or in flight and is a guttural mixture of calls.

The serin (*Serinus serinus*) is a tiny yellow-striped finch. In the wild, it lives in gardens, parks, vineyards and open woodlands, areas where the plants they feed upon are common. The call is a brief "tsooeet" and a harsh "chit-chit-chit;" its song, which it enerally delivers when posing at the top of a tree or telegraph pole is a guttural whistling or tinkling sound.

In addition to these species there are others that are also kept in cages: in the Mediterranean area these include the house sparrow (*Passer domesticus*), the blackbird (*Turdus merula*), mistle thrush (*Turdus viscivorus*), turtle dove (*Streptopelia turtur*), and the partridge (*Perdix perdix*). They are all easy to care for and can live in very small cages.

4. Management conflicts and environmental problems

4. Humanization that goes back to the ancient past

Few areas of the world have been so changed by humans over such a long time as the Mediterranean Basin, especially the eastern part. In fact, for millennia, humans have totally dominated the Mediterranean landscape.

The extraction of environmental resources

As is well known, the eastern Mediterranean was one of the first areas in the world where plants and animals were domesticated. The expansion of the Fertile Crescent agricultural complex led, as early as the Neolithic, to the complete destruction of the original vegetation in the low-lying areas and

222 Sandstone extraction; this is a modern sedimentary rock whose homogeneous texture makes it easy to work. The stone can easily be cut from the mass of rock in blocks. Quarries, like this one in the island of Favignana (one of the Isole Egadi, off Sicily) are common on many Mediterranean islands, and clearly show how humans in the Mediterranean have transformed, over the millennia, even the geomorphology of the landscape.
[Photo: Lluís Ferrés]

me fu una moria in molti luoghi e moziono migliaia di p....

223 An image of the victims of the Black Death in the 14th century Sercambi Codex kept at the Archivo di Stato in the Tuscan city of Lucca (Italy). Nobles, commoners, and Catholic dignitaries alike are harvested by the scythe of death, the Grim Reaper, epitomizing the illness that decimated the Mediterranean's population from the mid-14th to the mid-15th centuries (as well as the population of the rest of Europe and Asia). The Black Death is estimated to have killed about 25 million people in Europe, and it is easy to imagine the negative effect on population growth. [Photo: Scala]

river terraces most suitable for cultivation, followed in turn by intense soil erosion. About 5,000 years ago, in the Bronze Age, the domestication of some fruit trees meant intervention of farmers in shaping the landscape extended to sites at medium altitudes in the mountains, but the destruction of the original vegetation was not completed until the introduction in the third millennium B.P. of iron working, as this meant people could make tools to uproot woody plants and trees, and to create clearings in woodlands at higher altitudes to plant crops. Even where the slopes were steep and rocky, the smallest space suitable was planted with olives, almonds, or grape vines, and plots that could not be worked were grazed by sheep and goats.

The unfortunate combination of deforestation, fires, and overgrazing gradually led to the increasing aridity of many areas in the Mediterranean Basin, especially those that were already drier or less fertile. In fact, in the Mediterranean area, land and natural resource use models diversified and rapidly became more complicated. Studies of about 50 Neolithic Bronze Age settlements in central Italy revealed up to eight different resource use strategies, but all combined agricultural use of the best land near the settlement, the grazing of scrub, pastures, or dehesas a little further away, and the extraction of forest products from the steeper or less agriculturally valuable land. The differences between these strategies are related to the use as pasture of land nearer to, or further from, the settlement depending on whether the nearer land was agriculturally productive enough to justify herding the livestock the distance required.

Demographic pressure

Over its history the population of the Mediterranean Basin has grown neither continuously nor homogeneously. It has in fact undergone large fluctuations that have been unequal and related to changes in the model of exploitation. In the Neolithic, for example, there was a population explosion, shown by the traces of agricultural and stock-raising activities, and by their spread throughout the area, even to places that now seem totally unsuitable for agriculture or sites that must have been difficult to reach with the means of transport of the time. In the Cerdenya (Cerdaña) valley in the eastern Pyrenees, there were farmers and stock-raisers in the first half of the seventhy millennium B.P. In the Col de Vizzavona in Corsica and on the north face of the Aigoual in the Languedoc, there were human settlements at altitudes of more than 3,200 ft (1,000 m).

The first centuries of the present era , the height of the Roman Empire, saw another period of sustained population growth. Nevertheless, in the centuries after the barbarian invasions populations declined in many areas and stagnated in others, doubtless coinciding with some recovery in the area covered by forest. From the 10th to the 14th century there was some population growth, which was cut short from 1346 onwards by the Black Death, whose effects varied greatly from place to place. For example, while Milan was almost unscathed Pisa lost almost two thirds of its population and in many places populations did not recover to pre-epidemic levels until the 17th or 18th century. Since this last recovery, population growth has been virtually continuous (despite some local interruptions due to war and epidemics), although since the mid-19th cen-

224 **The aqueduct over the river Gard**, near Nîmes, Languedoc (France) is locally known as the Pont du Gard. It is 902 ft (275 m) long with a maximum height of 164 ft (50 m), and clearly shows the Romans knew how to transform areas like this as long ago as the 2100 B.P. This type of construction allowed large volumes of water to be transported effortlessly over long distances, meaning irrigation could be introduced and towns and villages could thrive. Mediterranean agricultural space and the urban space that is now being replaced by industrial space, thus go back a long time.
[Photo: Aisa]

tury growth has been mainly in urban areas, and in the 20th century it has been mainly in the south and east of the basin. Farming and grazing pressure on forests has varied with these population changes, causing the area under forest to expand and contract.

Agricultural and industrial activity

The political unification of the entire Mediterranean Basin under the Roman Empire led to the spread of cultivated species and varieties and the techniques for their cultivation from one place to another, as well as the colonization of extensive unploughed areas, especially in the Iberian Peninsula, Gaul, and North Africa. It also led to the creation of large agricultural production units, villas, and the specialization of some areas in certain products (Sicily and Africa specialized in growing wheat, and Baetica [Roman Andalusia] in olives, etc.). Roman agronomists also introduced the ancient classification of the Mediterranean space into three categories, *ager, saltus, et sylva*, in other words, cultivated land, pasture land, and the remaining spaces as woodland.

For Roman agronomists, these three spaces were complementary from an economic point of view; each supplied something to the production unit, whether it was the patrician landowner's villa worked by slave labor or a community of free peasants. This model of occupation of space has survived for a long

time in many areas and, where it is still maintained, has been very effective at preventing the deterioration of the productive base by absorbing disturbances of all types. The grid structure of the most mature Mediterranean systems (*sylva*) around the other systems, which become more and more simple (the matorrals and pasture forming the *saltus* and the cultivated land of the *ager*) the closer they are to the artificial, built space of the *villa* or *urbs*, has been compared to similar structures, recognized in other ecosystems such as the phytoplankton, that are subject to exploitation because of their nature, but which tend to return to a reticulate structure after temporary destructions by each disturbance.

Yet it seems clear that what we now consider as the Mediterranean landscape, a mosaic with an almost inseparable mixture of woodlands, matorrals, meadows, grazing land, cultivated plots, and inhabited areas, is in fact a remnant of the Roman period that has survived war, invasion, plagues, fires, and disasters of all types since the fall of the Roman Empire. It has persisted through the centuries of instability after the fall of the empire until the Industrial Revolution and the present day. In the Mediterranean, a complex and uninterrupted dynamic of local, short-term changes and high environmental diversity has been combined with some degree of long-term stability, although there has been accumulative soil degradation due to agricultural practices on difficult sites. The abandonment of cropland never leaves the soil in the same state as it was before it was ploughed, and the woodlands that regenerate never regain their former splendor.

The Industrial Revolution reached only a few parts of the Mediterranean Basin and was relatively late. Almost the only industrialization before 1800 worth mentioning was the production of cotton fabrics in Catalonia. In the 19th century, industrialization continued in Catalonia, accompanied by some areas of the Languedoc, Provence, and the northern regions of Italy. Even at the end of the 20th century the active agricultural population exceeds the industrial workforce in most countries on the Mediterranean's southern coastline, as well as in Turkey, Albania, and some parts of southern Spain and southern Italy. In Albania, Turkey and Morocco, more than half the workforce is still engaged in agriculture.

Transforming influences in the overseas mediterraneans

The general situation in the Chilean mediterranean shows no essential differences from that of the Mediterranean Basin, although its impact has not been so intense nor over such a long time, as a result of the low population until recently. The Californian mediterranean, with its more intensive use of the land based on irrigation and advanced technology, has also developed a model of land use reminiscent of the Mediterranean areas of the Old World, even though the urban areas are totally different. The Cape region and the Australian mediterraneans are very different and their land use has followed the logic of colonial settlement of previously empty areas, or those occupied only by very small groups of hunter-gatherers in Australia, and more-or-less nomadic shepherds in the Cape region. Only urban growth and the importance of the urban population in the overall territorial context appear to converge with the "Mediterranean" model of land occupation.

The current Mediterranean environment

Altogther, then this space has been greatly changed, and all that is left of the pre-Neolithic ecosystems is a few scattered patches in the most inaccessible areas. The flora is rich and diverse, but many species are threatened, the vegetation is continuously exploited and has to face unfavorable conditions, and the fauna in danger of extinction. In addition, to get a broad perspective of the environmental problems of the Mediterranean Basin, it is necessary to take into account other factors such as the omnipresent urban structure, the high population, and the intense activity on the territory as a whole.

To sum up, in the Mediterraneans the problems of environmental management are mainly based on an excessive population growth that is highly concentrated in urban centers dependent on a larger area with varied but limited resources. Resources that were subject to seasonal or year-to-year oscillations and shortages that arise from the biome's special characteristics (the water shortage in summer, when it is most important, highly variable harvests that depend on the year's climatic conditions, and disturbances such as autumn floods, hail, relatively high frequency of late frosts, etc.). The incorrect use of some of these resources, mainly water and forest resources, can indirectly worsen many of these problems and even create new ones (an increase in the frequency of fires, contamination of watercourses and aquifers, increased erosion, etc.).

4.2 The Mediterranean forestry areas

The former splendor of the forests

The relative agro-sylvo-pastoral balance mentioned above, does not mean that the entire Mediterranean Basin is lacking in environmental management problems. To the contrary, where this model of organizing space has never been implanted, or where it has been abandoned for an excessively long period, or is now feeling pressures due to changes in the activity of the population, the inherent fragility of the Mediterranean's natural environment is cruelly laid bare. Ancient cities, paradoxically located in what are now semi-deserts, ruins buried under meters of fluvial deposits transported from eroded slopes, ancient mosaics showing hunting scenes in places where the vegetation is now little more than a little sparse scrub, traces of seafaring civilizations in the middle of treeless regions, the lush forests that ancient travellers described that do not exist and could not grow there now; these all show the profound transformation the Mediterranean biome has undergone, sometimes attributed to hypothetical climatic changes that have led to increasing aridity over the last few millennia. There is no consensus on this point, but is seems clear that there was a period wetter than the present that finished about 5,000 years ago, and since then there have not been major climatic changes, and thus the problem of forest degradation is essentially due to human activities over the centuries.

The current situation of the Mediterranean's forests is that they are highly degraded, the price paid for a

Good and bad government

Effeti del buon governo in campagna (The effects of good government in the countryside), Ambrogio Lorenzetti (1285-1348), Palazzo Pubblico (Siena) [Scala]

If anywhere in the world best sums up and expresses the essence of the Mediterranean landscape, it is the region of Siena. Light, color, relief, vegetation, human settlement, land use, all appear to be to have reached a delicate balance between restraint and exaggeration, generally considered the main characteristic of the Mediterranean world. This balance was captured more than six centuries ago by the awareness and painting skills of the first European master of landscape painting, Ambrogio Lorenzetti, in the wonderful frescoes of the Sala della Pace of the Palazzo Publico de Siena.

And his images of the effects of good and bad management on the countryside are landscape paintings, especially the one dealing with good management. His artistic awareness captured all the the features that characterize exactly the Mediterranean landscape, all still on view in Sienna. The painter depicts a hilly landscape with little room for flat ground. A landscape that was almost entirely covered by crops where only a few areas (not even the furthest) appear deserted. There is always a house, a castle, a small village or some people reminding us that this landscape cannot be understood without taking human presence into account. If confirmation is needed, the images of the consequences of bad government show the same landscapes without human presence (except for a few soldiers and somebody being chased by two others with clearly evil intentions) and houses in ruins. Could anything else have shown so clearly that the Mediterranean landscape is so profoundly humanized that it cannot be conceived without human beings? Could anything else have shown so clearly the role of human beings, part of Mediterranean systems for millennia, and whose absence can only be associated with devastation and ruin?

Between 1338 and 1340, Ambrogio Lorenzetti completed his commission to paint fresco decorations for the new rooms of the Palazzo Publico de Siena that had been built to hold the meetings of the city government. The theme chosen was the allegory of the effect of good and bad government on the city and on the countryside. A conventional enough subject, exactly what would be expected from an official commission from the oligarchic government of Tuscany in the 14th century. But Ambrogio Lorenzetti was not a conventional painter. He had already finished paintings without a single saint or virgin, that lacked representations of individuals, battles or miracles. He painted things like a castle on the shore of a lake or a seaside city. He painted, in a word, landscapes.

Effetti del buon governo in campagna (The effects of good government in the countryside), Ambrogio Lorenzetti (1285-1348), Palazzo Pubblico (Siena) [Scala]

Beyond the cultivated areas it is just possible to make out some hills topped with matorral, and a few more covered with scattered trees that look like pines and holm oaks, and some forested valleys in the distance. It is also possible to make out a patch of forest that is lighter in color, probably fringing forest next to a mill. These steep slopes, hidden and sparsely populated valleys, and riverbanks were the last remnant of the Mediterranean's natural ecosystems.

Natural, perhaps, but used profitably. On one of the more clearly visible hills, covered with matorral there appears to be a dog chasing a rabbit towards a hunter's hide; elsewhere small flocks or individual animals appear to be grazing while one of the beasts of burden entering the city is loaded with firewood. The painter also knew how to convey the total interconnection of the society of the time and the Mediterranean space. In the background is the sea, with a castle defending the port, probably the port of Talamone. The citizens of Siena were obsessed by access to the sea, even if it meant crossing the unhealthy marshes of the Maremma region of Tuscany. This obsession is easy to understand in a time when the sea was the easiest communication route, and also because in the Mediterranean Basin no place is far from the sea.

The cultivated fields are totally dominated by cereals, presumably wheat. Wheat was the staple foodstuff and in the 14th century a city like Siena had to ensure its bread supply. This is why so many cereal fields were needed so close to the city, unless wheat was to be bought at a high price in the markets where surpluses existed. Showing beyond all doubt the importance that he and his fellow citizens attributed to wheat, the painter represented all the different tasks relating to wheat: ploughing, sowing, harvesting, gleaning, threshing, and the transport of sacks of grain to the mill.

Buon governo (detail)

Wheat is not the only crop visible in the frescoes in the Sala della Pace. Grape vines are present and so, to a lesser extent, are olive trees, and there are even some plots where it is possible to make out the presence of the "promiscuous cultivation." The most pleasing examples of this type of agriculture are still to be found in Tuscany. The artist also took care not leave out a small vegetable plot behind a house with a building that might well be a henhouse.

Commerce was the basis of Siena's prosperity, and appears in the fresco representing the effects of good government. Both the city and the countryside show intense traffic of merchandise. It is worth pointing out that there is not a hand-cart to be seen, as all the transport used beasts of burden—asses and mules. Horses were less common and the only ones to appear are the mounts of persons of high rank. Beasts of burden to transport goods and oxen to plough fields are shown as the basic sources of the exosomatic energy used by the people of Siena in the 16th century. The mill is the only technological device shown that delivered exosomatic energy for human use that was not derived from muscle power.

Also in the part corresponding to the city, Ambrogio Lorenzetti included exquisite details of humble (but significant) and typical features of the Mediterranean, such as the swallow's nest below a balcony, women taking hens or eggs to market, a shepherd with a flock of sheep, a carrier with animals laden with firewood, flowerpots on terraces and windowsills, and especially the role in Mediterranean cities of the streets, and to a lesser extent the plazas, as a public spaces in full view of the world and open to curiosity (to commerce in the broadest sense of the term). Even in the fresco showing the ruin and devastation caused by misgovernment, it is in the street that we see the corpses of the hung and the woman attacked or kidnapped by two thugs.

uon governo (detail)

Buon governo (detail)

Buon governo (detail)

225 These deforested hills in Crete were covered in forests in antiquity. The splendor of Minoan civilization relied on large inputs of timber for construction and shipbuilding, and of firewood as fuel, but this is difficult to reconcile with the sparse woodlands now covering most of the island. Even the farming and livestock installations built in recent centuries, like the one in the photo, seem out of place in today's sparse forests. This has occurred repeatedly throughout the Mediterranean Basin.
[Photo: Jordi Bartolomé]

few millennia of civilization—at least in the part of the biome found in the Mediterranean Basin, since intensive agricultural practice and exploitation did not reach the other mediterraneans until the recent colonization by Europeans. Ever since humans first started practicing agriculture, the best land has been cultivated, and forests have been banished to the worst ground, the steepest slopes, and the poorest soil. Over 2,000 years ago, Theophrastus referred to this quite logical practice (as did other classical agronomists). He recommended the using the best soil for grain production and leaving the worst soil for trees, including tree crops such as olives, carob, etc. This is clearly stated in the Latin maxim, *Quid est agricola, sylvae adversarius* (the farmer is the enemy of the forest).

The overseas mediterraneans' forests before and after colonization

In the Californian, Cape, and Chilean mediterraneans, the forests underwent great degradation after European colonization, which was accompanied by the introduction of crops and livestock as well as direct forest exploitation. Yet, with the possible exception of the Norte Chico in Chile, irreversible soil deterioration and landscape changes have never become as serious as in the Mediterranean Basin. In Australia, soil poverty and relatively low population densities have saved many forest areas, especially in the Western Australian region. In this case, the forests have only suffered an intense exploitation that has changed their appearance, as happened to the mag-

nificent jarrah (*Eucalyptus marginata*) forests the colonists found, although it is not possible to talk of degradation or major, irreversible changes.

The Chilean forest space

In the Chilean mediterranean, the arrival of the first Neolithic agriculturalists led to the felling of many, often wide, clearings in the central valley and this was the situation found by the Castilian colonists who fought the Mapuches for it. It seems the city of Santiago was founded in a region that had already been cleared of its forests, and it soon became difficult to obtain timber and firewood, and timber had to be brought from San Francisco el Monte (now Copiapó) and the Precordillera, and the Castilian authorities took measures to control felling. The testimonies of some writers from the early colonial period, such as Ovalle, described Santiago as being surrounded by "dense scrub;" this should doubtless be interpreted as meaning that there had already been an invasion of large areas by espinales of *Acacia caven* and other similar formations, genuinely comparable to the sparse scrublands of Castile and Estremadura. Outside the central valley, deforestation must have been less important, as it is known that the city of Coquimbo, at the northern edge of the Mediterranean region, was founded in the middle of a basin covered by plants belong to the myrtle family (Myrtaceae) and there are records that the forests of the La Dehesa region, nearer to the Andes, were still considered inexhaustible in the 18th century. In fact, apart from the central valley and the areas around major cities, the deforestation of the Chilean mediterranean did not really become important until the first half of the 19th century, especially after Chile became independent. At that time, heavy demand for timber and firewood around mining and metalworking industries devastated all the woody formations, such as the matorral and even the riverine forests, to such an extent that they have never recovered in the drier, more northerly, areas. To the south of Santiago, where no minable resources were found, it was cereal cultivation that grew at the expense of the matorral and forests, together with the spread of stock-raising, and this virtually eliminated forests from the lowlands and mid-mountain region, and now virtually none of Chile's natural Mediterranean forest is left.

The Cape forest areas

Many authors believe that there were never any forests in South Africa other than the Afro-montane formations of the southern slopes of the coastal mountains facing the Indian Ocean. Perhaps this is why, although

trees seemed an inexhaustible resource when the Cape colony was established (the shady side of Table Mountain has a relatively dense tree cover), only six years later felling of the *geelhout* (*Podocarpus*) was restricted and the patches of forest nearest to Cape Town were successively exploited, especially those in the Swellendam region (heavily exploited during the 18th century). Probably only the distance from Cape Town and the lack of a nearby secure port explains the survival to the present day of most of the Afro-montane forests in the Knysna region. Some authors believe that the tallest formations of the "fynbos," in spite of soil nutrient deficiencies, could have evolved into forest-type vegetation, if it had not been for the repeated forest fires since the arrival of the first Khoikhoi shepherds, about 2,000 years ago; also perhaps even, earlier due to the actions of Bushmen hunter-gatherers in search of bulbous plants.

226 *Espinal* **converted into grazing land in the Chilean mediterranean**. Stock-raising activity has left only a few individual specimens of the espino (*Acacia caven*) of the initial dense matorral. Without a doubt the landscape that the first Spanish colonists found consisted of much denser espinal than those found today. The espino is still abundant, but it has been heavily exploited to make charcoal.
[Photo: Lluís Ferrés]

227 Chaparral next to riverine oak forest near San Diego (California). This combination of plant formations can still be seen in the less modified lower areas of California's mountains and must have been widespread when the first Europeans arrived. The different colors of the chaparral correspond to dominance by different shrub species.
[Photo: Lluís Ferrés & Ramon Folch]

The Californian forest space

When the first missionaries arrived in California, the coastal area and the lower slopes of the surrounding mountain ranges appear to have been dominated by chaparral shrub formations, although the better soils must have had a large area of dehesas of the savannah oak, while plane trees (*Platanus racemosa*), poplars (*Populus fremontii*, known as the cottonwood), and willows (*Salix*) dominated fringing forests. The Indians who did not practice agriculture or stock-raising, regularly set fire to certain areas to clear spaces in the chaparral and thus improve hunting. On the western slopes of the Sierra Nevada, however, an altitudinal zonation was established that included, in the montane level, coniferous forests dominated by western yellow pine (*Pinus ponderosa*), white fir (*Abies concolor*), red California fir (*A. magnifica*), and lodgepole pine (*P. contorta*) together with some deciduous trees, such as California black oak (*Q. kelloggii*).

In strictly Mediterranean California, until the mid 19th century there was no forest management other than the ploughing of chaparral or other shrub formations for use in agriculture or extensive stock-raising. The gold rush led to large-scale, indiscriminate fellings in mining districts to obtain timber and firewood. As California's gold rush only lasted a short time (a little more than 20 years) it did not irreversibly damage forests, except in some valley bottoms that were treated especially harshly. It did, however, lead to a population boom that continued when the gold rush ended. More mouths to feed and a larger workforce meant increasingly large areas of varying suitability came under cultivation. Agricultural production also was diversified by the introduction of market gardening (where irrigation was possible) and vineyards on sunny slopes. More houses built in the forested areas of the mountains also meant greater exploitation pressure on forests, especially on privately owned areas. Yet at almost the same time the mountainous regions of California saw the first actions taken to protect natural spaces (specifically, forests) such as Yosemite Valley, or especially spectacular stands of trees such as the giant redwoods of Mariposa County. The world's first management plan for a natural park was drawn up in 1865 for what became in 1905 Yosemite National Park by the landscape architect Frederic L. Olmsted (1822-1903).

The Australian forest space

In Mediterranean Australia, European colonization did not start until the second quarter of the 19th century, and from the beginning was based on ploughing the land suitable for cultivation or extensive exploitation of areas suitable for grazing, and it did not pay especial attention to forest management in an area where trees were scarce anyway. The only forests that really deserved to be called forests (and still do) were the karri

228 Scattered patches of *Eucalyptus* thickets growing among flat wheatfields near Adelaide (South Australia). Poor soils and water shortages meant these areas could not be ploughed until recently. As the photo shows, however, all that now remains of the mallee eucalyptus are a few isolated fragments. [Photo: Lluís Ferrés & Ramon Folch]

(*Eucalyptus diversicolor*) and jarrah (*E. marginata*) in the southwest of Western Australia and those formed by other species of *Eucalyptus* (basically the "messmate stringybark," *E. obliqua* and the "brown stringybark," *E. baxteri*) in the highest parts of the Mount Lofty Ranges and the western part of Kangaroo Island. There were open woodlands, and even dehesas in the valleys and slopes of these mountains and in other relatively high areas of Southern Australia and the neighboring areas within the states of New South Wales and Victoria. Yet the lower-lying areas with a Mediterranean climate, in both the south of Western Australia and South Australia, were above all, the domain of the mallee, the characteristic Mediterranean eucalyptus scrub.

On the other hand, the first colonists considered that the spaces occupied by the mallee were especially inhospitable and unproductive places, as there was little surface water as the soils were generally sandy. However, colonization of the mallee began tentatively in the 1860s, after the development of technology appropriate for the treatment of this land and, between 1880 and 1886, 1.2 million acres (half a million ha) were ploughed up in the plains to the north of Adelaide. Although this slowed down at the turn of the century, after 1910 the area of land under the plough spread non-stop until the economic slump of the 1930s. This was the period when many regions came under cultivation, such as the area now known as the wheat belt in Western Australia.

In this context it is not surprising that the best specimens of the jarrah (*Eucalyptus marginata*) disappeared almost completely before the end of the 19th century. Its wood, known as Swan River mahogany, was exported to Britain, and was cut into paving blocks and used to pave some streets in London. Nor is it surprising, bearing in mind the predatory tradition of the management systems initially applied and the unusual characteristics of Australia's tree species, that the forest management techniques developed in the Australia's mediterranean (as in all Australian forests) are very different from those used in other parts of the world

The effect of livestock on forests

Grazing is a traditional activity in the Mediterranean Basin and was introduced into the Cape region about 2,000 years ago and into the other areas of the biome only two or three centuries ago. Together with the ploughing of land for agriculture, grazing has led to considerable forest degradation. For centuries, shepherds have burnt forests and matorral to encourage the appearance of tender shoots and herbaceous plants, with the consequent soil degradation that continued fires cause. Furthermore, in Mediterranean forests the livestock may graze both the herbaceous plants and the leaves and shoots of the trees. This seriously inhibits forest regeneration if the grazing load is excessive or if the combination of grazing and felling is deleterious.

The amount of grazing varies between the different areas of the biome, and as in the case of crops, the worst affected area is the Mediterranean Basin, where for centuries the rural economy has revolved around livestock, especially in the regions forming part of the Islamic world and the areas that have historically been influenced by Islam. The mentality derived from the nomadic lifestyle introduced by Islam (sometimes already present in populations before the arrival of Islam) considers large numbers of animals as a sign of wealth, even without taking into account the benefits provided (which, anyway, are always considered greater than the poor results of cultivating not very productive land). This view considers it unfitting to feed the animals with the fruits of one's labor, and considers stock-raising as no more than vigilance to protect the livestock from predators and guiding them to where they can find food for themselves.

The north of the Mediterranean Basin, for example the Iberian Peninsula, was for centuries (13th to early 19th century) an area where the main economic activity was the grazing and transhumance of millions of Merino sheep. This meant that forest management and even ploughing land for cultivation were subordinated to grazing. The situation was similar in the kingdom of Naples. In fact, in some Mediterranean forest environments only their exploitation for timber was allowed, but the abusive way in which it was practiced led to degradation. The combination of forestry exploitation and stock-raising is a delicate one, since an excessive livestock load or its inadequate distribution, or inappropriate combination with felling, may jeopardize regeneration of the forest.

Other characteristics of Mediterranean forest masses are their role in protecting the soil from erosion and regulating the water cycle. Good forest management ought to subordinate forest exploitation to protection from erosion and water regulation, as otherwise there is the risk of losing a non-renewable resource, soil, and worsening the problems of water supply in non-forest areas in the same watershed. Livestock production, soil protection, and water regulation are thus factors to be borne in mind in the sustainable use of Mediterranean forests and in the assessment of their production and profitability.

Today, the greatest problems are in the Magreb, where deforestation as a consequence of overgrazing is advancing rapidly and sometimes gives rise to sights as sorry as some places in the Middle Atlas where it is possible to see the dead trunks of immense cedars that have not been able to survive the lopping of their branches as fodder for goats. Valuable trees in landscape and forestry terms have been converted into a small amount of fodder for livestock, a few branches of firewood, and a magnificent dead trunk in the middle of the forest.

In the Chilean mediterranean, to the north of Santiago, as early as 1870 excessive grazing had eliminated almost all the native vegetation, and even now some areas of the Norte Chico region are little more than desert.

The supply of firewood and timber

Forest exploitation as such, when carried out for centuries in isolated forests in difficult conditions, also leads to degradation. Mediterranean woodlands, especially those in the Mediterranean Basin, have provided both timber for construction and fuelwood. Even now, in the poorest countries, fuelwood and charcoal form part of the daily life of their growing population, and 82% of wood produced in Greece is fuelwood, as is 75% of Turkey's production. In the Magreb, ploughing and overgrazing have reduced the area covered by holm oak woodlands by 75%. Yet to grasp the current status of the woodlands of the Mediterranean Basin, it is necessary to consider timber and fuelwood consumption by many civilizations over the centuries.

The following examples may suffice to give an idea of past requirements for forest products. In antiquity a large metalworking center consumed the production of about 990,000 acres (400,000 ha) of forest. Between the 16th century and the beginning of the age of steam and the first iron-hulled ships, building a single ship required 1,000 metric tons of good-quality timber, the annual production of about 2,500 acres (1,000 ha) of forest, and a ship's working life was usually no more than 15 or 20 years. In the 16th century, a single country such as Spain, maintained a fleet of 2,000 large ships, and the other Mediterranean seapowers, especially France and Turkey, also had large fleets. In these countries at that time, the Navy was in charge of the management of coastal forests and all forests close to watercourses suitable for transporting the trunks. Not all these forests were Mediterranean, and, in fact, the wood considered best for shipbuilding came from the surrounding mountains, such as the Pyrenees and Alps (where the trunks were cut and floated downstream). In the late 18th century, the price of firewood in Mediterranean coastal cities like Barcelona increased more than those of other consumer products, showing the shortage of forest products in the areas nearest the cities.

229 **The destruction of wooden battleships in the battle of Lepanto** (1571) in the idealized painting by the artist H. Letter at the National Maritime Museum, Greenwich (London). Other factors also caused forest destruction, but the need for timber for use in shipbuilding was a major factor in the degradation of Mediterranean forests. This became especially important in the Renaissance, because of the need for high quality timber and its selective effect on particular species or on especially tall or well-formed specimens (used for the boat's main masts).
[Photo: National Maritime Museum, London]

Centuries of navigation, shipbuilding and losses at sea, heating buildings and cooking food with firewood, building houses, extracting minerals and smelting metals, and felling relatively unproductive forests again and again have all contributed to shaping the current forestry scenario in the mediterraneans.

The effects of war

In addition to cultivation, grazing, and forest exploitation, there is a further factor that leads to the degradation of forests—war. To make war is to sink boats, to adopt scorched earth policies when retreating, to move large armies that consume timber and firewood, and to leave large areas without a rational long-term model of exploitation. Making war means, in a nutshell, upheaval in a region that is already fragile, like the Mediterranean, whose history is a series of invasions and retreats, empires that rise then fall, and episodes of devastation. In 2588 B.P., the Babylonians besieged the city of Lachish and destroyed the city walls by heating them for days with many tons of firewood, as shown by a several-meter-thick layer of ashes found in the excavations.

It is not, however, necessary to go so far back in time to find examples of forest destruction caused by war. During the dispute between Greece and Turkey in 1974 on Cyprus, incendiary bombs were deliberately used to destroy most of the best forests on Cyprus. Between these two events, there have been periodic devastations. Furthermore, wars and devastations have favored stock-raising over agriculture, as flocks of animals are goods that can be herded from areas of conflict and can then rapidly exploit the new areas they occupy.

Forests reduced to ashes

A further major problem affects Mediterranean forest and plantation management and conservation—forest fires. Although Mediterranean vegetation has partially adapted to forest fires because it has experienced this phenomenon for millennia, these fires have become a catastrophic, destructive force because of the anomalous increase in their frequency, relative to the intense human influence on the area.

It is a fact that, in addition to being spontaneously integrated into the natural dynamics of Mediterranean vegetation, fire has often been used by the local populations, perhaps most intensively by Australians and western Amerindians, who used it against insects, and as a tool to prepare land for crops, to increase the quantity and quality of grazing, and to reduce the risk of catastrophic spontaneous fires. Now things have changed. Fire continues to be used for these purposes only in Australia and to a lesser extent in California, while most of the western world has sought to eliminate fires. Forest fires are a double-edged weapon—half solution, half problem—in less humanized areas, but in densely populated areas they are clearly a problem and a major problem at that.

The mediterranean peoples and forest fires
In the mediterranean areas, humans have been clearly responsible for forest fires, whether accidentally or deliberately. Fire has been a traditional tool to

eliminate forests and matorral, whether to plant crops or to encourage the development of the short-lived pastures that grow on the vegetation-free soil. This incendiary activity has been important for 10,000–20,000 years, since the Neolithic revolution, although there are traces of the use of fire 100,000 years ago in the Mediterranean Basin and in the Cape mediterranean. The Chilean and Californian mediterraneans have also suffered the use of fire by the basically hunter-stockraiser ethnic groups that lived there. Intense humanization has lasted for different lengths of time in the different areas of the biome, from millennia in the eastern part of the Mediterranean Basin to 200-300 years in the other mediterranean regions. Anyway, anthropogenic fires have had an effect in addition to those of natural ones, and this is a serious environmental problem. In the Australian mediterranean, for example, scars and marks left by fires on the "black boys" (*Xanthorrhoea*, grass trees that sprout a new tuft of leaves after a fire) have shown that in one forest area there had been three fires in the 150 years before European colonization, while in the 150 years after British colonization there were 22 forest fires. In the Chilean Mediterranean region, the natural recurrence of fires every 270 years, has now decreased to 160 years as a result of humanization, which among other things has led to the massive introduction of more pyrophytic species, such as the Monterey pine (*Pinus radiata*), planted on more than 1.8 million acres (750,000 hectares).

Today, due to the high population densities of the Mediterranean regions, fire has become a scourge. In the Mediterranean Basin alone fire destroys more than 494,000 acres (200,000 ha) of vegetation every year. It is estimated that only a small percentage of all fires (about 2%) are caused by lightning, the normal origin of the natural fires in the biome. Anyway, human transformation of landscapes may have unexpected effects, paradoxically protecting the landscape from the fire that this transformation itself favors. For example, human territorial divisions impose physical barriers to fires, meaning there are more of them but they affect smaller areas. Forest fire is a very serious problem in the more developed countries of the biome, as the abandonment of many marginal agricultural and stock-raising areas, together with the increase in the urban population and its great mobility have all combined to create landscapes with inflammable matorral where almost nobody lives, but which many people cross. Forest fires, are thus becoming more common and destroying areas of forest and scrub, as well as affecting timber plantations, crops, and urbanized areas with serious economic and environmental losses.

In the north of California, the natural vegetation is controlled by the frequency of fires. Complete protection from fire favors forests, while periodic fires favor matorral, and frequent fires favor pastures. In the moister areas, where forests dominate, the tendency is to try and prevent all fires. In drier areas, where grazing is important, the preferred management techniques are the use of fire and the control of grazing. In southern California, however, the chaparral leaves no alternative: burned or not, it persists indefinitely, and this has been a serious problem for the cities and their surrounding areas, as the inhabitants like to be surrounded by natural vegetation, which they find aesthetically pleasing, even though it poses a certain degree of risk. In this case, preventive and controlled burning is not possible because the chaparral has no undergrowth that can be burned separately from a tree layer.

The forestry policies followed by the Californian authorities, based on making firebreaks and extinguishing fires when they occur, as in Europe, give good results. A 1983 study was published that used satellite images taken between 1972 and 1980 to compare the burnt areas in the south of the state of California with the adjacent burnt areas in the Mexican state of Baja California. In Baja California, control of fires is almost non-existent and for this reason there were more than twice as many fires, but each individual fire burned a smaller area as a result of the mosaic of plots differing in age and inflammability. The suppression of fire in the state of California had reduced the number of fires in the chaparral, but the fires were generally larger and more destructive, with the result that a similar proportion of the total chaparral was burned. This allowed them to draw the conclusion that in chaparral total suppression of fires may be counterproductive, as it opens up the possibility of later major conflagrations.

To sum up, every year in the Mediterranean area uncontrolled fires burn hundreds of thousands of hectares (1 hectare=2.5 acres), less than the number of hectares planted or reforested, with the consequent erosion and environmental degradation. The way to deal with fire is basically to take preventive measures and to have the most efficient fire-fighting services possible, but this tackles the problem after it has arisen without forestalling it. An effective policy against fires would consist of attacking the roots of the problem, that is to say by redesigning a territory that no longer functions because it has lost its ancestral organization without replacing with a truly functional new one. In this context, systematically favoring tree communities dominated by pyrophilic conifers does not seem a good approach, regardless of whether they are the

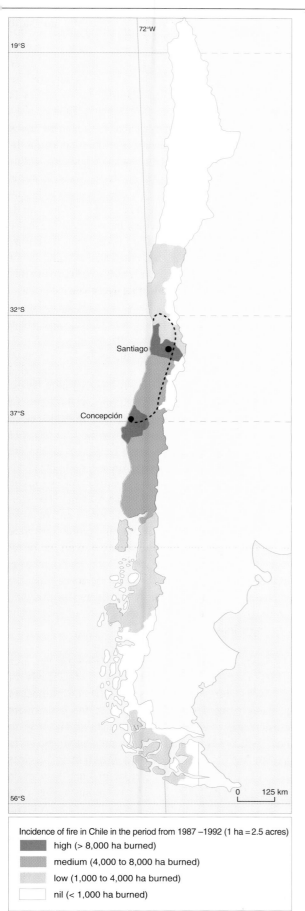

Incidence of fire in Chile in the period from 1987 –1992 (1 ha = 2.5 acres)

- ■ high (> 8,000 ha burned)
- ■ medium (4,000 to 8,000 ha burned)
- ▨ low (1,000 to 4,000 ha burned)
- □ nil (< 1,000 ha burned)

result of reforestation, timber plantations, or the spontaneous secondary forests arising as a consequence of the destruction of the sclerophyllous broadleaf forests.

The effects of repeated fires

The capacity of Mediterranean vegetation to regenerate after a fire is jeopardized when the frequency of fires at a single site is much greater than the rate that ecological mechanisms have been able to assimilat (the consequence of the rate of recurrence of natural fires). Thus, for example, a single plant cannot lose its aerial parts in a fire every year and then produce new shoots indefinitely. The ability to recover from a single fire does not confer the ability to recover from continuous additional fires.

In California, more than half of the species of the genera *Ceanothus* and *Arctostaphylos* do not sprout after a fire but germinate, and thus a change in the frequency of fires may greatly change the species composition. As an example, in a chaparral community that had not been affected by fire for 50 years, there was a reduction in the species of *Ceanothus* that sprout after fires, while there was an increase in the number of germinating species, such as *Prunus ilicifolia* and *Rhamnus ilicifolia*. Also in California, the savannah oak formations consisting of isolated trees scattered among grazing land are partly the result of frequent fires that do not damage large trees excessively; they degenerate if the frequency of fires diminishes, as this allows the invasion of matorral and some pines, meaning the fires are more destructive, leading in turn to the replacement of the tree community by a shrub formation accompanied by some pines.

In the chaparral, however, the measures taken against fires have meant that fire cycles with a frequency of less than 50 years seems to have been replaced by cycles with a longer periodicity, and so the fires are less frequent but more destructive. This is because the accumulation of biomass means the very high temperatures reached may increase mortality among individuals of resprouting species. In this case it appears that when the interval between fires is greater than 100 years, the rootstocks are often totally destroyed by the high temperatures and by loss of the ability to resprout that accompanies aging. At the same time, these fires favor the species that regenerate by germination, as these long periods lead to large accumulations of seeds in the soil and a parallel reduction in cover by species that resprout after fires. The reason for this is that their lifespan is limited, and the death of the old individuals opens gaps that remain empty or are occupied by new individuals less resistant to fire due to their youth and limited height.

230 **The occurrence of forest fires in Chile,** according to data provided by the Statistics Unit of the Fire Management Programme of CONAF (Chile). The map shows how the areas with most forest fires are concentrated within the Chilean Mediterranean (inside the dotted line) and the surrounding areas. Note that most forest fires in fact occur in the plantations of Monterey pine (*Pinus radiata*) and eucalyptus (*Eucalyptus*) that have become common since a government decree introduced grants for planting them.
[Drawing: Editrònica, from several sources]

231 **Fynbos on fire**, near Stellenbosch in the Cape region (South Africa) in the southern autumn of 1991. In the foreground is a specimen of *Protea nitida* whose height protects it to some extent from the fire.
[Photo: Colin Paterson-Jones]

It has been observed that about 90% of the dominant individuals in Cape fynbos communities that have not suffered a fire in 35 years belong to species that resprout after fire, such as those of the genus *Leucodendron*, without there being enough regeneration by germination to cover the empty spaces. In the continued absence of fire, species begin to appear that are the precursors of the "evergreen rain forest" that was formerly abundant in the Cape region, but if there

is a fire the community is dominated by species that reproduce by seeds. The seriousness of uncontrolled forest fires in the Mediterranean biome should not make us forget fire's natural role in sclerophyllous formations. Many Mediterranean shrub formations of species well adapted to fires (and whose growth cycles and reproduction are linked to fire) suffer aging and general failure to thrive if fires do not occur at the necessary frequency. Some shrub formations of the chaparral dominated by members of the Proteaceae become decrepit and do not regenerate if there are no regular fires, as a result of the aging and death of the plants and the total lack of new individuals. The same happens in forest communities, such as the California redwoods communities, in which a fire in the undergrowth is necessary to stimulate regeneration. In other forests, such as the jarrah, growing on poor soils in environments where the breakdown of leaf litter is slow, fire is necessary for the regular recycling of the minerals immobilized in the dead organic matter. In any case, not even the best fire prevention policies can completely eliminate them, so that the most effective policies to reduce the frequency of fires may be the most disastrous policies in the long run. The large accumulation of biomass resulting from the lack of fires leads to a gradual increase in the risk of a fire occurring, and when it occurs, the consequences are catastrophic because of its great size and heat.

Fighting fire with fire: the case of Australia

The European colonization of Australia is relatively recent. This, together with the low average popula-

232 **A controlled prophylactic fire** in Western Australia. These fires are set in the wet season and are closely supervised by the forestry and firefighting services. The aim of the fires is reduce the dead biomass that has accumulated in the undergrowth and to reduce the shrubby undergrowth. This reduces the load of plant fuel in the following dry season, when the risk of a spontaneous wildfire is greatest. This practice is common in the Australian and Californian mediterraneans and forms part of a forest management strategy that seeks to preserve the tree masses, but can only be used in sparsely populated or unpopulated areas that are flat enough to allow the forestry services to maneuver.
[Photo: Ramon Folch]

20–500 kW/m²
Low intensity fire, with
a frequency of 4-6 years.

500–1,700 kW/m²
Medium intensity fire, with a
frequency of 7-20 years.

1,700–3,500 kW/m²
Medium intensity fire,
with a frequency of
7-20 years, and some
trees are damaged if
the fire is severe.

3,500–7,000 kW/m²
High intensity fire with
a frequency of 21-50 years.
Above 7,000 kW/m²; the
fire becomes violent and the
trees are physically damaged.

60,000 kW/m²
or more Wildfire, with
a frequency of 200 years.
Rarely occurs, and only
in moist sclerophyllous
forests. It kills the trees
and breaks them.

Regeneration of species that
produce new shoots after fire.
This favors browsing animals,
such as the grey kangaroo
(*Macropus occidentalis*)
and the wallaby (*M. irma*).

Regeneration of species that
sprout after fire, and some
that regenerate from seed.
This favors browsing animals,
such as the grey kangaroo
(*Macropus occidentalis*)
and the wallaby (*M. irma*).

Favors some epiphytes
and some seeds adapted
to fire. This favors
the tammar
(*Macropus eugenii*).

Epiphytes grow on trees and
in the canopy. Dense regeneration
from seeds. Some species capable
of sprouting after a fire are killed.
These conditions favor the
mardo (*Antechinus flavipes*).

Artificial regeneration with
eucalyptus seedlings.

233 Diagrammatic representation of the effect of the different natural fire regimes that may occur in eucalyptus forests. The intensity of the fire is measured in kilowatts (kW/m²). The intensity of the fires deliberately set by humans is low (less than 500 kW/m²) while natural fires have values of 500 kW/m² or greater. Of course, the frequency in years given in the drawing is only approximate. The most intense fires only occur after the longest periods without fire and depend on climatic and other factors.
[Drawing: Jordi Corbera, from several sources]

tion of the Australian mediterraneans has allowed fire control techniques to develop that are based on the use of deliberate controlled fires in favorable times of year, just as the Aborigines used to do.

In fact, the first European colonists said that each summer the Aborigines systematically burnt their the territory by sectors, controlling the fire's spread as far as this was possible. As there was not enough fuel to burn every year, this allowed the existence of some burnt patches, together with others that were severely burnt and others that were little affected. After European settlement, this practice of setting fires was interrupted, and the accumulated fuel led to appearance of serious and uncontrollable outbreaks, often with the loss of life and damage to property. Recently there has been a return to controlled fires to diminish the fuel load, especially in the undergrowth: this is truly fighting fire with fire. To maintain the ecosystem perhaps it would be desirable to return to the Aboriginal practice of setting fires each year, as they did for thousands of years, but this would represent too high a cost.

In the jarrah forests, for example, the forest services have found that if the accumulated biomass in the undergrowth increases from 15 t/ha to 30 t/ha, the number of days when there a risk of a fire increases by a factor of 7.

So in the areas dominated by jarrah, the forest services burn the undergrowth, setting fires every 3-5 years, either from the ground or by starting small fires with incendiary bombs dropped from small planes. The territory is divided into parcels that are burnt in different years, making it a grid of forests in different stages of fuel accumulation in the undergrowth, which means the landscape is less inflammable, fires are more controllable and their negative effects are less severe. Controlled fires, set on spring days with suitable conditions of wind and humidity are a tool of proven effectiveness in the control of this forest disaster. Since the introduction of this practice in the west of Australia in 1953, the number of unintentional fires has not decreased, but the fires are less intense and have spread less.

Firefighting

Diffuser nozzle in use against forest fire
[N. Martinez / Index]

The main problem, certainly, is the availability of water. This is why before the days of four-wheel-drive tanker vehicles and planes to bombard fires with water, firefighting relied on many volunteers beating the burning scrub with branches cut on the spot and throwing buckets of water at the fire as fast as possible. A great deal of effort for very poor results, and even these limited results only in the areas where the accumulated fuel load is low, i.e. areas of matorral, since this method is clearly ineffective in fighting a major forest fire. Rather than seeking to put the fire out, this approach sought to stop the fire spreading by clearing firebreaks (gaps made by cutting down strips of trees in a hurry), or ensuring the fire did not cross existing barriers formed by gaps, such as roads, paths, and watercourses. The last resort was, and sometimes still is, a counterfire, a fire set in the right place in front of the fire, causing the first fire to advance towards the second, sucked in by the lower pressure created by the rising hot air; when the fire and the counterfire meet, both go out when they run out of fuel to burn.

Fighting forest fires is now like a war, with an air force, four-wheel-drive vehicles, and an almost military system of logistics based on effective communications and carefully studied tactics of positioning and attack. The only ammunition used in this war is, however, water. Water that is pumped or sprayed from the ground with conventional hoses. It seems simple. The problem is getting the water to the tops of inaccessible mountains, and flying precisely in the turbulent conditions associated with rising hot air when visibility is reduced by smoke. And then there is also the problem of finding water at the height of the Mediterranean summer.

Forest fires never retreat, because they can only burn what has not yet been burned, and forest fires thus always form a front that advances and is never compact and static. Immediately in front of the advancing fire there is a desiccation front— a strip of vegetation that the approaching heat is killing and drying out before it is consumed by the flames. Thus when a forest fire goes out, it always leaves a burned area of variable size, surrounded by a strip that is scorched or dead but not burned. Advancing and desiccation fronts, with a favorable wind, may move forward across plains or up slopes (it is more difficult for a fire to advance downhill because heat rises, and the flames and the desiccation front go up, not down) until a change in the direction of the wind, a gap the fire cannot cross, or rain brings the fire to a stop. Unless, of course, the firefighters put it out first, and they always have to base their strategy on the way the fire is actually behaving.

River bank of the Ebro. Northeast of the Iberian Peninsula [Ernest Costa]

Sign prohibiting making fires, Morocco [Josep Germain]

Sign prohibiting making fires, Spain [Iranzo / Index]

Sign prohibiting making fires, Catalonia [Josep Pedrol]

Firefighting strategy, however, depends on one essential factor—prompt action. This is because every forest fire starts as just a tiny fire, a clump of plants smoldering after a careless spark, a piece of glass focusing the sunshine, or one of the lightning strikes that have always caused natural fires. Starting as just a small fire, it may take 15-30 minutes to reach a diameter of just a few meters, when it is blown by the wind to form a moving wall of death. This is why the key to a successful firefighting policy is to locate small fires as soon as possible and to get to them. Once a forest fire has grown and formed a front, it is very difficult to stop, as it often forms an implacably moving front several kilometers long. This is why well-communicated observation towers able to spot and locate fires, and allow them to be triangulated as soon as possible, together with rapid intervention teams, are more effective than complicated systems, even though they are slower.

Fires often form important fronts, and then all these resources must be combined. Aerial methods are costly and dangerous, but are excellent for putting out fires that are difficult to reach on the ground. Large groups of people are essential to clear a firebreak or to prevent a fire from crossing a road. In any case, the use of hoses with a diffusor nozzle supplied by self-pumping cisterns, is the most common method of fighting fires. Good results are also obtained by adding fire-retardants to the water.

Yet nothing is as effective as prevention. Preventive prophylaxis involves the rapid organization of mobile firefighting teams and getting them to the right place at the right time so they can act immediately. Precautionary prophylaxis includes neutralizing pyromaniacs and reckless farmers who set fire to the stubble or edges of their fields whether conditions are suitable or not, as well as people who light fires to cook meals in the countryside or throw cigarette butts to the ground without putting them out first. But what is most needed is political prophylaxis—a change in forest management toward an emphasis on the maintenance of forests that are not excessively pyrophilous (unfortunately, the fastest-growing trees are pines and eucalyptus), or forests whose fuel load has been reduced (by preventive controlled fires, as in Australia), or even plots separated by fire-resistant barriers (stable firebreaks, ideally planted with plants that do not burn, instead of burned strips).

All these measures are necessary to fight against uncontrolled fires and against the catastrophe of wildfires —to prevent our forests being reduced to ashes.

234 Slopes with forest and matorral, in the pre-Andean interior of Mediterranean Chile. The Chilean Mediterranean's forest cover is typically discontinuous with areas of clearly xerophytic matorral, with many cacti, especially the quisco (*Trichocereus chiloensis*), in the very sunny areas.
[Photo: Lluís Ferrés & Ramon Folch]

Forest management

Unlike the situation in other biomes, such as the temperate biome in which forest management is relatively simple because it deals with relatively large and homogeneous forests, forests in the mediterranean biome are smaller and show a very high spatial heterogeneity, making their management much more complicated.

The state of the mediterranean forests

This problem is not so severe in the Australian areas, especially in the western area, as the relatively limited and recent humanization and the relatively smooth relief give rise to larger and more homogeneous forest masses, located on plains and suitable for exploiting as a normal tall forest, that is to say with regular clear cuts followed by spontaneous regeneration from seed. This is what happens with the forests of jarrah (*Eucalyptus marginata*), which now only have to face the problem of fires as large clear cuts have been replaced by selective thinning. Today, the exploitation model consists of clear-cutting small areas.

In the Chilean mediterranean, and especially in the Mediterranean Basin, the forest masses are discontinuous, with a highly variable species composition and a productivity that is highly influenced by the great diversity of conditions linked to the relief. The composition of the bedrock, the depth of the soil, the slope, and the exposure give rise to a forest mosaic in which productivity may change greatly in a distance of just a few dozen meters. In these conditions clear-cutting is not advisable, and more careful management is necessary. This careful management is based on thinning or selective felling, since this makes best use of the ability to regenerate from seed as well as the ability of many species to resprout from rootstocks. It gives rise to irregular forest masses—masses formed by trees of different ages and mixed in terms of regeneration, since there are trees that resprouted after the fire and others that grew from

germinated seed. The overall picture thus reveals a fragile forest that is unproductive and complicated, requiring considerable management effort in terms of both imagination and labor in return for rather limited profits, and this favors over-exploitation.

In the Mediterranean Basin, the area of the biome where these problems are worst, it is necessary to distinguish two very different areas. In the more economically developed countries of the northern shoreline, this century has seen the continued abandonment of marginal stock-raising and agricultural areas, at the same time as the population has become increasingly concentrated in the lowlands and on the coastline. This has given rise to an increase in the areas of forest, due to the regeneration of the vegetation in the abandoned areas or as a result of reforestation. The slow recovery and the non-existent or bad management of these recovered forest areas, however has not changed the general outlook at all with respect to productivity. In the low-lying and coastal areas the forests are suffering a progressive deterioration due to the high population density, leading to an increase in the number of fires and the appearance of problems caused by pollution.

The southern shoreline, especially in Turkey, Lebanon, and the Mahgreb, a quite different set of problems exists, because much of the rural population depends on forest resources for domestic use and even for small craft workshops. This, together with the tendency to plough new areas for crops and the maintenance of traditional grazing practices, is causing the progressive degeneration of the forests, which were already poor since they occur in the area of the biome with the harshest climate. In these countries, a household requires about 1 metric ton of fuelwood per year. The deficit in fuelwood is increasing and is further exacerbated by population growth.

Plantation and reforestation

The final result of all these problems and damaging actions is that the Mediterranean biome shows a

235 **Plantation of coppiced sweet chestnuts** on the volcanic lava of Etna (Sicily). The sweet chestnut (*Castanea sativa*) is a deciduous tree widespread throughout Mediterranean montane areas with acid soils, and may be grown as a forest or in plantations for wood. The tree provides fruit (chestnuts) and wood, used mainly for making furniture and barrels. The tree's ability to sprout after being cut to the ground, as shown in the photo, means it can be coppiced, which means new trees can be grown from the shoots that grow very quickly from the stem.
[Photo: Lluís Ferrés]

deficit in timber and even in fuelwood, in the less developed countries, such as those in the south of the Mediterranean Basin. Over the course of this century there have been many reforestation and plantation programmes to correct this progressive degradation of the forests.

It is, however, important to distinguish between reforestation and plantations. Reforestation is uncommon. It consists of the regeneration of degraded forest masses and seeks to recover the former landscape. Plantations often use exotic species intended for wood production. They should be considered as tree crops and not as forests. Planting trees for wood production is an economically profitable way of using land formerly used for farming or stock-raising. Only too frequently, however, plantations have replaced unproductive forests and scrub on difficult sites, resulting in soil degradation and the gradual loss of productivity. This type of management, with a view to immediate yield, will have to be abandoned in order to avoid making the degradation of the Mediterranean biome's forests even worse.

Mediterranean forest management should bear in mind that, in most of the biome, a living tree is worth much more than its wood. Almost all forest masses should be considered and managed as forests that protect the soil and regulate the water cycle. Unproductive forests and matorral should only be lightly exploited or not at all, since the profits they produce are minimal. The constant river surges and floods, so worrying in a biome where torrential rains are common, might be less intense and frequent if the land had a denser plant cover that effectively regulated the water cycle. Nor should the important role of mediterranean forests as recreation areas in more populated areas be underestimated, since it is one form of forest use that allows people to benefit without exploiting the forests directly. In the management of mediterranean forests, wood production should not prevent us from seeing the landscape. This is the only way to slow down and hopefully reverse thousands of years of degradation of the biome's forests.

The consequences of increasing atmospheric CO_2

Atmospheric carbon dioxide levels are now at their highest level for the last 160,000 years. According to analysis of ice samples taken in Antarctica, CO_2 concentrations were very stable at around 270-290 ppm between the 10th century and the early 19th century. Since then and up to the present, atmospheric CO_2 levels have increased continuously. This increase is due almost entirely to the growth of human populations, specifically industrial society. The human activities apparently responsible for this increase in CO_2 are the combustion of fossil fuels, cement production, and deforestation. All these affect the biosphere in general (greenhouse effect) and the vegetation in particular. This particularly affects, but is not unique to, mediterranean vegetation.

CO_2 levels and plant physiology

The average annual level of CO_2 in the atmosphere has risen over the last 130 years, and the rate of increase has risen sharply in the past 30-40 years. The rate of CO_2 emission due to the combustion of fossil fuels has followed a very similar pattern, although the annual increase in emissions has stabilized at 0.5-2.0% per year. If the tendency to CO_2 accumulation continues, CO_2 levels will double within the next century. Bearing in mind the residence time of CO_2 in the atmosphere and the current rates of fossil fuel use and deforestation, whatever reduction in emissions that is started immediately would not stabilise atmospheric concentrations for many years.

The increase in atmospheric CO_2 has a double effect on the vegetation. Carbon dioxide, together with other greenhouse gases, such as methane, nitrogen oxides, and chlorofluorocarbons will produce an increase of 4-7°F (2-4°C) in the average global temperature in the next 50-100 years. Furthermore, analogous changes are expected in rainfall patterns and other climatic variables, and these will be unequally distributed over the Earth. It seems that there will be very significant regional differences in the degree and type of change. Alterations in temperature and rainfall patterns as a result of the increase in the greenhouse effect will undoubtedly have a profound effect on plants and on ecosystems.

CO_2, in addition to its effect on climate, also has an effect on the growth and physiology of plants. These effects may in turn alter processes at the level of community and ecosystem. In most cases the CO_2 acts as a fertilizing agent. If the environmental conditions are optimal for growth, the rise in atmospheric CO_2 will increase photosynthesis at the level of the leaf, and as a consequence, the growth of the plant as a whole. Although this response is very rapid when plants are grown in atmospheres rich in CO_2, it has in some cases been shown that photosynthetic rates become adapted (accclimatized) to environments with high concentrations of CO_2. Thus, in the medium and long term, plant growth might not increase in response to the rise in atmospheric CO_2. Yet acclimatization does not appear to be a universal phenomenon, as other cases have shown a sustained increase in photosynthetic rates for many years. On the other hand, it is expected that the increase in atmospheric CO_2 will have little or no effect on ecosystems where primary production is highly limited by some essential factor, apparently the case in plant communities growing on serpentine-rich soils where nitrogen concentrations are very low.

Another direct effect of the increase in CO_2 is the reduction of the stomatal conductance—an increase in resistance to the loss of water—which leads in turn to a reduction of transpiration at the level of the leaf and an increase in the efficiency of water use. However, whether this increase will necessarily lead to a reduction in whole plant water use is still under debate.

Both the increase in plant growth and the reduction of stomatal conductance may be very important in the first phases of plant growth. Bearing in mind the exponential nature of the first stages of plant growth, small changes in growth rates may have large effects on competition for resources, and thus on community structure. Thus the increase in atmospheric CO_2 may alter the competitive relationships between species within a community. The base for these changes lies in the fact that each species responds in a different way to the increase in CO_2, as happens with other resources.

Mediterranean vegetation and the increase in atmospheric CO_2

Little is now known about the effects of the increase in atmospheric CO_2 on ecosystems with Mediterranean climates. Two of the factors that condition some of the more important adaptive characteristics of Mediterranean ecosystems are the scarcity of water, which in some systems lasts for up to seven months a year, and the recurrence of fires. In Mediterranean regions, the consequences of the increase in atmospheric CO_2 and its interaction with the factors of fire and water might result in important structural changes at the level of community and ecosystem. The increase in leaf-area and root production will interact with changes in transpiration, and this alters the system's patterns of water use and hydrology. On the other hand, the increase in the rates of growth and the accumulation of biomass might lead to a shortening of the interval between fires, and this might in turn have drastic effects on species' composition. The increase in biomass might also lead to an increase in fire duration and intensity, factors which play a very important role in the structure of the communities that develop after the fire. In general, the forecast is that ecosystems where water is the limiting factor, such as mediterranean ecosystems, will show a relatively large response to the increase in atmospheric CO_2 as a result of the increase in their resistance to water loss. If we also add to this the indirect effect on the cycle of fires, of such importance in many Mediterranean communities, the ecosystems in these regions may be expected to suffer major changes in the very near future. The climate change associated with CO_2 and other gases may further reinforce this. In the mediterranean region of California the first field research projects have recently started. The use of large computer-controlled growth cham-

236 Burning organic fuels and the increase in atmospheric CO_2 are clear in this view (1987) of Santiago de Chile. The city is badly affected by the retention, as a result of temperature inversion, of exhaust fumes derived from burning hydrocarbon fuels, and the photo shows how pollutants, such as CO_2 accumulate locally. In general, the increase of CO_2 in the Earth's atmosphere's is not so obvious, although its effects are equally important.
[Photo: Jaume Altadill]

bers allows the production of atmospheres that have been altered with respect to the external environment, making it possible to modify CO_2 levels, temperature and humidity. Only experiments that take into account the complex multifactorial interactions between plant physiology and the environment can help to predict the responses of ecosystems to increases in CO_2.

4.3 Mediterranean farming and stock-raising

If the situation of the forests forming the *sylva* is bad, those of the arable and grazing lands—the *ager* and *saltus*—is also ominous. Flourishing market gardens provide a stark contrast to unproductive dry-farming areas, pastures in mountain areas overrun by scrub, and deserts encroaching on dry grazing areas. The burden of history weighs heavily on the Mediterranean Basin: innovations have often been ill-considered or inadequately implemented in the overseas mediterraneans. Furthermore, scarce productive land has been squandered on urban or suburban functions.

The effects of agriculture on the environment

Mediterranean farming and stock-raising is characterized by its diversity. The contrast between lowland and upland areas mentioned above is mirrored by the contrast between irrigated and dry farming, and between large estates and smallholdings (there is a distinct lack of medium-sized properties). The contrast in all senses between large and small holdings is also typical of the Mediterranean. In many countries in the Mediterranean Basin significant damage has been caused by *latifundismo* (the ownership of large estates) and the cult of landownership has often been, and sometimes still is, more a phenomenon of status than of material enrichment. Many large land owners have permitted land erosion to occur, having failed to implement correct land use on their properties. Small-scale agricultural production, often in the form of market gardening, has also often contributed to the over-exploitation of the soil. Poor small holders, often unable to modernize their tools or to afford fertilizers, rarely obtain more than one subsistence crop a year. This situation was the cause as well as effect of the establishment of rural despotism in the Mediterranean and the cause of

237 The agricultural and stock-raising year in the Mediterranean is well depicted in these miniatures from a mid-15th century Italian manuscript book, kept at the Condé de Chantilly Museum (France). Each activity is portrayed in the month when it is typically performed, including pruning the rootstocks of vines, pressing grapes, taking pigs to snout for acorns in holm oak woodlands, sheepshearing, sowing grain, its harvest, and falconry. Farming and stock-raising had already totally transformed the Mediterranean landscape by the Middle Ages.
[Photo: Archiv für Kunst und Geschichte, Berlin]

238 **Modern industrial agriculture** has changed the face of the traditional Mediterranean agricultural landscape, as shown in this photo of plastic greenhouses covering the sandy hills of the Maresme region in northeastern Spain. Forced cultivation inside these greenhouses makes it possible to grow non-Mediterranean species and local crops out of season.
[Photo: Lluís Ferrés]

social conflict. In mountain areas, other than in the forestry exploitations already discussed, stock-raising is the predominant activity, especially sheep farming. In the southern and eastern parts of the Mediterranean Basin goats are also important, although in some areas of southern Europe and California, cattle are also important. Other than in highly urbanized areas, in some marginal deserts (southeast of the Iberian Peninsula, central and southern Tunisia), and in wetlands (the Maremma in Tuscany, the Languedoc coastal marshes), agricultural landscapes dominate the plains and middle mountain areas. The landscape is a patchwork of open cereal fields interspersed with fenced fields usually associated with livestock, and mixed farms. Typical Mediterranean mixed farming combines the cultivation of cereals, trees, and legumes; especially in its most intensive form, the irrigated *huerto* or market garden.

From ploughs to greenhouses

In the Mediterranean Basin, despite 10,000 years of agriculture, the plough has not fallen into disuse. It is precisely because the Mediterranean agricultural system is based on very delicate techniques such as irrigation, drainage, tree culture, and crop rotation, that it is more prone than other areas to harm from invasions, wars, and social conflicts. Temporary abandonment of the land can easily become permanent over the centuries as irrigation channels fall into disuse, the walls which support terraces disintegrate, and trees are cut down. Throughout history, Mediterranean agriculture, above all on the coast, has undergone many cycles of alternating thriving growth and complete ruin. Nevertheless, at any moment, any piece of land that appears to be abandoned, unproductive, or marginal

may be cultivated. Areas covered by scrub or secondary woodland are still being ploughed up wherever a favorable site, soil, and the availability of water permit the production of a crop—and if economic conditions determine it is valuable. Areas with sandy soils in the Maresme, near Barcelona (Catalonia), until recently covered by scrub and even by pine forests, have been ploughed up to grow strawberries. A century ago, grape vines escaped the phylloxera plague that swept Italy, the Iberian Peninsula and the Magrib in areas like cliffs and mountains; today they are returning to the Penedès or Priorat regions of Spain, to Chile, and to Australia. Some land is even being ploughed up so as to introduce crops under one of the newest agricultural technologies: industrial agricultural in greenhouses. In the mild Mediterranean climate this can be very productive if there is sufficient water, and fertilizers and pesticides are used generously.

Cereal fields

The few plains found in the Mediterranean Basin, although not on the same scale as the North American and eastern European plains, are also cereal production centers. The following are examples of this type of landscape: the Spanish Meseta, the Ebro and Guadalquivir Depressions, the Sardinian Campino, the interior of Sicily, the Tavoliere de Foggia in La Pulla in Italy, the Macedonian and Thessalian plains in Greece, the plains of Lydia, Phrygia and Pisidia in Turkey, the Bekaa in the Lebanon, and the inland plains of Syria. Typically, these areas possess poor soils and have very dry summers. The local population is usually concentrated in a few large towns and the predominant agricultural system tends to be based on leaving some land fallow. After the harvest livestock are pastured, often communally, in the stubble and fallow fields.

Over half of the agricultural land of mediterranean Chile is occupied by these type of systems, although property ownership tends to be organized differently. In the center of the region large estates comparable to the great estates of southern Spain and Italy are frequent. In the Norte Chico, community property systems, such as those found in Castile and Aragon, are common, and here livestock play an important role in land use. Areas to be cultivated are temporarily fenced off with branches, sticks, and especially the cactus *Trichocereus coquimbanus*, called *quisco*, while they are being exploited individually (6-8 years) by one of the communal farmers. When the land is no longer economically worthwhile, the farmer destroys the fences and the land reverts to communal pasture. This system of property ownership intensifies erosion in an area already at risk from erosion.

In California the first agricultural system to be introduced consisted of extensive livestock pasturing based on seasonal movements from mountain to plains, combined with cereal cultivation in the plains. However, owing to the agricultural changes introduced during the last few decades, land under cereals has been reduced to a third of the total area farmed and is now almost exclusively dedicated to the production of barley as fodder for milk cows in sheds.

The Australian mediterranean, as has already been mentioned, is much flatter than the Mediterranean Basin and has a system of production that is also dominated by the combination of wheat (or barley) and grazing, but on a very different scale. All properties are large and in general highly mechanized extensive cultivation of wheat alternates with clover (*Trifolium subterraneum*) production for fodder for wool-producing Merino sheep. Nevertheless, the landscape is not unlike the plains of Castile or other areas of the Mediterranean Basin, except for the strips of mallee between the fields. In the South African mediterranean property structures and levels of mechanization are similar, and cereal fields also occupy a large part of the cultivated land. In both Australia and South Africa cereal cultivation occupies a larger area in the more humid regions, where there is no summer drought, than in those with a Mediterranean climate.

Dry farming orchards and "promiscuous cultivation"
The low production of non-irrigated cereal cultivation combined with stock-raising in the plains of the Mediterranean Basin is somewhat mitigated in some areas by dry farming woody crops, especially olives and vines. In fact, olive groves and vineyards epitomize the Mediterranean agricultural landscape and differentiate it from neighboring regions. Fields are generally small, irregularly shaped, and molded to fit the intricacies of the landscape. Terraces are built if the gradient of the land is steep. Some fields are left open, while others are enclosed with stone walls or hedges. Some fields contain a single crop, and others combine rows of trees or grape vines with cereals and legumes in the free spaces.

Tuscany is perhaps the best example of this type of scenery which is found all over the Mediterranean Basin, from eastern Andalusia to Palestine, from Morocco to the coasts of Anatolia, from Provence to Cyprus, from the plains to mid-altitudes and even higher in the mountains. In Tuscany, as well as in Umbria, and other areas of central Italy, the isolated farms scattered throughout the plains and middle mountains alternate with the fairly large towns that are partly rural and partly urban, and often in a strategic location recalling the turbulent past of the city-states of this part of Italy. The plots are surrounded by shallow ditches with a ridge on one side on which a few trees, vines or both are planted. The fields may be planted with more trees or vines, or with cereals, fodder crops and tobacco. Vines are often trained around trees or specially placed poles, and in exceptionally windy areas cypress or pomegranate hedges act as windbreaks, thereby creating one of the richest and most diverse agricultural landscapes in the world. In the more humid plains, the landscape is organized around a network of canals and ditches which are used both for irrigation and for drainage, while on the slopes they are arranged following the curves of the contour thanks to the terraces constructed as a result of the efforts of many generations.

Apart from grape vines in California and Chile, and the olive in Chile, these crops have not been exported to the other mediterraneans, probably because the original colonizers of South Africa and Australia were not from the Mediterranean. Nevertheless, vines are currently being planted more and more in areas with a viticultural tradition such as the Napa Valley in California and the Talca area of Chile, as well as in some specialized areas of the Cape Province and southwest Australia.

The consequences of irrigation

Market gardening is an extreme case of agricultural intensification as a result of this irrigation of woody or mixed crops. Market gardens are usually found in low-lying plains irrigated by water from nearby mountains and are occupied by fruit trees, especially citrus fruits, and vegetables. The best example of this Mediterranean market garden landscape is probably the *Huerta de Valencia* along the banks of the River Júcar, in the Mediterranean overflowing of the Iberian Peninsula. The Mediterranean cultivator will make a market garden on the smallest site where there is enough water.

Irrigated areas

The most ancient systems of irrigation date back to Roman times, although the Romans did not develop techniques capable of bringing water to large areas of land. In the southern half of the Iberian Peninsula during the reign of the Caliphate of Cordoba there was a considerable increase in the amount of irrigated land. Records of the *Tribunal de las Aguas* (Water Court) of Valencia go back to 1016, although it seems to have been created in 960 when an even older system was reorganized. During Roman times and into the Middle Ages, irrigation in the Mediterranean Basin was practiced in the areas close to the large urban markets where the fruit and vegetables produce could be sold.

Only during exceptional droughts was water diverted to save cereal and legume crops, the main human foodstuffs produced by Mediterranean agriculture in classical times. Irrigation led to urban population growth and a change in eating habits, and this in turn caused, at various moments in the history of the Mediterranean Basin, a further increase in the area irrigated and its more intensive use. The changes introduced during the Muslim dominance of Al-Andalus have already been mentioned, but population growth in the 12th and 13th centuries meant that throughout the European shores of the Mediterranean both cities and monasteries sought to increase the area of land irrigated to improve yields of the basic crops, such as wheat, that were normally not irrigated. Even at the beginning of the 19th century, in the region of Murcia in southeast Spain known as the *huerta*, a quarter of the irrigated land was under cereals, basically wheat, a third was under alfalfa and other fodder crops, and another third in tree crops, mainly mulberries cultivated to feed their leaves to silkworms, while the rest was taken up by vegetables (3%), vines (1%), hemp, and other crops. Today, half the irrigated land is devoted to citrus fruit which, along with vegetable crops (16%), not only supply local and national markets, but are exported on a large scale. Today grape vines and hemp cultivation have disappeared completely, while cereals, principally maize, only occupy around 8% of the irrigated area.

239 Agricultural scene near Volterra, Tuscany (Italy). The photograph depicts a good example of the mixture of carefully tended fields, scattered woodland and other mixed habitats so typical of the humanized northern shores of the Mediterranean Basin. Apart from the vineyards, the crops, and the clumps of stone pine (*Pinus pinea*), the landscape's main feature is the rows of funeral cypresses (*Cupressus sempervirens*). This species from the eastern Mediterranean is commonly planted on roadsides, as a living memorial in cemeteries and religious sites, and as living windbreaks.
[Photo: Atlantide SDF / Bruce Coleman Limited]

240 Sprinkler irrigation has changed the agricultural landscape of many Mediterranean areas such as these fields in the Ebro delta. Formerly unproductive dry farming areas have become flourishing market gardens. Nevertheless, the large amounts of water needed by this system limits it to areas with plenty of groundwater or with plentiful surface water supplies.
[Photo: Xavier Ferrer]

241 A modern session of the *Tribunal de las Aguas* **(Water Court)** at the Apostles Door of Valencia Cathedral, February 1989. The tribunal has met every Thursday morning since the 10th century. It was first created in the period of Islamic domination, and was ratified in 1238 by the first Christian king, James I. It originally met in the mosque and then later in front of the cathedral door so the Muslims would not have to enter. It meets to resolve conflicts arising over the use of water from the Turia River in Valencia's 222,300 acres (90,000 ha) of market gardens. Ten centuries ago the first members of the Tribunal were democratically elected by those using irrigation water and the eight members have been making common law judgments since then without a single interruption. Their oral judgments take effect immediately and are almost always obeyed, only needing ratification by ordinary courts in the rare cases when their judgment is not respected. This remarkably durable example of medieval justice (in 1991 the Tribunal heard 14 cases) shows how important correct management of limited water supplies is considered in Mediterranean agriculture and its deep-rooted respect for the concept of justice and for customary practice.
[Photo: Lourdes Sogas]

The intensification of Mediterranean agriculture is nowhere better seen than in California. The transformations in land-use over the last two centuries have been really spectacular. Extensive grazing of sheep and cereal cultivation has given way to intensive fruit (both fresh and dried) and vegetable (mainly tomatoes, lettuces, and asparagus) production which currently occupies a third of all cultivated land in California and generates half the value of California's entire agricultural production. The main centers are around Los Angeles, the San Joaquin Valley (where citrus fruits are also important), and the coastal valley near San Francisco. They are relatively small enterprises by American standards and are highly specialized. Despite the high level of mechanization a large percentage of all Californian agricultural laborers find work here, including a large number of mainly Mexican seasonal workers.

On the other hand, in the southern hemisphere mediterraneans market gardening is far less important. In Chile, low demand from urban areas has meant that market gardening has only really developed around Santiago, Valparaiso, and a few other cities, although recently there has been an increase in fresh fruit production for export, out of season, to Europe and the United States. In the Cape and Australia market gardening is most common outside the areas with a Mediterranean climate, although it is present near the urban centers, as occurs in Chile.

Water management

Water management has played a very important role in agricultural transformations in the Mediterranean Basin over the last few decades. Water management has traditionally concerned the peoples living in these areas with their limited water resources and irregular rainfall, but the demand for water for all uses (not only agriculture) has grown exponentially. As mentioned above, the oldest documented irrigation projects date from Roman times but the technology of the time could not supply water to large areas of land. Even the substantial improvements in the Middle Ages were limited. The most spectacular increases have occurred in the last two centuries and were often associated with the drainage or filling in of wetlands to eliminate malaria, a disease endemic to the Mediterranean Basin and still not fully eradicated in the eastern and southern Mediterranean. These increases were largest in the Iberian Peninsula and Italy and more recently Israel, but also occurred in the other countries of the Mediterranean Basin, where there are still large projects to irrigate new land and to intensify cropping in sites that are already irrigated.

In Spain in 1858, there were 2 million acres (850,000 ha) of irrigated land, in 1900 2.5 million acres (1 million ha), in 1940, 3 million acres (1.2 million ha), and by 1990 this figure had reached 7 million acres (3 million ha), approximately 15% of all cultivated land. The percentages are even higher in Italy (18%, although a large part corresponds to the non-Mediterranean plain of the Po) and Israel (30%, again partially in extra-Mediterranean regions). In Greece the figure about 11%, while Turkey has less than 3%. This is in stark contrast to California where 6.9 million acres (2.8 million ha) is irrigated, over two-thirds of the cultivated land in the entire state, and all in Mediterranean areas except for the Colorado Valley. Irrigation is less important in the Australian mediterraneans.

Soil salinization

In the most arid parts of the Mediterranean Basin, the high levels of evapotranspiration cause salinization, i.e. the accumulation of salts in irrigated soils that lack adequate drainage. The risk is even higher when the irrigation water already has a high content of salts or when too much fertilizer is used. In irrigated plains on the coastline or in deltas, the aquifers are exploited (often over-exploited) for agriculture, industry, or domestic consumption, and this often causes salt water to enter the aquifers, a further cause of salinization. This happens even in areas whose rainfall levels should mean they are not at risk of salinization, such as the Po Delta and the Llobregat Delta (south of Barcelona). Salinization diminishes the osmotic potential of the water and provokes a physiological drought by decreasing the water available to the plant, and so agricultural productivity drops.

According to the Food and Agriculture Organization (FAO), a third of all irrigated land in Greece and Syria suffers from salinization, while in Tunisia 5% of the territory is threatened, in Algeria 3%, and in Morocco 1%. In terms of actual land surface under threat, the Nile Delta is most at risk. A third of the delta (an area with a climate somewhat more arid than a typical Mediterranean climate), is affected by salinization and over half the remaining area downstream from Cairo is also at risk.

Integrated agriculture and animal husbandry

Apart from water management, the main problems facing Mediterranean farming regions over the past 2,000 years have been soil erosion, and the integration of agriculture and animal hsubandry activities. In the Mediterranean Basin, agriculture and stock-raising are traditionally closely linked. Already in the Roman period, Cato considered a flock of sheep essential for an olive grove and recommended, in modern units of measurement, that a little over one and a half sheep per hectare would assure perfect fertilization with their dung. Artificial fertilizers have now largely replaced animal manure and draft animals are a thing of the past, but it is almost unthinkable that a peasant in the Mediterranean Basin will not have some farm animals. Often their fowl or pigs are the lasts signs, in the more developed regions of the Mediterranean Basin, of this activity by today's part-time farmers. However, these forms of animal husbandry can cause serious water pollution, precisely in areas with inherent water shortages.

Despite the traditionally complementary nature of agricultural exploitations and animal husbandry in the European Mediterranean Basin, there have been disputes between farmers and stock-raisers just as there are in other regions of the world. They have been especially important when institutions or the owners of large herds have moved their herds through the an area on their way to fresh grazing and on the way back. In the most developed parts of the European mediterranean, after years of over-grazing, the current problem is the opposite as insufficient pasturing in mountain pastures is favoring the regrowth and encroachment of scrub. Another problem is that the lack of stubble left by harvesting machines and the disappearance of weeds caused by herbicides. This means that what little grazing left for the herds may even be poisonous. In the south and east of the Mediterranean Basin, as well as in general in the most arid parts of other mediterraneans and in particular in the Norte Chico region of Chile, the situation is very different as grazing pressure on fragile ecosystems favors erosion and even desertification. A further contrast exists in Australia. Since the middle of the century, the increasing integration of sheep into wheat-growing areas has created a new type of mixed exploitation that has, in fact, tended to improve the fertility of previously poor soils by cultivating clover (*Trifolium subterraneum*) for forage.

Over-grazing

Shepherds who understand Mediterranean systems know how to use their resources completely and do

242 Overgrazing is a leading to desertification in the driest areas of the Mediterranean, especially in areas where agriculture is impossible and grazing pressure from goats is high. Plant life is almost eliminated everywhere livestock gathers, such as on seasonal migration routes or around water holes, shown clearly in these steep slopes in northeast Spain.
[Photo: Antoni Agelet]

not graze some areas to ensure that there will be some grass left in summer, however dry it may be, when grass production comes to an almost complete halt.

The management of Mediterranean pastures requires early grazing as soon as the autumn rains arrive, without waiting for the grass to grow. In this way livestock will consume the resources provided by the rapid-growing perennial plants, and thus encouraging the germination of annual species. If this early grazing is not performed, quick-growing perennials will hinder the development of annual species. In this biome delayed grazing is not as good a strategy as early grazing, because the life cycles of many species overlap. Rotational pasturing gives perennials a chance to regenerate but is not necessary in the mediterraneans as plant formations are dominated by annual species. Nevertheless, rotational grazing is advisable in Australian pastures where the plants are not adapted to pasturing and in California where the original pastures are dominated by perennial grasses.

In practice, intense and continual grazing pressure leads to degradation. The poor-quality species that

the livestock tend not to consume are favored and perennial species are gradually eliminated by exhausting their root-level soil reserves. In the Mediterranean Basin and other areas of the biome dominated by herbaceous plants from this region, there is what might be called the "Mediterranean pasturing paradox." The paradox is that the pre-adaptation of herbaceous species and pastures as a whole due to their long tradition of exposure to livestock grazing means that continuous intensive grazing, preferably by sheep and goats, actually favors the best plant species and prevents the invasion of pasture by shrubs. This only occurs, however, on sites at little risk of erosion, such as the pastures with low gradients (around 10-15%). If the gradient is 20-30% the pasture is at risk of erosion, and if it is over 40% it is at great risk, and little or no grazing pressure can be allowed, if major, irreversible degradation is to be avoided.

Thus Mediterranean herbaceous formations cannot support livestock grazing throughout the year, even if grazing pressure is light, because their growth ceases in the dry summer and the problem is aggravated by the winter pause due to the cold. Various strategies have been developed to overcome this problem. Summer drought is solved by the consumption of dry production left over in areas with high levels of production, by transhumance towards mountain pastures, or by the consumption of the leaves of trees and bushes, In winter, the solution is migration to warmer areas or to feed livestock on the leaves of trees and shrubs, often from pruned tree crops, or on acorns and chestnuts. In both winter and summer there is a increasing tendency to supplement grazing with animal feeds or with agricultural products. Anyway, stock-raising in Mediterranean regions is inextricably linked to agricultural and forestry activities, as the livestock eat the remains and some of the products of agriculture, forests, and scrublands.

Pastures may be improved in a variety of ways. The use of controlled burning to reduce scrub cover is being studied and can be combined with the seeding of selected herbaceous species to increase the production and fertilization of the soil. Other methods include the use of fertilizers and selected herbicides, irrigation, crop and pasturing rotation, and the implementation of pasturing practices which distribute grazing pressure over space and time. Nevertheless, a satisfactory model for pasturing in Mediterranean regions which can determine the loads that the system will support and optimum grazing frequency has yet to be found. These elements must also be compatible with forestry exploitation since in the past in the Mediterranean biome there have been constant clashes between forestry and stock-raising interests.

Over-grazing in forests and scrublands, not only hinders regeneration of woody species, but also leads to the increasing rarity or elimination of the most palatable plants and allows species less suitable as animal feed to dominate. In the Chilean mediterranean, for example, over-grazing has almost eliminated *Atriplex repanda*, a very productive and palatable bush that resists disease and provides good grazing. A more adequate distribution of grazing pressure may help to avoid this sort of problem. For example, a grazing model based on high livestock density and low frequency is preferable to a low-density, high-frequency model. High-density, low-frequency models force the animals to eat even the least palatable species, thus preventing the highly selective grazing that leads to the disappearance of the most palatable species in high-density, low-frequency grazing models.

Underuse of shrubby vegetation for grazing

Mediterranean shrub formations, often only exploited for food by livestock, normally have a high density of shrubs that hinders and even prevents grazing. Some clearing is thus required if livestock are to graze these areas. There have been experiments in maquis and garrigue formations in the western Mediterranean Basin, namely Corsica and southern France, that consist of a thorough clearing or total elimination of the lowest shrub layer. This material is then ground up and spread over the ground, not taken away, in order to improve soil fertility. This technique has created shrub communities with a herbaceous production ten times greater than the original that are also more accessible for livestock. This new structure can be maintained indefinitely by planning an adequate grazing load that prevents the regeneration of the low shrub layer.

Experiments like these are very useful, as they make it possible to plan improvements in matorral use, which would be very welcome in some of the poorer areas of the biome. In North Africa, for example, 60-80% of the benefits obtained from shrub communities are a product of pasturing and this yield could easily be improved with better management. This would also prevent degradation of woody formations by over-grazing or practices as absurd as cutting almost all the branches off

specimens of the Atlas cedar (*Cedrus atlantica*) to feed to goats, a practice that kills magnificent trees, but does not even use their wood, leaving the trunks to rot in the forest.

It has been shown repeatedly that maintaining a certain density of trees and shrubs in pastures is a stabilizing factor. This is because production is diversified, soils are protected and improved, maximum soil temperatures and evapotranspiration are reduced, more water is made available to the herbaceous layer and it responds by extending its vegetative period in comparison with unshaded areas. The price paid for an overall reduction in herbaceous production is more than compensated by the increased stability. Under the best conditions, such as in Californian oak woods in valley bottoms, it has been shown that herbaceous production increases spectacularly if the tree stratum is removed. However, production declines over time, and after fifteen years the negative effects of eliminating the tree cover are apparent.

Over the last few years in a number of areas in the biome, experimental replantations with shrubs for forage have been initiated in order to recover degraded zones and to establish stable and productive grazing areas. In the Chilean mediterranean these replantations have used *Atriplex repanda*, a good shrub for forage, and other species of *Atriplex* and *Acacia* in certain areas of North Africa and the eastern Mediterranean Basin. In the driest and most degraded areas trials have even been attempted with species of cactus that livestock can eat. Centuries of grazing have degraded much of the Mediterranean biome, and this type of solution, combined with the sensible livestock management of the remaining well-preserved tree masses is the only path to sustainable grazing of matorral. This question is very important given that most Mediterranean forests and shrublands are more suitable for stock-raising than for silviculture.

Inadequate dehesa management

The *dehesa* system entered a crisis point in the early 1960s, and in many areas the transformation was so profound that it led to the disappearance of the dehesa system of production. In the past this system was supported by two economic pillars that have now disappeared—the intensive use of poorly paid human labor (which discouraged mechanization), and the use of draft animals for power. Currently, the tendency is to mechanize and intensify stock-raising, a profound change to the system that now tends towards heavily subsidized extensive monocultures, and a reduction in livestock diversification.

This leads to the elimination of trees in cultivated areas and the invasion of pastures by matorral as a result of reduced grazing pressure. Management systems based on the search for maximum production simplify these systems, but also make them less self-sufficient and more dependent on external resources. In effect, these systems may be more productive, but they are less stable. Today, some people are pressing for a return to management systems closer to traditional practices that are more closely attuned to environmental limitations and that are more stable and independent, with all the advantages and disadvantages that this implies. The export and adaptation of this model to other regions of the biome has been suggested. The idea is to establish integrated agro-sylvo-pastoral regimes that require very little external input instead of just simple wooded pastures. Experiments have been carried out with umbrella pine (*Pinus pinea*) and in Mediterranean Chile similar plantations of the soapbark tree (*Quillaja saponaria*) have been proposed.

As has already been noted in reference to the general management of pastures, the value of these traditional systems is only just beginning to be appreciated. In some aspects they can be improved, although in terms of their overall integration into the environment they are difficult to beat. In practice, awareness of the limitations of these traditional systems can only help stability and is far more realistic than their replacement by agricultural and livestock systems that are highly productive but highly simplified and clearly dependent on external supplies of energy and resources.

Erosion and soil exhaustion

The rugged terrain, steep slopes, and rainfall patterns are the main reasons why much of the Mediterranean Basin (and of the Californian and Chilean mediterraneans) is at great risk of soil erosion. A further contributing factor is the elimination of plant cover to prepare land suitable for agriculture and grazing. Thus, traditional agricultural practice has had to deal with the problem of erosion since time immemorial and has developed several soil conservation strategies. These strategies include the conversion of pastures into dehesas, the construction and maintenance of terraces, restrictions on livestock movement through the area or confining it to improve soils by manuring, the channelling and protection of water courses, and many more. However, a number of other widespread practices which are not nearly so well-adapted to the biome also exist and have caused serious erosion, above all in the drier parts of the mediterraneans. These include regular burning to obtain better grazing, the uprooting of woody plants for firewood, and using heavy machinery to combine two or more small terraces into a single larger terrace with a steeper, and often unprotected, slope.

Particularly in the north of the Mediterranean Basin and increasingly in the south, agriculture is becoming

243 Erosion after the loss of tree cover, seen here in loamy soils in Bages (Catalonia). The cultivation of the relatively flat lands of this area has caused the disappearance of the Aleppo pine (*Pinus halepensis*) that retained the soil on steep slopes. The friable texture of the substrate has led to the soil being sculpted into active "badlands" that threaten the future of cultivation in this area.
[Photo: Josep Maria Barres]

increasingly dependent on the industrial sector. Machinery, chemical fertilizers and pesticides may increase yields in the short term, but in the longer term they only serve to exhaust the soil and degrade its structure, due to the limited return of organic material to the soil (previously supplied by animal dung). This continuous structural degradation is making the soil more fragile and vulnerable to erosion, so that torrential autumn rains and strong winds bear away the fine topsoil. An estimated 247,000 acres (100,000 ha) of agricultural land are lost in Mediterranean areas of Algeria, Morocco, and Tunisia for this reason every year. The transformation and commercialization of agricultural products is now, however, clearly dependent on the industrial sector.

In one way or another, the introduction into Mediterranean agriculture of the logic of increasing production at any cost and maximizing short-term profitability has profoundly changed rational soil use and agricultural systems in general by creating a division between the sites (and farmers) able to adapt to these requirements and those unable to do so. Those sites able to adapt include, some of the most productive land in the world in monetary terms, and landowners in these areas are coming to resemble typical businessmen (vineyard and wine producers in the Penedès or Chianti, market gardeners in Rousillon or Murcia, orange producers in Valencia, Morocco and Israel). The sites that have been unable to adapt are gradually becoming marginalized and are at risk of abandonment in the northern countries of the basin while in the southern and eastern countries these sites are in danger of desertification and over-exploitation.

Pollution by agriculture and stock-raising

Pollution is normally associated with industrial and urban activity, but modern agriculture and stock-raising are also important sources of contamination. The main pollutants from these sources are fertilizers, pesticides, and the residues of olive oil production.

Fertilizers and pesticides

Increasingly intensive cultivation, especially of irrigated land, based on massive usage of chemical fertilizers and pesticides can cause the eutrophication of surface water and the contamination of aquifers. The citrus orchards of Valencia have an estimated fertilizer use of 2.5 metric tons/ha per year, more than half of which is nitrogen-containing fertilizer. Excess fertilizer not used by plants may permeate the soil and enter the aquifers. Nitrate input into aquifers is estimated at 575 kg/ha per year from citrus orchards and 840 kg/ha per year from vegetable crops. Market gardening areas in Castellón de la Plana in the north of the Autonomous Community of Valencia have been checking water supplies since 1976. Even then nitrate concentrations of over 100 ppm (parts per million) had been reached in some wells used for irrigation and for drinking water. In December 1983 levels as

244 Aerial fumigation of crops in California. The massive and systematic use of agricultural pesticides represents an increasingly worrying interference in the ecological mechanisms regulating agricultural systems. *[Photo: AGE Fotostock]*

high as 450 ppm were found in some wells, while most of the aquifer had levels of over 200 ppm. To give a better idea of just what these figures mean, the World Health Organization (WHO) recommends a limit of 10 ppm in drinking water (especially for children) as higher levels of nitrates can increase the incidence of methemoglobinemia.

Fertilizers not taken up by plants mix with liquid waste from pig and poultry farms, as well as from urban centers, causing eutrophication, especially in lakes and reservoirs. This phenomenon is not restricted to Mediterranean environments. However, the Santa Ana River, southeast of Los Angeles, has the unenviable distinction of having the highest levels of nitrate of any river yet studied.

Data on pesticides are scarce and scattered. Pesticide use rose steadily until 1980 and seems to have stabilized, at least in countries on the Mediterranean's northern shoreline. The use of some of the more stable pesticides showing bioaccumulation, such as DDT, has even been banned in some countries. Pesticides are particularly worrying in rice fields in coastal wetlands such as the Ebro and Po Deltas, the Camargue, the Albufera coastal lagoon near Valencia and the Guadalquivir saltmarshes. In these areas pesticides are sprayed indiscriminately from small airplanes on areas which border on important wetlands (lakes in deltas, the Albufera itself, and the Coto Doñana reserve), key points on bird migration routes from northern and central Europe to sub-Saharan Africa. By 1972, in the Albufera wetland in Valencia, for example, lindane concentrations had reached 0.39 mg/l and DDT concentrations rose to 0.49 mg/l. Pesticide levels also tend to be very high in any area where attempts have been made to intensify production, whatever the product. This is perhaps most severe in areas with hothouses and similar installations.

Olive oil-mill waste

One type of pollution that is characteristically if not exclusively Mediterranean is the waste from the pressing of olives for oil. Everyone knows oil is extracted from olives, yet quantitatively olives produce a lot of liquid waste (mainly *alpechines*), some solid waste (*orujo*), and a little olive oil. The liquid waste is a particularly harmful pollutant, although it is virtually unknown outside the Mediterranean Basin.

In an average year, around 88 lbs (40 kg) of olives can be harvested from each of the 100 or so olive trees found per hectare (1 ha=2.5 acres). Harvesting takes place in winter when the olives are perfectly mature and the oil easiest to extract. The oil is

obtained traditionally by mills or by more modern systems of continuous extraction. The traditional method consists of milling the olives into a runny pulp which is then placed upon circular mats called *capachos*. The mats are piled up and submitted to pressure that makes the oil and other liquids drip out, leaving the solid waste, the *orujo*, between the mats. The oil is separated from the liquid fraction by decanting, often after gentle washing with hot water, leaving the useless lower fraction, the liquid waste called *alpechines*.

Approximately 2 lbs (1 kg) of olives yields 9 oz (250 g) of virgin olive oil, 11 oz (320 g) of solid waste, and 15 oz (430 g) of liquid waste, ignoring any water that may have been added during the extraction process. The liquid waste still contains a little oil, about 0.6-0.7%, while the rest is water and solids. The solid waste contains 6% oil and 27% water, with solids making up the rest. In continuous extraction, sometimes known commercially as *pyralisis*, the process is similar but more mechanized, and the liquid fractions are separated by centrifuge. About 1 quart (1 liter) of water is added for every 2 lbs (1 kg) of olives to make the pulp more fluid. The end result is 9 oz (250 g) of oil, 17.5 oz (500 g) of solid waste, and 44 oz (1,250 g) of liquid waste. The liquid waste is more diluted and only contains 0.3% oil, while the solid waste is also more diluted, and contains 3.5% oil and 50% water in addition to the solid wastes.

The oil obtained by cold pressing is known as extra virgin oil if the olives' quality and ripeness means the oil produced has an acidity of less than a 1° (ideally less than 0.5°). If the acidity is over 3°, the oil is not fit for consumption and must be refined physically and chemically to eliminate the acidity, a process that also removes the oil's color and aroma. The refined oil, transparent and insipid, is used to reduce the acidity of some virgin oils and the resultant product is called pure olive oil. The oil in the liquid waste is decanted, and the oil content of the solid waste is extracted with organic solvents. The oils obtained by these means are of very low quality and need to be refined and then blended with virgin oils.

After removing the water and oil, the solid waste make a good fuel with a caloric value of 2,800 kcal/kg. It is used in industry, for example in kilns for pottery, for domestic heating, and is also used to power the installations where the olives are processed.

The liquid wastes, on the other hand, are totally useless and their high content of organic matter makes them a serious problem. When potential pollution is measured in terms of the biochemical oxygen demand (BOD) —the amount of oxygen used to decompose the organic material it contains—the liquid wastes are 100 times more contaminating than normal urban sewage. In the Mediterranean Basin about 1,590 million gallons (6,000 million liters) of these liquid wastes are produced annually. Treatment, when carried out, consists of storage in ponds away from urban centers where they slowly decompose. The problem is that these liquid wastes are often discharged without any treatment and seriously contaminate watercourses.

Mediterranean agriculture today

As the final report of the UNEP's Blue Plan for the Mediterranean Basin points out, agriculture is more important in the region than its percentage (in most cases less than 20%) of the gross national product (GNP) of the various countries of the Mediterranean Basin would seem to indicate. In fact, in the southern and eastern countries of the Basin almost half of the population is employed in agriculture. Nevertheless, the population growth and the living and working conditions this leads to, is causing massive emigration to the countries of the northern shoreline and the rest of Europe. The countries of the Mediterranean Basin devote between 80 and 85% of their regulated water resources and around a third of their land to agriculture.

Population growth has, however, been faster than growth in agricultural production. This is especially true on the southern shoreline and in countries like those of the Mahgreb. Countries that only a few decades ago contributed to their trade balance by exporting agricultural products (Morocco, for example, exported cereals until the early 1970s) now no longer produce enough food for their own populations. Extensive dry farming has probably reached its limit in terms of area covered and is often unproductive (even less than 10 quintals [equivalent to 100 kg] per hectare in Morocco, Anatolia, and Al-Jazirah in Syria). Intensive agriculture is patchy and mainly in irrigated areas, and has caused problems, such as increased erosion and desertification as a result of intensive or inappropriate use of machinery or pollution by fertilizers and pesticides.

The expansion of irrigation comes into conflict with alternative uses of water, such as industry and domestic consumption. This is made even worse by the low water quality of many watersheds and aquifers that are now polluted. For an irrigation pro-

245 **Dried waste from an olive mill** in northeast Spain. Residues from Mediterranean olive mills are potent organic pollutants, although they are not poisonous. Their breakdown gives rise a very high BOD (biological oxygen demand) and no satisfactory treatment or elimination system has yet been found. [Photo: Josep Loaso]

ject to be truly successful, it must not only include the network of supply and drainage channels, but also reach the plots to be irrigated. Many things have to be taken into account—the methods of using water, the overall situation of the watershed, feeding the reservoirs (soil erosion caused by deforestation can quickly silt up reservoirs), and discharges of pollutants which can reduce water quality.

All these factors have driven the authors of the UNEP report to take a pessimistic point of view. This pessimism is due to the fact that forecasts indicate greater pressures on natural resources due to population growth in the south and east of the Mediterranean Basin and the inadequate growth in agricultural production in both the north and south of the Basin. The agricultural policies in operation in Mediterranean countries over the years to come will be crucial for the sustainable development of the area.

In California, Australia, and the Cape the situation is completely different. These regions are much more productive and produce large surpluses in relation to size of the local population, and thus export much of their agricultural production. The situation in Chile's mediterranean region is more similar to the less developed regions of the northern shores of the Mediterranean Basin, but here the agricultural sector is integrated into a national economy of great diversity in terms of renewable natural resources, the result of the country being a long thin strip.

The architecture of hillsides

Homer narrates in the *Odyssey* that Ulysses found his father Laertes "in the vineyard on the terraces", thus confirming that the mountains of Thessaly had been terraced into slopes, and that terraced slopes have been a feature of the Mediterranean agricultural landscape for at least 2,500 years. It is now believed that some of the agricultural terraces that surround Mycenae were already there in the time Atreus, Agamemnon, and Electra (or the real persons that governed, rather than these probably mythical figures), and this would mean Mediterranean terrace cultivation methods are more than 3,000 years old. This is not surprising. In view of the Mediterranean's relief, terraces must be as old as agriculture.

The very scarce references to terrace cultivation in classical texts—whether Greek, Roman or Arabic—indicate in effect that terraces must have been very common. The texts do not mention them because they were taken for granted. Homer mentions that the islands of Cos and Lemnos in their day were densely populated, which would have been impossible in these rocky islands unless agricultural supplies were ensured from terraced fields (it is only necessary to see the island of

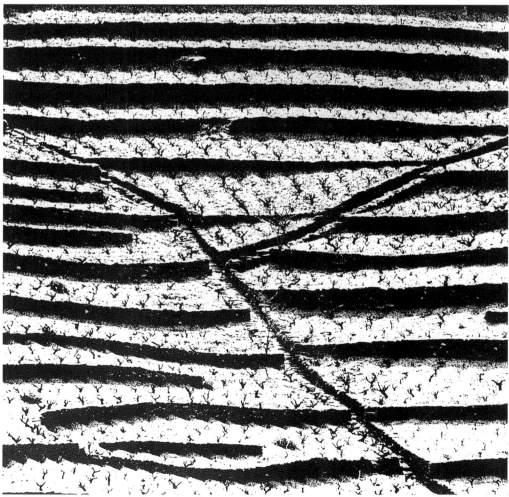

Terraced vineyard showing "crowsfoot" drainage system in Albares (Pyrenees) [Xavier Miserachs]

Calci, which like Rhodes is now totally terraced, to grasp this). Later, in the Middle Ages, Muslim treatises on agriculture written in Al-Andalus gave precise instructions on the construction and preservation of the terraces, and landscape engravings made in Liguria in the 16th and 17th centuries show mountains with terraces. What exactly was grown on these terraces, why are they so widespread in the Mediterranean landscape, and how were they built?

Cabin wall built of schist [Teresa Franquesa]

Steps up the wall built using limestone [Ernest Costa]

The grape vine and the olive are the most typical crops grown on Mediterranean terraces. The difficulty of working the land and the virtual impossibility of irrigation explain the choice of these two species, both highly resistant perennials, together with the fact that their products, wine and olive oil, command high prices. It was in fact increasing demand for wine that explains the last and largest expansion of terracing in the 19th century, the period when they were most widespread, at least in the western Mediterranean. Thus, almost all the hillsides of the coastline of the Iberian Peninsula, Balearic Islands, Provence, Liguria, Corsica, and Sardinia were terraced up to altitudes of 1,640 ft (500 m) or more. Now, after most of the terraces have been abandoned and are now covered in matorral, these gigantic flights of steps are occasionally revealed by forest fires. And they really were titanic works of engineering, that required the investment of millions of working days of backbreaking labor by thousands and thousands of peasants over generations and generations. Millions of working days spent constructing kilometers and kilometers of dry-stone walls that follow the contours, and then filling in the resulting flat surface with earth brought uphill from lower ground, or at least with earth from the hillside, cleaned of stones and leveled. Spurred by their necessity, the peasants thus managed to turn the entire slope into flat areas, creating narrow strips of flat arable land, superimposed on a steep landscape with poor soils. Quite a feat!

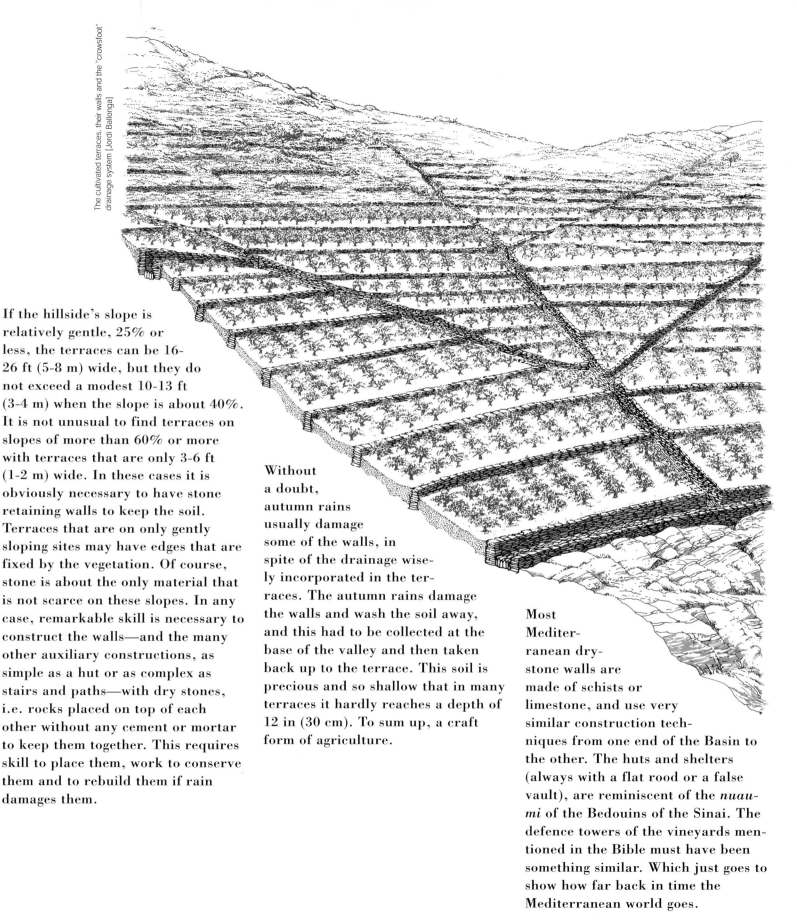

If the hillside's slope is relatively gentle, 25% or less, the terraces can be 16-26 ft (5-8 m) wide, but they do not exceed a modest 10-13 ft (3-4 m) when the slope is about 40%. It is not unusual to find terraces on slopes of more than 60% or more with terraces that are only 3-6 ft (1-2 m) wide. In these cases it is obviously necessary to have stone retaining walls to keep the soil. Terraces that are on only gently sloping sites may have edges that are fixed by the vegetation. Of course, stone is about the only material that is not scarce on these slopes. In any case, remarkable skill is necessary to construct the walls—and the many other auxiliary constructions, as simple as a hut or as complex as stairs and paths—with dry stones, i.e. rocks placed on top of each other without any cement or mortar to keep them together. This requires skill to place them, work to conserve them and to rebuild them if rain damages them.

Without a doubt, autumn rains usually damage some of the walls, in spite of the drainage wisely incorporated in the terraces. The autumn rains damage the walls and wash the soil away, and this had to be collected at the base of the valley and then taken back up to the terrace. This soil is precious and so shallow that in many terraces it hardly reaches a depth of 12 in (30 cm). To sum up, a craft form of agriculture.

Most Mediterranean dry-stone walls are made of schists or limestone, and use very similar construction techniques from one end of the Basin to the other. The huts and shelters (always with a flat rood or a false vault), are reminiscent of the *nuaumi* of the Bedouins of the Sinai. The defence towers of the vineyards mentioned in the Bible must have been something similar. Which just goes to show how far back in time the Mediterranean world goes.

246 Lammergeier (Gypaetus barbatus). The loss of wild areas and decreasing food supplies, partly caused by the lack of animal carcasses that once provided these great vultures with their food, have reduced lammergeier numbers greatly. In the north of the Mediterranean Basin lammergeiers are decreasing and there is evidence that they have disappeared from traditional nesting areas. There are also signs that they are no longer present in other mountainous areas that fulfil all the requirements of breeding areas, although breeding has never been recorded there. Nevertheless, programs designed to provide food and protection for birds of prey have increased the lammergeier's chance of survival.
[Photo: Josep Maria Barres]

4.4 Wildlife in retreat

Mediterranean-type ecosystems, and especially those within the Mediterranean Basin, have always been a highly suitable setting for human settlements. Throughout history, many different cultures have fought over the valuable sites suitable for agriculture, and they are among the most highly modified of all terrestrial system. Their fauna is in direct competition with humans, and the animals have been greatly affected.

Hunting and direct impact

In the Old World, the first inhabitants of the Mediterranean Basin were predominantly hunters or hunter-gatherers. During the Palaeolithic human impact was slight since humans, as predators, in general did no more than keep most other animal populations in check. One of the most important consequences, however, was the eradication of animals competing with humans (for shelter and food, mutual predation, etc.). One species which was eliminated rapidly from Mediterranean ecosystems and from its whole range was the grizzly bear (*Ursus spelaeus*).

Later on and right up to historical times, humans continued wiping out all competing species, especially predators and the large animals, that caused greater problems—species that were not particularly abundant around the Mediterranean Sea and in Asia Minor. The relatively low density of human populations, inefficient hunting techniques, and the limited economic use of animal carcasses meant these animals retreated to sparsely inhabited regions, normally mountainous areas. Yet only over the last three or four centuries of profound changes in human society (industrial and agricultural revolutions, overseas trade, major wars, etc.) have these extinctions become widespread in Old World Mediterranean systems. The most important factors were undoubtedly the invention of firearms (directly through hunting and indirectly through the reduction or disappearance of prey) and the use of poisons.

Thus, Mediterranean systems were deprived forever of the lion (*Panthera leo*) then found in Greece, Anatolia and Asia Minor as well as North Africa, the serval (*Felis serval*), and the cheetah (*Acinonyx jubatus*). Generally, though, most other large predators were not exterminated completely and today hang on precariously in a few small areas. For example, between 50 and 100 leopards (*Panthera pardus*) survive in the Atlas mountains, the brown bear (*Ursus arctos*) has disappeared totally from Africa and in total around 200 individuals cling on in the Cantabrian cordillera, the Pyrenees, Abruzzi Mountains, Greece and Albania, as well as an unknown number in Turkey. Only 1,000 Iberian lynxes (*Felis pardina*) survive in the Iberian Peninsula, most of them in the south (it once lived in France), while 200 European lynxes (*F. lynx*) struggle to maintain a foothold in Albania, Kosovo, and Greece, while its status in Anatolia in unknown. Wolves (*Canis lupus*) hang on, while barely 80 pairs of lammergeier (*Gypaetuss barbatus*) survive in the western Mediterranean (Pyrenees and Corsica), along with fewer than 10 pairs in Crete, and a small but unknown number in Greece and Turkey. Less than 100 breeding pairs of the imperial eagle (*Aquila heliaca*), a typically Mediterranean eagle, remain. If these figures are compared with the size of the Mediterranean ecosystem, the full drama of the current situation is plain. Animals that could once be counted in their thousands, can now almost be counted on the fingers of a single hand.

Not only the predators, but also many large herbivores have also become extinct or declined in numbers. Artiodactyla (even-toed ungulates) have been largely forced to take refuge in isolated mountainous regions. Hunting is the most obvious direct cause, but the effects of pastoralism and livestock activity were equally decisive, taking the form of competition for pastures, the reduction in favorable biotopes and the transmission of disease as a consequence of increased livestock grazing. Some Artiodactyla such as ibex (*Capra pyrenaica*) are endemic to the Iberian Peninsula. In the mid-20th century fewer than 800 individuals were left alive, while a mere 90 specimens of the wild goat or pasang of Southwest Asia (*C. aegagrus*) still survive on Crete, with around 1,000 in Greece and Turkey. For this type of animal, these populations are too small to be viable and they are at great risk of becoming extinct. It also should be noted that practically all the species of equids from Asia Minor have become extinct—mainly the Asiatic wild-asses or chigetai (*Equus hemionus*), although a few may be left in Syria, and the mountain gazelle (*Gazella gazella*) and Persian gazelle (*G. subgutturosa*). However, in comparison with past populations, the North African members of the Artiodactyla have probably been affected more than any other group and many species such as the scimitar-horned oryx (*Oryx dammah*), addax (*Addax nasomaculatus*), dor-

247 Imperial eagle (*Aquila heliaca*). Large raptors at the top of the food chain suffer most from environmental problems such as demographic pressure, loss of habitat, and contamination of food resources. The imperial eagle has been especially affected and numbers in the Mediterranean Basin have been low for many years. Once far more abundant, it is thought that this eagle (a symbol of one of the Evangelists) is at risk of imminent extinction. [Photo: José Luis González Grande / Bruce Coleman Limited]

248 Ibex (*Capra pyrenaica*) in the Puertos de Beceite, in the northeast of Spain. Proof of the widespread distribution of goats throughout the Mediterranean Basin is the large number of place names associated with goats. Mountain goats are one of the most typical animals of Mediterranean mountains, although in the past they were also found in the plains as archeological remains and cave paintings prove, having been forced in the past to retreat to mountain areas by human presence. In various places in the Mediterranean Basin successful reintroduction programs have been carried out, as, for example, in the Puertos de Beceite in the western Mediterranean where the ibex is now recovering.
[Photo: Oriol Alamany]

cas gazelle (*Gazella dorcas*), Cuvier's gazelle (*G. cuvieri*), slender-horned gazelle (*G. leptoceros*), and Barbary sheep (*Ammotragus lervia*) are either extinct or almost extinct.

The most abundant Artiodactyla today in northern and central Europe are red deer (*Cervus elaphus*) and roe deer (*Capreolus capreolus*), yet in Europe's Mediterranean ecosystems they have been practically wiped out, even in undisturbed, poorly populated areas. Cultural differences have often been blamed for this situation; nevertheless, a closer look at the ecology of these two species reveals that Mediterranean systems' lower production of plant material that herbivores can eat plays a fundamental role. As a result, population densities in natural conditions tend to be only a half or a quarter of those in central Europe. Reproductive efficiency is also significantly lower in Mediterranean biomes. Herein lies the greater fragility of the Mediterranean megafauna, today far more important than that in central Europe for game hunting.

These profound changes have not occurred in the other Mediterranean ecosystems (California, Chile, South Africa and Australia) where human pre-industrial and industrial settlements are more recent. Yet there have been serious losses. Thus the South African mediterranean has lost many large mammals,

victims largely of the expansion of agriculture, stock-raising, over-hunting, and the human population explosion. Notable examples include the Cape hunting dog (*Lycaon pictus*), spotted hyena (*Crocuta crocuta*), serval, lion, cheetah, desert lynx or caracal (*Felis caracal*), black rhinoceros (*Diceros bicornis*), white-tailed gnu (*Connochaetus gnou*), and two strictly Mediterranean species, the bontebok (*Damaliscus dorcas*) and the blue buck (*Hippotragus leucophaeus*). The communities of large fauna in the Cape's Mediterranean ecosystems have been the most affected in the entire African continent.

The Californian mediterranean's wildlife communities have been less affected, although there are notable absentees such as the wolf (*Canis lupus*), brown bear (*Ursus arctos*), kit fox (*Vulpes velox*), pronghorn (*Antilocapra americana*), elk (*Cervus canadiensis*), bald eagle (*Haliaeetus leucocephalus*), and California condor (*Gymnogyps californianus*). Nevertheless, some important species, such as the mountain lion (*Felis concolor*), bobcat (*Felis rufus*), whitetail deer (*Odocoileus virginianus*), and the mule deer (*O. hemionus*), are still fairly abundant.

In Chile fewer large species have become extinct although some, such as Geoffroy's cat (*Felis* [=*Leopardus*] *geoffroyi*), huiñas (*Felis guigna*), the mountain lion (*F. concolor*), the guanaco (*Lama guanicoe*), and the huemul (*Hippocamelus bisulcus*), have become increasingly rare over the last 100-150 years.

Lastly, in Australia many species have suffered a decline, such as the hare-wallaby (*Lagorchestes leporides*), toolache wallaby (*Macropus greyi*), and numbat (*Myrmecobius fasciatus*).

Habitat transformation

Whereas the extinction of larger species is generally considered to be a direct result of hunting by humans, many smaller species have been severely affected by various types of changes, such as modifications to their habitats. Everywhere, Mediterranean systems have been transformed into cultivated land (vineyards, olive groves, orange orchards, wheat fields, etc.) or into pasture (largely for sheep and goats). Undoubtedly, as a result the number of wild species associated with forests, scrub and chaparral habitats has been reduced and many have been replaced by other species better adapted to agricultural habitats. Mechanization has led to increasingly large holdings and this has totally negative effect as the boundaries between properties, often hedges and bramble patches etc., which acted as ecotones (which were rich in species) and as refugia for wildlife, have disappeared. This process can be seen clearly in European and North African small game species, for example the red-legged partridge (*Alectoris rufa*), and in other animals such as various species of tortoise (*Testudo*). The use of crop varieties with earlier or later cycles has modified the phenology and life cycle of many wild species. A good example of this is Montagu's harrier (*Circus pygargus*), a bird of prey that used to nest in scrub, steppes, and wetlands but which has become a typical inhabitant of dry-farmed cereal plains. Earlier harvests (late May or early June) and the use of mechanical harvesters, which crush their nests, are two of the main reasons for this bird's decline in the Mediterranean. This and other species have also been much affected by the conversion of dry farming areas into irrigated farmland. The case of the great bustard (*Otis tarda*) is worth mentioning, as it initially benefitted from the conversion of forests into dry-farmed cereal plains, but more recently it has suffered the effects of hunting and the conversion of its habitat into irrigated farmland. This surprising phenomenon also benefitted many species from North Africa (often steppe

species), but they too are now under threat (sandgrouse [*Pterocles*], bustards [*Otis*], Dupont's lark [*Chersophilus duponti*], etc.).

The increase in the use of insecticides, snail-killing chemicals, and other artificial products has reduced the amount of food available for insectivorous species, and leads to the input of synthetic products that may cause the bioaccumulation of certain pollutants (PCBs, DDT, lindane, dieldrin, aldrin, heavy metals, etc.). In large quantities, these substances directly kill most animals, but smaller doses may cause alterations, such as lower reproduction, infertility, and immunodepressive effects. As they are neither easily eliminated nor biodegradable, they persist in ecosystems. This then leads to bioaccumulation or biomagnification —the animals at the top of the food chain accumulate higher levels of these toxins and are the most affected. Species affected include the Algerian hedgehog (*Atelerix* [=*Aethechinus*] *algirus*), insectivorous birds, and bats, and many other birds, such as falcons (*Falco*) and little owls (*Athene noctua*). In California and Europe, the result of this process was that some birds produced egg shells that were too thin for the embryo to develop.

Fresh and salt marshes have suffered more habitat transformation than most biotopes. Wetlands have been drained for cultivation, to combat certain diseases such as cholera and malaria, or simply for the exploitation of underground water supplies. Apart from tampering with existing physical conditions in wetlands, many marsh areas, once oases in arid areas or dry farming areas, have been reduced considerably in size with negative effects on the fauna (fish, amphibians, waterbirds, etc.). A good example is the California salamander (*Ambystona californiense*) which is threatened by the reduction of its wetland habitat and breeding sites. The behavior of many migratory species that use wetlands as stop-over or wintering areas has also been affected. This has become obvious in the biomes that migratory species originate from, as well as in their eventual destinations, and not just in the Mediterranean biome. White-headed duck (*Oxyura leucocephala*) populations in the western Mediterranean (Iberian Peninsula, Algeria, and Tunisia, for example), had plummeted by the early 1980s to a mere 300 individuals. Yet many species have adapted to the irregularity and unpredictability of water levels in Mediterranean wetlands, such as the greater flamingo (*Phoenicopterus ruber*). Despite possessing certain traditional breeding

249 The white-headed duck (*Oxyura leucocephala*) has never been bred in captivity. Of Asiatic origin, this duck is seen occasionally in the Mediterranean Basin, above all in the west, and has become increasingly rare owing to the destruction of wetlands. It is a very localized breeder and the few pairs that are left are only found in certain lagoons and often not in other, similar lagoons in the vicinity. These diving ducks are characterized by their almost completely white head and sky-blue bill.
[Photo: Eckart Pott / Bruce Coleman Limited]

sites (the Camargue [southern France], the Fuentepiedra Lagoon, the Doñana Reserve [southern Spain], and a number of places in Magrib [North Africa]), they will sometimes reproduce in large flocks (thousands of pairs) in newly colonized areas where conditions are temporarily ideal. Afterwards these flamingoes may not breed again for many years or even centuries. The Ebro Delta is a case in point—due to problems with water levels in the Fuentepiedra Lagoon, flamingoes bred there in 1992 for the first time since the Middle Ages.

Large quantities of persistent pollutants also accumulate in wetlands and diseases may break out, such as botulism (frequent in ducks and geese). Many animals are susceptible to pesticides (there were many cases in the 1960s and 1970s of thousands of birds dying after a pesticide application), and two cases are very clear. The otter (*Lutra lutra*) disappeared within a space of ten years from all cultivated areas in low and middle altitude habitats in western European Mediterranean biomes. The bald ibis (*Geronticus eremita*), once found in Asia Minor, most of Europe, and in North Africa, had been reduced by 1989 to a single pair in Turkey and 78 pairs in Morocco, a wild population that is far less numerous than the 700 captive individuals present in the world's zoos.

The expansion of anthropophilous fauna

Yet not all the fauna of Mediterranean natural systems have become scarcer, since when some species lose ground, others gain it. We have already commented on the expansion of steppe birds at the expense of Mediterranean forest species, obviously accompanied by replacements in other taxonomic groups. Animals that benefitted include mammals, such as the Iberian root vole (*Microtus duodecimcostatus*) and the Mediterranean rat (*Mus spicilegus* [=*M. spretus*]), amphibians such as natterjack toads (*Bufo calamita*) and reptiles, such as geckos. In general, the animals of deforested, humanized open spaces have benefitted from changes in Mediterranean biotopes, while forest and wetland species have declined. Among Old World birds, some of the most favored groups include larks, *Sylvia* warblers, pipits (*Anthus*), wheatears (*Oenanthe*), Mediterranean gamebirds (*Alectoris* and *Coturnix*), and various birds of prey such as short-toed eagles (*Circaetus gallicus*), harriers (*Circus*), eagles (*Aquila* and *Hieraaetus*), and falcons (*Falco*).

Species which are at home in and around human settlements have also been favored. The worldwide expansion of the brown rat (*Rattus norvegicus*) and house mouse (*Mus musculus*), especially in all the world's Mediterranean ecosystems, has been well documented. House sparrows (*Passer domesticus*) and starlings (*Sturnus*) have also spread widely. Europe is now witnessing a northward expansion of the spotless starling (*S. unicolor*) and a southward expansion of the common starling (*S. vulgaris*), induced by changes wrought by humans. In the case of the gregarious common starling, however, ethological changes are also partly responsible: there are now many roosting places in urban parks and spaces with trees, where hundreds of thousands of birds congregate. The birds can roost peacefully in these places given the absence of hunting, the structure of the environment, and the more favorable climatic conditions (greenhouse effect). By day they feed outside the city and return in the evening, providing city dwellers with the spectacular sight of massive flocks of returning birds forming in the city skies in preparation for their entry *en masse* into roosting sites. It is worth mentioning that predators such as the peregrine falcon (*Falco peregrinus*) prey on these large concentrations of birds and in Barcelona, for example, the peregrine is now very much a part of the urban fauna.

Another species which has benefitted from human expansion in natural Mediterranean systems is the wild boar (*Sus scrofa*). This animal was not particularly abundant in the Iberian Peninsula in the 1940s and 1950s due to predation by wolves, the formerly greater human presence in the countryside, the smaller amount of forested land, and the type of forestry exploitation practiced (charcoal burning, extraction, and clearance of undergrowth, and grazing inside forests). The number of wild boars increased between 1950 and 1970 and they spread to recolonize many areas where they had been absent for centuries, thus increasing both its population density and range.

Other species that have taken advantage of greater human presence in Mediterranean ecosystems are wild canids which are generalists as regards their ecological requirements. They have also benefitted from the decrease in their natural predators (top predators such as the wolf, lynx, large eagles and owls) and the increase in food and waste originating from human settlements. They reproduce efficiently and tolerate human presence well and have become the commonest predators in simplified and humanized Mediterranean systems, reaching densities they would not normally achieve under natural conditions. The best examples are the fox (*Vulpes vulpes*) in Mediterranean regions throughout the Holarctic region, the coyote (*Canis latrans*) in California, and the Asiatic jackal (*C. aureus*) in North Africa, Asia Minor, and Greece.

The spread of animal diseases

From what we have seen, the Mediterranean's original fauna has clearly been transformed. There has been a particularly large decline in the numbers of predators, especially top predators (wolves, bears, big cats, large birds of prey, etc.). Together with certain human activities, this decline has generally favored the proliferation of certain species of primary consumers and enabled them to reach high population

250 **The expansion of black-headed gulls (*Larus cachinnans*), linked to human activity**, has been very noticeable over the last few years. They are, for example, frequent visitors to garbage dumps with organic waste. They once only frequented the coast but are now found far inland. Black-headed gulls follow rivers, traveling from garbage dump to garbage dump, and have become scavengers, no longer feeding exclusively on fish.
[Photo: Josep Maria Barres]

densities. For health reasons this situation is worrying, and the use of wild species or closely related breeds has opened the door to various pathogens, many of which are new to these ecosystems.

The case of the rabbit (*Oryctolagus cuniculus*) is well known. This Iberian species is widely distributed throughout the Old World, and has been affected by two different viruses provoked or spread by humans: myxomatosis and a viral haemorrhagic fever. Rabbit populations have, as a result, crashed everywhere.

The effects of the decline in rabbit populations has been most marked in those natural systems in the Mediterranean Basin that included the rabbit. It was once the principal prey of many carnivores such as the Iberian lynx (*Felis pardina*), wild cat (*Felis silvestris*), mongoose (*Herpestes ichneumon*), imperial, golden, and Bonelli's eagles (*Aquila adalberti*, *A. chrysaetos*, and *Hieraaetus fasciatus*), eagle owl (*Bubo bubo*), and birds of prey such as Egyptian vulture (*Neophron percnopterus*), to name but a few. The biological efficiency (the number and survival of eggs, hatchlings and adults) of these species has declined considerably, which is very negative for their continued survival. Moreover, the reduction of rabbit numbers has deprived many human hunters of their favorite prey, and many have taken to hunting other species, even those that are protected or in danger of extinction. Other animals are accused of having reduced rabbit numbers when humans are in fact responsible (unsuitable reforestation techniques, transfer of viral strains from one area to another, hybridization with domestic species, poor sanitary conditions in rabbit farms, etc.).

Another clear example of the spread of disease has occurred in the ibex (*Capra pyrenaica*). In the Sierra de Cazorla National Hunting Reserve in Spain this wild goat's population fell victim to mange (*Sarcoptes scabiei*), and in a mere two years (1989-1991) numbers plummeted from around 10,000 to just over 300 individuals. The generalist species discussed above also sometimes suffer from disease, and fox populations in western Europe are also affected by mange. In all these cases, it is humans that have altered the equilibrium between animals and the diseases and parasites with which they have coevolved. The situation of rabbits and ibexes is especially serious as these species are at the base of the trophic webs on which most predators depend (lynx, wild cat, eagles, Egyptian vultures, owls, other birds of prey, etc.). Mortality has increased and breeding success decreased in all these predator species. Furthermore, Mediterranean biomes, as one of the world's most densely populated ecosystems, are undergoing rapid modifications. Agricultural and livestock systems are changing due to production surpluses, irrigation, abandonded land, the creation of standards for consumer products *denominacion de origen*, etc. This will all lead to a series of changes whose effects cannot always be foreseen.

Hunting control and the recovery of animal populations

Hunting, together with many other factors, has pushed many species towards complete or virtual extinction, but in the 1960s many countries banned the hunting of many formerly hunted species. Protection is now becoming more widespread for species such as Iberian lynx, leopard, crane, great and little bustards, pin-tailed sandgrouse, and stone curlew. Other, formerly disdained species are now widely hunted and considered as fair game, such as many species formerly considered pests (magpies, crows, jackdaws, and ravens, for example), and other animals that compete for small game (carnivorous mammals and birds of prey), and species shot purely for fun are now hunted more than ever. Species taken for sport include vast numbers of passerine birds caught in traps or by liming, often a way of initiating children into the world of hunting.

In urban areas, throughout Mediterranean Europe and the rest of Europe, there is a tendency to look upon hunting as a negative aspect of human behavior, but in rural areas the situation is in fact quite the opposite. This negative perception is largely the result of the effects of the high population density of humans (and thus of hunters), as it is clear that the current scarcity of natural predators means that hunting is necessary to keep the populations of some wild species under control. In most of the Mediterranean Basin the current situation is fairly worrying, and is particularly serious in Greece, Turkey, and North Africa where hunting has little social or economic value and is almost totally unregulated. Wild populations of North African game and non-game species are continually falling as a result of the use of methods which are frequently totally unacceptable (laying of poison, cruel and unselective trapping, illegal forms of hunting, etc.).

The effects of the law

In central and northern Europe, where hunting is administered according to Germanic law (the ownership of the animals present is linked to ownership of the land, and thus to a responsible person), but in southern Europe hunting is generally poorly organized and controlled. Roman law separates land ownership from ownership of hunting rights, considering game as *res nullius* (the concern of no-one), and thus belonging to everybody. This consistently negative factor has totally prevented any control of hunting, and in many areas game populations have decreased since "first come, first served" is the only rule governing hunting. Only the most abundant species (often, like rabbits and hares, considered pests) in areas with low human populations have tolerated this

251 Skinned fox corpses (***Vulpes vulpes***) in Isona (Pallars, Catalonia). Predators have historically been considered as vermin and persecuted by hunters and farmers in various ways (traps, snares, and poisons). Their bodies are displayed publicly as proof of the efficiency of the gamekeepers charged with protecting game. This macabre sight, which seems almost to be designed as a warning to other foxes, may in fact be a covert way of trading furs using the excuse of the damage caused by foxes.

[Photo: Antoni Agelet]

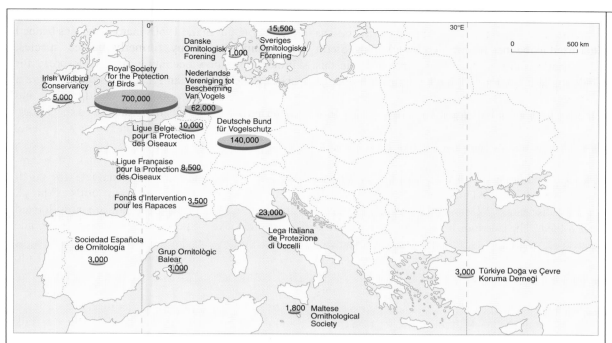

252 **Members of European ornithological societies** in 1989 (the figure for Spain refers to 1992). Humans have always felt a fondness for birds, shown by the number of ornithological societies, their age, and high membership. This love of birds is so great that when a rare bird is spotted thousands of people may travel to see it.
[Diagram and map: Editrònica with data provided by Xavier Ferrer]

3. HUMANS IN SCLEROPHYLLOUS FORMATIONS

pressure without problems. This became evident in the 19th and above all in the 20th century when hunting became very popular in the Mediterranean and firearms began to be used. In Islamic countries in the south and east of the Basin, the world view and system of property rights associated with Islam have led to a model of game hunting that is very similar to that in European Mediterranean countries, although with the addition of persistent feudal values.

The recovery of game species as a whole varies from country to country and essentially depends on legislative developments. In Spain, the 1970 Hunting Act was a giant step towards controlling and organizing hunting. Apart from the National Hunting Reserves (very efficient ways of aiding the recovery of big game species), the Act established private hunting areas with owners who thus could be held responsible for their functioning. This person was responsible for deciding who

could hunt or not, reducing the pressure on game, and the system also meant that it was the owners of hunting areas who gained or lost from the correct or incorrect management of natural resources, in this case game.

A similar model is the French, also based on Roman law, and hunting zones usually correspond to municipal boundaries. In each municipality, it is the local hunting society, generally well organized in hunting federations at the level of the Department, that controls hunting. Results vary and are very much influenced by the high population density of France.

Whether due to inadequate legislation or inability to enforce existing legislation, or both, the situation in other European Mediterranean countries and North Africa is totally different to that in France and Spain. In Italy, especially in the North, over-emotional hunters more concerned

253 **A sign indicating a private hunting area** in Catalonia under the watchful eyes of a robin (*Erithacus rubecula*). Throughout human history the role of hunting has been constantly changing, and over the last few centuries hunting has been a sport rather than a way of obtaining animal resources.
[Photo: Oriol Alamany]

254 The California condor (*Gymnogyps californianus*) is an example of a species —once all but extinct in the wild— that is now recovering thanks to reintroduction programs. Recovery programs often involve the cooperation of several countries, since wildlife does not recognize the existence of human frontiers. Rivers, large habitats, and birds that can travel large distances can only be protected if different countries collaborate.
[Photo: Jeff Foott Productions / Bruce Coleman Limited]

about for the size of the day's bag than the environment have led to a sad dearth of animals. Hunters even patrol motorway verges for the few animals, often repopulated only for hunting, that still remain. Since the 1970s Italian hunters have had to travel further afield (Iberian Peninsula, Balkans, France, and Eastern Europe) to find game to hunt.

Hunting is regulated in the New World mediterraneans. Most European settlers came from cultures with Germanic law (Holland and Britain), and so in much of East Africa hunting is strictly organized and managed (above all since 1950).

Nevertheless, despite these regulations there are still conflicts between the management of wildlife and agricultural and livestock interests. Wildlife populations are somewhat better conserved in California: in spite of the initial impact of white settlers, hunting is generally well controlled and populations recovery of certain species is often promoted by the Wildlife Service or sometimes by private concerns.

The recovery of fauna

Fortunately attitudes toward wildlife are changing and many factors that used to operate (poisoning, lack of planning or legislation, etc.) are now less widespread. Great efforts are being made to aid the recovery of certain species. In the Western European, Californian, Chilean and South African mediterranean, herbivore and bird of prey populations are recovering, although the trend is still negative in the eastern Mediterranean Basin (Albania, Greece, Asia Minor) and in North Africa. The following species have all benefitted from genuinely effective local, national or international recovery programmes set up to safeguard animal populations: lammergeiers, griffon vultures (*Gyps fulvus*), black vultures (*Aegypius monachus*), wolves, artiodactyla, especially the ibex with a current-day population of 25,000 individuals, roe and red deer in Western Europe, bald ibis in Israel, leopards in South Africa, and pronghorns and California condors (once again flying free in the wild) in the United States.

Although many species have in fact recovered somewhat in recent years, much time and energy is having to be spent on other highly endangered species (brown bear, lynx, Iberian imperial eagle, gazelles, black rhinoceros and bald ibis). It is very worrying that populations of some species as ecologically important as the rabbit and the cottontail are in a very bad state. The outlook in Mediterranean ecosystems throughout the world is fairly disheartening, but a change is visible in the rate of regression of animal species, many of which have had to modify their ecology and behavior.

4.5 A Mediterranean of naked brickwork

The landscapes of the Mediterranean have attracted and enchanted travelers and invaders from many different origins throughout history, and have often been associated with values of balance and harmony. The countless contemporary visitors also tend to perceive the Mediterranean as little more than a decorative backdrop; the July sun accompanied by the sea lapping at the sand on the beach or breaking gently against the rocks, rugged landscapes that change greatly at each bend in the road, a natural vegetation with large open spaces and few but impenetrable, forests. All this leads visitors to think the area's nature is lush, or even generous. Visitors are less likely to grasp, even though it is quite clear, that the Mediterranean is in fact an unrewarding place to live; stones lie just below the pretty flowers, terraced slopes built with the sweat of generations of peasants may collapse and be lost in a few minutes of torrential rains, vegetable gardens can be washed away by a flood, and the forests can burn like tinder. The Mediterranean landscape has more stones and bare ground than vegetation, more ruins than gardens, and more urban space occupied by concrete plazas than green spaces. It is a landscape of "naked brickwork," where the disasters perpetrated by tourist development have disfigured the entire coastline.

255 Humans have been living in the Mediterranean for millennia. Over this entire period, humans have modified the landscape, adapting it to their needs and extracting whatever they have thought necessary. Many places have been continuously inhabited, such as the village in the photo (Lindos on the island of Rhodes) and humans have left their stamp. The photo shows the remains of a Greek temple, a medieval fortification, and modern buildings —some unfinished—together with some signs of the use of natural resources, such as stone pens for sheep.
[Photo: Lluís Ferrés]

The difficult healing of landscape wounds

The entire Mediterranean landscape, even the areas that appear most "natural," reveals traces of human toil. It has all become artificial, to a greater or lesser extent and has been for a greater or lesser period of time. First impressions to the contrary, the Mediterranean has never been a paradise created for human enjoyment. Places that might seem "paradise" have had to be built entirely from scratch, often with greater effort than required in other areas, and its maintenance requires just as much effort, because in

256 **The large seaport of Naples has grown in a way that reflects the growth of other Mediterranean cities.** It was founded by the Greeks —the "modern" *Neapolis*— and it has felt the influence of its Roman allies, the Catalans, the Spanish, and population growth, especially in the 17th, 18th, and 20th centuries. Naples has always been an important center for the economic life of the surrounding regions.
[Photo: AGE Fotostock]

this world paradises are always precarious, and in the Mediterranean they are especially fragile. The shallow soils break down easily when exposed to the impact of the violent rains as a result of discontinuous plant cover. The soil escapes downhill in the runoff and leaves the rock increasingly bare. Forests that might protect the soil occupy increasingly small and marginal spaces, and without suitable forest management, seem to be waiting for a spark to set them ablaze. Natural succession in Mediterranean areas is extremely slow, and even slower where soil is lost through erosion, and therefore the Mediterranean landscape consists almost entirely of transitory communities (maquis, scrub, garigue, meadows, and wasteland) and actively cultivated, fallow or definitively abandoned ground. Less than 5% of the entire basin is occupied by forests. In these conditions, soil erosion can eliminate, in next to no time, a layer of soil that has taken thousands of years to accumulate. Therefore, cultivated spaces require imaginative solutions to protect their soil from erosion or to remedy it. The natural plant cover, which both depends on the topsoil and protects it, should be maintained in uncultivated areas, since it retains the topsoil. The areas disturbed by major public works (motorways, railways, reservoirs, etc.) require artificial replanting as energetically as possible. The Mediterranean world has a landscape of "naked brickwork" where every terrace wall and every small clump of vegetation is a precious shield against erosion and soil loss.

Uncontrolled housing development

The last few decades have added a new threat, urban growth, to those that have long faced Mediterranean soils, although it is not strictly new, all that is new is its scale. This includes urban growth, and the spread of second homes and tourist accommodation and facilities. This new interpretation of a "naked brickwork" landscape is only too literally landscapes covered in unfinished brickwork.

Urban growth

The growth of the Mediterranean cities, without reaching the extremes of some metropolis in sub-Saharan Africa or eastern and southern Asia, has been very high for the last few decades. Growth has been rapid in both the Mediterranean Basin and the other Mediterranean areas. Little more than a century ago, around 1880, not a single city Mediterranean city had a million inhabitants, and the only one approaching this figure was Istanbul, then the capital of the Ottoman Empire, with 875,000 inhabitants. No other city exceeded half a million, and only Naples (475,000),

Madrid (400,000), Lisbon (300,000), Rome (275,000), and Barcelona (250,000) had a quarter of a million or more inhabitants. Santiago de Chile and San Francisco, both with about 150,000 inhabitants, were the largest cities of the overseas Mediterranean areas. The only other cities in California were Sacramento, with 17,000 inhabitants and Oakland, with 15,500; San Jose, Los Angeles and San Diego had fewer than 10,000 inhabitants. Cape Town, together with its suburbs, had about 50,000 inhabitants. In Australia, Adelaide had a population of about 100,000 and Perth was an insignificant settlement with a population of less than 10,000. Fourteen conurbations in the Mediterranean Basin and another 10 in the overseas Mediterranean areas have more than a million inhabitants, and the urban area of Los Angeles has a population of more than 13 million. All this urban growth has been at the expense of the land surrounding the urban centres, regardless of whether they were productive agricultural soils or relatively well preserved natural vegetation. The area around the Yerba Buena Mission, the site of the founding of San Francisco, was next to some of the first land in California to be ploughed. The area that is now occupied by the Olympic Village in Barcelona, which was recovered from the factory districts built in the last century, was still covered in the mid-18th century by wetlands and salt marshes.

In some of the Mediterranean urban areas, this growth seems to be running out of steam, or at least, moderating. This seems to be the case in general in the north of the Mediterranean Basin since the mid 1970s. But this growth seems to be accelerating in the south of the Basin and in the overseas Mediterranean areas. In the metropolitan area of Cape Town, in the 1980s, new "informal" settlements have sprung up, such as Khayelitsha, where every day an estimated hundred new arrivals join the more than half a million inhabitants.

The quality of the construction work, when growth is so rapid and the population involved has a low purchasing power, is often dreadful, whether because the newcomers have reused materials to build precarious shacks or because constructors take advantage of the need for housing and use low quality materials and finishes. Many buildings put up in different urban areas in Spain, especially Catalonia, in the boom years of the 1960s have now been discovered to have beams made with aluminous cement; recently this has caused failures in several buildings, within twenty years of construction, in the Barcelona metropolitan area. Entire blocks have had to be demolished and their inhabitants rehoused.

The spread of second homes near the city

Furthermore, in the more developed areas, when growth and the increasing density of the urban spaces make the city almost uninhabitable, part of the population acquires the desire, rapidly converted to a need by massive publicity, to have a second home to spend the weekend and the holidays. Nature, for "townies" may just as well be a vineyard or a pinewood, and for many owners of peripheral forest or unproductive cropland within an hour away of the city by car, this obsession among city dwellers means a chance to sell otherwise low value property at nearly urban prices. In the areas around the medium and large cities in the richer regions of the Mediterranean areas, this desire has been one of the most important factors in the loss of soil and the destruction of some of the last remaining forests.

This is combined, especially in some metropolitan regions of southwest Europe, with the introduction of second residences that are used for purposes closer to tourism. In fact, in these areas, housing development for second homes are inseparable from tourist developments. Only in the areas furthest from the large urban centers is it possible to find purely tourist developments, which however threaten to occupy the entire coastline of the whole Mediterranean, or at least the western Mediterranean and some of the islands.

The Blue Plan for the Mediterranean estimated in 1986 that tourist accommodation occupied an area of 850 square miles (2,200 km^2) in the countries of the Mediterranean coastline (and twice this, including their infrastructure and facilities)—more than half of it in coastal regions. Approximately one third of all international tourism has as its destination somewhere in the Mediterranean Basin, in an annual summer migration in which more than 115 million people leave their homes and head for the coast, not including the internal tourism in the Mediterranean countries, estimated at about 100 million people.

The tourist explosion

The increasing urbanization and the higher living standards of the European population (urban growth leads to the desire to get away from a city environment that is often oppressive and degraded, while higher living standards allow part of family income to be devoted to leisure after covering basic needs), the development of the means of transport—both individual and collective—and the social organization of work (paid holidays, more free time at weekends) have all encouraged, since the 1950s, the

257 A heavily used Medi-terranean beach in Barcelona. In the Mediterranean's sunny beaches tourism is a resource that has contributed to the economic development of many areas but has also given rise to many new prob-lems. Mass tourism degrades the landscape and the temp-tation to get rich quick leads people to abandon local and traditional means of exploit-ing resources without provid-ing lasting, solid alternatives. [Photo: Jordi Todó / Tavisa]

growth of tourism and the implantation of second homes in western Europe. Beginning on the coast of Provence (la Côte Azur) and of Liguria (the Riviera), the tourist phenomenon in the Medi-terranean grew spectacularly after the recon-struction of western Europe following the Second World War. It started in Majorca and the Catalan coast (in the Costa Brava) and then spread to the rest of Spain's Mediterranean coastline, the Adriatic coast of Italy, and then to all the coastline, develop-ing later in Africa.

The impact of tourism and second homes on the nat-ural and social environment are so similar that they can be treated together. In fact, the most noticeable difference between them is the periodicity of the use of the accommodation and facilities, one or two days a week in the case of second residences, or two or three months a year in the case of tourist lodgings and facilities. The impact on the natural environ-ment varies greatly, as both the introduction of sec-ond homes and of tourism may start from a pre-existing village (reusing buildings or land already dedicated to human dwellings) or from landscapes formerly occupied by crops or natural ecosystems. Especially in periods of uncontrolled growth, prop-erty developers have often sought to bring the "townie" or tourist into such close contact with "nature" that they choose the last remaining wood-ed areas for development.

A forest that has been developed soon ceases to be a forest; light enters clearings along tracks and around houses, eliminating the shade-loving plants of the understorey and favoring the sun-loving plants of edges and scrub. Together with human presence, this prevents any possible natural regeneration, and ensures the forest will disappear when the existing trees die, if they are not felled before then. As a place in the country, a second home and a tourist flat have become mass consumption goods for many Europeans, the quality of the construction and of the infrastructure of many development has declined seriously, and in many cases the final result appears unfinished, provisional and degraded, not unlike some of the marginal districts in the big cities that many of the temporary inhabitants are escaping from. At the opposite end of the scale, the promoters who seek to offer high-quality tourist accommodation or second residences, for clients with higher purchasing power, search out the most beautiful sites, and then change them drastically. This has happened with the "marinas" that have occupied one after another of the already scarce wetlands of the Mediterranean coastline or some deluxe developments on the eastern coast of Sardinia (Costa Smeralda), until recently one of the areas of the Mediterranean coastline with the best preserved natural vegetation.

Yet the indirect effects on the environment of tourism and second residences is as great as, or greater than, the direct impact. The consumption of water increases several-fold, right at the height of the summer dry season, due to the presence of an increased population with wasteful water use habits. The number and size of forest and woodland fires has also increased several-fold, due to the presence of many people in natural areas, many of whom know little about the characteristics of Mediterranean ecosystems and do not realize the danger of some of their actions. The beach vegetation of most major summer resorts has already been totally destroyed.

And then there are the social effects, such as the abrupt abandonment of traditional lifestyles and values, and the adoption of new social values. These often give rise to conflicts and tension with the risk of social destabilization and loss of a cultural identity, usually reduced to stereotyped cliches for tourist consumption or blatantly falsified in response to the stereotypes the tourists are looking for.

Water, in short supply

The climatic conditions, the density of the population, and the presence of economic activities that are highly dependent on the availability of water mean that this resource is crucial in most of the Mediterranean areas, especially in the Mediterranean Basin and California. In both these areas, water resources are highly limited in comparison to demand.

In addition to the permanent demand, which is more or less fixed or subject to limited oscillations, water for human consumption and for industrial use, irrigation, and tourism, is subject to large seasonal demands precisely at the time of year when the volume of flow in the rivers is lowest and least reliable (because of the dry summer conditions). In the arid areas to the south, such as the Sahel in Tunisia, which lack permanent watercourses, or on islands,

258 Community well in Turkey. Water shortage in the Mediterranean Basin has led to the development of different water use strategies. Drinking water is shared in many towns and villages, and is greatly valued because it is so scarce.
[Photo: Lluís Ferrés]

especially on the smaller ones with permeable sub-soil, the over-exploitation of sub-soil aquifers is almost inevitable and will lead in the short or long term to the exhaustion of this water resource. Pollution of surface water and even subsoil aquifers by the discharges of population centers, industries, and irrigation activities, makes the water deficit worse in many watersheds, leaving the water unfit for human consumption.

The problems of the management of water in the Mediterranean are due its incorrect usage, rather than its deterioration. Of course, not all water resources are regulated, but regulation of rivers by dams does not solve every problem, and sometimes creates new ones. Good arable land is flooded in order to irrigate much larger areas that are not always of proven quality and the hydraulic infrastructure used is not always adequate and may thus lead to the salinization of the newly irrigated land. The retention of silt and sediment in reservoirs may have consequences in the lowlands and in the delta plains, at the same time as the reservoir silts up and loses the ability to regulate the flow of water. Bear in mind that Mediterranean rivers may carry, especially after torrential rains, up to 15% or 20% of solid materials (30% solid particles was measured in the River Medjerda in Tunisia in autumn, 1973). It is estimated that the reservoirs in Algeria lose 2-3% of their capacity every year. But deltas, like the Ebro Delta whose entire watershed is highly regulated, are losing ground, eroded away by the sea due to the greatly reduced amount of sediments arriving.

The Mediterranean rivers, whose volume of flow is always low, are indiscriminately used as sewers to discharge untreated residues of all types, even in more developed areas. In some more developed areas, in addition to these organic residues from the sewage of population centers, there are also discharges from farming and stock-raising, as well as often highly toxic industrial discharges. Catalonia is one of the most industrialized regions of the Mediterranean Basin and the River Besòs, whose mouth is a little to the northeast of the city of Barcelona, shows some of the most extreme cases of water pollution ever recorded—99.25% of water flowing in the Ripoll, a small tributary of the Besòs, every day in the summer low water period was untreated discharge from population centers and industries until a few years ago.

Both the north and the south of the Mediterranean Basin will be forced in the medium term to introduce more careful water management in order to meet demand, both by improving the efficiency of the use of the water (drip irrigation, recycling of industrial discharges, saving water in the home, etc.) and by returning it to the environment (treatment and reuse of discharges). Even with more efficient use of surface resources, foreseeable growth in the next 25 years (based on economic and population growth comparable to those of the last 25 years) will make it inevitable that regions with a sharp imbalance between supply and demand will have to turn to artificial resorts. These include desalination of seawater (already essential in some islands like Formentera or Lampedusa and some of the areas in the east and south of the Mediterranean basin) or transfer of water from other watersheds with a surplus, a solution viable only for areas in the north of the basin that are near watersheds fed by the more abundant rains of central Europe.

This solution, transfer, has been adopted in California, where the population growth and the increasingly intensive irrigation of one of the most developed regions of the world has only been possible thanks to the use of the excess water from the basin of the Colorado River. At the beginning of the 20th century, part of the water from the lower Colorado was transferred to Imperial Valley in southern California, and since 1940, the Los Angeles region has been supplied by an aqueduct almost 250 mi (400 km) long.

The new Mediterranean landscape

From a landscape that was "unfinished" due to the natural conditions, the Mediterranean Basin's landscape has changed in little more than a quarter of a century to a landscape full of "unfinished brickwork" in the most literal meaning of the term. Dividing walls, piles of rubble, empty plots, peeling paintwork, crooked signs, bulldozers flattening walls or carving roads through forests, the arms of (building) cranes rising above the pines are all now more characteristic of the coastal Mediterranean landscape than cork oaks, olive groves or lavender. Even the most prestigious spaces, precisely because they are famous and visited by many people, are spoiled by their setting of soft drink and fast food stands, shops selling souvenirs (often made far away and unrelated to local culture or to the monument or site visited) and premises dedicated to entertainment. From the Acropolis in Athens to the Alhambra in Granada, from Djerb to Santorini, from the area around Doñana National Park to the Hyères Islands tourism-linked speculation has killed the goose that lays the golden eggs with its peculiarly gross sense of aesthetics and total indifference to the natural and cultural values that are the basis of its success in the Mediterranean Basin.

4
Protected areas and biosphere reserves in the sclerophyllous formations

1. The world's protected sclerophyllous formations

1.1 Some general considerations

It general it can be said that there are protected areas of one type or another in all the Mediterraneans. Out the biome's total area of about 400,000 square miles (1.02 million km²), the protected areas account for 32 million acres (13 million hectares), i.e. about 1.3% of the total. The prolonged and uninterrupted interaction between human beings and the environment that has taken place in the Mediterranean Basin, mean that some of the oldest protected areas are in this area. Protected and sacred forests are known to have existed in Roman times, and since the Middle Ages there have been hunting reserves and forests. Awareness of the lack of protected areas has been most recent in South Africa and has led to the creation of new reserves in the Cape region where all the fynbos vegetation is considered to be endangered.

Yet as elsewhere on the planet, in the Mediterranean area strictly protected areas (with the aim of total protection from all human influence) coexist with areas that are loosely protected, and even other areas whose management involves the protection of some sections and the productive use, in an ecologically rational manner, of others. Strictly protected areas include all the classical national parks, natural parks,

and reserves, while the second group consists of the biosphere reserves as defined by UNESCO (volume 1, "The protection of species against the danger of extinction"). The preserved areas of the Mediterranean Basin are especially dedicated to the maintenance of systems that are little or highly modified artficially, and which are often close to human dwellings and in harmonious interaction with humans for centuries. The situation is different in the South African and Australian mediterraneans, where the vulnerable native vegetation has been seriously damaged by human activity, especially over the last two centuries, and projects to restore these habitats seek first of all to reduce human interference.

1.2 Protected parks and areas

Protected areas in the Mediterranean Basin

Classical literary sources such as the Old Testament, the writings of Theophrastus (372-287 B.C.), and the works of Pliny the Elder (A.D. 23-79) provide much information about the ancient Mediterranean landscape, the distribution of forests, and their exploitation. These testimonies are thus contemporary to the

259 Tsitsikamma National Park, South Africa, with a heath in the foreground and sclerophyllous formations in the middle distance. The first protected areas known as national parks were created in the United States more than a century ago; the concept of protected space soon spread throughout the world, and there are now protected spaces in most countries. As natural parks were created to protect nature and also for human leisure—sometimes in total contradiction—it became necessary to create other forms of protection, but none was totally satisfactory, as the human species was not considered as an integral part of nature. This is why the concept of the biosphere reserve was born in 1971, which included human beings and their activities as integral elements of the area in question. Areas were not fenced off and controlled exploitation was permitted. At the same time, an international network of reserves was established to exchange experiences and research results.
[Photo: Colin Paterson-Jones]

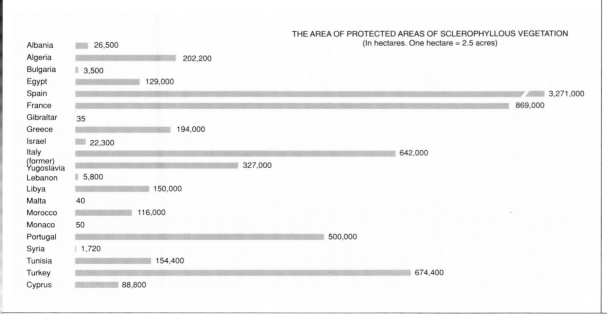

THE AREA OF PROTECTED AREAS OF SCLEROPHYLLOUS VEGETATION
(In hectares. One hectare = 2.5 acres)

Country	Area
Albania	26,500
Algeria	202,200
Bulgaria	3,500
Egypt	129,000
Spain	3,271,000
France	869,000
Gibraltar	35
Greece	194,000
Israel	22,300
Italy	642,000
(former) Yugoslavia	327,000
Lebanon	5,800
Libya	150,000
Malta	40
Morocco	116,000
Monaco	50
Portugal	500,000
Syria	1,720
Tunisia	154,400
Turkey	674,400
Cyprus	88,800

260 The area of protected areas of sclerophyllous vegetation in the countries of the Mediterranean Basin, including all the different protective statutes (natural parks, biosphere reserves, etc.).
[Drawing: Editrònica, on the basis of data provided by the author]

ancient silviculture practices that regulated the plantations of trees and the protection of forests and matorrals. Over a period of centuries, sacred areas, forests, and hunting reserves were established and maintained. Many of the areas that are now national parks or protected areas have their origins in such areas, such as Doñana, Ichkeul, and Mount Olympus.

In all the Mediterranean areas there have been regional initiatives to establish networks of protected areas. Some of the most complex programs have been drawn up precisely for the Mediterranean Basin, where a total of 26 countries have to reach agreement for there to be consensus on the priority actions for Mediterranean cooperation. The current initiatives follow the directives of the European Union, the World Bank, UNESCO's MAB Program, and the United Nations' Environment Program (UNEP), which is reflected in their support for the Barcelona Convention (1975), an agreement between the countries of the Mediterranean coastline. The protected areas now represent 12.8% of the basin, i.e. about 18.3 million acres (7.4 million hectares) out of a total of 223,880 square miles (580,000 km²) of sclerophyllous Mediterranean biome. The largest protected area in the region is the Corsican Regional Natural Park, which covers 741,000 acres (300,000 ha), followed by the natural parks of the sierras de Cazorla, Segura and Las Villas, covering 530,000 acres (214,300 ha) in Andalusia, and then the Velebit Mountains covering 494,000 acres (200,000 ha) in Croatia.

Protected areas in the South African mediterranean

The southern tip of Africa is known for its very high floristic diversity and for the large number of endemic species that grow there, especially in the southwestern tip of the Cape Province, whose unique flora makes it one of the world's six floral kingdoms, and has a special sclerophyllous vegetation (*fynbos*) that covers 26,600 square miles (69,000 km²), 5.72% of the area of the Republic of South Africa (see "The Cape Kingdom").

Conservation has a long history in South Africa. The first signs of concern arose bacause of the the regression of the Afro-montane forests in the Cape Province in the first half of the 19th century. The first protective measures were taken in 1856.

In the early 1980s a resolution was adopted that recognized the importance of the coastal fynbos and a committee was established to investigate different proposals to create nature conservation areas. The resulting report presented a summary of the state of the area and the strategies necessary to fulfil the conservation objectives. The report also proposed the creation of a fynbos project that prepared, by means of a set of working groups, detailed reports on particular sites, to preserve the most valuable areas on the basis of complete information on the sites in question. Thus, the priority was to carry out conservation actions where they were most necessary, and conservation mechanisms were installed in all these sites. Even so, images taken in 1981 by the Landsat satellite revealed that 34% of the natural fynbos vegetation had been destroyed by diverse human activities, such as intensive agriculture and urbanization. The greatest losses were among the coastal fynbos, 47% of which was lost.

Almost 15% of the area of fynbos in South Africa enjoys some form of protection. Both in number and in area, most of the protected area is in the mountains. Only 10% of these areas, representing only 5% of the total area of the biome, consists of coastal fynbos and other types of lowland vegetation. The fynbos is protected on the basis of a series of zones classified into different categories: state-owned conservation areas, semi-state conservation areas, areas protected by private bodies, and South African Defence Force areas. Even so, there are many areas without protection. On the other hand, the South Africa's difficult relations with the United Nations until their democratic transition impeded the creation of biosphere reserves in that country; finally in 1998 Kogelberg was cataloged as a Biosphere Reserve. Perhaps the largest protected areas within the Cape Region center of endemism are the basins of the Hawequas (447 square miles [1,159 km²]) and the Cederberg (488 square miles [1,264 km²]). The rest of the protected area consists of small but well-managed sites, that are good representatives of the fynbos. Although there are many reserves of fynbos in the lowlands, most are so small that their long-term viability is doubtful.

Most of the approximately 580 nature reserves in South Africa are concentrated in the relatively narrow strip between the coastline and the inland plateau. Almost all the reserves have been created in the last 30 years, and only five cover more than 247,000 acres (100,000 ha). And 64% of the reserves are isolated from each other. Yet more than 50% of the land surrounding these reserves is so highly artificial that it is almost impossible for the flora and fauna to emigrate to neighboring reserves. Only 88 of these 580 South African nature reserves have lists, mostly incomplete, of the species that live there and only 28 have reason-

261 The Bontebok National Park in South Africa owes its name to the bontebok (*Damaliscus dorcas dorcas*), the animal it was set up to protect. This park fulfills all the criteria established by the United Nations for national parks. Firstly, its ecosystems have not been exploited by humans, secondly, the flora, fauna and geomorphology are of enormous scientific interest (the renosterveld, shown in the photograph; see section 1.3 of this volume), thirdly, the Park is run by state authorities, and lastly, educational, cultural, and recreational visits are permitted. The type of protection given to natural areas varies from one country to another according to the perceived necessities and priorities, as well as the financial, institutional and legislative efforts that each country has chosen to or has been able to implement. Equally, the great variety of natural formations found in protected areas means that management strategies cannot necessarily be compared. Hence the variety of different names used throughout the world for protected natural areas and management policies.
[Photo: Anthony Bannister / NHPA]

262 The Cape Mountain Zebra (*Equus zebra zebra*) is one of the most important animals in the De Hoop Reserve, a protected area which embraces various ecosystems including the typically South African fynbos. Only 600 individuals of this finely marked zebra remain; its narrower stripes and slighter build separate it from other members of its genus. It is found scattered throughout many South African nature reserves.
[Photo: N.J. Dennis / NHPA]

263 Eucalyptus forest typical of the pre-dune systems of Nambung National Park, in Western Australia, created in 1968 and which covers an area of 43,253 acres (17,500 ha). The photo shows an area dominated by the manna gum (*Eucalyptus viminalis*), also called stringybark because of the long strips of bark that are shed from the trunk was taken in the moist sclerophyllous formations of the Brindabeld Ranges. The landscape is highly homogeneous throughout the formation, with a low herbaceous layer, a shrub layer of ferns, and a tree layer of eucalyptus.
[Photo: Wayne Lowler / Auscape International]

ably thorough lists. There is a clear lack of information, and especially in the lists for the small reserves in the Cape. The protected areas of the Cape Mediterranean region include, on the one hand, the Bontebok National Park with coastal renosterveld, the basin of the Groot Swartberg, the Tsitsikamma Coast National Park, and the De Hoop Natural Reserve, which represent, if only partially, the Cape fynbos; and, on the other hand, the Storms River National Reserve (within the Tsitsikamma National Park), the Goendal Natural Area, the basin of the Groot Winterhoek, and the basin of the Hawequas, which represent areas with "false" karoo (areas formerly covered with grassland but now transformed through grazing pressures to a karroid shrubland). Other National Parks in the Mediterranean climatic region are the Wilderness National Park (the lake area) and the West Coast National Park.

The Protected areas of the Australian Mediterraneans

The first national park in Australia was created as a royal park south of Sydney in 1879, and the early 20th century saw the progressive development of the system of national parks and reserves. In 1990 there were about 100 million acres (40.78 million hectares) of protected ground, about 5.3% of Australia's surface area. The Australian Federal Government carried out a study of the continent's biological resources, with the intention of documenting the entire flora and fauna, including the mediterraneans in south and southwestern Australia that cover areas of 36,285 mi^2 (94,000 km^2) and 60,600 mi^2 (157, 000 km^2) respectively.

The protection of the Australian mediterranean biome, which is located in the states of South Australia and Western Australia, depends on the state legislation. The legislation on protected areas in Western Australia was enacted with the 1933 Land Act that allowed the Governor to set aside land for public use. The National Parks and Nature Conservation Authority, in accordance with the 1984 Conservation and Land Management Act, develops strategies to stimulate public respect for nature and the protection of the environment. In South Australia, the body responsible is the National Parks and Wildlife Service, within the Department of Environment of Planning. In parallel, the Woods and Forests Department within the Ministry of Forests is responsible for the administration of the forest reserves.

The Mediterranean areas are represented in almost 200 of the protected areas in Australia. They make up a total of 10 million acres (4 million ha), approximately 10% of the total protected area in Australia. Two thirds of the protected area within the Mediterranean areas in Australia, more than 6.2 million acres (2.5 million ha), are in western Australia. These include about 40 national parks that alone account for more than half of the total area and a hundred or so nature reserves, that cover approximately one million ha. On the other hand, the roughly 70 protected zones in South Australia cover more than 15 million acres (6 million ha), a quarter of which is in eight national parks, and a little more than half in 43 conservation parks.

In Western Australia, most of the national parks were created in the early the 1970s, except for the oldest, Yanchep, created in 1905 and covering 6,915 acres (2,799 ha), the Stirling Range, declared a national park in 1913 and covering 285,700 acres (115,661 ha), Neerbup, created in 1945 and covering 2,675 acres (1,082 ha), and Cap Le Grand, created in 1948 and covering 77,530 acres (31,390 ha). Within the sclerophyllous area, the largest protected areas are the Cape Arid National Park, covering 690,155 acres (279,415 ha), and the Jilbadji Nature Reserve, covering 515,900 acres (208,866 ha). The average area of the reserves is 4,940 acres (2,000 ha) the national parks vary in size from 2,470 acres (1,000 ha) to 98,800 acres (40,000 ha).

In the South Australia mediterranean, the oldest national park is the Flinders Chase, which was created in 1919, while the oldest conservation park is Fairview created in 1960. The largest protected areas in the state (outside the Mediterranean area) are the Nullarbor National Park, covering 572,800 acres (231,900 ha), and the Ngarkat conservation park, covering 513,615 acres (207,941 ha).

The protected areas of the Chilean mediterranean

The first natural park in Chile was created in 1926, in order to protect its natural resources and beauty. Tourism was allowed, as long as it did not threaten the model of life of the local people. Between 1935 and 1945, 12 more parks were created with the aim of protecting the endemic flora and fauna, but the management plans were not drawn up until the early 1970s detailing the infrastructure and the research and educational projects required by the protected areas. By 1990 there were 30 national parks, 36 nature reserves, and 10 natural monuments, covering a total of 33.6 million acres (13.6 million ha), 18% of the country. Now, CONAF (National Forestry Corporation) is responsible for the administration the network of protected areas by means of the National System of State-

Protected Wild Areas (SNASPE). The different definitions used for the different categories of protection are: national parks, national reserves, and natural monuments. CONAF is developing a management system for each of the SNASPE protected areas.

The Chilean mediterranean covers an area of 8,878 square miles (23,000 km²), about 3% of the country. Within this, the 1990 United Nations List of National Parks and Protected Areas recognises eight areas which cover a total of 363,110 acres (147,008 ha), i.e. 1.37% of the total protected area in the Neotropical kingdom. This corresponds to 6% of the Chilean mediterranean, or 0.2% of the country's total area. Among the protected areas there are two national parks with a total area of 44,358 acres (17,959 ha), eight national reserves that cover 315,011 acres (127,535 ha), and five national monuments covering 6,815 acres (2,759 ha). The first declared protected areas were: the national monument of Cerro Nielol in 1939, the Fray Jorge National Park in 1941, and the national reserve of Lake Peñuelas in 1952.

Among the recommendations of the 1985 report by the CONAF's Department of Protected Forest Areas titled *Ecological Representation of the National System of Wild Areas Protected by the State in Relation to the Udvardy Classification* it was stated that according to

265 The composite *Coreopsis gigantea* is endemic to Santa Cruz Island, one of the Channel Islands (California), a site with some of the last representatives of the Mediterranean vegetation of North America, including 80 endemic species. The Channel Islands have never been joined to the mainland, but it is thought that when sea levels were lower, the four northernmost islands formed a single island.
[Photo: Stephen Krasemann / NHPA]

the criteria of the IUCN, within the frontiers of Chile there was an important biogeographic unit, the Mediterranean, that was not adequately protected. In 1985, the CONAF started a basic system of classification of the native Chilean vegetation in order to understand better the country's plant formations and ecosystems and to ensure that protection under SNASPE was effective. The system recognized 8 ecological regions, 17 subregions, and 83 different plant formations.

In 1986, it was realized that in Chile there were many different forms of protection in the different regions, and some ecosystems were not even considered by the SNASPE. Almost 82% of the protected areas were concentrated in the Aisén and Magellanes regions in the far south of the country. The conclusion was reached that the sclerophyllous matorral, the steppe, and the Patagonian desert were not sufficiently protected. Special priority was given to the inclusion in the SNASPE of these under-represented ecosystems. These included the Chilean mediterranean which was progressively deteriorating. And so, several new protected areas were declared in the Mediterranean area, including the national reserve Laguna Torca covering 1,492 acres/604 ha (1986), the national reserve of the Pampa de Tamarugal covering 248,605 acres/100,650 ha (1988), and the natural monument of Isla Cachagua (1989).

These protected areas play an important role in leisure and recreation within the country. In 1988, the Chilean national system of protected areas recorded a total of 520,000 visitors, of whom 280,000 visited the national parks. In the Chilean Mediterranean the most visited areas are the natural monument of Cerro Niclol, with 130,000 tourists, and the national reserve of Río Clarillo with 70,000 visitors.

Protected areas in the California Mediterranean

In the United States, the administrative systems governing protected areas are very complex. The conservation legislation of the protected areas is regulated at both the state and federal level. Then, at the state level, there are different areas that are protected by local or regional administrations. The conservation of parks did not officially begin until June 30, 1864 when President Abraham Lincoln signed the document ceding Yosemite Valley and the Mariposa Grove of giant redwoods to California for "public use, resort, and recreation." Shortly afterwards, on March 1, 1872 Yellowstone was declared a national park, the first in the world.

The State of California is typical of the North American mediterranean, and specifically the chaparral formation, which covers about 135,100 square miles (350,000 km²) of hills and rolling but not very high sites of which 32,810-38,600 square miles (85,000-100,000 km²) correspond to sclerophyllous forest. The dominant genera of woody plant in this chaparral, such as *Adenostoma*, *Ceanothus*, *Heteromeles* and *Rhus*, are not usually found in the other regions with a Mediterranean climate or they do not play such an important role. The proportion of endemic plants in California is very high. There are more than 25 endemic species in the chaparral, some of which are considered scarce or in danger of extinction within the state according to the California Native Plant Society.

In California the chaparral is highly protected. Within the state there are almost 100 federally protected areas. These include the following 6 national parks: Channel Islands National Park (covering 249,438 acres [100,987 ha] and created in 1980), Kings Canyon (covering 462,060 acres [187,069 ha] and created in 1940), Lassen Volcanic (covering 119,284 acres [48,293 ha] and created in 1916), Redwood National Park (covering 103,740 acres [42,000 ha] and created in 1968), Sequoia (covering 402,894 acres [163,115 ha] and created in 1890) and Yosemite (covering 761,449 acres [308,279 ha] and created in 1890). There are also 20 national wildlife reserves, 4 national monuments, and 47 fauna protection areas and 3 national leisure areas. In fact, more than 10% of the Mediterranean vegetation is within protected areas.

2. UNESCO's biosphere reserves in sclerophyllous formations

2.1 The Mediterranean biosphere reserves

The biosphere reserves declared by UNESCO under their MAB (Man and Biosphere) Programme represent the latest generation of protected areas (see section 4.1.1 of this volume and, especially, the insert "The protection of species against the danger of extinction" of volume 1). The designation "biosphere reserve" is, above all, concerned with overall management and not simply with the protection of a natural area since the latter is a logical consequence of the former. The declaration of a biosphere reserve implies the preservation of virgin natural areas as well as a guarantee that economic activities within the reserve will continue to be viable. The fulfilment of these goals depends on opportune and ecologically sound management policies. The 34 biosphere reserves with matorral and Mediterranean sclerophyllous forest formations are distributed throughout 15 countries. More than 4.9 million acres (two million hectares) of the Mediterranean biome (more than the 0.5% of the total biome) form part of a biosphere reserve: more than two million acres (833,000 ha) in the Mediterranean Basin; 671,100 acres (271,700 ha) in western and southern Australia; almost 77,065 acres (31,200 ha) in Chile; and two million acres (893,300 ha) in California. In the sclerophyllous fynbos zone of the Cape in South Africa there is the Kogelberg Biosphere Reserve. The first Mediterranean biosphere reserves, created in 1976, were those of the Channel Islands and San Joaquin and San Dimas in California. Of all the biosphere reserves in mediterraneans, that of the Channel Islands National Park is one of the largest (one million acres [500,000 ha]), followed by the Californian Central Coast, and the Fitzgerald River National Park in Western Australia. Smallest is the Miramare marine park (148 acres or [60 ha]) near Trieste in Italy. Links have been established between different Mediterranean biosphere reserves, the most important being the relationship between the Californian and Chilean biosphere reserves and the relationships established between all of the Mediterranean Basin reserves at the Workshops on Mediterranean Biosphere Reserves held in 1986 in Florac in the Park Headquarters of the Cevennes Natural Park Biosphere Reserve (France).

2.2 The Mediterranean Basin biosphere reserves

Within the sclerophyllous biogeographical formations of the Mediterranean Basin, 25 biosphere reserves exist in ten countries of southern Europe, western Asia, and North Africa. Not all of these 25 reserves can be considered entirely as examples of the Mediterranean biome as some represent small examples of mountain biomes, an may embrace marine meadows and coastal matorrals or extend as far as deciduous forests and alpine pastures. The first seven Mediterranean biosphere reserves were created in 1977: two in Tunisia, two in France, one in Spain (Montseny), one in Italy, and one in Croatia. The Montseny was designated as a protected area in 1928, one of the first areas of the Mediterranean Basin to be declared a reserve.

Spain possesses the greatest concentration of biosphere reserves in the Mediterranean Basin. Of the 15 Spanish biosphere reserves, 11 are sclerophyllous and Mediterranean: the Sierras de Cazorla and Segura, Doñana, Grazalema, *Mancha Húmeda*, Odiel Marshes, Montseny, Sierra Nevada, the upper reaches of the Manzanares River, the island of Minorca, Cabo de Gata-Níjar and the Sierra de las Nieves. Other countries in the Mediterranean Basin such as Algeria (2), Croatia, France (3), Greece (2), Italy (2), Portugal, Tunisia (2), and Israel also contain biosphere reserves. Between the largest there are the Sierras de Cazorla and Segura and the Sierra Nevada, both covering 469,300 acres (190,000 ha). As mentioned above, the smallest is the Miramare marine park in Italy. The total surface area of biosphere reserves in the Mediterranean Basin excels 2.5 million acres (990,000 ha).

Under the auspices of UNESCO's MAB Pro-gramme, some countries have begun work on projects to create biosphere reserves which as yet have not borne fruit. For example, in 1980 Libya invited consultants from UNESCO to give guidance on the setting up of a national park and biosphere reserve in Kouf in the Jabal al-Akhdar. Lists of possible parks have also been drawn up for countries such as Egypt where there are currently three biosphere reserves which are, moreover, not within the Mediterranean biome.

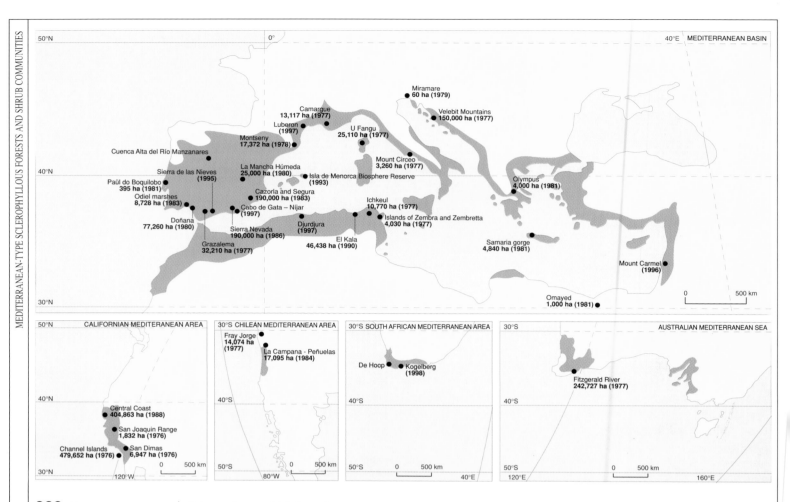

266 Mediterranean biosphere reserves included in the United Nations list of national parks and protected areas with surface areas and declaration dates. (In hectares, 1 hectare= 2.5 acres.) The De Hoop Natural Reserve has been included although strictly speaking it is not a biosphere reserve. Some of the reserves included not only represent the sclerophyllous formations dealt with in this volume. Instead, they represent oro-biomes or mountain biomes, or in other words, biomes within the Mediterranean climatic area that are also heavily influenced by another important physical factor such as the proximity to the coast (the Camargue) or altitude (the Montseny).

[Maps: Editrònica, from data supplied by UNESCO]

Others places which have been proposed as biosphere reserves include the Bentael National Park, proposed by the permanent UNESCO delegation in the Lebanon, and various others selected by the MAB Programme committee of Tunisia. Many of these proposals were made within the framework of the Mediterranean Biosphere Reserve Action Plan, first formulated during the I Workshop on Reserves held in Florac (France) in September 1986. Their claims were further advanced at the second meeting in Montesquiu (Catalonia) and in the joint meeting of the III MAB Workshop on Mediterranean Biosphere Reserves and the first IUCN-CNPPA Workshop on Protected Areas in North Africa and the Near East held in 1992 in Tunis.

Many areas of today's biosphere reserves have a mythical and historical significance for Mediterranean people. Mount Olympus springs to mind as the legendary home of the Greek gods, as does the Samaria Gorge, a stronghold of Christianity during the Ottoman occupation of Greece, and once said to be inhabited by nymphs.

Recently discovered cave paintings in the Cuevas de Pardis near Segura and the Iberian settlements from 4000 B.P. at Galera and Orcera and at Baños de la Marrana show that the Cazorla and Segura Mountains have been inhabited since prehistoric times. The El Ichkeul National Park Biosphere Reserve in Tunisia has been a hunting reserve for many years and its buffaloes are thought to be descendants of the herds present in Carthaginian times which were subsequently preserved for hunting. The Zembra Reserve, also in Tunisia, with its Phoenician tombs and Roman ruins has an equally interesting historical past. The palaeomediterranean, Illyrian, Roman, and Slavic remains in the Velebit Mountains Biosphere Reserve in Croatia are of great archeological importance and this region has also been declared of great ethnological interest owing to its fine rural architecture.

The 700 years of history behind the Doñana Biosphere Reserve are more modest in comparison. The Doñana Reserve was the favorite hunting ground of Philip IV in the 17th century, Philip V in the 18th century, and Alfonso XII in the 20th century. The forests in the Italian Reserve of Circeo are the remains of the Agro Pontino forest that in days gone by covered large areas of the Italian coastline and that were believed to be the magical kingdom of the witch Circe who turned Ulysses's companions into swine.

267 **The U Fangu Valley in NW Corsica**, carved out from rhyolite deposits which have filled an ancient granitic canal, was declared a Biosphere Reserve in 1977. It is principally important for the typical sclerophyllous Mediterranean vegetation, consisting of a dense holm oak forest and the different types of matorrals form part of the succession to a forest formation. Some of the most interesting animals include the moufflon (*Ovis musimon*), wild boar (*Sus scrofa*), various species of deer, and a number of very important bird populations. [Photo: Ernest Costa]

268 **The Montseny mountains in the Catalan Pre-Coastal Range in NE Iberian Peninsula,** that reaches 3,323 ft (1,700 m), are characterized by a wealth of different plant formations. The photograph shows the typical Mediterranean holm oak (*Quercus ilex*) forest which covers the slopes of Tagamanent, one of the principal peaks (3,461 ft or 1,055 m) of the SW of the Montseny mountains. [Photo: Oriol Alamany]

269 Mediterranean landscape in the Sierra de Caillo in the Grazalema Biosphere Reserve in the extreme south of the Iberian Peninsula. Most of the Reserve experiences a Mediterranean climatic regime, although the peculiarities of the climate and relief means that a much more varied vegetation with a large number of endemic forms has been able to evolve. Apart from the typically Mediterranean holm oaks and carob trees, the Reserve also contains Spanish fir, cork oak, and deciduous oak forests, as well as numerous rock and fluvial communities. However, human action has severely altered the whole face of the Reserve, and in the more arid zones the vegetation has been degraded by centuries of grazing by goats, fire, as well as by firewood extraction and agriculture.

[Photo: Oriol Alamany]

The Grazalema Biosphere Reserve

The Grazalema Biosphere Reserve in the south of the Iberian Peninsula contains an imposing karstic massif reaching 5,425 ft (1,654 m) with a complete altitudinal zonation of Mediterranean vegetation and exceptional botanical richness. Grazalema is home to splendid thermophilous matorral, spiny cushion summit communities, as well as various forest formations including holm oak (*Quercus ilex*) forests and Spanish fir (*Abies pinsapo*) forests. This fir is endemic to the Grazalema and the Serranía de Ronda. Grazalema also holds important bird of prey populations, and good number of large mammals. The griffon vulture (*Gyps fulvus*) colonies are among the largest in Europe.

Characteristics and natural riches

The Grazalema Natural Park lies within the provinces of Cádiz and Málaga, almost equidistant between the cities of Seville, Cádiz and Málaga, and covers an area of 127,687 acres (51,695 ha). It is the western-most massif in the Cordillera Baetic and its nucleus is made up of spectacular rocky ridges (Sierra del Pinar) which stand out from the flatter surrounding countryside.

The mountains of Grazalema act as a barrier to the wet winds which blow in off the nearby sea (Gulf of Cádiz, 50 mi or 80 km away). The clouds cool quickly as they rise over the mountains and produce considerable rainfall, making the Grazalema region one of the wettest areas of Europe. Annually, Grazalema receives an average of 98 in (2,500 mm) of rain, although in some years this figure exceeds 156 in (4,000 mm). Despite these rainfall levels, this area is without doubt part of the Mediterranean biome and exhibits the typical irregularity of Mediterranean systems. In some years only 39 in (1,000 mm) of rain falls, and the seasonal distribution of rainfall is characteristically Mediterranean with torrential rain in winter and a long dry summer with temperatures reaching 104°F (40°C).

The substratum is composed principally of Lias, Jurassic, and Cretaceous limestones, while in the basal areas clays and Triassic gypsums appear along with irregularly distributed El Aljibe sandstones typical of the Algeciras region. The predominance of limestones and the singular amount of rain have combined to create an abrupt and complicated landscape, rich in geomorphological formations produced by karstic erosion. The landscape is full of scarred rocks, steep slopes, cliffs and vertical-sided gorges (Garganta Verde and Garganta Seca). To the east and southeast, the sandstones have created an undulating landscape with low hills that contrast greatly with the impressive peaks of the limestone massif.

Surface water in the Grazalema Mountains drains into the River Guadalete to the north, the River Majaceite to the southwest and the Guadiaro to the southeast. However, in the interior of the mountains the solubility of the limestone rocks in rainwater and the contact between materials of different degrees of permeability have created a highly original hydrological system. The almost total absence of superficial water courses means that there are very few river valleys. On the other hand, there are numerous depressions (dolines) without any apparent outlet for water. In fact, the water penetrates through cracks and fractures, dissolving the rock, and circulates in subterranean rivers through a vast network of galleries, often full of stalactites and stalagmites (the Yedra, la Rajada, la Ermita and la Pileta caverns) connected to the outside world by well-known potholes (Villaluenga, el Cacao). Many springs occur where underground rivers meet less permeable strata. For example, the River Gaduares is swallowed up by the Hundidero Cave and reappears 2.5 mi (4 km) away and flows into the large Gato Cave where it joins the River Guadiaro. The orographic and climatic peculiarities of the Grazalema Mountains have generated various microenvironments and an incredibly rich flora: over 1,200 species of vascular plants have been identified. The different vegetation strata are showcases of many different Mediterranean plant communities, from the thermo-Mediterranean, right through to the supra-Mediterranean.

The mountain range from 820 ft (250 m), the lowest point, to the high point of Torreón (over 5,250 ft/1,600 m) in the Sierra del Pinar. The predominant species, *Quercus ballota*, is ecologically very flexible and is found at different altitudes alongside the plants that are more specific to each altitudinal community. In the low-lying areas up to about 2,950 ft (900 m) dominated by low Mediterranean scrub, the holm oak appears with the carob (*Ceratonia siliqua*), wild olive (*Olea europaea* var. *sylvestris*), *Rhamnus lycioides oleoides*, mastic tree (*Pistacia lentiscus*), mock-privet (*Phillyrea angustifolia*) and *Smilax aspera*. These communities are still well preserved in some areas (Monte de las Encinas, Los Laureles), although a large part of the lowland holm oak forests has been converted into dehesas or replaced by olive groves, wheat fields, and pastures. There are areas of garrigue with kermes oak (*Quercus coccifera*), *Ulex parviflorus*, grey-leaved cistus (*Cistus albidus*), the European fan palm (*Chamaerops humilis*), and other areas of matorral with rosemary (*Rosmarinus officinalis*) and *Anthyllis cytisoides*. On the acid sandstone soils venerable cork oaks (*Quercus suber*) grow, sometimes mixed in with a few holm oaks and an understorey characterized by a large number of heathers (Spanish heath [*Erica australis*], tree heath [*E. arborea*], *E. scoparia* and *E. umbellatum*), and cistus species such as *Cistus ladanifer, C. monspeliensis, C. populifolius, C. salviifolius, C. crispus*. Other common species include ling (*Calluna vulgaris*), bracken (*Pteridium aquilinum*), wood sage (*Teucrium scorodonia* ssp. *baeticum*), *Genista tridens*, and *G. monspessulana* [=*G. candicans*]. More humid oak and mixed deciduous and conifer forests, as well as the extraordinary Spanish fir forests appear between 2,950-4,950 ft (900-1,400 m). The Lusitanian oak (*Quercus faginea*) occupies the lower parts of this zone where the environment and soils are more humid, and it is often found mixed in with holm oaks, cork oaks, Spanish firs, and with the typical bushes such as hawthorn (*Crataegus monogyna*), sloe (*Prunus spinosa*), and wild roses (*Rosa*) found on the edges of deciduous forests. A few oaks (*Quercus canariensis*), common in the neighboring region of Algeciras, grow in the Barrida Gorge near Ubrique.

270 **Spanish fir (*Abies pinsapo*) forest**. This fir is found in the mountains of Grazalema, Ronda, and Las Nieves in the pre-Baetic sierras, and represents the most valuable plant formation in the Grazalema Biosphere Reserve. The forest occupies 989 acres (400 ha) and has been expanding and recovering since the Reserve was created. Some authors consider the fir to be endemic to the southern Iberian mountain ranges, while others consider that it is only a variety of the firs found in the Mahgreb. [Photo: M. Rafa / Arxiu Alamany]

271 The poppy *Papaver rupifragum* is almost only found in the Grazalema area. It lives in shady, rather humid cracks in calcareous rocks at around 2,953 ft (900 m). The reason why many plant species are restricted to these southern Iberian mountains is related to its special climatic features and relief which distinguish it from the surrounding areas, which are generally totally Mediterranean, making it the ecological equivalent of an island. Rainfall is abundant, falling above all in autumn and winter, and is combined with the typically high temperatures of the southern Mediterranean which can reach 104°F (40°C) for a few days at the height of the summer.
[Photo: Abelardo Aparicio]

The Spanish fir (*Abies pinsapo*), endemic to the Sierra de Grazalema and Serranía de Ronda and undoubtedly the most valuable floristic element of the Reserve, grows on the northern, more Atlantic slopes of the Sierra del Pinar. It is found between 2,950 and 5,250 feet (900 and 1600 m), although the densest patches are concentrated between 3,280 and 4,590 ft (1,000 and 1,400 m) where it has formed a forest of 988 acres (400 ha), the best surviving community today of this species. The gloomy atmosphere of the forest floor and the acid leaf-mold prevents the formation of an understory, although some species such as spurge laurel (*Daphne laureola*), hellebore (*Helleborus foetidus*),

ivy (*Hedera helix*), honeysuckle (*Lonicera etrusca*), and helleborine (*Cephalanthera rubra*) can survive. In humid hollows at lower altitudes, the Spanish fir mixes in with oak, whereas at higher levels it penetrates into communities of maple (*Acer monspessulanum*) and whitebeam (*Sorbus aria*). It can also survive alongside holm oak in crevices and depressions in the karstic areas, as well as with the low cushion formations at the summit of *Erinacea anthyllis*, *Vella spinosa*, *Bupleurum spinosum*, *Ptilotrichum spinosum*, and *Arenaria racemosa*. On the south-facing slopes of the Sierra del Pinar, Phoenician juniper (*Juniperus phoenicea*) is dominant and a few scattered holm oaks and Spanish firs descend as low as 2,296 ft (700 m). The rupicolous (growing on and around rocks) vegetation is very impoprtant due to the abundance and diversity of rocky habitats. Many interesting species are found, some of which belong to more ancient floras. Especially worthy of mention are *Papaver rupifragum*, a poppy endemic to Grazalema, *Saxifraga boissieri*, endemic to the Sierra de Grazalema and Serranía de Ronda, and *Silene andryalifolia* and *Ionopsidium prolongoi*, Baetic-Mauritanian endemics. Along permanent water courses there are still magnificent riverbank communities of narrow-leaved ashs (*Fraxinus angustifolia*), smooth elm (*Ulmus minor*), white poplar (*Populus alba*), black poplar (*Populus nigra*), and willow (*Salix alba, S. fragilis* and *S. atrocinerea*). In dry river beds, there are oleander formations (*Nerium oleander*), reeds (*Phragmites australis*), and bulrushes (*Typha dominguensis*). In a few areas, the presence of saline water has allowed halophyte communities of glasswort *Salicornia ramosissima*, sea spurrey (*Spergularia marina*), and beaked tassel-pondweed (*Ruppia maritima*) to flourish. The variety of habitats and the wide range of trophic resources offered by the plant communities opf the Park allow a very diverse fauna to survive. Moreover, various migration routes between Europe and Africa pass through Grazalema. In total, 136 species of birds (sedentary and migrant), 40 mammals, 14 reptiles, 10 amphibians, and three fish have been recorded in the Park. The forests are home to abundant large herbivores such as the roe deer (*Capreolus capreolus*), here at the southern limit of its range, and the red deer (*Cervus elaphus*), a recently reintroduced species. Among the smaller mammals, northern hedgehogs (*Erinaceus europaeus*), common white-toothed shrews (*Crocidura russula*), pipistrelle bats (*Pipistrellus pipistrellus*) and greater horseshoe bats (*Rhinolophus ferrumequinum*) are some of the more typical species, as is the fire salamander (*Salamandra salamandra*) among the amphibians. Birds are everywhere: phytophagous species such as the wood pigeon (*Columba palumbus*), turtle dove (*Streptopelia turtur*), chaffinch (*Fringilla coelebs*) and greenfinch (*Carduelis chloris*) live alongside insectivores such as the coal, blue, great and crested tits (*Parus ater, P. caeruleus, P. major* and *P. cristatus*), golden oriole (*Oriolus oriolus*), green woodpecker (*Picus viridis*), great spotted woodpecker (*Picoides* [=*Dendrocopos*] *major*) and short-toed treecreeper (*Certhia brachydactyla*). Forest predators include the rare and endangered wild cat (*Felis silvestris*), Egyptian mongoose (*Herpestes ichneumon*), European genet (*Genetta genetta*), weasel (*Mustela nivalis*), and polecat (*Mustela putorius*), traditionally domesticated and used for catching rabbits. Higher up the forest trophic chain, there are a number of raptors such as the tawny owl (*Strix aluco*), buzzard (*Buteo buteo*), sparrowhawk (*Accipter nisus*), and booted eagle (*Hieraaetus pennatus*), as well as the horseshoe snake (*Coluber hippocrepis*). The jay (*Garrulus glandarius*) and wild boar (*Sus scrofa*), neither very common, are omnivores which are closely linked to forest environments. Rocky areas and cliffs offer few trophic resources but are, on the other hand, excellent biotopes for breeding and shelter. The ibex (*Capra hispanica*) is one of the most interesting species in Grazalema, and one of the most evident animals in the first level of the trophic chain. The commonest insectivores are birds such as the common and spectacular alpine swift (*Tachymarpis* [=*Apus*] *melba*), crag martin (*Ptyonoprogne* [=*Hirundo*] *rupestris*), black wheatear (*Oenanthe leucura*), blue rock thrush (*Monticola solitarius*), kestrel (*Falco tinnunculus*), and little owl (*Athene noctua*); the latter two species are raptors which supplement their diet of small mammals with large insects. Among the predators, the area is very rich in birds of prey such as the golden eagle (*Aquila chrysaetos*), Bonelli's eagle (*Hieraaetus fasciatus*), eagle owl (*Bubo bubo*), and peregrine falcon (*Falco peregrinus*) and there are good populations of beech martens (*Martes foina*) and Lataste's viper (*Vipera latasti*). Finally, all the Park's scavengers, including the Egyptian vulture (*Neophron percnopterus*) and one of the biggest griffon vulture (*Gyps fulvus*) colonies in Europe, breed there on the cliffs. Many of the predators which live in and around the forest and rocky zones hunt in the maquis and garrigue, as well as in other open spaces. There are many types of small finches such as the linnet (*Carduelis cannabina*), goldfinch (*Carduelis carduelis*) and serin (*Serinus serinus*), as well as rabbit (*Oryctolagus cuniculus*), Cape hare (*Lepus capensis*), and partridge (*Alectoris rufa*), much bigger prey. All feed on the rich and diverse plant matter of the shrub communities. Many insectivores feed in the matorral: the Algerian sand racer (*Psammodromus algirus*) and spiny-footed lizard (*Acanthodactylus erythrurus*), as well as the subalpine

272 **The griffon vulture** (*Gyps fulvus*) is a gregarious bird that breeds in large colonies on ledges or in caves on cliffs, preferably on sunny calcareous rock faces such as those found frequently throughout the Grazalema Reserve. The population of griffon vultures has declined greatly in the Iberian Peninsula, above all due to a loss of habitat. Nevertheless, a number of recuperation programs are currently underway and many are proving to be successful thanks in no small part to the support given by the public.
[Photo: Oriol Alamany]

warbler (*Sylvia cantillans*), Dartford warbler (*Sylvia undata*), northern wheatear (*Oenanthe oenanthe*), black-eared wheatear (*Oenanthe hispanica*) and stonechat (*Saxicola torquata*) are all common. One of the most common predators in the matorral is the ocellated lizard (*Lacerta lepida*). The few rivers and streams which flow in the Park form a biotope characterized by the presence of water and soft herbaceous plants. They contain fish such as the barbel (*Barbus bocagei*), amphibians such as the sharp-ribbed salamander (*Pleurodeles waltl*) and tree frog (*Hyla meridionalis*), reptiles, such as *Mauremys caspica*, and

273 Zahara in the Grazalema mountains, like other villages within biosphere reserves, has been inhabited for many centuries and many of its popular traditions are still alive today. Humans from across the centuries have left their mark in many areas of the Park and Paleolithic, Roman, Visigothic, and Muslim remains, as well as cave paintings (especially in the Pileta cave), have been found. The Muslim inheritance has been immortalized in the name *Zahara* or flower which, according to legend, was the name of Mohammed's daughter. In June, on the religious holiday of Corpus Christi, the village holds its annual *fiestas* and the local population covers the streets with flowers that fill the village air with perfume and color.
[Photo: Oriol Alamany]

mammals such as Mediterranean water shrew (*Neomys anomalus*) and Etruscan shrew (*Suncus etruscus*). Among the many birds, there are white, grey, and yellow wagtail (*Motacilla alba, M. cinerea* and *M. flava*), chiffchaff (*Phylloscopus collybita*), Bonelli's warbler (*P. bonelli*), wood warbler (*P. sibilatrix*), melodious warbler (*Hippolais polyglotta*), olivaceous warbler (*H. pallida*), Sardinian warbler (*Sylvia melanocephala*), Orphean warbler (*S. hortensis*), blackcap (*S. atricapilla*), house martin (*Delichon urbica*) and sand martin (*Riparia riparia*). The most typical predator of fluvial environments is the otter (*Lutra lutra*).

Reserve management and problem solving

The interesting archeological remains discovered in the Park are evidence that the area was densely populated in prehistoric times. In addition to many lesser finds, the most important are the Palaeolithic cave paintings in the Pileta and Gato caves and the extraordinary Veredilla cave complex with its wonderful Neolithic red pottery dating from the 7th millennium B.P., Iberian and, above all, Roman remains, including the vestiges of towns (Ocurris [Ubrique], Iptuci [Prado del Rey], and Lacibula [Grazalema]), Roman roads, aqueducts, necropolises, villas and water tanks, as well as coins and pottery have been found. There are also important medieval remains from the Visigothic period and from the Muslim civilization which held on in the area until 1485 (the upper parts of Benaocaz, fortifications in Zahara, Aznalmara, Cardela, and Audita). Today over 30,000 people live in the area, principally in the attractive white villages (*pueblos blancos*) which still preserve the old urban structures and local architectural forms. These villages, spread out against the dark background of the hillsides, are an essential part of the Grazalema landscape. Economic activities were closely linked to the natural environment and consisted of stock-raising, agriculture, sylviculture, and the traditional manufacture of cloth and leather goods, activities which survived until fairly recently along with the traditional way of life, customs, and religious holidays (Corpus Christi in Zahara and el Gastor, reenactments of the battles between Moors and Christians in Benamahoma, etc.).

During the second half of this century, the socio-economic structure of these mountain regions has had to face a series of difficulties which has caused many young people to move away and, in a more general sense, has provoked an economic decline in the area. Nevertheless, some traditonal economic activities still persist. Cork and carob beans are still collected, dry-farming continues, and extensive stock-raising is practiced in the forests which have been transformed into *dehesas*. Small game is still shot and honey, cheese and artesanal sausages are produced. The traditional production of crafted leather goods in Ubrique, which is said to have imported seal furs from Russia, has evolved towards a thriving industry based on goat skins. Recently, an attempt has been made to revive the traditional manufacture of blankets in Grazalema. Clean springs and rivers have enabled trout farms to be set up and coarse fishing is popular. Tourism, attracted by the natural, cultural and, scenic splendors of the area, is beginning to be seen as an important facet in sustainable local economic development. Grazalema was declared a Biosphere Reserve in 1977, and on December 18, 1984, the Environmental Protection Agency of the Andalusian Regional Government (*Agencia de Medio Ambiente* [*AMA*]) declared the area a Natural Park. Since then, it has formed a part of the network of protected areas in Andalusia whose aims are four-fold: conservation, education, recreation, and socio-economic growth. The Park is managed by a Director appointed by AMA. The Management Plan, ratified on December 27, 1988, establishes a management framework for the Park which includes the delimitation of areas and the different activities that can be carried out therein. This is a necessary step towards the fulfillment of the Park's objectives. Three categories of areas have been defined. Firstly, the

Reserve itself which includes the most valuable ecosystems, strictly protected, where the only permitted activities are the investigation and study of wildlife. Secondly, there is an extensively managed area where primary economic activities compatible with the stability of natural systems and recreational visits are allowed. Finally, there is an intensively managed area consisting of areas which have been highly altered by economic activities and which the Plan proposes to regulate and manage.

The main areas of environmental management involve the monitoring of animal populations through studies of nesting and breeding; fire prevention; regeneration of the plant cover; and forest conservation. The strict protection of the Spanish fir has been proved to be working as the Spanish fir forest is currently expanding. The vegetation on karstic slopes, after years of pasturing by goats, cutting for firewood, and deliberate fires, is also recovering well. The principal problems facing the Reserve are poaching and forest fires.

The Plan has also given priority to certain lines of research which are supervised by a scientific committee. The use of the Park for scientific work is considerable although more could be carried out. The AMA has, above all, encouraged the study of the fauna of the Park (situation of the roe deer, censuses of ibex and birds of prey), while universities and other bodies are studying the vegetation, geology, and landscape.

The Plan has created a management structure (*Gerencia de Promoción*) whose aim is to promote economic development in local communities by optimizing existing activities and promoting new ideas through grants from outside the Park and technical help for individual concerns. A lot of work has still to be done to maintain the human population of the area and to involve local inhabitants in the future of the Reserve. This policy is very promising as it is very important to establish effective mechanisms that will encourage coordination with local people and their participation in the Park. One pioneering project is the establishment of a cooperative to promote green tourism and adventure tourism. Planning for public access will concentrate on information and interpretation, environmental education, and civil protection. Currently there is basic infrastructure for visitors with centers in the villages of El Bosque and Grazalema, although the information center is only open all year in the latter village. Various information brochures and routes have been published and a small botanical garden is being built that will contain the most characteristic plant species of the area. Nevertheless, the interpretation and educational programs are just beginning

274 The Phoenician juniper (*Juniperus phoenicea*) is a palaearctic and circum-Mediterranean species well adapted to arid climates and windy places. It prefers calcareous rocks and dry, rocky locations, although it will grow on any substratum. It grows very slowly and survives for many years.

[Photo: Oriol Alamany]

and will have to be enhanced with better facilities (an interpretation center and a historical-ethnographic museum are to be opened) and more staff.

Given the surface area of the Park, the number of visitors is low (1.7 visitors per acre or 0.7 visitors per hectare per year). However, one of the principal problems facing the Reserve is the negative impact of the public (high concentrations of visitors in small areas, fires, camping and other uncontrolled activities). The boom in adventure sports may also have a negative effect on certain animal populations (hang-gliding disturbing breeding birds of prey). The villages of the area

still need to be provided with better sewage and waste disposal systems to prevent pollution, especially serious in the case of the contamination from the leather good factories in Ubrique. Equally urgent is better control over construction to stop the growth of dangerous illegal housing complexes and the disfiguring of the traditional architectural styles of many villages.

The case of the El Kala Biosphere Reserve

The complex of the El Kala Biosphere Reserve in Algeria is considered one of the three most important wetlands in the Mediterranean Basin, and is notable for its sclerophyllous forests and other ecosystems that contain plants that are nationally and internationally rare. Its natural cork oak forests contain the last specimens in Algeria of a variety of maritime pine (*Pinus pinaster renoui*) and the alder grove in Nicha Rirhia (in the center of the reserve and approximately 2,625 ft [800 m] long) is one of the largest and least deteriorated riverine forests in North Africa and contains many species that need humidity, such as the abundant royal fern *Osmunda regalis*. The Kala is also important as the last refuge of the Barbary red deer (*Cervus elaphus barbatus*) in the Mahgreb and a refuge for the otter (*Lutra lutra*) and the caracal (*Felis caracal*).

Natural characteristics and values

The National Park of El Kala existed before the Biosphere Reserve and is located on the Mediterranean coast of northeast Algeria, near the frontier with Tunisia in the *wilaya* (district) of al-Tarf near the city of el-Kala, 311 mi (500 km) east of Algiers. Its climate corresponds to the limit between the humid and sub-humid Mediterranean bioclimatic zone. The average temperature is 47.4°F (8.5°C) in the coldest months and reaches 86°F (30°C) in the hottest months. The average yearly rainfall varies between 34.3 in and 46.4 in (879 mm and 1,191 mm) and depends on the altitude and the distance from the mountains. El Kala was accepted as a Biosphere Reserve in 1990 with a total area of 235,733 acres (95,438 ha). The central area occupies 45,730 acres (18,514 ha), the buffer zone occupies 138,649 acres (56,133 ha), and the transition zone 4,424 acres (1,791 ha). The peripheral area occupies a further 46,930 acres (19,000 ha). When it was declared a national park on July 23, 1983 it received total legal protection. Until then and during the French colonial period is was known as the "Réserve de la Calle." The park includes the natural reserve of lake Oubeira and the Lake Tonga and Mélah hunting reserves, included in the Ramsar Convention on November 4, 1983. The park is situated on the plain and the outlying hills of the Great Tell Atlas, specifically the mountains of Kroumirie which reach an altitude of 663 ft (202 m). The rocky layer of the low land consists mainly of an alternation of Tertiary and Quaternary sandstones and clays, although in the hills there are also acidic Oligocene rocks. The relief is gentle, reaching 656 ft (200 m) in Djebel Ghorra, and between the coast and the mountains there is a well-developed dune system, running parallel to the coast and reaching a height of 581 ft (177 m). The filtration of water in the soil from the dunes allows the formation of fringing forest dominated by the alder (*Alnus glutinosa*), and very shallow lakes and marshes surrounded by woody hills. Some of these small lakes and marshes are refuges for overwintering migratory birds. Lakes Oubeira and Tonga consist of enclosed basins of fresh water 19.7-39.4 in (0.5-1 m) in depth with abundant vegetation and limited supply of water. Lake Mélah is a saltwater lagoon connected to the sea.

The plant communities include large areas of primary forest, maquis, garrigue, highly degraded croplands as well as meadows of riparian or marsh vegetation. The dominant ecosystem is sclerophyllous Mediterranean forest, which occupies 71.4% of the Biosphere Reserve. In El Kala it is possible to find seven or eight forest communities, dominated by cork oak (*Quercus suber*), Lusitanian oak (*Q. faginea*), garrigues, forests of maritime pine (*Pinus pinaster*), maquis of mastic tree (*Pistacia lentiscus*) and wild olive (*Olea europaea sylvestris*), riparian vegetation, and groves of alder. The most widespread forest community is the cork oak with undergrowth dominated by different species of heath (*Erica scoparia*, *E. cinerea*, *E. tinnaria*, *E. arborea*), strawberry tree (*Arbutus unedo*), myrtle (*Myrtus communis*) and mock-privet (*Phillyrea angustifolia*). Groves of *Quercus faginea* are only found in the areas at altitudes greater than 2,952 ft (900 m).

The plant dominating the coastal juniper groves are juniper (*Juniperus oxycedrus*), savin or Phoenician juniper (*J. phoenicea*), kermes oak (*Quercus coccifera*) and *Cistus salviifolius*, which may develop into thermophilous matorral with mastic tree (*Pistacia lentiscus*), olive (*Olea europaea var. sylvestris*), strawberry tree (*Arbutus unedo*), mock-privet (*Phillyrea angustifolia*), genista (*Genista tricuspidata*), bracken (*Pteridium aquilinum*), and myrtle (*Myrtus communis*). The noteworthy species of the garrigue include sage (*Phlomis bovei*), *Halimium halimifolium*, *Coronilla valentina*, and violet (*Viola sylvestris*). The kermes oak may reach a height of one or two meters and is found all along the coast in exposed and highly degraded biotopes near human settlements; it is the dominant species on the coastline between Cape Rosa, El Kala and Cape Roux.

On the coast there are two types of pine forest; those of Aleppo pine (*Pinus halepensis*) and mixed forests of Aleppo pine and maritime pine (*P. pinaster*). In the coastal dunes these communities also include the Phoenician juniper (*Juniperus phoenicea*), *Jasminum fruticans*, *Daphne gnidium*, *Rhamnus alaternus* and genista (*Genista aspalathoides*).

The riverine forests are dominated by communities of alder (*Alnus glutinosa*) and ash (*Fraxinus angustifolius*), tall forests that apparently represent the least deteriorated sites of the Biosphere Reserve, following the network of river channels. There are also other associated species, such as willow (*Salix pedicellata*), elm (*Ulmus minor*), oleander (*Nerium oleander*), and white poplar (*Populus alba*). In the transition between the lake and land zones there are alder groves with populations of willows (*Salix alba* and *S. cinerea*).

In the levels with the deepest water table, the undergrowth beneath the alders includes shrubs like the Mediterranean buckthorn (*Rhamnus* [=*Frangula*] *alnus*), brambles (*Rubus ulmifolius*) and laurel (*Laurus nobilis*), climbing plants such as ivy (*Hedera helix*) and the wild vine (*Vitis vinifera*) and plants endemic to Algeria, such as the *Hypericum afrum*, *Campanula lata* and *Solenopsis* [=*Laurentia*] *bicolor*.

The species dominating the vegetation submerged in the marshes and pools in of El Kala are pondweed (*Potamogeton*) in the eutrophic Oubeira lake, and the tasselweed (*Ruppia spiralis*) in the brackish Lake Mélah. In the freshwater lakes, the helophytic vegetation consists of beds of club-rushes, reeds, and reedmaces, with club-rush (*Scirpus lacustris*), common reed (*Phragmites australis* [=*communis*]), and the narrow-leaved bulrush (*Typha angustifolia*) together with variable patches of yellow flag iris (*Iris pseudacorus*) accompanied by the white water-lily (*Nymphaea alba*) and *Sparganium erectum*.

The endangered plants found in El Kala include, in Lake Oubeira, *Polygonum senegalense* and *Paspalidium obtusifolium*; and in lake Tonga, cord-grass (*Spartina patens*), duckweed (*Lemna trisulca*), white water-lily (*Nymphaea alba*), lesser spearwort (*Ranunculus flammula*), *Cardamine parviflora*, and the water chestnut (*Trapa bispinosa*).

The low areas of the Biosphere Reserve are very important feeding and breeding areas for the local bird fauna. The cork oaks and shrub communities near the lakes contain wood pigeon· (*Columba*

275 **The desert lynx or caracal (*Felis caracal*)** has always been much appreciated and assimilated by human cultures. The Egyptians revered it as a mythical figure and often depicted the svelte form of its body and eyes. In India, it was domesticated to hunt small mammals. Being found all over Africa, it has often mistakenly been called the African lynx. However, it is also common throughout the Arabian Peninsula and in Asia as far as the Aral Sea and NW India. It is equally at home climbing trees in the open savannah as in sandy deserts. One of its African strongholds is the Algerian El Kala reserve, although this photograph was taken in the Zebra Mountain National Park in South Africa.
[Photo: Anthony Bannister / NHPA]

palumbus), turtle dove (*Streptopelia turtur*), black kite (*Milvus migrans*), green woodpecker (*Picus viridis*), great spotted woodpecker (*Picoides* [=*Dendrocopos*] *major*), wryneck (*Jynx torquillo*), Cetti's warbler (*Cettia cetti*), blackcap (*Sylvia atricapillla*), Bonelli's warbler (*Phylloscopus bonelli*), and serin (*Serinus serinus*). The aquatic fauna includes many wigeons (*Anas penelope*), up to 9,000 pochards (*Aythya ferina*), about 12,000 tufted ducks (*A. fuligula*), and about 5,000 coots (*Fulica atra*). The low areas are also important hunting grounds for the osprey (*Pandion haliaetus*) and hundreds of marsh harrier (*Circus aeruginosus*) that overwinter in the region.

The mammals include wild boar (*Sus scrofa*), otter (*Lutra lutra*), caracal (*Felis caracal*) and the Barbary red deer (*Cervus elaphus barbarus*), which is found among the cork oaks and the Lusitanian oaks. In the past the fauna of the forest areas was much richer, and in the late 18th century abbot Poiret commented on the presence of Barbary leopards and lions in the forests close to the former French fortification of Vieux Calle and Lake Mélah.

In the El Kala region there are some endemic and sub-endemic species of Mediterranean bird, some endemic to Algeria itself, such as a subspecies of the Yelkouan shearwater (*Puffinus puffinus yelkouan*), the shag

(*Phalacrocorax aristotelis demarestii*) and a sub-species of the purple gallinule (*Porphyrio porphyrio porphyrio*). The Bou Redim marshes house one of the most important heron colonies in the Mediterranean Basin, with squacco heron (*Ardeola ralloides*), grey heron (*Ardea cinerea*), purple heron (*Ardea purpurea*) and night heron (*Nycticorax nycticorax*).

Other areas in the El Kala park are occupied by grazing land, plantations of white poplar, thickets of brambles, fields subject to flooding, and croplands that as a whole represent a large part of the protected area.

Management and problems

Human beings have lived in this area since antiquity. Archeological remains include Neolithic megaliths and dolmens in the Djebel Ghorra, Roman ruins and the remains of a French fortress from the 17th century, Vieux Calle, where at least 400 people died from malaria in 1679. There are also several monuments in the park built to commemorate battles that took place in Algeria's war of independence. Throughout the park there are many scattered villages and hamlets with traditional houses made of adobe mud walls. The total estimated population in 1990 was 67,246 people, and their livestock grazed a large area of land within the park. The rural economy is based on the cultivation of cereals and fruit trees, stock-raising, and small industry. Agricultural land covers about 37,050 acres (15,000 hectares) within the park. The park was created to protect the unusual marsh complexes, and their fauna, flora, water resources and historic monuments, as well as the region's typical landscape and lifestyles. In collaboration with University College, London, studies and a draft management plan were proposed, and the plan has been followed since 1986. Recently UNEP and the World Bank funded a scheme to develop this management plan. The Biosphere Reserve is divided into five areas. The first is an integral reserve; the second is a protected, or wild area; the third area allows limited growth; the fourth area acts as buffer zone; and the last area is a peripheral transition area.

Although the El Kala Biosphere Reserve is considered as a protected site, it is inhabited and exploited, as tourism, industry, and housing developments are permitted within the park. The activities that are tolerated vary from one area to another. The first area's unique and unusual characteristics mean it can only be used for scientific research; the second area (natural or wild) allows no development, even the construction of main roads; in the third area there are sections where modest expansion is permitted, although it is regulated; the fourth area is intended to buffer the first and second areas against possible external impacts and camping is allowed in it; in the fifth zone, the peripheral areas, all types of construction are allowed. The park authorities consider it very important to maintain human activities, such as agriculture and aquiculture, and thus grazing, silviculture, and fishing are allowed. The exploitation of natural resources is controlled and hunting is prohibited. The park's administration is the responsibility of a resident director and a number of permanent staff under the control of a management committee with representatives from 11 ministries and the local authorities. Recently, there have been ecological improvements to the Tonga and Fetzara lakes (which were dredged in the 1930s), including bringing back into operation the former sluices to bring water back to the lakes.

The park's educational activities consist of guided visits round the shores of Lake Tonga. The usual complement of researchers consists of about 100 Algerians and about ten foreign. The most complete study of the park was performed by Bougacelli and a group of

276 Linnaeus named the shag (*Phalacrocorax aristotelis*—a common bird in the Mediterranean—after Aristotle. It normally breeds on inaccessible ledges and in sheltered caves up to 328 ft (100 m) up on sea cliffs. Linnaeus' homage to Aristotle was in recognition of the Greek philosopher's writings on nature: of his 50 known works on zoology 25 have survived and include descriptions of 140 species of birds.
[Photo: M. Rafa / Arxiu Alamany]

research students in 1976. The resources available to the researchers include a field research station, a meteorological station, and experimental stations to plant trees and raise fish. The Algerian National Office of Forest Studies has drawn up vegetation maps, and maps of the park's limits and zones. The major environmental problems are the degradation of the forests due to grazing and forest fires, and construction of roads through the forest, hunting pressure on the shores of the lakes, the drainage of wetlands, the extraction of ground water, and dredging.

Other problems include agricultural exploitation, excessive human occupation, illegal hunting, the accelerated increase in semi-natural habitats, such as the vegetation of wetlands, and the loss of the coastline. Other changes have occurred in the park's characteristics with the gradual loss of the traditional culture, shown in the replacement of the typical mud-straw huts by modern European-style cement houses.

2.3 The biosphere reserves in the Australian mediterraneans

The two Mediterraneans in the Australian continent have a single biosphere reserve, the Fitzgerald River National Park, which was established in April 1978. The Fitzgerald River reserve is on the coast, and rises from sea-level to a height of 1,499 ft (457 m). Its relief is as highly varied as are its flora and fauna. Its climate is mild with a rainfall varying between 24.6 in (630 mm) a year on the coast and 14.6 in (375 mm) inland. The minimum annual temperature is 49.3°F (9.6°C) and a maximum of 74.3°F (23.5°C).

The Fitzgerald River Biosphere Reserve

The Fitzgerald River Biosphere Reserve occupies a site full of contrasts, on the southwest coast of Australia, 124 mi (200 km) east of the city of Albany, that covers an area of approximately 494,000 acres (200,000 ha), representing the Mediterranean biogeographic province of Western Australia.

Natural characteristics and values
The Fitzgerald River National Park is one of the world's two biosphere reserves that illustrates the

effective integration of human beings with their environment. (The other is the Cévennes National Park in France, whose landscape has been modelled by human beings.)

The reserve has many species of animals and plants, some of which are in danger of extinction while others have very restricted distributions. The vegetation is very diverse and the flora includes 1,750 species, 75 of which are endemic to the reserve or a part of it, and 204 species have features that make their conservation a priority. This means that 20% of the plant species growing in Western Australia are represented in the reserve. This rich flora is a sample of diversity of the the biotopes and complex evolutionary history of the area in which the reserve occurs, and which in many cases has paralleled that of the South African floras. The landscape is divided between forests of mallee, scrub, ponds, agricultural areas, farms, and hedges. The fauna is also the richest in southwestern Australia, as it consists of some species adapted to semi-arid climates and others adapted to wet environments. There are five species of rare and endangered mammals, four species of bird that need protection, and one rare species of reptile.

The area was declared a Biosphere Reserve in April 1978. Shortly before, on January 19, 1973, the area had been declared a national park of the first category, covering 599,726 acres (242,804 ha) and with the possibility of adding a further 125,970 acres (51,000 ha). The Biosphere Reserve is located to the west of the southern coast of the Western Australia, 248 mi (400 km) southeast of Perth and 93 mi (150 km) north of Albany. It stretches from near Hopetoun and Ravensthorpe in the east to Brener Bay in the west. The Gairdner River which flows into Gordon Inlet, forms part of the park's western limits, and the Phillips River which flows into Culham Inlet, forms part of the eastern boundary.

The region's climate is typical of the southern part of Australia. The summer season between November and March tends to be temperate, with hot days and cool nights. Most of the annual rainfall is in the winter season, when the days are cooler, and there may be frosts at night inland. Downpours and thunderstorms are common throughout the year. The annual rainfall varies between 24.6 in (630 mm) on the coast and 14.6 in (375 mm) inland.

The topography includes wide, deep valleys produced by erosive action of rivers, rolling sandy plains and rugged mountains reaching an altitude of 1,499 ft (457 m). There are also some of the most impressive coastal landscapes of the southwestern coast of Australia. The sandy plains overlie a substrate of laterite crossed by many streams. The tributaries of the Fitzgerald, Gairdner, Phillips and Hamersley rivers run southeast through very narrow valleys. The Hamersley and the Fitzgerald rivers have formed cliffs of splendidly colored spongiolite (a sedimentary rock of marine origin). Both these rivers flow into very wide estuaries that are often separated from the open sea by sandbanks that may stabilize as a result of the plant colonization. The park is crossed by a series of ranges of isolated of hills with some spectacular metamorphic peaks of pre-Cambrian quartzite. To the east of the Fitzgerald River the ridges continue straight into the sea, and so the coast has some very steep slopes and cliffs. The west of the estuary of the Fitzgerald River is characterized by long beaches edged by dunes.

As a result of its exceptionally rich flora, the Fitzgerald River Biosphere Reserve is a very rich group of habitats, including forests, matorral, and mallee scrub. If introduced plants are included, the total number of plant species reaches 2,400. The plants that most dominate the landscape are different species of eucalyptus, as they grow in both the wet areas and the semi-arid ones. The shrub species of eucalyptus in the mallee are typical of the drier areas of the Mediterranean climate and replace the forests of the wetter areas.

It is possible to identify 12 major plant communities. The park is dominated by very open mallee, true mallee, and shrubland. The "kwongan," a coastal heath typical of southwestern Australia, is more common in the more exposed areas of coastline, while the forests are mainly found on soils formed on serpentine on riversides and in marshy areas. The mallee ecosystems have a shrubby undergrowth, while the more open mallee scrub is on the colonized, more fertile brown soils and appears to grow only in the areas with Mediterranean climates. In the open mallee formations just 10 or 12 species of shrubby eucalyptus dominate the ecosystem. They have broad, sclerophyllous leaves and are about 10 ft (3 m) in height. The undergrowth normally consists of herbaceous plants, although in the more arid areas there are some semi-succulent chenopods.

277 **Dunes cover much of the surface of the Fitzgerald River Biosphere Reserve.** The most structured ones are on lateritic soils at the mouth of the Hamersley River. [Photo: Reg Morrison / Auscape International]

278 The flowers and buds of one of the eucalyptus of the Australian mallee, *Eucalyptus incrassata*, a shrubby species that does not grow higher than 7 ft (2 m).
[Photo: C. Andrew Henley / Auscape International]

In the mallee vegetation there are six major plant formations; the white mallee (*Eucalyptus diversifolia*); that of *E. incrassata* and broombush (*Melaleuca uncinata*); that of *E. incrassata* and porcupine grass (*Triodia irritans*); on the most fertile soils the formation of *Eucalyptus behriana*, the red mallee (*E. socialis*) and *E. dumosa*; and the formation consisting of *E. oleosa* and *Triodia irritans*. In the Biosphere Reserve the structure of the mallee scrub tends to be dispersed, with scattered shrubby eucalyptus (including *E. diversifolia*) and a lower layer of sclerophyllous matorral 10-79 in (25 cm-2 m) in height. The most representative plant families in the area are: the Myrtaceae, represented by the fringed myrtle (*Calythrix involucrata*) and *Leptospermum laevigatum*; the Proteaceae, represented by seven species of the genus *Hakea*; as well as Leguminosae, Asteraceae, and Epacridaceae. The more prominent species include banksia (*Banksia coccinea*), with its spectacular scarlet inflorescence and leaves similar to the kermes oak (*Quercus coccifera*) of the garrigues of the Mediterranean Basin, and the royal hakea (*Hakea victoriana*) with its variegated leaves.

The wealth of the flora in the Fitzgerald River Biosphere Reserve is clearly reflected in the great diversity of its biotopes and in the complexity of its evolutionary history. Recent research shows the dominant eucalypts in the mallee still maintain a rhythm of growth in the summer that dates from the Tertiary, when this part of Australia enjoyed a sub-tropical climate in which the seasonal variation in climatic conditions were quite unlike the current Mediterranean regime with its dry summer period. This is why these trees only grow in the summer, the season that is now the driest part of the year in this part of Australia but which was the rainy season in the Tertiary before Australia separated from the remains of the ancient continent of Gondwana.

The Biosphere Reserve's great variety of biotopes means there is also rich birdlife. There are 175 species of bird, including the endangered ground parakeet (*Pezoporus wallicus*). The birds of the Australian mediterraneans account for 36% of Australia's terrestrial bird species, including the honey eaters (Meliphagidae), psittacids, such as parrots and cockatoos (*Calyptorhynchus*), thornbills (Acanthazidae), whistlers (*Pachycephala* and monarchs [*Monarcha*]), fly-catching muscicapids, wrens (*Malurus*), pied butcher bird (*Cracticus nigrogularis*), and pardalote (*Pardalotus*). The coastal area between the Fitzgerald River and Two People Bay near Albany is the natural habitat of four bird species in danger of extinction, including the western bristlebird (*Dasyornis longirostris*) and the western whipbird (*Psophodes nigrogularis*). Two more birds worth mentioning are the Australian pelican (*Pelecanus conspicillatus*) and the mallee fowl (*Leipoa ocellata*). The ground parakeet (*Pezoporus wallicus*) has a dark green plumage, and is a noisy inhabitant of the scrub, the only region where it is found. Surveys show that in the 1920s only two or three nests were found. The species now only lives in the wild in two areas of Western Australia: one is the Cape Arid Natural Park, in an area of 6 square miles (15 km²) and the other area is in the Fitzgerald River Biosphere Reserve, where there is only a single very small population of just 100-200 individuals.

279 The mallee fowl (*Leipoa ocellata*) lives in areas of dry vegetation in semi-arid areas of south and southeast Australia. It is one of the three curious mound builder or incubator birds found in the Australian continent. At the beginning of the southern winter (June), the pair prepares an incubator consisting of a depression that the birds fill with decomposing vegetable matter. The female lays her eggs inside the mound in September and the heat from the decomposing plants along with the heat from the sun is sufficient to incubate the eggs until they hatch. The chicks are capable of looking after themselves from the moment they hatch. In the photograph, taken by Frank Park, a pair of mallee fowls are working on their nest depression.
[Photo: Frank Park / ANT / NHPA]

A total of 21 species of native mammal are now known, including the very rare marsupial called the dibbler (*Parantechinus apicalis*) formerly thought to be extinct, and the Australian sea lion (*Neophoca cinerea*). However, in the last 80 years several species of mammal have in fact become extinct. The relatively common animals in the reserve's areas of mallee include the grey kangaroo (*Macropus fuliginosus*, *M. giganteus*) and black-gloved wallaby (*M. irma*), and the tiny honey possum (*Tarsipes rostratus*). Another possum, *Trichosurus vulpecula*, lives in the taller mallee located near to watercourse. There are 41 species of reptile in the reserve including lizards (Scincidae), and there are 11 species of frog (Sphenophrinidae). Most of the mallee's amphibians are normally found in the permanent waters of rivers and lakes. Some of the nest-forming frogs, however, reproduce in temporary pools and later spend the dry summer season in their nests. These include *Lymnodynastes dumerili*, *Neobatrachus centralis*, and *N. pictus*. There is little information on invertebrates, but it is known that there are occasional spectacular shows of handsome coleopterans, jewel beetles, members of the Buprestidae.

Management and problems

The Aboriginal population was decimated by European colonists and there are almost no traces of them left. After their disappearance, there have been almost no permanent inhabitants inside the area that is now the Biosphere Reserve, although remains found include many small mines, a 19th century telegraph line, rabbit traps, farms that failed, and the remains of fishing activities. The only permanent residents in the center of the reserve are the park's wardens. In the outer areas there are farms and other constructions. Near the park there are four settlements with fewer than 300 inhabitants, and some

cultivated areas. Access to the reserve is restricted for tourists and visitors. There are only two camping centers and a small number of nature routes. Within the park, in Twerput, there is a local group that works in the Field Studies Centre. Important research incudes broad ecological studies of a Mediterranean ecosystem virtually undisturbed by human action, and studies of the biogeographical

280 The honey possum (*Tarsipes rostratus*), a small marsupial from the Australian mediterranean, feeds on nectar and pollen but not honey, in this case from a flower of a member of the Proteaceae. Lightly built and resembling a shrew, this opossum uses its prehensile tail and the first opposable toe of its hind feet to grip branches when it needs its front feet for feeding. Phylogenetically it is an isolated species and it is thought that it must have evolved over 20 million years ago when large open spaces covered with flowering plants were common. The photograph was taken in the Fitzgerald River Reserve where this species is one of the most common marsupials.
[Photo: Reg Morrison / Auscape International]

281 **The Barren Ranges run into the sea just west of the mouth of the Fitzgerald River**. Formed by pre-Cambrian quartzites that cross the park from west to east, they terminate in spectacular coastal cliffs. The vegetation of this small range (37 mi [60 km] long) is a mosaic of heaths and mallee areas mixed with riverine forests and marshes. Of the 600 species of plant in the area, 60 are endemic.
[Photo: Reg Morrison / Auscape International]

and interspecific relationships between the fauna of southern Australia and the rest of the former continent of Gondwana.

The park's location and size and its relatively untouched state makes it a very suitable place for the study of the world's climate. Research projects are now underway studying the relief, hydrology, soil, and socioeconomic relations with a cooperation area around the reserve, whose authorities have organized a database that any researcher may access. There are also points of contact between the local community and the visiting researchers, offering accommodation, guides, specialist assistance, and advice.

In order to manage the area well, studies have been performed on the ecology and behavior of the rarest species of the reserve, for example the ground parakeet (*Pezoporus wallicus*). In the case of this species, monitoring the individuals requires them to be caught, using large nets in order to fix a radio emitter that allows them to be monitored at a distance. It is also becoming more and more necessary to understand the impact of several factors on the reserve, such as the effects of mallee fires on the flora and fauna, the spread of the deadly fungal dis-

ease caused by *Phytophthora cinnamomi*, and tourist activities. The wardens have selected and inventoried 72 permanent sites for the study of their flora and fauna. A geological map has been drawn up a scale of 1:250,000, and doctoral theses have been written on a wide range of subjects, including the structure of the quartzites.

There have been no known major changes in the natural ecosystems as a result of human action, except for changes in the fire regime. Yet the fact that during the last three decades many upstream areas supplying the reserve's watercourses have been brought into cultivation has caused an increase in agricultural contamination by chlorides, phosphates, and silt. Tourism, in turn, has significantly contributed to further erosion of the paths in the coastal areas. In 1990 a management plan came into force. Most of the management tasks that have been carried out so far relate to forest fires or tourist access. The area's has been divided with a peripheral buffer system that is 656 ft (200 m) wide to protect it from fires. The local management is the responsibility of the Regional Manager of the Albany Department of Conservation and Land Management, and the Fitzgerald River MAB Project in Hopetoun.

The Fitzgerald River MAB Project started in September 1988 with a study of a 66,690 acre (27,000 ha) watershed east of Hopetoun that sought to remove the causes of soil degradation, mainly salinization and flooding, without restricting itself to considering each use of the area individually. The idea was to work with a group of farmers, who each contributed 500 Australian dollars, and to draw up outline action plans to rehabilitate the soil through cultivation.

Eight farms in the project area drew up partial plans based on the project's objectives. Where appropriate perennial plants were used to fix the soil, the fences were moved to make them coincide with the drainage lines or the adequate soil types, and the shrubs left were protected to avoid degradation. Most of the commercial perennial plants were introduced species, including legumines such as *Cytisus proliferus* and alfalfa (*Medicago sativa*); grasses such as meadow fescue (*Festuca pratensis*), rye (*Secale cereale*) and lovegrass (*Eragrostis tenella*); and trees such as pistachio (*Pistacia vera*), grown for its nuts and some eucalypts, such as the blue gum (*Eucalyptus saligna*) from southeast Australia, grown for their wood.

Even so, the approximately 2,300 native wild plant species of the zone have great potential for the project's objectives. One example of this is (*Eucalyptus spathulata*), known locally as the swamp mallee, which is remarkable for its tolerance of flooding and moderate salinity, and also because it is an excellent source of eucalyptus oil. A group of farmers on the southern coast, near Needilup, is now is organizing itself so as to replant this local tree in areas at risk of flooding and to build a factory to extract the essence. Another species under investigation is *Melaleuca alernifolia*, known as the tea tree, which is originally from New South Wales and produces an essential oil with antifungal properties that is used in medicine.

2.4 The biosphere reserves in the Chilean Mediterranean

In the biogeographical provinces of the Chilean Mediterranean there are only two biosphere reserves, covering a total of 76,987 acres (31,169 ha). The oldest, the National Park of Fray Jorge y Las Chinchillas, covers an area of 34,763 acres (14,074 ha) and was declared a reserve in 1977. The other, the La Campana-Peñuelas Biosphere Reserve covers 42,225 acres (17,095 ha), and was established in 1984.

General considerations and overall assessment

The biosphere reserves are located in the sclerophyllous complex of the Andean coast and include very diverse landscapes, with peaks reaching 2,952-7,288 ft (900-2,222 m) in La Campana, to the mountains forming the Coastal range, which run through Fray Jorge to reach 2,499 ft (762 m) above sea level. Las Chinchillas is at an altitude of 1,230-1,401 ft (375-427 m) above sea level. In both reserves, the soil is almost totally sandy, with a total lack of nitrogen. The average annual temperature varies between 59.9°F and 66.2°F (15.5°C and 19°C) and rainfall is 8-15 in (215-375 mm) per year. Fray Jorge Park has a modified desert climate. The average monthly temperature is 57.9°F (14.4°C), and the wind blows from the southwest all year long, apart from the winter when it blows from the north. When the humid coastal breezes blow from the sea to the coastal mountains almost permanent clouds form around the summits and they are usually accompanied by rainfall. The waters of the Peñuelas lake are always fresh and the area includes marshy areas, flowing waters and some meadows that sometimes flood.

The Fray Jorge and Las Chinchillas Biosphere Reserve

The Fray Jorge and Las Chinchillas Biosphere Reserve covers an area of 34,763 acres (14,074 ha), and consists of a series of protected areas that are discontinuous, some separated by distances of a hundred of kilometres. The Biosphere Reserve includes the Fray Jorge National Park and Las Chinchillas National Reserve. The climate is Mediterranean, and the vegetation of this part of central Chile is represented by sclerophyllous matorral; also present is the Fray Jorge cloud forest, also called temperate rainforest. These forests are called "Valdivian," and are green oases surrounded by semi-arid sites, and they are of importance because they conserve, in isolated locations, some plants that are genuine living fossils more characteristic of more southerly rainforests. The Biosphere Reserve contains nearly all the most representative animals of the Chilean mediterranean's fauna, and Las Chinchillas contains the only known population in the wild of the endangered chinchilla (*Chinchilla laniger*).

Natural characteristics and values

Fray Jorge was declared a national park for tourism-related purposes on March 29, 1941, and after some modifications on June 1, 1967 it was accepted as a Biosphere Reserve in June 1977. Las Chinchillas was created as a national reserve, and as an extension of the Fray Jorge Biosphere Reserve, on November 30, 1983.

282 Coastal sclerophyllous matorral growing in the Fray Jorge y Las Chinchillas Biosphere Reserve in an area with one of the few patches of the site's original vegetation. In the foreground are clumps of cardon (*Puya chilensis*). [Photo: Juan A. Fernández / Incafo]

The Fray Jorge National Park has an area of 24,599 acres (9,959 ha), between the national parks of Fray Jorge, Talinari, and Punta del Viento. It is located in Chile's Administrative District IV, in the province of Limari and the municipality of Ovalle, more than 280 mi (450 km) north of Santiago and 7 mi (11 km) south of La Serena. Las Chinchillas covers 10,4446 acres (4,229 ha), and is 124 mi (200 km) to the south in a mountainous area 9 mi (15 km) northeast of Illapel. It is also the administrative district IV. The reserve forms part of the coastal Andean complex and runs from the river Elqui in the north to Aconcagua in the south, and goes from sea level to 2,500 ft (762 m) in Fray Jorge, and from 1,230 ft to 4,681 ft (375 m to 1,427 m) in Las Chinchillas. The Fray Jorge park runs along the Coastal range. In the reserve there are two types of relief; coastal plains and the mountainous interior. In the coastal plains of Fray Jorge there are terraces of marine or pluviomarine origin. There are no permanent rivers or streams in the park but there are springs. The mountainous area of Las Chinchillas has a relief crossed by small rolling hills and the valleys of the basins of the tributaries of the River Auco, including the El Cobre, El Pollo, Torca Chillán, Las Mollacas, Las Yeguas and Las Gredas rivers. The soils are mainly sandy, deficient in nitrogen, and formed from Cretaceous rocks and sediments. In Fray Jorge, the climate ranges from sub-humid Mediterranean to modified desert. The average monthly temperature is 57.9°F (14.4°C), with a maximum of 65.5°F (18.6°C). The winds are usually from the southwest, except in winter when they blow from the north. As the moist coastal breezes reach the Coastal range, the summits of the mountains are almost permanently covered in clouds, with abundant mist and rainfall. In Las Chinchillas, the average annual temperature is 59.9°F (15.5°C), and the rainfall is about 8 in (215 mm) per year at an altitude of 1,230 ft (375 m) above sea level.

The Biosphere Reserve is located in an area adjoining Mediterranean matorral and semi-desert steppes on one side, and tree and shrub formations of the Coastal range on the other. The coastal communities of matorral have been highly modified by development and stock-raising, although they are still rich in endemics. These communities reach as far as La Serena, and form a zone of transition between desert and matorral, and they have annual rainfall of 5-17.5 in (127- 449 mm) per year. Perennial sclerophyllous plants gradually become scarcer to the north and are replaced by typical desert species. About 95% of the plants of the matorral are endemic. In the sclerophyllous forests of matorral there is a mixture of tropical elements with temperate elements from the southern hemisphere. From an evolutionary perspective its flora, and the more arid transition communities with the northern desert, show clear affinities with tropical and subtropical floras, like the Tertiary elements in the Californian chaparral.

The typical plant communities of matorral consist of evergreen sclerophyllous shrubs between 3 ft and 10 ft (1 and 3 m) in height, such as the litre (*Lithrea caustica*) and soapbark tree (*Quillaja saponaria*) which are the dominant species. Other shrubby (and sometimes tree) species include the canelilla (*Cryptocarya alba*), corontillo (*Escallonia pulverulenta*), *Kaganeckia oblonga* and *Colliguaja odorifera*. Three genera of asteraceae (*Baccharis*, *Haplopappus*, *Senecio*) occasionally form diverse associations of low shrubs. The herbaceous cover of the mature communities of matorral, unlike those in the Californian chaparral is abundant and can reach 40%.

Yet the plant communities in the reserve are very diverse. At an altitude of 1,476 ft (450 m), at the park's northern tip, the woody vegetation is restricted to patches dominated by *Aextoxicon punctatum* and *Mirceugenia correaefolia* together with *Rhaphithamnus spinosus*, separated by areas of shrub and herbaceous vegetation dominated by *Lythrum hyssopifolium*, *Distichlis spicata*, *Haplopappus foliosus*, *Berberis*, *Kageneckia oblonga* and *Fuchsia lycioides*. Above altitudes of 1,640 ft (500 m), there are typical associations of winter's bark (*Drimys winteri*) with *Aextoxicon punctatum*. In contrast, on the coastline the strip of woody vegetation is dominated by *Adesmia angustifolia*, *Proustia pungens*, and the bromeliad *Puya chilensis*. *Cassia stipulacea* and *Porlieria chilensis* are the most characteristic constituents of the vegetation up to an altitude of 492-656 ft (150-200 m).

In Fray Jorge and in Talinay there are the best-known examples of the relict forests of the late Tertiary and early Quaternary. These are temperate moist forests where the moisture-laden winds arriving from the sea are intercepted by the slopes of the coastal range. They condense and form cool mists that allow many Valdivian plants to grow far to the north of their range. The existence of cloud forest in San Jorge is analogous to the southern groups of redwoods (*Sequoia sempervirens*) in California. These relict species in Fray Jorge, are essentially southern species whose center of distribution is 621 mi (1,000 km) to the south, and include forest communities of *Acaena ovalifolia*, *Aextoxicon punctatum*, *Azara microphylla*, winter's bark (*Drimys winteri*), *Dryopsis glechomoides*, *Griselinia scandens*, *Gunnera chilensis*, *Mitraria coccinea*, *Sarmienta repens*, and *Urtica magellanica*. Some scattered sites have several ferns of zones that are intermediate

283 The cardon (*Puya chilensis*) forms part of the typical coastal vegetation of Chile's Mediterranean area. This bromeliad belongs to a family with many terrestrial or epiphytic species that grow in areas ranging from tropical forests to altitudes of more than 13,123 ft (4000 m) in the Andes. The photo was taken in the Concón area of Valparaiso. The accumulation of water between the leaves allows the development of communities, called *phytotelmic*, found only in these moist microenvironments. [Photo: Xavier de Sostoa & Xavier Ferrer]

between cool and temperate, which are at the northern limit of their range in the Fray Jorge park. The vegetation characteristic of the east of the reserve consists of semi-desert shrub formations dominated by the *Gutierrezia paniculata* and *Chuquiraga ulcina*. Other important trees include the incense tree (*Flourensia thurifera*), *Proustia pungens*, and *Adesmia bedwelli*. In the more arid areas to the north the vegetation consists of cacti such as quisco (*Trichocereus chiloensis*) and *Eulychnia acida*, mixed in with *Adesmia angustifolia* and *Cassia stipulacea*. In Las Chinchillas there are six plant communities, dominated respectively, by *Coliguaja odorifera* and *Proustia pungens* on the southern faces; by *Flourensia thurifera* and *Bridgesia incisiifolia* on the northern slopes; and by *Adesmia microphylla* on the others.

The park contains most of the species typical of Chile's Mediterranean fauna. Of the 60-100 species of bird, many are normally found there or just over-winter there. The bird community of the matorral is

284 The only surviving population of the chinchilla (*Chinchilla laniger*) in the wild lives in the Las Chinchillas Biosphere Reserve. Native to the Chilean and Peruvian Andes, this species was once common in dry rocky mountainous areas between 9,842 and 16,404 ft (3,000 and 5,000 m). Its natural populations were decimated by the fur trade hunted for its long, smooth coat; today, it is bred in specialized farms.
[Photo: Gérard Lacz / NHPA]

very rich and diverse, and has common species, such as the Chilean tinamou (*Nothoprocta perdicaria*), unprotected and often overhunted, dove (*Zenaidura auriculata*), meadowlark (*Sturnella loyca*), thrush (*Turdus falcklandii*), finch (*Diuca diuca*), mockingbird (*Mimus thenca*), and blackbird (*Curaeus curaeus*). There are also swallows (*Tachycineta meyeni* [=*Hirundo leucopyga*]), hawk (*Parabuteo unicinctus*), barnowl (*Tyto alba*), striped woodpecker (*Picoides lignarius*), burrowing owl (*Speotyto cunicularia*) and *Oreopholus ruficollis*. The population of the Chilean pigeon (*Columba araucana*) was almost eliminated by an epidemic in 1956, but by 1972 the survivors had managed to recolonize several areas of forest. The bird fauna of Las Chinchillas also includes the Chilean tinamou (*Nothoprocta perdicaria*), the Andean condor (*Vultur gryphus*), harrier (*Circus cinereus*), burrowing owl (*Speotyto cunicularia*), American black vulture (*Coragyps atratus*), white-throated caracara (*Phalcoboenus albogularis*), dove (*Columbina picui*), firecrown (*Sephanoides sephanoides* [=*S. galeritus*]), *Pteroptochos megapodius*, *Anairetes parulus*, and meadowlark (*Sturnella loyca*).

There are few mammals left in Fray Jorge, the most notable are South American fox (*Pseudalopex*

[=*Dusicyon*] *culpaeus*), small grison (*Galictis cuja*), and hog-nosed skunk (*Conepatus chinga*). The guanacos (*Lama guanicoe*) and chinchilla (*Chinchilla laniger*) are now extinct in the reserve, although there is still a population of chinchillas in the Las Chinchillas reserve. In the same area there are mouse opossum (*Marmosa elegans*), *Octodon degus*, small grison (*Galictis cuja*), puma (*Felis concolor*), and pampas cat (*F. colocolo*). In Fray Jorge there are many rodents but few amphibians, which are abundant in the areas of cloud forest. There are various species of reptile typical of the Chilean and coastal area, the most notable being the iguana (*Calopistes maculatus*). The Valdivian forest contains several species that are at the northernmost limit of their ranges. The more important members of the introduced exotic fauna include the California quail (*Lophortyx californica*) and the European hare (*Lepus europaeus*).

Management and problems

In 1974 a management plan was drawn up that lays down the aims of the Fray Jorge National Park. Zonation, of great importance in planning the reserve, includes an area that is inaccessible, a pristine area, one for intensive use, another for extensive use and another for special use. Bearing in mind that most of the agricultural space is in the Chilean biogeographical province, the management takes into account the traditional uses of the area by the local human population. The management plan lays down the detailed planning for camping areas, accommodation, and a tourist information and environmental education center.

There is no agriculture, intensive stock-raising, or forest exploitation, although some livestock has been introduced from neighboring areas. However, in Las Chinchillas there has been overgrazing, the felling of trees, and coal mining. Although the entire area is protected, erosion has not been halted, livestock is still grazed in the meadows and there are motorways and railway lines crossing the area.

The most important current research projects are basically related to the management of the district. Research priorities include the study of the microclimate, plant succession, the natural regeneration of the desert area, and an evolutionary study of the forest and its degradation. There are also studies of the fauna and of the general ecology, and the possibility of reintroducing species that used to live in the district, such as the guanaco (*Lama guanicoe*) and the Chinchilla (*Chinchilla laniger*). In Fray Jorge the facilities are still limited, but in Las Chinchillas there is good access and accommodations for scientists.

2.5 The biosphere reserves in the Californian Mediterranean

The Californian Mediterranean has four biosphere reserves (the experimental forest of San Dimas, the experimental dehesa of San Joaquin, the Central California Coast, and the Channel Islands), which cover a total of about 2 million acres (893,294 ha), ranging in size from 4,446 acres (1,800 ha) to about 1 million acres (480,000 ha). Three were approved in June 1976, and the fourth, the Central California Coast Biosphere Reserve, in November 1988. Of these sites, the one that has been protected for longest is the San Dimas Experimental Forest. It was established as a reserve by order of the Chief of the Forest Service on March 28, 1934. The Channel Islands Biosphere Reserve was declared a national monument in 1938, and on March 5, 1980 it was extended and declared a national park. The Californian biosphere reserves are mostly concentrated on the Pacific coastline in an area stretching from the south of Santa Barbara County to the north of the San Francisco Bay, as well as an area 50 mi (80 km) to the northeast of Los Angeles and a sector of the central part of the Sierra Nevada.

General considerations and overall assessment

The relief of these areas ranges from the coastal plains and even part of the sea, such as the national marine sanctuary of Gulf Farallon, to the heights of the Sierra Nevada. The limits of the Marine sanctuary reach six nautical miles into the eastern Pacific Ocean in the Central California Coast Biosphere Reserve. The Channel Islands Biosphere Reserve includes the five islands in the California Channel (Anacapa, San Miguel, Santa Barbara, Santa Cruz, and Santa Rosa), as well as an area of six nautical miles from each island. In the hills of the San Joaquin Experimental Dehesa Biosphere Reserve there are many outcrops of granitic rocks. The Channel Islands reserve includes representative examples of the world's largest coastal terraces, several marine caves, steep coastlines, beaches, mountains, and valleys. The San Dimas Experimental Forest Biosphere Reserve includes two watersheds on the slopes of the San Gabriel Mountains, one of which, the basin of the California River, descends to Los Angeles, while the other, the San Gabriel River, flows into the Pacific Ocean.

285 Aerial photo of one of the five islands forming the Channel Islands Biosphere Reserve, USA. The island's plant cover has suffered from human pressure in the form of agriculture, stock raising, and many accidental fires. *[Photo: Stephen Krasemann / NHPA]*

The area's climate is Mediterranean. Most of the annual precipitation occurs in winter in the form of rain, although on the coast humidity may be high in summer. The rainfall varies between 19.5 in and 26 in (500 mm and 678 mm) at an altitude of 2,395 ft (730 m), and 95% of this falls between October and April. The temperatures are between 68°F and 86°F (20°C and 30°C) in summer and 28°F to 46°F (-2°C to 8°C) in winter on the Channel Islands. These temperatures can reach 81°F to 102°F (27°C to 39°C) in July in San Joaquin and in the San Dimas experimental forest, where the annual precipitation lies between 19 in (483 mm) in San Joaquin and 26 in (678 mm) in San Dimas.

The Central California Coast Biosphere Reserve

The Central California Coast Biosphere Reserve lies to the west of San Francisco Bay and north of the city. It consists of the coastal area and an area six nautical miles into the Pacific Ocean, including the Farallon Gulf and some parts of the bay and the main islands, Alcatraz and Angel.

Natural characteristics and values

The Biosphere Reserve is especially interesting for many different reasons: the vegetation is virtually untouched, with coastal meadows, islands, freshwater wetlands, forests and moist coastal areas. It also houses some of the largest and most diverse populations of marine birds and pinnipeds on the western coast of North America below Alaska, and the largest seabird breeding sites in the mainland United States. There are seven species that are classified as endangered in the United States, four of which are on the State of California's list of protected species and an indeterminate number of species covered by federal laws protecting marine mammals.

The area was declared a Biosphere Reserve in November 1988, and covers over one million acres (404,863 ha), 974,620 acres (394,583 ha) of which are on dry land, and 948 square nautical miles in the sea. The zone includes and surrounds several especially important protected areas: the coast of Point Reyes, the Golden Gate Recreational Area, the National Marine Sanctuary of Farallon Gulf and the wildlife reserve of the Farallon Islands, the western part of Marin Peninsula (Marin Municipal Water District) and the Californian state parks of Tomales and Samuel P. Taylor. The extensions proposed for the years 1992-1993 include the Audubon Canyon Ranch, the Bodega Marine Reserve, the Cordell Bank Marine Sanctuary and the Jasper Ridge Biological Reserve. The first area of the reserve to be protected was Point Reyes, designated a National Seashore on September 13, 1962, followed by the Golden Gate recreational area, whose legislation was approved in 1972. The Farallon Gulf Marine Sanctuary was declared a protected area in 1981, under section III of the law governing the protection, research and marine sanctuaries.

The Biosphere Reserve's zonation is complex, and includes 76,0766 acres (30,800 ha) that are strictly protected to maintain as them relatively isolated and undisturbed. There is a 41,990-acre (17,000-ha) area where some stock-raising activities are permitted. Fishing is permitted in the waters of the Farallon Gulf, but only using traditional methods. Broadly speaking, the different areas are distributed in a central area, a buffer area of 877,566 acres (355,290 ha), and a transition area of 22,410 acres (9,073 ha). The extension proposed for 1992–1993 would add another four additional units, adding 2,984 acres (1,208 ha)—2,347 acres (950 ha) in the central area of the reserve and 630 acres (255 ha) in the buffer zone and 397 square nautical miles of sea (all in the core area).

This Biosphere Reserve is best example in the United States of the Californian sclerophyllous biogeographical province, with 20 of the terrestrial communities of the biome. One of its unusual characteristics is that it covers the entire spectrum from terrestrial zones to estuary and sea, and includes representatives of all the elements of the biogeographical unit it belongs to. There are more than 674 different plants in the area it covers, making it a center of endemic species of great importance at the level of the State of California.

The terrestrial part consists of protected areas that are natural, semi-natural and grazed, in some acres controlled fires are still permitted in order to maintain biodiversity. Within the reserve it is possible to admire excellent examples of natural or modified areas, from the virgin areas of coast redwoods and the intact chaparral communities to the modified coastal meadows or the forests of Douglas fir. The region's landscape is typically open, with mountain chains, which reach 2,394 ft (730 m), running to the coastline and giving rise to rocky promontories, coastal terraces, dunes and lagoons.

The marine area is formed by a series of low granitic islands located at the edge of the continental platform. Then the sea bottom falls away to a depth of 4,260 ft (1,300 m) between Farallon escarpment and the Santa Clara Basin. This zone is well known as being tectonically active, especially along the San

286 A pair of brown pelicans (*Pelecanus occidentalis*) in a mangrove swamp on the island of Anacapa in the Californian Channel Islands Reserve. This pelican, the only one of its genus which completely submerges itself when fishing, dives powerfully into the water when it spots a fish. Once it has caught its prey, it returns to the surface and empties the water from its mouth pouch and swallows its catch. Remarkably, the water taken up in its pouch weighs more than the bird itself. The brown pelican is the most common pelican and has a world population of over one million birds. Despite the fall in numbers in North America in the 1960s due to pesticide contamination, over 30,000 pairs still breed in the Gulf of California. A variety of different pelicans are found commonly in all mediterraneans: in South Africa, the white pelican (*P. onocrotalus*); in Australia, the Australian brown pelican (*P. conspicillatus*); in California, the North American white pelican (*P. erythrorhynchos*) and the *P. occidentalis*; and two species in the Mediterranean Basin, the white pelican (*P. onocrotalus*) and the Dalmatian pelican (*P. crispus*), the only species of Mediterranean pelican that is endangered.
[Photo: David S. Boyer / National Geographic Society]

Andreas Fault, the border between the Pacific and North American plates. The fault crosses the Biosphere Reserve together with other smaller connected faults. The area is basically formed of sedimentary rocks with some intrusive mesozoic rocks, and Tertiary granites and basalts. To the west of the San Andreas fault there are also basalts and metamorphic rocks, mainly in the San Francisco formation. The marine habitats range from intertidal zones and pelagic and deep ocean habitats representative of the Oregon marine province of the northeast Pacific. Offshore, there are extensive beds of *Laminaria* with a very diverse marine fauna. Normally, at least five species of pinniped are found there, together with 17 species of cetaceans, and because this area is an important migratory route in the migration season a large number of both sedentary and migrant birds are observed. There are also some historic sites, such as Alcatraz Island, Fort Point, Fort Funston, West Fort Miley and Fort Mason.

The region's climate is typically Mediterranean, with dry summers and very rainy, wet, winters. In the summer months the coast is typically covered in sea mists. Rainfall varies from 20 to 27 in (500–700 mm) at sea level and 79 in (2,000 mm) at an altitude of 2,395 ft (730 m). The average temperatures are 66.9°F (19.4°C) for the warmest month and 42.9°F (6.1°C) for the coldest month. This part of the Pacific coastline is bathed by cold water that circulates from Punta Concepción to British Columbia, known as the California Current.

Throughout the Biosphere Reserve, the vegetation is typical of the Californian mediterranean. It can be divided into three broad categories, the coastal chaparral, the coastal shrub communities and the mixed forests of conifers and broadleaves. The chaparral includes protected areas and others that are grazed, and is dominated by chamiso (*Adenostoma fasciculatum*), manzanita (*Arctostaphylos*), redroot (*Ceanothus*) and chinquapin (*Castanopsis*). The coastal shrub communities are represented by sagebrush that basically consists of Californian sagebrush (*Artemisia californica*), and locally the white sage (*Salvia apiana*). The communities of mixed forest grow in specially protected areas, where controlled fires are allowed in order to encourage new growth. The forests contain a mixture of broadleaf trees and needle-leaved trees, they are dominated by species such as the coast redwood (*Sequoia sempervirens*), mixed with sclerophyllous species, such as the tanbark oak (*Lithocarpus densiflora*), California live oak (*Quercus agrifolia*), Douglas fir (*Pseudotsuga menziesii*), and madrona (*Arbutus menziesii*), a member of the same genus as the strawberry tree that grows in the Mediterranean (*A. unedo*). On the coast there are meadows, grazing land with grasses, and saltmarshes dominated by succulent plants and glasswort (*Salicornia*). Other land communities include riverbank and meadow communities growing on serpentine substrates. More than 80% of the remaining wetlands in California are within the limits of the Biosphere Reserve, an exceptional area for the migratory birds that travel along the Pacific coastline. In the marine part of the reserve, the subtidal and intertidal areas have dense beds of large seaweeds,

287 Coastal vegetation with arum lilies (*Zantedeschia palustris*) from the South African Mediterranean, in the San Francisco Bay, in the Central Coast Biosphere Reserve. The vegetation of the California coast Biosphere Reserve is typically Mediterranean, although in some moister areas there are plants typical of swamps.
[Photo: James P. Blair / National Geographic Society]

kelps, dominated by the laminarian algal genera *Nereocystis* and *Macrocystis*, and meadows of marine flowering plants.

The notable plants in danger of extinction in the United States include Raven's manzanita (*Arctostaphylos hookeri ravenii*), and the California state list of endangered species includes Bolinas Ridge (*Ceanothus masonii*) and clarkia (*Clarkia francisca*), plants whose distribution has shrunk due to human use and abuse over the centuries. There are a further 11 species on the state list that might be found in the Biosphere Reserve, including the Tamalpais manzanita (*Arctostaphylos montana*), Bolinas manzanita (*A. virgata*) and the San Francisco gum plant (*Grindelia maritima*).

The fauna of the Biosphere Reserve, except perhaps for the cetaceans and pinnipeds, has been decimated by their direct exploitation and use of their habitats by humans and by their proximity to major urban centers.

Some land animals, such as the San Francisco strangling snake (*Tramnophis sirtalis tatrataenia*), the marsh mouse (*Reithrodontomys raviventris*), peregrine falcon (*Falco peregrinus*), the Blue Mission butterfly (*Icaricia icarioides missionensis*), spotted bay butterfly (*Euphydras edita bayensis*) and Saint Bruno's Magic butterfly (*Incisalia mossii bayensis*) are in danger of extinction at both the state and federal level.

In the sea, the fauna is richer than on land; there are up to 22 marine mammals, including five species of pinniped, such as the California sea lion (*Zalophus californianus*), Steller's sea lion (*Eumetopias jubatus*), the northern elephant seal (*Mirounga angustirostris*) and the common seal (*Phoca vitulina*). There are also 17 species of cetacean, most of which pass through the reserve on their migration. These interesting cetaceans include the grey whale (*Eschrichtius robustus*), humpback whale (*Megaptera novaeangliae*), common porpoise (*Phocoena phocoena*), and Dall's porpoise dolphin (*Phocoenoides dalli*). Although it seems beyond doubt that the marine fauna is very varied, the research performed cannot reasonably be considered complete, partly due to the difficulty of obtaining samples from the deep sea. In the subtidal and pelagic environments there are more than 20 species of very common fish, including several species of salmon (*Oncorhynchus*), redfish (*Sebastes*) and several species of serranidae. The grey whale (*Eschrichtius robustus*), sea otter (*Enhydra lutris*), the Californian freshwater shrimp (*Syncaris pacifica*), and *Eucocyclogolius newberryi* are some of the species on the state or federal lists of coastal or marine species in danger of extinction.

The reserve's bird fauna includes more than 123 species of aquatic birds. The Farallon Islands and the cliffs at Point Reyes are the sites chosen by large breeding colonies and by migratory populations, such as the colonies of brown pelicans (*Pelecanus occidentalis*) and some notable birds of prey such as the bald

288 The island of Alcatraz and the Golden Gate Bridge in San Francisco Bay, within the California Central Coast Biosphere Reserve. When the Spaniards reached the region, they occupied the island and called it the island of Los Alcatraces (Gannet Island). In 1851 it passed to the United States of America and was fortified, and a prison was built that was in operation until 1963, when it was closed because its usefulness did not compensate for the high maintenance cost.
[Photo: David S. Boyer / National Geographic Society]

eagle (*Haliaeetus leucocephalus*) and the peregrine falcon (*Falco peregrinus*). Marine and wetland species on state and national protection lists include the brown pelican (*Pelecanus occidentalis*) and the clapper rail (*Rallus longirostris obsoletus*)

Management and problems

The area is of great cultural importance, as it includes a former Indian settlement, the remains of a Russian colony, a Spanish colony, and one of the pioneer settlements of the gold rush and the American Civil War. Archaeological excavations have uncovered more than 100 sites of interest in Point Reyes National Seashore, and the remains of settlements proving the existence of communities of Miwok Indians.

The area is also rich in more recent historical and cultural remains, including traces left by the English explorers of the period of Queen Elizabeth I of England, Mexican ranchers, Asiatic and European gold prospectors, and by the continual flow of immigrants from Europe. In 1579, the area was explored by Sir Francis Drake's fleet led by the Golden Hind. Seven years later the galleon under Sebastián Rodríguez Cermeño was shipwrecked off San Francisco. This ship has already been found, and so the area is also of interest for the old boats conserved at the bottom of the bay, including the schooner *C.A. Thayer*, the steamship *Wapama*, the lighter *Alma*, and the boat *Jeremiah O'Brien* built in 1943. Other notable historical constructions include the grim former prison-island of Alcatraz, several military forts, such as Fort Point, Fort Funston, West Fort Miley, Fort Mason, and other

and many costal gun batteries, such as the Chamberlin, famous for its 95,000 pound guns, dated 1906.

The region has a total of six million inhabitants concentrated around the bay, although nobody lives in the center of the reserve, in the buffer zone or in the zone of influence. Yet controlled grazing is permitted within the reserve. Every year there are almost a million visitors attracted by the museums, galleries, boat excursions, the bathing on the beaches, nature routes, leisure areas, campsites, recreational parks and the horse riding. Popular activities also include salmon fishing, whale watching, and sea excursions along the coast. In some cases, large-scale commercial fishing is allowed for human consumption, including salmon (*Oncorhynchus*), redfish (*Sebastes*), serranids, ling (*Molva*), and jurels (*Trachurus*).

The Biosphere Reserve is organized into independent units. Thus, the Farallon Gulf Marine Sanctuary belongs to the NOAA (National Oceanic and Atmospheric Administration); the Farallon fauna reserve belongs to the United States Fish and Wildlife Service; Point Reyes National Seashore and the Golden Gate Recreation Area belong to the National Park Service; and the Tamales Bay and Samuel P. Taylor State Parks are the responsibility of the California State Parks and Recreation Department. The Biosphere Reserve is divided into a series of management and multiple use areas, in some of which restoration projects are underway or planned. The core area is scrupulously protected, while some of the buffer areas are often used for recreational activities or

289 **Sports activities in the Golden Gate recreational area** in San Francisco Bay, within the Central California Coast Biosphere Reserve. With the help of two other people, the pilot of this hang-glider is running towards the edge of a cliff, which has been reinforced with tree trunks on its edge to prevent erosion. To the right of the photo is the beach and the Pacific Ocean.
[Photo: James P. Blair / National Geographic Society]

grazing. Activities allowed include angling, commercial fishing and aquaculture, controlled grazing and burning of matorral, tourist development, and military maneuvres. Experimental restoration projects are also repopulating the coastal dunes with natural vegetation.

The center of the reserve is protected by the master-plan for the Farallon Gulf Marine Sanctuary, which was prepared in 1987 by the NOAA, of the U.S. Commerce Department. The adminstration plan for the marine sanctuary gives special priority to research projects to be carried out over periods of more than ten years. These include of basic studies on populations and habitats, their distribution, and other basic aspects that are still not well understood, control studies of the most representative species and habitats and analytical studies to determine the causes of the environmental impacts. The management plan includes an action plan with a program in three parts: resource protection, interpretation, and education, and research. The plan's priority is to ensure improved protection of the environmental resources of the Mediterranean vegetation and the marine environments, and to ensure the multiple and compatible use of the oceanic area, to increase public awareness and support, and lastly, the promotion of management-related research programs. The main agency responsible for the wardening the reserve is the California State Department of Fish and Game. The National Park Service collaborates in the ordinary administration and in the development and implementation of the educational program. NOAA has cooperation agreements with both bodies, they have a base in the Golden Gate National Recreation Area to coordinate the participation of each agency in the reserve's administration. There are also programs that involve the participation of several agencies in the Point Reyes National Seashore and California's state parks.

Perhaps the most serious of the menaces threatening the Biosphere Reserve are the high levels of water and air pollution, and the potential danger of oil spills. The beaches and dune areas receive too many visitors, and fishing is still not sufficiently controlled. Over the last few years more and more seabirds and marine mammals have been trapped in the nets of local fishermen.

Over the last 100 years a great deal of research has been carried out in the reserve. Scientific research has always been encouraged, especially when the results might help to solve major management problems. Many activities have been based on comparative ecological studies of the chaparral and herbaceous communities with the convergent communities that exist in Chile and in other parts of the world. These include studies by H. Mooney on plants and E. Fuentes on lizards. A major study has been performed without interruption since 1850 of the migratory birds that pass over San Francisco Bay and along the coastal Pacific route, that has been the basis of ornithological research over the last century and a half, and is now controlled from the bird observatory at Point Reyes, which was built in 1972. Other institutions undertaking research include the National Fisheries Service Tiburton Laboratory, the Bodega Bay Marine Laboratory and the University of California Long Marine Laboratory. Yet more institutions participate in studies of the region, such as the University of California (at Berkeley, Davis, and Santa Cruz), California State University (at San Francisco, Hayward, and Sonoma) and World College West.

Bibliography

This bibliography includes general works used for basic reference. It thus includes general works on geography, climate, soils, wildlife, plant life, etc., as well as more specific works on the Mediterranean biome, either as a whole or separately.

BALCELLS, E. (ed.) (1977). *Estudio integrado y multidisciplinario de la dehesa salmantina. Estudio Fisiográfico-Descriptivo* (2nd and 3rd fascicules). Contribución a proyectos UNESCO-MAB. Centro de Edafología y Biología Aplicada de Salamanca / Centro Pirenaico de Biología Experimental.

BARBOUR, M. AND W.D. BILLINGS (eds.) (1988). *North American Terrestrial Vegetation.* Cambridge: Cambridge University Press. 434 p.

BARIGOZZI, C. (eds.) (1986). *The origin and domestication of cultivated plants.* New York: Elsevier Scientific Publishing Company. 218 p.

BARRACLOUGH, G. AND N. STONE (eds.) (1989). *The Times Atlas of World History.* Times Books Ltd., London. 358 p.

BARRAU, J. (1983). *Les hommes et leurs aliments.* Temps Actuels, Paris. 378 p.

BEARD, J.S. (1990). *The plant life of Western Australia.* Kangaroo Press, Sydney. 319 p.

BERTRANPETIT, J. AND L.L. CAVALLI-SFORZA (1991). "A genetic reconstruction of the history of the population of the Iberian Peninsula." *Annals of Human Genetics,* 55: 51- 67.

BLACK-MICHAUD, J. (1986). *Sheep and land: the economics of power in a tribal society.* Cambridge: Cambridge University Press.

BOARDMAN, J. (1988). *The greeks overseas.* Thames and Hudson, London. 288 p.

BONILLA, L. (1975). *Breve historia de la técnica y del trabajo.* Ed. Istmo, Madrid. 292 p.

BONTE, P. AND M. IZARD (1991). *Dictionnaire de l'ethnologie et de l'anthropologie.* Presses Universitaires de France, Paris. 755 p.

BOOMSMA, C.D. AND N.B. LEWIS. *The native forest and woodland vegetation of South Australia.* Woods and Forests Department, South Australia. 313 p.

BORDE, J. AND R. SANTRANA-AGUILAR (1980). *Le Chili. La terre et les hommes.* Éditions du Centre National de la Recherche Scientifique, Paris. 252 p.

BOSERUP, E. (1981). *Population and technological change. A study in longterm trends.* University of Chicago, Chicago.

BOZON, P. (1983). *Géographie mondiale de l'élevage.* Librairies techniques, Paris. 256 p.

BRAQUE, R. (1988). *Biogéographie des Continents.* Masson et Cie., Paris. 470 p.

BRAUDEL, F. AND G. DUBY (1986). *La Méditerranée. Les hommes et l'héritage.* Flammarion, Paris. 217 p.

BRAUDEL, F. (1985). *La Méditerranée. L'espace et l'histoire.* Flammarion, Paris. 223 p.

BRIMBLECOMBE, P. AND C. PFISTER (eds.) (1990) *The Silent Countdown. Essays in European Environmental History.* New York: Springer-Verlag.

BRISTOW, D. AND G. CUBITT (1988). *The Natural History of Southern Africa.* Cape Town: Struik Publishers.

BRÜCHER, H. (1989). *Useful Plants of Neotropical Origin and their Wild Relatives.* New York: Springer-Verlag.

BUCKLEY, R. (eds.) (1992). *The Mediterranean. Paradise under Pressure.* Cheltenham, England: European Schoolbooks Publishing Limited.

CAMPOS PALACÍN, P. (1984). *Economía y energía de la dehesa extremeña.* Madrid: Instituto de Estudios Agrarios, Pesqueros y Alimentarios.

CAMPOS PALACÍN, P. (1984). *Evolución y perspectiva de la dehesa extremeña.* Universidad Complutense de Madrid.

CANDOLLE, A. (1896). *Origine des plantes cultivées.* Paris: Félix Alcan Éditeur.

CAVALLI-SFORZA, L.L. (1981). "Human Evolution and Nutrition." In: *Food, Nutrition and Evolution,* New York.

CAVALLI-SFORZA, L.L. (1992). "Genes, pueblos y lenguas." *Investigación y Ciencia,* 184: 4-11.

CEE (1989). "Informe de la Comisión de las Comunidades Europeas al Consejo, al Parlamento y al Comité Económico y Social sobre protección del medio ambiente en la región Mediterránea." *Revista de Derecho Ambiental,* 3:79-92.

CHALINE, J. (1972). *El cuaternario.* Madrid: Akal.

CHARLET, P. AND J. BOUGLER (1979). *Les races locales et leur devenir. In: Utilization par les ruminants des pâturages d'altitude et parcours méditerranéens.* Versailles.

CLOUDSLEY-THOMPSON, J.L. (1978). *Animal Migration.* London: Orbis.

CLOUDSLEY-THOMPSON, J.L. (1979). *El hombre y la biología de las tierras áridas.* Barcelona: H. Blume.

CLUTTON-BROCK, J. (1987). *A Natural History of Domesticated Mammals.* Cambridge: Cambridge University Press.

CNRS (ed.) (1961). *Le peuplement des îles mediterranéennes et le probleme de l'insularité.* Paris: CNRS.

COLE, S. (1970). *The Neolithic Revolution.* London: British Museum.

COLOM, G. (1964). *El medio y la vida en las Baleares.* Palma: Fundación Juan March.

CONTRERAS, T.D., C.J. GASTÓ AND G.F. COSSIO (eds.) (1986). *Ecosistemas pastorales de la zona mediterránea árida de Chile. I. Estudio de las comunidades agrícolas de Carquindaño y Yerba Loca del secano costero de la región de Coquimbo.* Montevideo: Oficina Regional de Ciencia y Tecnologia de la UNESCO para América Latina y el Caribe.

COWLING, R. (ed.) (1992). *The Ecology of Fynbos. Nutrients, Fire and Diversity.* Oxford: Oxford University Press.

CRAMP, S. (1977). *Handbook of the Birds of Europe, the Middle East and North Africa: The Birds of the Western Palearctic.* Oxford: Oxford University Press.

CROCKER, R.L., A. KEAST AND C.S. CHRISTIAN (eds.) (1959). *Biogeography and Ecology in Australia.* The Hague: W. Junk.

CROSBY, A.W. (1988). *Imperialismo ecológico. La expansión biológica de Europa, 900-1900.* Barcelona: Editorial Crítica.

DATURI, A. AND C. VIOLANI (1985). *África del Sur. Enciclopedia de la Naturaleza,* ADENA/WORLD WILDLIFE FUND, vol. 17. Debate/Itaca/Círculo, Madrid.

DEBEIR, J.C., J.P. DELÉAGE AND D. HÉMERY (1986). *Les servitudes de la puissance. Une histoire de l'énergie.* Paris: Flammarion.

DEBUSCHE, M., S. RAMBAL AND J. LEPART (1987). *Le changement de l'occupation des terres en région méditerranée humide: évaluation des consequences hydrologiques.* Acta oecologica, 8: 317-332.

DELAMARE DEBOUTEVILLE, C. AND E. RAPOPORT (eds.) (1968). *Biologie de l'Amérique Australe.* Paris: C.N.R.S.

DELANO, C. (1979). *Western Mediterranean Europe. A Historical Geography of Italy, Spain and Southern France since the Neolithic.* London: Academic Press

DI CASTRI, F., D.W. GOODALL AND R.L. SPECHT (eds.) (1981). *Mediterranean-Type Schrublands.* Amsterdam: Elsevier Scientifing Publishing Company.

DI CASTRI, F. AND H.A. MOONEY (eds.) (1973). *Mediterranean Type Ecosystems. Origin and Structure.* London: Chapman and Hall Limited/Springer Verlag.

DREGNE, H.E. (1976). *Soils of Arid Regions.* Amsterdam: Elsevier Scientific Publishing Company.

EPSTEIN, H. (eds.) (1971). *The Origin of Domestic Animals in Africa (vol. I and II).* New York: Africana Publishing Corporation.

ERICKSON, R., A.S. GEORGE, N.G. MARCHANT AND M.K. MORCOMBE (1986). *Flower and Plants of Western Australia.* Sydney: Reed.

ESCARRÉ, A., C. GRACIA, F. RODÀ AND J. TERRADAS (1984). "Ecología del bosque esclerófilo mediterráneo." *Investigación y Ciencia,* 95: 69-79.

FACCHINI, F. (1990). *El origen del hombre. Introducción a la Paleontología.* Madrid: Aguilar.

FAO (1985). *Soil Map of the European Comunities 1:1000 000.* Luxembourg: CEC.

FAO (1992). *Cultivos marginados. Otra perspectiva de 1492.*

FAO-UNESCO (1988). *Soil Map of the World. Revised Legend.* Rome: FAO.

FELDMAN, M. AND E.R. SEARS (1981). "Los recursos genéticos del trigo silvestre." *Investigación y Ciencia,* 54: 50-61.

FEREMBACH, D., C. SUSANNE AND M.C. CHAMLA (1987). *L'homme, son évolution, sa diversité. Manuel d'anthropologie physique.* Paris: CNRS.

FILLAT, F., J.M. GARCÍA RUÍZ, P. MONTSERRAT AND L. VILAR (1984). *Els pasturatges. Funcionalisme i aprofitament dels ecosistemes pastorals.* Barcelona: Servei de Medi Ambient de la Diputació de Barcelona.

FOLCH, R. (ed.) (1981). *Història Natural dels Països Catalans. 15 vols.* Barcelona: Enciclopèdia Catalana.

FOLCH, R. (1988). *Natura, ús o abús? Llibre Blanc de la Gestió de la Natura als Països Catalans.* Barcelona: Barcino.

FOLCH, R., L. FERRÉS AND M. MONGE (1991). *Mediterrània. L'home i els ecosistemes mediterranis al llarg de l'any.* Barcelona: Enciclopèdia Catalana.

FRAISSINET, M., B. MASSA AND M. MILONE (1985). *El Mediterráneo. Enciclopedia de la Naturaleza,* Madrid: ADENA/WORLD WILDLIFE FUND, vol. 1, Debate/Itaca/Círculo.

GARCÍA MARTÍN, P. AND J.M. SÁNCHEZ BENITO (1986). *Contribución a la historia de la transhumancia en España.* Madrid: Ministerio de Agricultura, Pesca y Alimentación.

GARRABOU, R. AND J. PUJOL (1988). "La especialización de la agricultura mediterránea y la crisis." In: *La crisis agraria de fines del siglo XIX.* Barcelona.

GIACOMA, C. (1985). *Australia. Enciclopedia de la Naturaleza.* Madrid: ADENA/WORLD WILDLIFE FUND, vol. 24, Debate/Itaca/Círculo.

GLANTZ, W.E. (1977). *Comparative ecology of small mammals communities in California and Chile.* Doctoral Thesis. University of California.

GRAMKRELIDZE, T.V. AND V.V. IVANOV (1990). "La protohistoria de las lenguas indoeuropeas." *Investigación y Ciencia,* 164: 80-87. Barcelona.

GREENPEACE AND PASTOR, X. (1991). *El Mediterráneo.* Debate, Madrid.

GRENÓN, M. AND M. BATISSE (eds.) (1988). *El Plan Azul: El futuro de la Cuenca Mediterránea.* Madrid: Ministerio de Obras Públicas y Transporte.

GRIGG, D.B. (1974). *The agricultural systems of the world.* Cambridge: Cambridge University Press.

GROVES, R.H. AND F. DI CASTRI (eds.) (1991). *Biogeography of Mediterranean Invasions.* Cambridge: Cambridge University Press.

GUIDONI, E. (1978). *La città europea. Formazione e significato dal IV all'XI secolo.* Milan: Electa.

HARANT, H. AND D. JARRY (1963). *Guide du naturaliste dans le Midi de la France.* Neuchâtel: Délachaux and Niestlé.

HARLAN, J.R. (1987). *Les plantes cultivées et l'homme.* Paris: Presses Universitaires de France.

HARRIS, M. (1987). *Introducción a la Antropología General.* Madrid: Alianza.

HASSAN, F.A. (1981). *Demographic Archaeology.* New York: Academic Press.

HAUDRICOURT, A.G. AND L. HÉDIN (1987). *L'Homme et les Plantes Cultivés.* Paris: Éditions A.M. Métailié.

HAUDRICOURT, A.G. (1988). *La technologie science humaine. Recherches d'histoire et d'ethnologie des techniques.* Paris: Éditions de la Maison des Sciences de l'Homme.

HEADY, H.F. (1971). *La explotación de los pastizales de secano.* Saragossa: Ed. Acribia.

HEIZER, R.F. AND A.B. ELSASSER (1980). *The natural World of the California Indians.* Berkeley: University of California Press.

HOYT, E. (1992). *Conservando los Parientes Silvestres de las Plantas Cultivadas.* Mexico: Addison-Wesley Iberoamericana.

HUECK, K. (1978). *Los bosques de Sudamérica. Ecología, composición e importancia económica.* Deutsche Gesellschaft für Technische Zusammenarbeit (GTZ) GmbH., Eschborn.

HUENNEKE, L.F. AND H.A. MOONEY (eds.) (1989). *Grassland Structure and Function.* California Annual Grassland. Dordrecht: Kluwer Academic Publishers.

IUCN (1990). *1990 United Nations List of Nationals Parks and Protected Areas.* Gland, Switzerland: IUCN.

JOFFRE, R., B. HUBERT AND M. MEURET (1991) *Les systèmes agro-sylvo-pastoraux méditerranéens.* Paris: Dossier MAB UNESCO.

JOHNSON, H. (1985). *The World Atlas of Wine.* R. D. Press.

KEITH, P. AND M.C. PALGRAVE (1987). *Everyone's Guide to Trees of South Africa.* Johannesburg: CNA.

KLEIN, R.G. (1989). *The Hman Career. Human Biological and Cultural Origins.* Chicago: University of Chicago Press.

KRUGER, F.J., D.T. MITCHELL AND J.U.M. JARVIS (1983). *Mediterranean-Type Ecosystems. The Role of Nutrients.* Berlin: Springer-Verlag.

LABORATOIRE DE PRÉHISTOIRE DU MUSÉE DE L'HOMME ET ACTION CULTURELLE MUNICIPALE DE LA VILLE DE NICE (eds.) (1982). *Origine et évolution de l'homme.* Paris.

LAGUNA SANZ, E. (1986). *Historia del Merino.* Madrid: Ministerio de Agricultura, Pesca y Alimentación.

LANGANEY, A. (1988). *Les Hommes.* Armand Colin.

LAVOCAT, R. (ed.) (1966). *Atlas de Préhistoire.* Paris.

LEAKEY, R.E. (1981). *La formación de la humanidad.* Barcelona: Ed. Serbal.

LEBEAU, R. (1983). *Grandes modelos de estructuras agrarias en el mundo.* Barcelona: Ed. Vicens-Vives.

LICHARDUS, J. AND M. LICHARDUS-ITTEN (1987). *La protohistoria de Europa.* Barcelona: Ed. Labor.

LIEUTAGHI, P. (1991). *La plante compagne: pratique et imaginaire de la flore sauvage en Europe Occidental.* Conservatoire et Jardin botaniques de Genève, Alimentarium/Musée d'histoire naturelle de Neuchâtel, Vevey - Neuchâtel, Geneva.

LIGHTON, C. (1973). *Cape Floral Kingdom.* Cape Town: Juta and Company.

LOPEZ LINAGE, J. (1989). *Agricultores, botánicos y manufactureros en el siglo XVIII.* Ministerio de Agricultura, Pesca y Alimentación, Madrid.

LOUSSERT, R. AND G. BROUSSE (1980). *El olivo.* Madrid: Ed. Mundi Prensa.

MANETTI, O. AND V. TOSONOTTI (1984). *Scienza del maiale. Tecniche di allevamento, transformazione e utilizzazione.* Bologna: Edagricole.

MARCHAND, H. AND et al. (1990). *Les Forêts Méditérranéenes. Enjeux et perspectives.* Paris: Economica.

MARGALEF, R. (1982). *Ecología.* Barcelona: Ed. Omega.

MARGALEF, R. (1986). *Limnología.* Barcelona: Ed. Omega.

MARGARIS, N.S. AND S.A. MONEY (eds.) (1981). *Components of Productivity of Mediterranean-Climate Regions - Basic and Applied Aspects.* The Hague.

MATHUR, H.S. (1988). *Essentials of Biogeography.* Jaipur: Pointer Publishers.

MAURIZIO, A. (1932). *Histoire de l'Alimentation Végétale.* Paris: Payot.

MCKELL, C.M. (ed.) (1988). *The Biology and Utilization of Shrubs.* New York: Academic Press.

MCNEILL, W.H. (1989). *Plagues and Peoples.* New York: Doubleday.

MEIER, H.M.E. (1978). *Enciclopedia sistemática agropecuaria.* Barcelona: Aedos.

MENÉNDEZ FERNÁNDEZ, M. (1986). "La aparición del Neolítico en Oriente Próximo." *Revista de Arqueología,* 7:10-19.

MILLER, P.C. (ed.) (1981). *Resource Use by Chaparral and Matorral.* Berlin: Springer-Verlag.

MINELLI, M.P. AND A. MINELLI (1984). *El canguro y los animales de Australia.* León: Ed. Everest.

MINELLI, M.P. AND A. MINELLI (1984). *El ciervo y los animales de Europa.* León: Ed. Everest.

MINELLI, M.P. AND A. MINELLI (1984). *La llama y los animales de América del Sur.* León: Ed. Everest.

MOLINA VÁZQUEZ, F. (ed.) (1988). *Reservas de la Biosfera en Andalucía.* Sevilla: Junta de Andalucía.

MOLNAR, S. (1983). *Human Variation. Races, Types, and Ethnic Groups.* Englewood Cliffs, NJ: Prentice Hall.

MONTOYA, J.M. (1983). *Pastoralismo mediterráneo.* Ministerio de Agricultura, Pesca y Alimentación. ICONA, Monographs 25.

MOONEY, H.A. (1977). *Convergent Evolution in Chile and California Mediterranean Climate Ecosystems.* Stroudsburg: Dowden, Hutchinson and Ross.

NOBLE, J.C., P.C. JOSS AND G.K. JONES (eds.) (1990). "The Mallee Lands. A conservation perspective." *Proceedings of the National Mallee Conference,* Adelaide, April 1989. Australia: CSIRO.

NOWAK, R.M. (1991). *Walker's Mammals of the World.* Baltimore: The Johns Hopkins University Press.

OADES, J.M., D.G. LEWIS AND K. NORRISH (1981). *Red-brown earth of Australia.* Adelaide: CSIRO.

PARRA, F. (1987). *Monte mediterráneo. Enciclopedia de la naturaleza de España.* Madrid: Debate.

PITTE, J.R. (1989). *Histoire du paysage français. Tome I. Le Sacré: de la Préhistoire au XVe siècle.* Tallandier, s.l.

POIRIER, J. (ed.) (1972). *Ethnologie Régionale.* Paris.

POISSONET, P., F. ROMANE, M.A. AUSTIN, E. VAN DER MAAREL AND W. SCHMIDT (eds.) (1987). *Vegetation Dynamics in Grasslands, Heathlands and Mediterranean Ligneous Formations.* The Hague: Dr. W. Junk Publishers.

POUNDS, N.J.G. (1974). *An Economic History of Medieval Europe.* London: Longman.

POUNDS, N.J.G. (1989). *Heart and Home. A History of Material Culture.* Bloomington, IN: Indiana University Press.

QUEZEL, P., R. TOMASELLI AND R. MORANDINI (1977). *Mediterranean forest and maquis: ecology, conservation and management.* Paris: UNESCO.

RAINE, P. (1990). *Mediterranean Wildlife.* London: Harrap-Columbus.

RAMADE, F. et al. (1990). *Consérvation des Écosystèmes Méditerranéens.* Paris: Economica.

RAYNAUT, C. (ed.) (1983). *Milieu naturel, techniques, rapports sociaux.* Paris: Editions du CNRS.

REED, C.A. (1977). *Origins of Agriculture.* The Hague - Paris.

RECHER, H.F., D. LUNNEY AND I. DUNN (eds.) (1986). *A Natural Legacy.* New York: Pergamon Press.

REICHE, K. (1937). *Geografía Botánica de Chile. Tomo II. La flora de Chile.* Imprenta Universitaria, Santiago de Chile.

RENFREW, C. (1989). "Orígenes de la lenguas indoeuropeas." *Investigación y Ciencia,* 159: 82-91.

RENFREW, C. (1990). *Arqueología y lenguaje.* Barcelona: Editorial Crítica.

REYNOLDS, P.R. (1983). "La agricultura en la Edad de Hierro." *Mundo Científico,* 2(14): 484-493.

RIDE, W.D.L. (1970). *A Guide to the Native Mammals of Australia.* Melbourne: Oxford University Press.

RODRÍGUEZ DE LA FUENTE, F. (1975). *Fauna Ibérica.* Barcelona: Salvat.

RODRÍGUEZ, R., O. MATTHEI AND M. QUEZADA (1983). *Flora arbórea de Chile.* Chile: Editorial de la Universidad de Concepción.

ROLLI, K. (1991). *Plantes d'Afrique du Nord.* Deutsche Gesellschaft für Technische Zusammenarbeit (GTZ) GmbH., Eschborn.

ROUGERIE, G. (1988). *Géographie de la Biosphère.* Paris: Armand Colin.

SÁNCHEZ-MONGE, E., A. MANUEL, A. BERMEJO, J. SALAZAR, J. DEL CAÑIZO AND D. VIDAL (1962). *El trigo.* Madrid: Ministerio de Agricultura.

SELL, J. AND F. KROPF (1990). *Propriétés et caractéristiques des essences de bois.* Le Mont, Switzerland: Lignum.

SERVENTY, V. AND C. SERVENTY (1981). *Australian Wildlife.* Adelaide: Rigby.

SNAYDON, R.W. (ed.) (1987). *Managed Grasslands (Analytical studies).* Amsterdam: Elsevier.

TAKHTAJAN, A. (1986). *Floristic Regions of the World.* Berkeley: University of California Press.

TAMARO, D. (1968). *Tratado de fruticultura.* Buenos Aires: Gustavo Gili.

TERMIER, H.G. (1973). *Trama geológica de la historia humana.* Barcelona: Editorial Labor.

THIAULT, M. (1979). "Réflexions à partir de quelques aspects bioclimatiques." In: *Utilization par les ruminants des pâturages d'altitude et parcours méditerranéens,* Versailles.

THIRGOOD, J.V. (1981). *Man and the Mediterranean Forest.* London: Academic Press.

THROWER, N.J.W. AND D.E. BRADBURY (eds.) (1977). *Chile-California Mediterranean Scrub Atlas.A Comparative Analysis.* Stroudsburg, PA: Dowden, Hutchinson and Ross.

TINDALE, N.B. (1974). *Aboriginal Tribes of Australia. Their Terrain, Environmental Controls, Distribution, Limits and Proper Names.* Berkeley: University of California Press.

TOUS, J. AND I. BATLLE (1990). *El algarrobo.* Madrid: Ediciones Mundi Prensa.

TOUSSAINT-SAMAT, M. (1987). *Histoire Naturelle et Morale de la Nourriture.* Paris: Bordas.

TRABAUD, L. (1971). *Les combustibles végétaux dans le département de l'Hérault.* Montpelier: CEPE Louis Emberger.

VALVERDE, J.A. (1967). *Estructura de una comunidad mediterránea de vertebrados terrestres.* Madrid: CSIC, Consejo Superior de Investigaciones Científicas.

WALCHER, D. AND N. KRETCHMER (eds) (1981).*Food, Nutrition and Evolution.* New York.

WALTER, H. (1976). *Vegetació i Climes del Món.* Barcelona: Departament de Botànica, Facultat de Biologia, Universitat de Barcelona.

WALTER, H. AND S.W. BRECKLE (1985). *Ecological Systems of the Biosphere. 1. Ecological Principles in Global Perspective.* Berlin: Springer-Verlag.

WERGER, M.J.A. (ed.) (1978). *Biogeography and Ecology of Southern Africa.* The Hague: W. Junk.

WILLIAMS, W.D. (1983). *Life in Inland Waters.* Melbourne: Blackwell Scientific Publications.

WINKLER, A.J. (1965). *Viticultura.* Mexico: Compañía Editorial Continental.

Authorship and source of the illustrations

Pictures and maps:

- Anna Maria Ferrer, 56
- Editrònica, 18; 32; 34; 46; 47; 53; 65; 102; 109; 127; 129; 130; 176; 189; 212; 238; 247; 252; 254; 285; 295; 306; 308; 310; 343; 377; 387; 394
- Eugeni Sierra, 82; 186
- Jordi Ballonga, 366; 369
- Jordi Corbera, 101; 131; 345
- Lluís Sanz, 134
- Marisa Bendala, 326
- Marisa Bendala / ECSA, 126
- Miquel Alonso / ECSA, 159
- Román Montull / ECSA, 269

Photographs:

- Abelardo Aparicio, 398
- Adolf de Sostoa and Xavier Ferrer, 84; 413
- Adrian P. Davies / Bruce Coleman Limited, 105
- AGE Fotostock, 41; 215; 255; 265; 287; 295; 311; 323; 346; 347; 363; 380
- Aisa, 207; 276; 330
- Ajuntament de Barcelona / BIMA, 190; 191; 193
- Anthony Bannister / NHPA, 389; 404
- Antoni Agelet, 309; 360; 376
- Archiv für Kunst und Geschichte (Berlin), 262; 354
- Archivio di Stato di Foggia / Foto d'Autore / Ariston de Miticocchio G.N., 315
- Archivo General de Indias (Seville) / Arenas, 208; 209; 210
- Archivo General de Simancas (Valladolid), 313
- Atlantide SDF / Bruce Coleman Limited, 357
- Biblioteca de Catalunya (Barcelona), 218; 230; 237; 250
- South African Library (Cape Town), 38; 39; 41; 199
- Bibliothèque Nationale de France (Paris), 202
- C. Andrew Henley / Auscape International, 408
- Carol Hughes / Bruce Coleman Limited, 73
- C.C. Lockwood / Bruce Coleman Limited, 236
- Centre de Documentació i d'Animació de la Cultura Catalana (Perpignan) / J.L. Valls, 270
- César L. Barrio Amorós, 118
- Claudio del Río / Fotobanco, 228
- CNRI / Science Photo Library / AGE Fotostock, 213
- Colin Paterson-Jones, 36; 79; 94; 97; 149; 344; 387
- Colla Swart / ABPL, 39
- David S. Boyer / National Geographic Society, 417; 419
- Denis and Therese O'Bryne / ANT / NHPA, 325
- E.A. Janes / NHPA, 117
- Eckart Pott / Bruce Coleman Limited, 374
- Eric Crichton / Bruce Coleman Limited, 63; 72; 251
- Erich Lessing / Archiv für Kunst und Geschichte (Berlin), 182; 183
- Ernest Costa, 23; 26; 45; 48; 66; 75; 76; 89; 92; 93; 162; 163; 208; 217; 225; 227; 233; 235; 243; 246; 255; 266; 272; 273; 274; 278; 283; 303; 307; 312; 313; 314; 315; 347; 364; 368; 395
- Erwin and Peggy Bauer / Bruce Coleman, 294
- Felix Labhardt / Bruce Coleman Limited, 124
- Field Museum of Natural History (Chicago), 197
- Firo Foto, 235; 250
- Francesc Muntada, 75; 256
- Frank Park / ANT / NHPA, 409
- G. Llop / ECSA, 301
- Frieder Sauer / Bruce Coleman Limited, 155
- Fritz Prenzel / Bruce Coleman Limited, 31
- Genin Andrada / Cover / Zardoya, 317
- Gérard Lacz / NHPA, 414
- Graphische Sammlung Albertina (Vienna), 114
- H.J. Deacon, 173
- Hans Reinhardt / Bruce Coleman Limited, 125; 136
- Herbert Kranawetter / Bruce Coleman Limited, 219
- Índex, 190; 220; 296
- Institut Agrícola Català de Sant Isidre / ECSA, 270
- Iranzo / Índex, 347
- Israel Antiquities Authority (Jerusalem), 172
- Jacana, 117
- Jaime Álvarez / Fotobanco, 194
- Jaime Plaza van Roon / Auscape International, 69
- James P. Blair / National Geographic Society, 418; 420
- James Simon / Bruce Coleman Limited, 77
- Jan Taylor / Bruce Coleman Limited, 95
- Jane Burton / Bruce Coleman Limited, 115
- Jaume Altadill, 58; 72; 158; 242; 353

- Jaume Gual, 180
- Javier Andrada, 50; 112; 113; 145; 157; 299
- Jean-Paul Ferrero / Auscape International, 49; 62; 148; 300
- Jeff Foott Productions / Bruce Coleman Limited, 143; 293; 378
- Jen and Des Bartlett / Bruce Coleman Limited, 132
- J. Enric Molina, 33; 201
- J.H. Brackenbury / Bruce Coleman Limited, 150
- Jim Brandenburg / Minden Pictures, 140
- J. Myrdal / Índex, 234
- Joan Biosca, 17; 20
- Joaquim Reberté and Montserrat Guillamon, 80; 142; 324
- Joe Cornish / Photothèque Stone International, 263
- Jordi Bartolomé, 75; 107; 165; 222; 233; 239; 289; 290; 336
- Jordi Camardons, 305
- Jordi Gumí / Firo Foto, 288
- Jordi Gumí, 177; 254
- Jordi Muntaner, 120
- Jordi Vidal, 29; 96; 105; 180; 186; 235; 241; 264; 266; 267; 277; 317; 391
- José Luis González Grande / Bruce Coleman Limited, 138; 371
- José Luis Rodríguez, 111
- Josep Germain, 224; 240; 347
- Josep Loaso, 236; 263; 365
- Josep Maria Barres, 27; 86; 121; 143; 151; 229; 259; 261; 264; 273; 274; 319; 362; 370; 375
- Josep Pedrol, 52; 88; 267; 275; 347
- Josso / CNRI (Paris), 212
- Juan A. Fernández / Incafo, 412
- Jules Cowan / Bruce Coleman Limited, 72
- Kim Taylor / Bruce Coleman Limited, 123; 147
- Larry Minden / Minden Pictures, 35; 316
- Lourdes Sogas, 24; 187; 358
- Lluís Ferrés and Ramon Folch, 338; 339; 350
- Lluís Ferrés, 22; 36; 59; 81; 94; 98; 167; 171; 227; 261; 319; 328; 337; 351; 355; 379; 383
- Lluís Giralt and Joan Reyes / Departament d'Agricultura, Ramaderia i Pesca / Generalitat de Catalunya, 268; 269
- Manuel Ballesteros, 289
- Martí Domínguez, 288; 290; 291
- Mary Evans Picture Library, 125
- Matt Jones / Auscape International, 25
- Miquel Monge, 302
- Montserrat Comelles, 159
- M. Rafa / Arxiu Alamany, 397; 405
- Musée d'Orsay / Réunion des Musées Nationaux (Paris), 178
- Museu del Perfum (Barcelona) / Albert Masó, 219; 220
- Museu del Perfum (Barcelona) / Jordi Vidal, 220
- Museum of the History of Science, University of Oxford, 188
- National Maritime Museum (London), 341
- NHPA / Silvestris Fotoservice, 124
- Nigel Blake / Bruce Coleman Limited, 125
- Nigel Dennis / ABPL, 45; 61
- Norman Owen Tomalin / Bruce Coleman Limited, 206
- N. Martínez / Índex, 346
- N.J. Dennis / NHPA, 389
- Oriol Alamany, 36; 54; 67; 99; 119; 139; 216; 244; 372; 377; 395; 396; 399; 400; 402
- Phoebe Hearst Museum of Anthropology, University of California (Berkeley), 209
- Rafael Vela, 160
- Rambol / ECSA, 178; 179
- Ramon Folch, 51; 78; 271; 277; 279; 344
- Ramon Manent, 191; 192; 193
- Ramon Vallejo, 27; 30; 32
- Reg Morrison / Auscape International, 69; 73; 156; 284; 407; 408; 410
- Rod Williams / Bruce Coleman Limited, 300
- Romano Cagnoni / Zardoya, 232
- R. Wanscheidt / Bruce Coleman Limited, 253
- Sánchez-Durán / AISA, 251
- Sandro Prato / Bruce Coleman Limited, 286
- Scala, 322; 329; 332; 333; 334; 335
- Sebastián Bellón, 280; 281; 282
- S. Fiore / Firo Foto, 280
- Southern Book Publishers / Auriol Batten (Johannesburg), 39; 40; 71
- Stephen Krasemann / NHPA, 392; 415
- Tavisa, 21; 382
- Teresa Franquesa, 59; 103; 109; 166; 368
- The Ancient Art and Architecture Collection (California), 70; 185; 201
- The Huntington Library (San Marino, California), 174; 196; 207
- Unitat d'Ecologia, Universitat Autònoma de Barcelona, 90
- Wayne Lowler / Auscape International, 98; 153; 226; 390
- Xavier Ferrer and Adolf de Sostoa, 60
- Xavier Ferrer, 292; 316; 323; 358
- Xavier Miserachs / Firo Foto, 191
- Xavier Miserachs, 365
- Zev Radovan (Jerusalem), 177; 181

Indexes

Species' index

This index contains the scientific and common names of the species mentioned in the text. The number refers to the page or pages where the name appears in the main text. Page numbers in italics refer to illustrations.

Thematic index